Computer Design Aids for VLSI Circuits

NATO ASI Series

Advanced Science Institutes Series

A Series presenting the results of activities sponsored by the NATO Science Committee, which aims at the dissemination of advanced scientific and technological knowledge, with a view to strengthening links between scientific communities.

The Series is published by an international board of publishers in conjunction with the NATO Scientific Affairs Division

A	Life Sciences	Plenum Publishing Corporation
B	Physics	London and New York
C	Mathematical and Physical Sciences	D. Reidel Publishing Company Dordrecht and Boston
D	Behavioural and Social Sciences	Martinus Nijhoff Publishers The Hague/Boston/Lancaster
E	Applied Sciences	
F	Computer and Systems Sciences	Springer-Verlag Berlin/Heidelberg/New York
G	Ecological Sciences	

Series E: Applied Sciences – No. 48

Computer Design Aids for VLSI Circuits

edited by

P. Antognetti
Istituto di Elettrotecnica
University of Genova
Italy

D.O. Pederson
EECS Department
University of California
Berkeley, California, USA

H. de Man
Laboratory ESAT
Catholic University of Leuven
Heverlee, Belgium

1986 **Martinus Nijhoff Publishers**
Dordrecht / Boston / Lancaster
Published in cooperation with NATO Scientific Affairs Division

Proceedings of the NATO Advanced Study Institute on Computer Design Aids for VLSI Circuits, Urbino, Italy, July 21-August 1, 1980

ISBN 978-94-011-8008-5   ISBN 978-94-011-8006-1 (eBook)
DOI 10.1007/978-94-011-8006-1

Distributors for the United States and Canada: Kluwer Boston, Inc., 190 Old Derby Street, Hingham, MA 02043, USA

Distributors for all other countries: Kluwer Academic Publishers Group, Distribution Center, P.O. Box 322, 3300 AH Dordrecht, The Netherlands

First edition 1981
Second printing 1984
Third printing 1986

FOREWORD

The Nato Advanced Study Institute on "Computer Design Aids for VLSI Circuits" was held from July 21 to August 1, 1980 at Sogesta, Urbino, Italy. Sixty-three carefully chosen professionals were invited to participate in this institute together with 12 lecturers and 7 assistants. The 63 participants were selected from a group of almost 140 applicants. Each had the background to learn effectively the set of computer IC design aids which were presented. Each also had individual expertise in at least one of the topics of the Institute.

The Institute was designed to provide hands-on type of experience rather than consisting of solely lecture and discussion. Each morning, detailed presentations were made concerning the critical algorithms that are used in the various types of computer IC design aids. Each afternoon a lengthy period was used to provide the participants with direct access to the computer programs. In addition to using the programs, the individual could, if his expertise was sufficient, make modifications of and extensions to the programs, or establish limitations of these present aids. The interest in this hands-on activity was very high and many participants worked with the programs every free hour.

The editors would like to thank the Direction of SOGESTA for the excellent facilities, Mr. R. Riccioni of the SOGESTA Computer Center and Mr. M. Vanzi of the University of Genova for enabling all the programs to run smoothly on the set date.

P.Antognetti
D.O.Pederson   Urbino, Summer 1980.
H. De man

TABLE OF CONTENTS

VIII

COMPUTER AIDS IN INTEGRATED CIRCUITS DESIGN

Donald O. Pederson

Department of Electrical Engineering and Computer
Sciences and the Electronics Research Laboratory
University of California, Berkeley, California 94720

Today's VLSI and LSI Circuits

A very large-scale integrated (VLSI) circuit is commonly defined as a single chip which contains more than 100,000 devices.
This year (1980) several industrial electronic firms are designing IC's with device counts of more than 100,000. These circuits
are either in initial design or are possibly in initial prototype
production. Generally, the NMOS technology is being used with a
minimum feature size in the 1-2 micron range. The design methodology and the computer aids used in the VLSI design, in general,
are those that are also being used in today's LSI production or
are relatively modest extensions of these methodologies and
computer aids for design.

As is typical when a design is attempted with new size
restrictions and with extensions of processing technology, the
design proceeds slowly because there is a great interaction between the design and layout of the circuit functions and the
development of the fabrication technology. Usually the design
team, which will be quite large for VLSI, involves digital and
signal processing systems specialists, circuit designers, layout
specialists, testing experts, design aids guru's, and of course
the processing experts. For new types of circuits it is
common that a mixed design strategy will be used, i.e., a mixture
of different design methodologies will be used. A particular
problem in the development of computer design aids concerns the
fact that with these new developments or the new extensions in
the fabrication and processing technology there is little adequate modeling information on the process or the electronic devices.

Today's production LSI circuits can be represented by the 8086 and the 68000 microcomputer circuits initially developed by Intel and Motorola, respectively. The device count of the former is approximately 30,000; for the latter, it is about 34,000 (67,000 sites). The development of these circuits commenced in late 1976; thus, the technology used and the critical feature dimensions were those of late 1976, but probably modified before the design freeze to 1977 technology. The total development time for these circuits was of the order of 32 to 36 months. The design teams numbered between 20 and 40 and reached a peak during the layout phase. The initial layout, before subsequent corrections, took of the order from 60 to 100 man-months of effort. The development cost was of the order of 10 to 20 million dollars.

As is typical, even though the technology for these LSI's was reasonably well established, there was a lag in the development of the necessary modeling information for the computer design aids. None-the-less, circuit and logic simulation were used effectively, the former to develop and evaluate circuits of the order of 100 transistors, which involved basic building blocks used in the design. Logic simulation was used on several of the building blocks of the chips as well as larger portions in order to verify the logical design as well as to develop through fault simulation adequate test patterns for the IC testers. The layout was mostly accompanied by hand either initially on paper or directly into a graphical terminal; i.e., automatic design or synthesis procedures were not used extensively. In the base of development on mylar, there was a subsequent digitization. For both, interactive computer correction was used. Design-rule checkers were used extensively to establish that layout design rules were verified. Finally, in both cases a breadboard of the circuit or system was made or attempted. For the 68,000, the breadboard consisted of MSI and LSI CMOS parts.

The following data illustrate the increase of design, layout, verification and digitizing time. The early Z-80 microprocessor contained approximately 8,000 devices and took approximately 80-man weeks of effort for the design, layout, etc. The recently developed Z8000 microcomputer contains approximately 25,000 devices and took approximately 165-man weeks of effort to design. The 8086, mentioned above, has 30,000 devices and had a corresponding design time of 600 man-weeks. The total design and fabrication time from initial concept to manufactured product was approximately 350-man weeks for the Z-8000 microcomputer, approximately twice the design time.

Design Methodologies and Computer Design Aids

In integrated circuit design very distinct and different design methodologies can be used which have a large influence on the type of effective computer design aids that can be employed in the design. For example, in the polycell (standard-cell) design methodology a set of cells are developed usually with constant height, and definite dc supply ports. The cell information concerning detailed layout and electrical and logical functional descriptions are located in the computer design system library. The subsequent design problem is to place the cells appropriately in avenues and to route the signal connections between the cells. As brought out later, over the past decade effective virtually automatic placement and routing computer programs have been developed to achieve rapid designs.

In the gate-array approach, an array of modules is established where each module consists of basic devices and gates in SSI or MSI configurations. The design methodology is to achieve the signal path connections usually using two or more levels of interconnect capability. The connection layers achieve at each module location of the array the desired MSI function and as well as the connections between these modules to achieve the overall chip function. Again, placement and routing schemes can be used to achieve the overall design in a rapid manner. Typically, the design time for these two approaches are of the order of ten times faster than the conventional hand approach which sometimes is labelled the data-path approach.

In the data-path appraoch, a set of basic circuits is also developed initially. However, there is freedom in changing the performance and arrangements of these standard circuits to serve particular needs. For example, larger buffers might be needed in order to drive a long interconnection path. Design-rule checking and other design verification tools must be more effective in order to establish the necessary designs in the data-path approach.

Another design approach is commonly called the building-block approach in which the blocks are very regular structures, such as PLAs. For this type of regular structure in the building blocks, it is necessary to spend significant design time on the effective partitioning of the overall design. Programs are available for logic minimization to achieve small regular structures. Routing programs are needed to aid in the development of the complete chip.

As indicated above, for today's VLSI circuits it is common that a mixture of design methodologies are used. Thus in the development of a big circuit, one might use regular structures like PLAs whenever possible and effective. One might also use some building blocks developed with some poly-cell and placement routing schemes. Finally, certain very critical random logic circuits might be done with a data-path approach. A particular problem with the mixed approach is to achieve adequate design verification and to develop the necessary input test patterns to test the chip coming out of fabrication. It is because of the severe problem of verification and testing that more and more structured design approaches are used. These limit the freedom of the designer in order to obtain more easily a verifiable and testable circuit.

In all aspects the design of large integrated circuits is done in a hierarchical approach. We start with small known cir-cuits, whether they be standard cells or basic building block structures. These are developed and designed and verified very carefully. These are used then as building blocks in the buildup of the design. The buildup will commonly follow the same parti-tioning procedure that was used in the initial top-down parti-tioning which established the necessary subcomponents of the chip.

As the above indicates, different design aids can be des-cribed and developed for different design methodologies. In this Advanced Study Institute, a decision was made to concentrate on most of the features common to all of these methodologies. In the next section a design flow is proposed for a typical inte-grated circuit that might be used in a specific product. From this design flow, the computer aids that can be used in the integrated circuit design are identified.

An IC Design Procedure

A flowgraph of a possible integrated circuit design proce-dure is shown in Figs. 1a and b. The flowgraph starts from a product or system which would include the integrated circuit of interest, and the flowgraph ends with a ready-to-use, tested, functioning integrated circuit. This flowgraph has been pre-pared to illustrate the possible computer aids that are used or can be used in an integrated circuit design process. In follow-ing sections, the status of these design aids is presented together with an indication of the direction of present develop-ment activities. Reference to individual programs and contribu-tions is deferred to the chapters that follow.

The IC design process usually starts with an initial (tentative) definition of the data-processing or signal-processing function to be achieved. At the same time, an initial (tentative) choice of an IC processing technology is made. Finally, an initial (tentative) choice is made concerning the electronic device type to be employed as well as the basic circuit technology to be used. These three choices lead to three early study phases of the IC design process.

As shown in Fig. 1a, the initial choice of an IC technology leads to investigations with either actual test devices or through process simulation to determine the characteristics of the process and the characteristics and parameter values of the devices and the test circuits. For the circuits, it is necessary to propose basic cell or circuit configurations and to proceed through the electrical and geometrical layout design of the circuits. (This initial design of the circuit family will usually be used later in the complete VLSI design.)

For the initial IC function investigation, it is still common today to initiate a hardware simulation of the proposed circuit function using an older technology in order to study possible subpartitioning of the IC function as well as to provide a vehicle to check the circuit performance with the outside elements, i.e., the programming and interfacing necessary to use this IC in a complete system. For large integrated circuits, the hardware simulation constitutes a major design and realization effort on its own. Further, the reliability aspects of the hardware realization can be most troublesome. It is not uncommon for the development of the hardware simulation to take almost as long as the design and layout of the IC itself. Alternately, and as shown in Fig. 1a, functional simulation, utilizing higher-level computer languages often referred to as design languages, can be used. This type of program is in development and is coming into use for IC design.

The choice of the IC technology, the characterization and design of devices in basic circuits, and the establishment of the detailed data-processing or signal-processing function of the chip together with necessary partitioning involve iterative processes. If each of these aspects is successful (and to achieve this there will inevitably be interaction between these three design aspects), we achieve, as shown in Fig. 1a, a) a detailed process specification, b) the basic design rules and the basic cells and circuits for the IC, and c) the detailed specification of the performance of the integrated circuit and its major subfunctions. For the last, information will also be available that should lead to the development of successful test procedures for a final processed integrated circuit.

In Fig. 1b are shown the details of the design of the complete IC itself. The so-called top-down design procedure on the right of Fig. 1a leads to the individual subfunctions that must be designed. For the design of each subfunction there will generally be heavy use of IC simulation to investigate the behavior and interaction of the basic circuits as they are combined and laid out geometrically in the mask arrangements. Logic simulation will be used first for logic verification (true-value-simulation) in order to verify that the actual design achieves the desired logic or signal-processing behavior of the subfunction. At this stage one must think in terms of possible faults and, through the use of fault simulation, to develop final test input patterns for the complete IC to use on the processed chip. These developments naturally are based upon the initial information that was developed in the subpartitioning and concurrent functional simulation in earlier top-down design procedure.

The testing of a VLSI circuit is one of the major aspects and one of the most troublesome at the present time. Over the past decade, new structured design procedures have been investigated at several locations on three continents. This structured design (design for testibility) involves a constraint on the designer to use particular basic circuits and combinatorial procedures. With a slight penalty in circuit complexity and in mask area, a design is achieved which can be reduced, by a sequence of input signals to the chip, to the equivalent of a set of combinatorial logic circuits or separate signal-processing circuits. For complete ICs the input test patterns can be readily established and 100 percent testing possible. This topic is controversial, however, but it is one that is given major attention in this Institute.

To return to Fig. 1b, once the layout of a section of the integrated circuit has been achieved it is necessary to check the design rules. In the past, this aspect has been a very time consuming and tedious procedure. Now computer aids, developed in the last four or five years, have made this a manageable task. It should be noted that the design rule check (DRC) is done usually on smaller portions of the design as they evolve. Almost all DRC's increase their computer usage time by the function n^x where n is a measure of circuit size complexity and x is usually no smaller than 1.5. This is a first example, but only the first, of the "Tyranny of Increasing Size." To avoid this, a hierarchical design check is often used. Once a sub-circuit has been found to be design-rule safe, it is defined as an entity. In further checking, the software treats the

entity as such and it's internal features are not reinvestiga-
ted by the DRC.

Another DR procedure which is stressed in this Institute is
to include automatically the design rules and minimum geometry
constraints as the layout proceeds. This can be done with new
computer programs which are just now being developed and put into
IC design use.

As indicated in Fig. 1b, it is often desirable to simulate
a complete design layout in order to verify the circuit function
behavior of the complete chip and to check out the test patterns
that have been developed. Because of the size of today's integrated circuits, this complete simulation, even at the logic-
simulation level, is sometimes avoided because of the large cost
of computer usage. New mixed-mode (mixed-level) simulation
programs which have the ability to provide a hierarchical simulation are now being developed to make such design verification
possible.

With the design completed, including the test procedure, it
is possible to proceed to mask generation and IC processing. As
noted in Fig. 1b, the complete design procedure of the chip itself is almost never a one-way bottom-up procedure, but involves
iteration as simulation and verification show that performance
objectives and specifications have not been achieved.

Computer Aids for Circuit Design

The computer aids that can be used in the complete VLSI
design procedure include those shown in Table 1. In our present
Institute, a concentration is made on the aids for the actual
circuit design, layout, and testing. In the chapters that
follow, process simulation, device characterization and modeling
and design verification including design-rule checking and circuit and logic extraction are not ignored but are included
implicitly in the topics of other chapters.

A list of IC simulation programs is shown in Table 1. Logic
simulation has developed to a high state of perfection and heavy
use particularly for digital systems containing basic circuit
blocks, e.g., gate-level IC's. It has been found, however, for
LSI and VLSI circuits involving particularly MOS dynamic logic
technology that existing logic simulators are limited and consume
large computer time. Even for logic verification (true-value
verification) it is often not possible or economic to simulate
adequately even a large portion of an IC. Further, a great many

modifications have had to be made to the earlier logic simulators, and some are not usable at all. In the past few years new logic simulators have been developed at several locations which have been specifically developed with the new LSI and VLSI technologies in mind.

In a similar sense the earlier logic simulators based on existing gate-level circuit blocks for digital system design have not been found to be suitable to simulate the various faults that might occur in the new technologies. Here again, new logic simulators have been designed with this aspect in mind.

Functional simulation has also been developed to a significant degree for digital systems but again its use for VLSI has demanded new developments and extentions. Functional simulation can be accomplished in several ways. One can start with a given logic simulator and introduce models of larger protions of the logic, i.e., combinations of gates to represent larger logic functions. Higher-level languages can also be introduced to produce the functional simulation. Alternately, canonical circuit forms such as PLAs, ROMs, etc., can be described functionally rather than at the detailed electrical or logic levels. Finally, one can propose a top-down functional language which has an input and output format which describes the logical aspects of the blocks. Researchers developing these types of design aids for IC design are combining logic and functional simulation into programs for mixed-mode (mixed-level) simulation.

Circuit simulation is in an advanced state for IC design and is used significantly. There are several aspects that are important to keep in mind when we think in terms of the development of this type of design aid. In circuit simulation, it has been necessary to distinguish in the design of the programs between algorithmic optimality versus the overall effectiveness of the simulator. In the program SPICE, for example, if one uses an "optimum" algorithm for each portion of the simulation task, it has not been possible to achieve a reliable or even functioninng circuit simulator. SPICE constitutes an engineering choice of "suitable" algorithms that work together most effectively. This type of consideration also holds for most of the other IC design aids.

Another important aspect in the development of a design aid is that it cannot be based exclusively on theory or on earlier results and experience. Only extensive use will find the problems and the inevitable bugs. Another important aspect that became evident in the decade of development of circuit simulators is the need for very friednly input and output languages and features.

One must start with the needs of the designer and not with the desires and idiosyncrasies of the simulator developer. Finally, an important aspect that has been crucial to the success of circuit simulation for IC design has been an open users' community. Having an open, nonproprietary users' community enables one to share the effectiveness of the program, new methods to extend the program and of course means to exchange information on problems and bugs.

Circuit simulation in spite of its effectiveness can only be used on relatively small circuits. The following example indicates the problem. Typically, a circuit simulator used on computers comparable to the 370/168 consume cpu time in the simulation of the order of 1 ms per device-iteration. If we choose an example of 10,000 devices which has 1 signal input and 2 clock inputs, if we choose a computer run of 1,000 time points, which is small for a circuit of this size, and if we assume that the typical simulator uses 3 iterations per time point, the cpu time is approximately 10^5 seconds, somewhat over a continuous day of computer run time. This length of run time (and cost) is of course unacceptable if we think in terms of the circuit simulator being used as an iterative design aid.

The above circuit simulation example implicitly assumes that circuit simulation run time goes up linearly with circuit size, i.e., $n^{1.0}$. Early simulators provided n^2 while the best of the new versions have the property $n^{1.2}$. Here is another example of the problem of "The Tyranny of Increasing Size."

Industrial experience indicates that typical circuit simulation runs for IC design use consume approximately 15 minutes cpu time per run and involve transistor counts of from tens to several hundreds. However, there are many examples from industrial practice in which an integrated circuit after test has been found inexplicitly to not perform as expected. Critical paths of large integrated circuits have geen investigated with detailed circuit simulation. Single run cpu times of 20 to 30 hours are not uncommon. Fortunately, in most of these cases, the circuit simulation has given the designer an opportunity to isolate the problems and to achieve a final successful design.

In the past several years, several techniques have been investigated to extend circuit simulation to larger circuits. In the first of these, a macromodeling technique has been used. Rather than modeling the electronic behavior of individual devices, a basic portion of circuitry in the IC is modeled at its terminals. This circuit model, in place of device models, is included in the circuit simulation. For both analog and digital circuits it has been found possible to decrease the cpu analysis time by a factor of 10 to 20 while maintaining a high degree of accuracy of simulation (external to the macromodels).

In macromodeling, it is also possible to model noncritical portions of a circuit in a cruder, less accurate manner while at the critical portions of the circuit simulation is used at the detailed device-level. The technique is to relax the need for accurate simulation in those portions of a circuit which are not critical nor under attention. For this type of simulation, it is highly desirable to have an input language such that the designer/ user can change easily the description of a portion of the circuit from noncritical to critical levels of modeling.

Another extension technique for circuit simulation also involves the relaxation of accuracy and goes under the name of timing simulation. In timing simulation, we reduce analysis time and cost by using simplified circuit analysis techniques and also by using basic circuit-level macromodels. In the analysis scheme, matrix inversion is minimized by purposely omitting from consideration (what are hoped) minor feedback paths. For example, in a MOS device which is used in an inverter mode, one ignores the capacitive feedback from the output node to the input node. One assumes that this feedback is not significant and can be ignored. Thus the matrix inversion of circuits of these elements becomes a simple evaluation, node by node, through the circuit matrix. Of course, there is inaccuracy introduced, but particularly for digital circuits, the errors introduced by the simplication are not accumulative.

One of the first of these timing simulators to be developed was the MOTIS Program, developed at Bell Labs. Very large logic circuits can be evaluated and test patterns developed with this program. In the evaluation of complete ICs involving thousands of gates, computer run times for this for this program have been as long as 100 cpu hours on a Harris/6 computer. The group that produced this program has continued to develop and evolve this program and have now also included a logic simulation capability. This is a second type of mixed-mode simulation.

In mixed-mode (mixed-level) simualtion, a combination of several simulation capabilities are included. It is possible to include detailed electrical-level circuit simulation, timing simulation, logic simulation, and functional simulation in one program. For such a total simulation, it is planned that the designer/user will choose a critical path and will input detailed device-level circuit simulation for this path. This critical path will be imbedded in circuitry which will be simulated using timing simulation, for which a cruder calculation of timing delays is sufficient. Finally, all of this circuitry will be imbedded in the remainder of the circuit for which a logic simulation is used to simulate the proper sequencing and logic operation of the total circuit. Possibly portions of the remainder will be simulated at a functional level, particularly where large regular structures such as ROMs and PLAs are involved.

In mixed-mode simulation it has been found that one cannot use a set of existing circuit, timing, logic, and functional simulators and combine them with suitable translators to achieve a common input language and common output capabilities. Such a combination becomes entirely too cumbersome and too slow and costly for effective design use or even design evaluation. Instead it has been necessary to develop these new simulators from the beginning with particularly attention to a common data language and modeling, with great care being given to the communication aspects between the different levels of the simulation of the circuit. Another important point has been to make sure that the user will be able to input easily a natural circuit partitioning and that he can very easily shift a portion of the circuit from one type of description to another as he changes his investigation of critical paths. Finally, it has been important to develop these multi-level (mixed-mode) simulators so that they make use of event-driven procedures in order to achieve rapid, very effective computer usage.

Computer Aids for IC Layout

There are three principal methods used in the layout of integrated circuits. The first involves a sequence of hand drawing, the repeated use of basic circuit structures (the use of paper dolls), the digitizing of this information using a minicomputer, and finally the use of an interactive computer-graphic terminal to correct and update the hand-drawn layout. With the exception of only a few industrial concerns, this is the technique which is used today. This procedure demands a very large man-effort to design LSI and VLSI chips. W. Lattin has shown that the time to design and lay out random logic proceeds at the rate of only 10 devices per day per person. Therefore, for a LSI circuit of 10^4 devices with 30% random logic, the amount of effort for designing, laying out, and verifying the performance of the random logic in this chip involves 1 to 2 man years of effort. If design efforts increases only linearly with circuit size and linear dimensions, a VLSI with a device count of 10^5 to 10^6 per circuit will demand a design effort exceeding 100 man years.

Another existing layout technology involves (semi) automatic placement of standard cells and the subsequent (semi) automatic routing of the interconnections between these cells. At several electronic system houses, programs or program packages have been developed to achieve this placement and routing (P&R). Circuit layout and designs have been achieved which are approximately a factor of 10 faster than the hand-drawing technique (this improvement factor includes the overhead in the development of the standard-circuit library). Usually the layouts from the P&R programs consume slightly more silicon area than for the other technique. Because of the cost to produce parallel efforts for large IC's, exact figures are not available. It is estimated that the area penalty is of the order of 10 to 30 percent.

The use of the term (semi) automatic implies that some hand work is necessary. Usually a complete solution of the layout problem cannot be achieved by these P&R programs. Typically 95-98 percent of the layout can be completed. The last portion must be done by hand and constitutes of course a significant problem.

The new category of computer aids for IC layout involves an approach where automatic compaction algorithms are used and the input is made in symbolic form. The symbolic input is, in a sense, a caricature of a pictorial diagram of the layout. One uses the experience and/or intuition of the circuit designer to input into the computer the relative geometry of a portion of an IC. The computer program is written to include automatically all significant design rules and minimum geometry constraints. This is followed by the use of an automatic compaction routine. Usually an iterative scheme is followed; first, a horizontal compaction is used, then a vertical compaction and so forth, until no further compaction is possible. If the compaction is sufficiently rapid, the intermediate steps of the layout procedure can be displayed to the designer and he can quickly establish where the compaction algorithms are having difficulty. The process can be stopped at this point and a return to a symbolic representation can be made with geometric adjustments to achieve a better design. If the computer can supply design information to the designer/user in less than 1 minute, and preferably within a few seconds, very effective and rapid layout and design are possible.

An essential feature with this type of layout technique is that there must be a hierarchical procedure. After a portion of the circuit has been laid out, it is declared as an entity and the computer algorithms do not deal with it further except at its edges. For a hierarchical scheme, it should then be possible to move to greater and greater complexity in a chip layout without the penalty of circuit size tyranny.

Design Verification

As noted earlier excellent programs have now been developed for design rule-checking (DRC). Nonetheless, the computer execution time for a typical DRC goes up as the 1.5 power of the number of critical geometrical features. Thus individual portions of the circuit must be evaluated, preferably initially and as automatically as possible, as the design proceeds.

Another important aspect of design verification is the automatic extraction of the circuit and logic schematics from the finished mask set and the automatic determination of the performance of the extracted circuits. Excellent results have been achieved for MOS circuits and for some types of bipolar circuits. For circuit extraction, node-connection information is obtained together with the automatic determination of parasitic and device parameters and characteristics. In a few cases, for specific logic technology, it is possible to move from the circuit schematic to a logic description. The input test patterns and output response patterns which were evolved earlier in the design can be then used to check the design. To achieve an integrated set of design aids, automatic design verification is essential.

IC Test Pattern Generation

As menioned earlier, testing remains a most significant problem in the actual design and layout of the IC. Although programs are available to establish test patterns, successful complete results are usually available only for combinatorial circuits. For general sequential circuits and for mixed, analog digital circuits, there continue to be significant problems in developing automatic test procedures. At the present, a very large amount of hand work is necessary to program the testers for the actual IC's.

Logic simulators can be used to aid in developing a test of the logic function of the circuit and also in introducing faults, particularly stuck-at faults, to develop the effective input patterns to show outputs which reveal these faults.

As mentioned earlier, it appears that we must use new structured design procedures together with regular circuit subfunctions. Both new circuitry types and design constraints must be used so that the final circuit can be isolated into effective combinatorial sections to achieve 100 percent and rapid testing.

Another area that is becoming increasingly important with respect to test concerns onboard test generation. Circuitry will be introduced to or utilized within the IC itself to make internal logical checks. As chips get larger and more complex, but external pin count does not increase correspondingly, it will be more and more necessary to design the IC to include internal onboard testing, particularly of those portions that will not be available readily to external pins.

TABLE 1

COMPUTER-AIDS FOR IC DESIGN

SIMULATION

    PROCESS

    DEVICE (CHARACTERIZATION AND MODELING)

    FUNCTIONAL

    LOGIC FUNCTION VERIFICATION

      FAULT SIMULATION

    TIMING

    CIRCUIT (DESIGN AND CHARACTERIZATION)

      MIXED MODE

LAYOUT

    DRAW, PAPER DOLL, DIGITIZE AND CORRECT

    AUTOMATIC PLACEMENT AND ROUTING

    INTERACTIVE, SYMBOLIC AND COMPACTION

DESIGN VERIFICATION

    DESIGN-RULE CHECKING

    CIRCUIT EXTRACTION AND VERIFICATION

    LOGIC EXTRACTION AND VERIFICATION

TEST PATTERN GENERATION

16

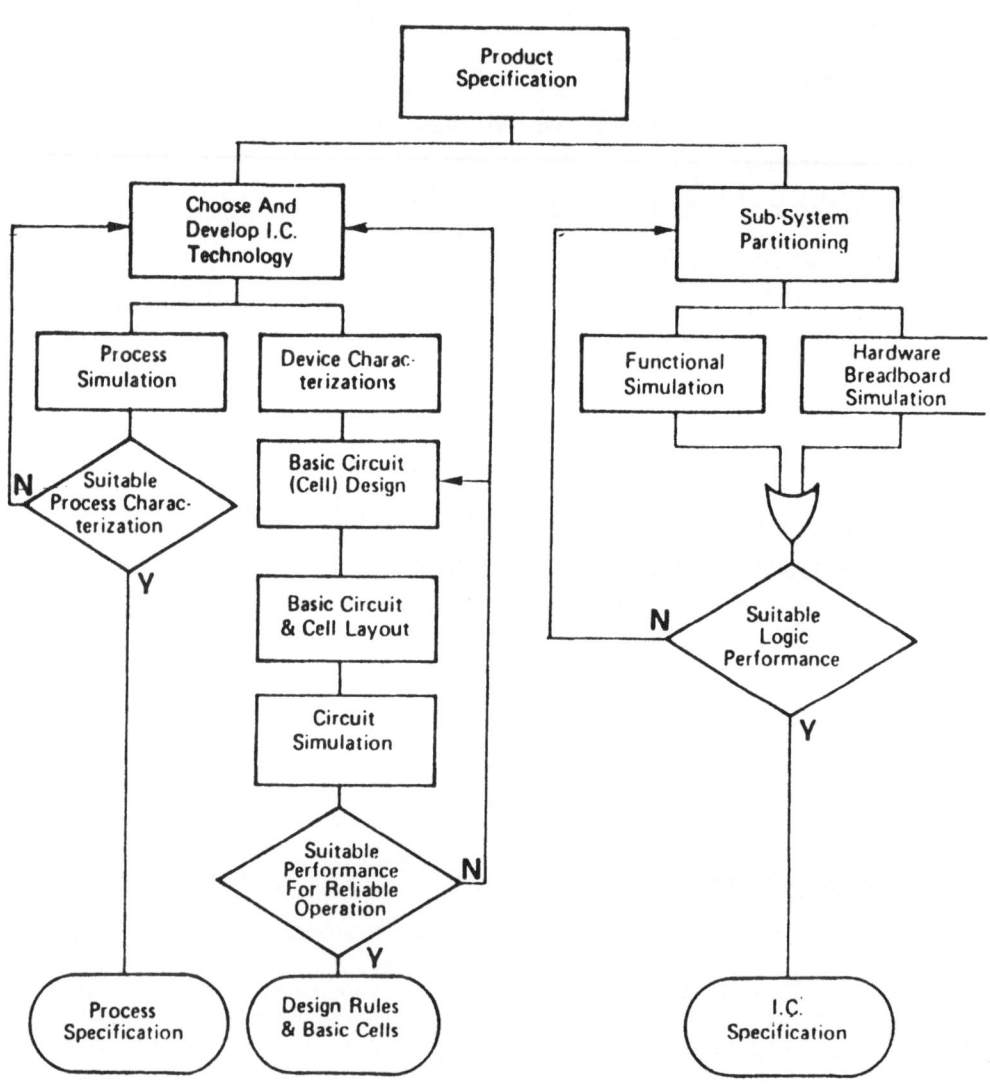

(a)

Fig. 1

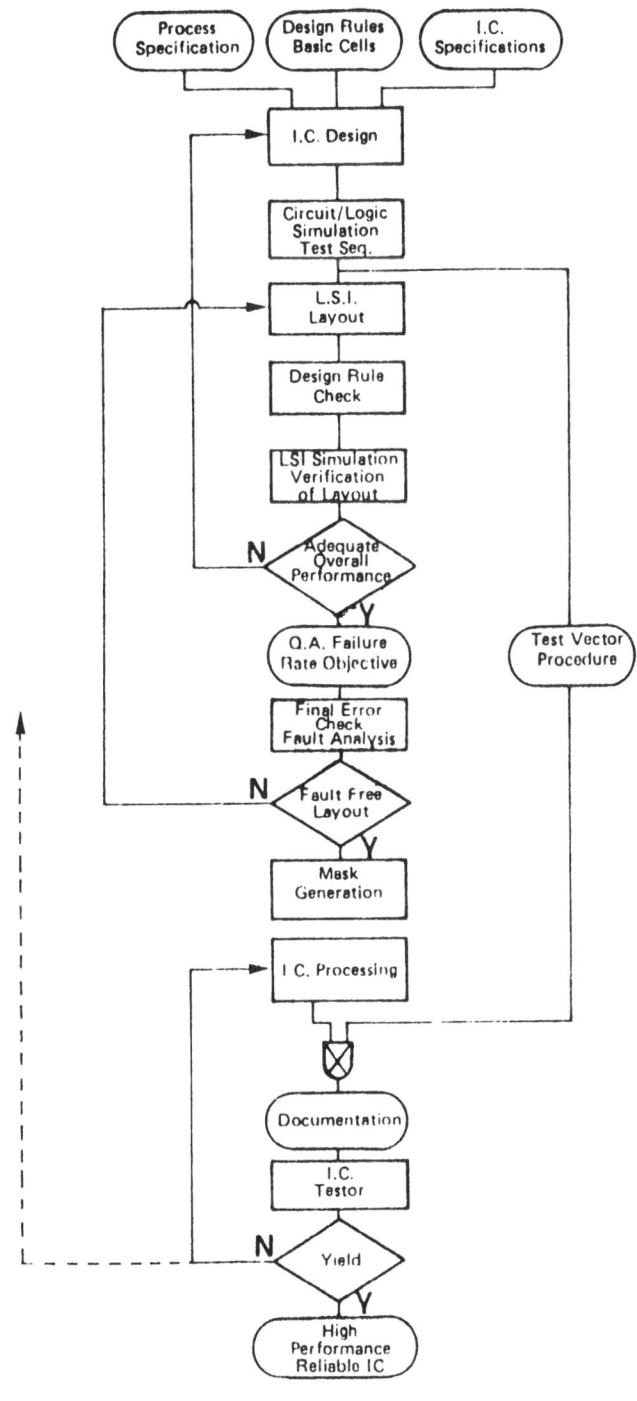

(b)
Figure 1

CIRCUIT SIMULATION

A. L. Sangiovanni-Vincentelli

IBM T. J. Watson Research Center
Yorktown Heights, NY 10598, USA[1]

1. INTRODUCTION

Circuit simulation programs have almost completely replaced the traditional breadboard or fabrication and testing of integrated circuits as means of verifying design acceptability. In fact, a breadboard may give results which have small resemblance to the manufactured circuit performances due to the completely different nature of parasitic components. Fabrication and testing of an integrated circuit for verifying a design is very expensive and time consuming. Moreover, extensive probing is not possible and modification of circuit components to determine a better design is practically unfeasible. On the other hand, circuit simulators such as SPICE [1] and ASTAP [2] give very accurate prediction of circuit performances and provide information impossible to obtain from laboratory measurements. As a measure of the use of circuit simulators we offer the following data:
1- At a major IC house SPICE is run about 10000 times per month.
2- At the University of California, Berkeley, SPICE is accessed about 60000 times per year.

---

[1] On leave from the Department of Electrical Engineering and Computer Sciences, University of California, Berkeley.

3- At IBM East Fishkill facility during December 1977 more than 150 ASTAP jobs per working day were submitted and about 40 hours per working day of CPU of IBM 370/168 were spent for ASTAP runs.

Even though circuits with hundreds of active devices are commonly simulated,circuit simulators are too expensive at present to perform the analysis of a complete VLSI circuit containing more than ten thousand devices (at Bell Laboratories a circuit with 3000 active devices has been analyzed on a CRAY computer in 1/3 of an hour). However circuit simulators still find their application in VLSI circuit design when analog voltage levels are important to verify a design of a part of the entire circuit or when tightly coupled feedback loops need to be taken into account.

In this chapter the basic algorithms upon which circuit simulators are built are reviewed, some implementation issues are discussed and the solution adopted in the simulation program SPICE are described. In particular, Section 2 is dedicated to the formulation of the circuit equations. Section 3 deals with algorithms for the solution of linear algebraic equations. Section 4 introduces algorithm and data structures for handling sparse matrices. Section 5 presents algorithms for the DC analysis of nonlinear circuits. Section 6 discusses algorithms for the transient analysis of dynamical circuits. Finally Section 7 presents the time and circuit size limitations of widely used circuit simulators such as SPICE2.

2. FORMULATION OF CIRCUIT EQUATIONS

2.1 Introduction

Most circuit simulators require the user to enter the description of the circuit (s)he intends to analyze. This description must contain information about the components and their interconnections. This information is then "transformed" by the circuit simulators into a set of equations. The topology of the circuit determines two sets of equations: Kirchoff Current Law (KCL) involving branch currents and Kirchoff Voltage Law (KVL) involving branch and node voltages [3]. The electrical behavior of the components of the circuit are described by a mathematical model. In most of the circuit simulators this mathematical model is expressed in terms of ideal elements.

2.2 Ideal Elements

In general the ideal elements are divided into two-terminal ideal elements,e.g.resistors, capacitors, inductors and independent sources, and two-port ideal elements,e.g. controlled sources and coupled inductors. The branch voltages and the branch currents at the component terminals are measured according to the associated reference directions [3]as shown in Figure 1. The ideal elements we consider here are shown in Table 1.

An ideal two terminal resistor is characterized by a curve in the branch voltage versus branch current plane. If this curve can be represented as a function of the branch current(voltage), it is said to be current(voltage) controlled. If the curve is a straight line passing through the origin, the resistor is said to be linear; otherwise it is said to be nonlinear.

An ideal two terminal capacitor is cha. .cterized by a curve in the charge versus branch voltage plane. If this curve can be represented by a function of the branch voltage(charge), it is said to be voltage(charge) controlled. If the curve is a straight line passing through the origin the capacitor is said to be linear; otherwise it is said to be nonlinear. For the sake of simplicity in this chapter we consider only voltage-controlled capacitors.

An ideal two terminal inductor is characterized by a curve in the flux versus branch current plane. If this curve can be expressed as a function of the branch current(flux) ,it is said to be current(flux) controlled. If the curve is a straight line passing through the origin , the inductor is said to be linear, otherwise is said to be nonlinear. For the sake of simplicity in this chapter we consider only current controlled inductors.

An ideal independent source can be either a voltage or a current source. An independent voltage source has a branch voltage which is prescribed a priori and is independent of the branch current. An independent current source has a branch current which is prescribed a priori and is independent of the branch voltage.

22

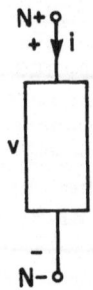

Figure 1.
Associated Reference Directions

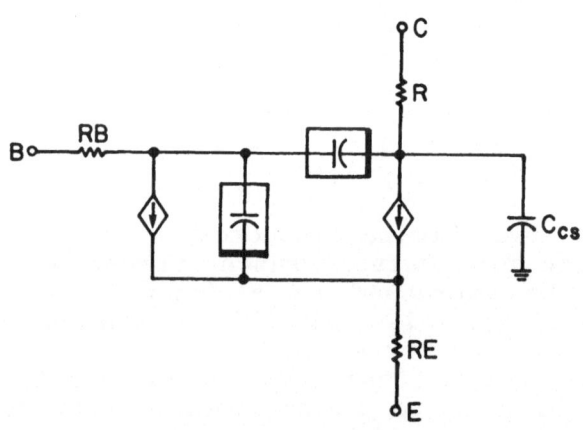

Figure 2.
The SPICE2 BJT model in terms of ideal elements.

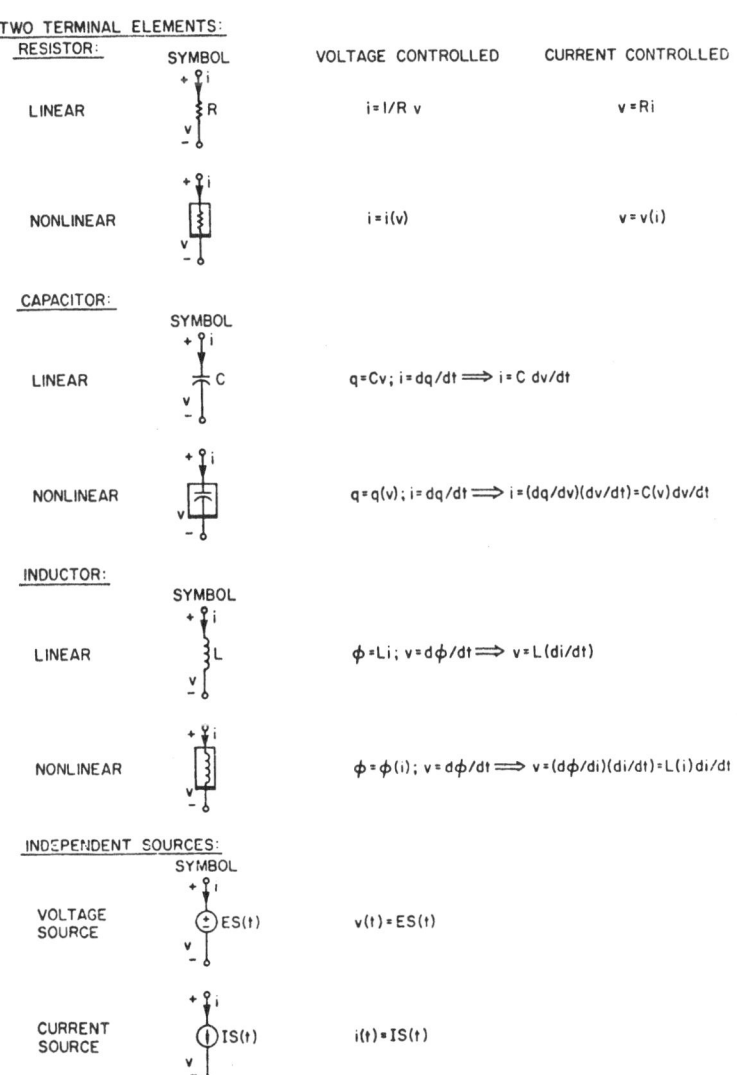

TWO TERMINAL ELEMENTS:

RESISTOR:	SYMBOL	VOLTAGE CONTROLLED	CURRENT CONTROLLED
LINEAR	R	$i = 1/R\, v$	$v = Ri$
NONLINEAR		$i = i(v)$	$v = v(i)$

CAPACITOR:

	SYMBOL		
LINEAR	C	$q = Cv;\ i = dq/dt \Longrightarrow i = C\, dv/dt$	
NONLINEAR		$q = q(v);\ i = dq/dt \Longrightarrow i = (dq/dv)(dv/dt) = C(v)\, dv/dt$	

INDUCTOR:

	SYMBOL		
LINEAR	L	$\phi = Li;\ v = d\phi/dt \Longrightarrow v = L(di/dt)$	
NONLINEAR		$\phi = \phi(i);\ v = d\phi/dt \Longrightarrow v = (d\phi/di)(di/dt) = L(i)\, di/dt$	

INDEPENDENT SOURCES:

	SYMBOL		
VOLTAGE SOURCE	ES(t)	$v(t) = ES(t)$	
CURRENT SOURCE	IS(t)	$i(t) = IS(t)$	

Table 1.a
Two Terminal Ideal Elements

24

TWO PORT ELEMENTS:
CONTROLLED SOURCES

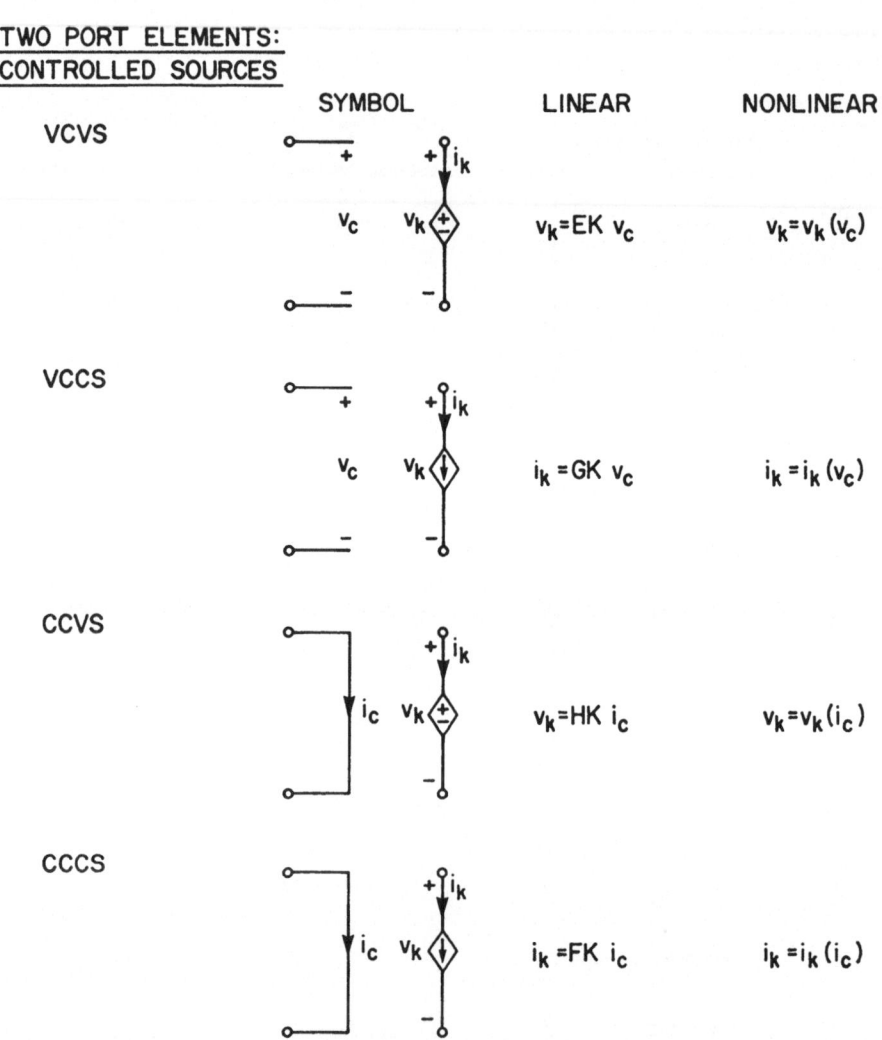

	SYMBOL	LINEAR	NONLINEAR
VCVS		$v_k = EK\, v_c$	$v_k = v_k(v_c)$
VCCS		$i_k = GK\, v_c$	$i_k = i_k(v_c)$
CCVS		$v_k = HK\, i_c$	$v_k = v_k(i_c)$
CCCS		$i_k = FK\, i_c$	$i_k = i_k(i_c)$

Table 1.b
Two Port Ideal Elements

An ideal controlled source is a two-port element. There are four kinds of controlled sources : Voltage Controlled Voltage Sources(VCVS), Voltage Controlled Current Sources(VCCS), Current Controlled Voltage Sources(CCVS) and Current Controlled Current Sources(CCCS). A VCVS has a branch voltage on port 2 which is a function of the branch voltage of port 1 and is independent of the port 2 current. Port 1 is an open circuit. The other controlled sources are defined similarly. (See Table 1 for the details.) An ideal controlled source is said to be linear if the curve in the controlled variable versus controlling variable plane is a straight line passing through the origin.

Controlled sources, two terminal resistors and independent sources are called resistive elements since their behavior is characterized by an algebraic equation involving branch voltages and branch currents. Capacitors and inductors are called energy storage elements.

In general, any mathematical model of an electronic device can be represented by an interconnection of these ideal elements. In Figure 2 the BJT model implemented in SPICE is represented in terms of the ideal elements.

2.3 Kirchoff Voltage and Current Laws

KCL and KVL can be written in several forms. The form we use in this chapter is the simplest : KCL will be written for the nodes of the circuit to be analyzed , and KVL will be written to specify the relation between node and branch voltages. For example, consider the simple circuit shown in Figure 3. The reference directions specified on the branches are the positive directions of the branch currents. KCL at the nodes of the circuit are :

node 1 : $i_1 + i_2 + i_3 = 0$

node 2 : $-i_3 + i_4 - i_5 - i_6 = 0$ $\hspace{4cm}$ (2.1)

node 3 : $i_6 + i_8 = 0$

node 4 : $i_7 - i_8 = 0$

The relations between node and branch voltages are :

branch 1 : $v_1 - e_1 = 0$

branch 2 : $v_2 - e_1 = 0$

branch 3 : $v_3 - e_1 + e_2 = 0$

branch 4 : $v_4 - e_2 = 0$

branch 5 : $v_5 + e_2 = 0$ (2.2)

branch 6 : $v_6 - e_3 + e_2 = 0$

branch 7 : $v_7 - e_4 = 0$

branch 8 : $v_8 - e_3 + e_4 = 0$

The advantage of using this form of KVL and KCL is in the fact that they can be assembled by inspection from the data given by the user. For example, in SPICE the data for a two terminal resistor are given in the following format:

RNAME N1 N2 RVALUE

where RNAME is the name of the resistor, N1 and N2 are the numbers of the nodes connected by the resistor and RVALUE is the numerical value of its resistance. The only information needed to assemble KCL and KVL in the above form is the nodes that are connected to each branch, which are obviously present in the data line for the resistor.

KVL, KCL and branch equations completely characterize the electrical behavior of the circuit. In general, when energy storage elements and nonlinear resistive elements are present in the circuit, the circuit equations are a set of mixed algebraic, linear and nonlinear, and ordinary differential equations. Unfortunately, in general, there are no closed form solutions to these equations and we must resort to numerical methods for their solution. We shall see in the next sections that the application of numerical methods to circuit analysis transforms the original analysis into a sequence of analysis of linear resistive circuits. Therefore , methods for the analysis of linear resistive circuits play a key role in circuit analysis programs.

2.4 The Sparse Tableau Analysis Method

The Sparse Tableau Analysis (STA) method was proposed in [4] and is used in a modified form in the
program ASTAP [2]. It consists of writing KCL and KVL
in the form previously introduced and appending the
branch equations to these equations. For the circuit
shown in Figure 3,STA equations are listed in Figure 4.
The reader with some knowledge of graph theory can
easily identify the reduced incidence matrix of the
linear graph associated to the circuit in KCL and KVL.
In fact, KCL and KVL can be written by making use of
the reduced incidence matrix A as follows : $Ai=0$ and
$v-A^te=0$, where i is the vector of branch currents, v is
the vector of branch voltages, e is the vector of node
voltages and A^t denotes the transpose of A. In general
STA equations for a linear resistive circuit with n+1
nodes and b branches assumes the following format :

$$\begin{bmatrix} A & 0 & 0 \\ 0 & I & -A^t \\ K_i & K_v & 0 \end{bmatrix} \begin{bmatrix} i \\ v \\ e \end{bmatrix} = \begin{bmatrix} 0 \\ 0 \\ S \end{bmatrix} \qquad (2.3)$$

where K_i and K_v are bxb submatrices containing the
coefficients of the branch equations and S is the vector of indipendent current and voltage sources. The
coefficient matrix of the STA equations has dimensions
(2b+n)x(2b+n) since we have n node equations, b KVL
equations and b branch equations in 2b+n unknowns (b
branch currents,b branch voltages and n node voltages).
The main advantages of STA are :

1-STA can be applied to any circuit containing all the
basic resistive ideal elements listed in Table 1.

2-STA can be assembled by inspection from the input
data.

3-The STA coefficient matrix is very sparse, i.e. many
entries are zero. As we shall discuss later, sparsity
is a very important structural property to take advantage of in order to be able to analyze large circuits.

The basic difficulty in implementing STA is in the
use of fairly sophisticated algorithms and programming
techniques in order to fully exploit the sparsity of

28

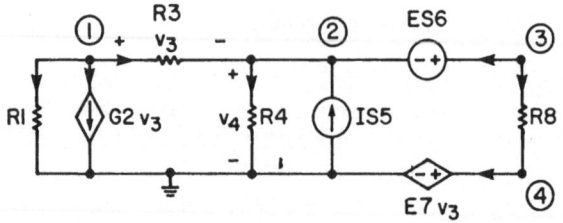

Figure 3.
A linear resistive circuit.

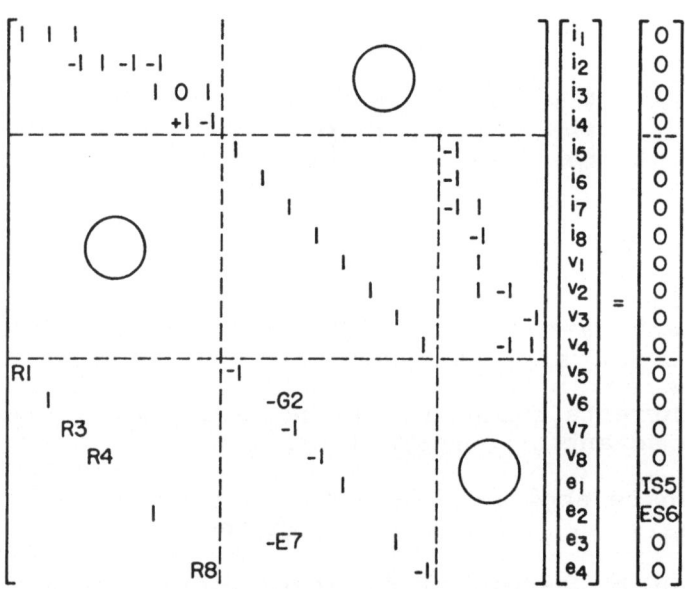

Figure 4.
The Sparse Tableau Equations for the circuit of Fig. 3.

the coefficient matrix. Moreover, the solution of the STA equations provides all the node voltages, all the branch currents and all the branch voltages. In general a designer needs only a small subset of this information. In order to save the computation involved in solving the STA equations when only a subset of the unknowns is required as output, the programming complexity must be further increased. These implementation problems caused people to look for methods which share most of the nice features of STA but are easier to implement.

2.5 Modified Nodal Analysis

Modified Nodal Analysis (MNA), pioneered by McCalla, Nagel and Rohrer, was subsequently formalized by Ho et al. in [5] It can be considered as a generalization of Nodal Analysis (NA). Before describing MNA, let us briefly review NA.

Nodal Analysis has been used in several simulation programs,such as MSINC [6] and consists of the following logical steps. KCL at the nodes of the circuit are written first. Then the branch equations are used to express the branch currents in terms of the branch voltages and current of independent current sources. Finally, KVL are used to express the branch voltages in terms of the node voltages. For the example in Figure 5,we have:

node 1: $i_1 + i_2 + i_3 = 0$ $\qquad$ (2.4)

node 2: $-i_3 + i_4 - i_5 = 0$.

Then

node 1: $1/R1\ v_1 + G2\ v_2 + 1/R3\ v_3 = 0$ $\qquad$ (2.5)

node 2: $-1/R3\ v_3 + 1/R4\ v_4 = IS5$

and finally

node 1: $1/R1\ e_1 + G2(e_1 - e_2) + 1/R3(e_1 - e_2) = 0$

node 2: $-1/R3(e_1 - e_2) + 1/R4\ e_2 = IS5$ $\qquad$ (2.6)

In matrix form we have :

$$Y_n \ e = \begin{bmatrix} 1/R1+G2+1/R3 & -G2-1/R3 \\ -1/R3 & 1/R3+1/R4 \end{bmatrix} \begin{bmatrix} e_1 \\ e_2 \end{bmatrix} = \begin{bmatrix} 0 \\ IS5 \end{bmatrix} = IS \quad (2.7)$$

where the coefficient matrix Y_n is called the nodal admittance matrix and IS is called the right hand side vector or simply the right hand side. Y_n is an nxn matrix for a circuit with n+1 nodes , since we have n node equations in n unknowns, the node voltages. It is important to note that Y_n and IS can be assembled by inspection from the input data. For example suppose that the input processor of a circuit simulator based on NA receives a data line from the user terminal describing a two terminal resistor:

RK N+ N- RKVALUE

We know that its branch current will appear in two node equations, the one relative to node N+ and the one relative to N-. If we perform the substitutions outlined above, we see that RK contributes to four entries in Y_n as shown in Figure 6. Therefore the input processor will only add to the four locations shown in Figure the value of the admittance of RK with the appropriate sign. Similarly it is possible to see that an independent current source will contribute to the right hand side only as shown in Figure 7. A little more complicated is the case of a VCCS. The input line for a VCCS in SPICE is

GK N+ N- NC+ NC- GKVALUE

where N+ and N- are the positive node and the negative node of GK,NC+ and NC- are the positive and negative nodes of the controlling branch. By following the procedure outlined above, it is easy to see that the VCCS contributes to four entries of Y_n as shown in Figure 8.

The advantages of NA are :

1-The circuit equations can be assembled by inspection.

2-Y_n has nonzero diagonal entries and is often diagonally dominant [3].This is an important feature to simplify the algorithms used in the solution of linear algebraic equations.

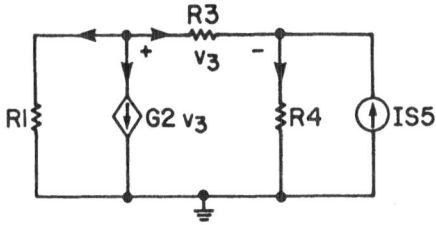

Figure 5.
A linear resistive circuit.

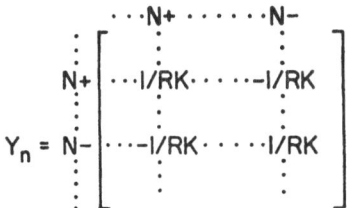

Figure 6.
The NA pattern of a two terminal resistor.

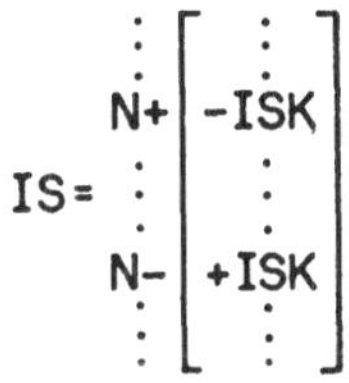

$$IS = \begin{array}{c} \vdots \\ N+ \\ \vdots \\ N- \\ \vdots \end{array} \begin{bmatrix} \vdots \\ -ISK \\ \vdots \\ +ISK \\ \vdots \end{bmatrix}$$

Figure 7.
The NA pattern of an independent current source.

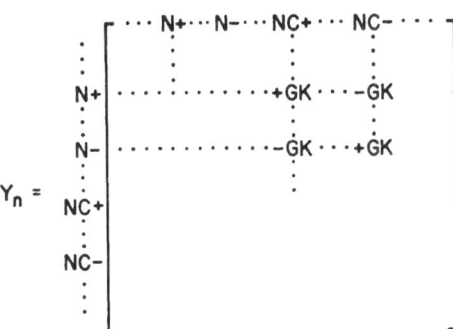

Figure 8.
The NA pattern of a VCCS.

3-Even though Y_n is not as sparse as the coefficient matrix of STA, it is sparse. In fact, the number of nonzero elements in row m of Y_n is approximately equal to the number of nodes connected to m by some branch and in usual electronic circuits, there is an average number of three nodes connected to any one node.

Unfortunately, NA cannot be applied to circuits containing independent voltage sources, CCCS, CCVS and VCVS(strictly speaking there are ways of modifying the circuit so that nodal analysis can still be applied, but they are cumbersome and require preprocessing. For these reasons they will not be considered here). In fact, for these elements the substitution of branch currents with branch voltages in the node equations cannot be accomplished. Another drawback of NA is that, if a branch current is needed as output by the user, some postprocessing is necessary to compute the branch current from the node voltages which are obtained as solution of the NA equations. MNA has been introduced to cope with these problems associated with NA. MNA is the analysis method used in SPICE. MNA follows the logical steps of NA as close as possible. For the circuit of Figure 3, MNA equations can be derived as follows. KCL at the nodes of the circuit are written first :

node 1: $i_1 + i_2 + i_3 = 0$

node 2: $-i_3 + i_4 - i_5 - i_6 = 0$ $\hspace{4cm}$ (2.8)

node 3: $i_6 + i_8 = 0$

node 4: $i_7 - i_8 = 0$

Then, we substitute as many branch currents as we can with branch voltages using the branch equations :

node 1: $1/R1\ v_1 + G2\ v_3 + 1/R3\ v_3 = 0$

node 2: $-1/R3\ v_3 + 1/R4\ v_4 - i_6 = IS5$

node 3: $i_6 + 1/R8\ v_8 = 0$ $\hspace{4cm}$ (2.9)

node 4: $-1/R8\ v_8 + i_7 = 0$

Note that we cannot substitute i_6 and i_7 with branch voltages because of the branch equations of the corresponding elements. Next we append the branch equations we did not use in the previous step:

branch 6: $v_6 = ES6$

branch 7: $v_7 - E7\, v_3 = 0$ $\hspace{4cm}$ (2.10)

Finally we substitute all the branch voltages with the node voltages using KVL.

node 1: $1/R1\, e_1 + G2(e_1 - e_2) + 1/R3(e_1 - e_2) = 0$

node 2: $-1/R3(e_1 - e_2) + 1/R4\, e_2 - i_6 = IS5$

node 3: $i_6 + 1/R8(e_3 - e_4) = 0$

node 4: $-1/R8(e_3 - e_4) + i_7 = 0$ $\hspace{3cm}$ (2.11)

branch 6: $e_3 - e_2 = ES6$

branch 7: $e_3 - E7(e_1 - e_2) = 0.$

We end up for this circuit with a sytem of six equations, four node equations plus the two branch equations of the elements for which the branch equations do not provide a branch current versus branch voltage relation, the "bad" elements, in six unknowns, the node voltages plus the currents of the "bad" elements. In matrix form we have :

$$\begin{bmatrix} Y_n & A \\ B & C \end{bmatrix}\begin{bmatrix} e \\ i' \end{bmatrix} =$$

$$\begin{bmatrix} 1/R1+G2+1/R3 & -G2-1/R3 & 0 & 0 & 0 & 0 \\ -1/R3 & 1/R3+1/R4 & 0 & 0 & -1 & 0 \\ 0 & 0 & 1/R8 & -1/R8 & 1 & 0 \\ 0 & 0 & -1/R8 & 1/R8 & 0 & 1 \\ 0 & -1 & 1 & 0 & 0 & 0 \\ E7 & -E7 & 0 & 1 & 0 & 0 \end{bmatrix}\begin{bmatrix} e_1 \\ e_2 \\ e_3 \\ e_4 \\ i_6 \\ i_7 \end{bmatrix}$$

$$= \begin{bmatrix} 0 \\ IS5 \\ 0 \\ 0 \\ ES6 \\ 0 \end{bmatrix} \hspace{2cm} (2.12)$$

where Y_n is the submatrix corresponding to the node
admittance matrix of the circuit obtained from the
original one by removing the "bad" elements. The coefficient matrix and the right hand side of the MNA equations can be assembled by inspection in a way similar
to the one followed to obtain the NA equations coefficient matrix. Two terminal resistors, independent
current sources and VCCS contributes to entries in the
Y_n submatrix and in the right hand side in the same way
as in the NA equations. VCVS and independent voltage
sources contribute to entries in the MNA equations
coefficient matrix as shown in Figure 9. For CCCS and
CCVS, an additional branch is automatically added to
the circuit. The branch is a short circuit, i.e. a zero
independent voltage source and is added in series to
the element whose current is the controlling current of
the CCCS or CCVS. This addition is usually performed by
the designer before entering the description of the
circuit. In SPICE the input data required for a CCCS is

FK N+ N- VNAM FVALUE

where the current in the CCCS flows from N+ to N- and
VNAM is the name of the voltage source through which
the controlling current flows. The patterns generated
by the CCCS and the CCVS are shown in Figure 10. Note
that the solution of the MNA equations provides the
value of some currents in the circuit, e.g. the currents through the voltage sources of the circuit. Now,
if we are interested in some branch current which is
not the current through a voltage source, we can still
arrange the input to the program, so that the branch
current of interest will be computed. In fact, if we
add a zero shared independent voltage source in series
with the element whose current we are interested in,
the program will append to the vector of unknowns the
current of the new voltage source. This is the strategy followed by SPICE users.

The main features of MNA are :

1-The coefficient matrix and the right hand side of the
MNA equations can be assembled by inspection.

2-MNA can be applied to any circuit with elements in
the ideal element list of Table 1.

3-MNA yields a coefficient matrix which has a zero-
nonzero pattern close to Y_n of nodal analysis. There-

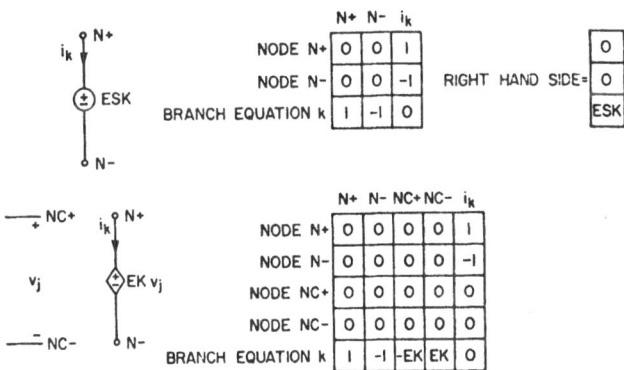

Figure 9.
The condensed MNA patterns for independent
voltage source and VCVS.

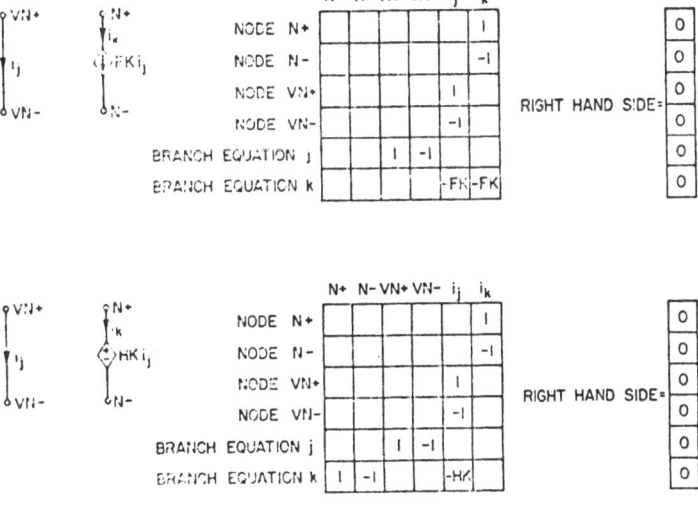

Figure 10.
The condensed MNA patterns for CCCS and CCVS.

fore, the coefficient matrix is sparse and has most of its diagonal entries different from zero. Moreover, most of the principal minors are diagonally dominant in many practical circuits.

3. ALGORITHMS FOR THE SOLUTION OF LINEAR ALGEBRAIC EQUATIONS

3.1 Introduction

Assuming that the circuit equations have been assembled in one of the forms we discussed previously, we have now to solve them in the most efficient way. In this chapter, we denote the system of equations to solve as

$$Ax = b \qquad (3.1)$$

where $A = [a_{ij}]$ is a nonsingular nxn matrix of real numbers, x is a vector of n unknowns and b is the right hand side vector. The methods for solving (3.1) belong to two classes : direct methods and iterative methods. By direct methods we mean algorithms which can find the exact solution of (3.1) in a finite number of steps if the computations are carried with infinite precision. By iterative methods we mean algorithms which produce a sequence of approximate solutions of (3.1) hopefully converging to the exact solution of (3.1). Circuit simulators use direct methods because of their reliability. In fact iterative methods converge to the exact solution of (3.1) only if certain rather strong conditions on A are satisfied. Among direct methods the most well known is perhaps Cramer's rule, but this method is too costly to be considered in circuit simulators. In fact the number of long operations (divisions and multiplications) required is of order n! Even though we know that Gaussian Elimination (GE) is not optimal, [8] i.e. there are other methods which require less operations to produce the solution of a linear system of algebraic equations, GE is still the most efficient algorithm available today for the practical solution of systems of linear algebraic equations.

3.2 Gaussian Elimination

The idea behind GE is to eliminate the unknowns in a systematic way, so that an equivalent system, whose coefficient matrix is triangular, is produced. A triangular system can be easily solved. In fact, let Ux=b be a system to solve where U is an upper triangular matrix with nonzero diagonal elements :

$$u_{11}x_1+u_{12}x_2+\ldots+u_{1n}x_n=b_1$$

$$u_{22}x_2+\ldots+u_{2n}x_n=b_2$$

$$u_{nn}x_n=b_n$$

We can compute $x_n,x_{n-1},\ldots,x_1$ almost by inspection from

$$x_n = b_n/u_{nn}$$

$$x_{n-1}=(b_{n-1}-u_{n-1,n}x_n)/u_{n-1,n-1}$$

$$\begin{matrix} \cdot & & \cdot \\ \cdot & & \cdot \\ \cdot & & \cdot \end{matrix}$$

$$x_1 = (b_1-u_{12}x_2-\ldots-u_{1n}x_n)/u_{11}$$

or in a compact form we have :

$$x_j = (b_j - \sum_{k=n}^{j-1} u_{jk}x_k)/u_{jj}, \quad j = n,\ldots,1 \tag{3.2}$$

where $\sum_{k=n}^{n-1}$ is zero. This solution process is called <u>back substitution</u>. If we have a triangular system $Lx=b$ to solve, where L is a lower triangular matrix, then a similar solution process yields :

$$x_j = (b_j - \sum_{k=1}^{j-1} l_{jk}x_k)/l_{jj}, \quad j = 1,\ldots,n \tag{3.3}$$

This solution process is called <u>forward substitution</u>. From (3.2) and (3.3), we see that the number of divisions and multiplications needed to solve a triangular system are respectively n and of the order of $1/2\, n^2$. Now it is clear why GE transforms the original system into one which is triangular. Consider now the system ot equations :

$$a_{11}x_1+a_{12}x_2+\ldots+a_{1n}x_n=b_1$$
$$a_{21}x_1+a_{22}x_2+\ldots+a_{2n}x_n=b_2 \qquad\qquad (3.4)$$

$$a_{n1}x_1+a_{n2}x_2+\ldots+a_{nn}x_n=b_n$$

If a_{11} is different from zero, we can eliminate x_1 from the last n-1 equations by solving the first equation for x_1 and replacing it with this expression. Now, this operation corresponds to subtracting from the i-th equation the first equation multiplied by a multiplier denoted by $l_{i1} = a_{i1}/a_{11}$, i=2,...,n. Then the last n-1 equations become

$$a_{22}^{(2)}x_2 + \ldots + a_{n2}^{(2)}x_n = b_2^{(2)}$$
$$\vdots \qquad\qquad \vdots \qquad\qquad \vdots$$
$$a_{n2}^{(2)}x_2 + \ldots + a_{nn}^{(2)}x_n = b_n^{(2)}$$

where

$$a_{ij}^{(2)} = a_{ij}-l_{i1}a_{1j}, \quad i = 2,\ldots,n; \quad j = 2,\ldots,n.$$

$$b_i^{(2)} = b_i-l_{i1}b_1 \quad , \quad i = 2,\ldots,n.$$

Now we have a system of n-1 equations in n-1 unknowns. If a_{22} is different from 0, we can eliminate x_2 from the last n-2 equations in a similar way. Thus we obtain a system of n-3 equations in n-3 unknowns. Setting the multiplier $l_{i2}=a_{i2}/a_{22}$, the coefficients of the reduced system are :

$$a_{ij}^{(3)} = a_{ij}^{(2)}-l_{i2}a_{2j}^{(2)}, \quad i = 3\ldots,n; j = 3,\ldots n.$$

$$b_i^{(3)} = b_i^{(2)}-l_{i2}b_2^{(2)} \quad , \quad i = 3,\ldots,n.$$

We can repeat this process until after n-1 steps we obtain

$$a_{nn}^{(n)} x_n = b_n^{(n)}$$

Of course, to be able to complete these steps we need to have $a_{ii} = 0, i = 3, \ldots, n-1$. Now x_n can be computed by inspection. All the other unknowns can be obtained by back substitution. In fact, if we collect the first equation at each step, we have :

$$a_{11}^{(1)} x_1 + a_{12}^{(1)} x_2 + \ldots + a_{1n}^{(1)} x_n = b_1^{(1)}$$

$$a_{22}^{(2)} x_2 + \ldots + a_{2n}^{(2)} x_n = b_2^{(2)}$$

$$. \quad a_{nn}^{(n)} x_n = b_n^{(n)}$$

where $a_{1j}^{(1)} = a_{1j}$ and $b_1^{(1)} = b_1$, and we see that to obtain x we need to solve an upper triangular system. The n-1 steps followed to obtain the upper triangular system can be summarized as : at step k, the coefficients of the reduced system of dimension n-k+1 are transformed according to

$$a_{ij} = a_{ij}^{(k)} - 1_{ik} a_{kj}^{(k)}, i = k + 1, \ldots, n; j = k + 1, \ldots, n. \qquad (3.5)$$

where

$$1_{ik} = a_{ik}^{(k)} / a_{kk}^{(k)} \qquad (3.6)$$

The coefficients $a_{kk}^{(k)}$ are called <u>pivots</u>.

It is easy to see from (3.5) and (3.6) that n divisions and approximately $n^3/3 + n^2/2 - n/3$ multiplications are needed to reduce (3.3) to upper triangular form. Since a triangular system can be solved with $n^2/2$ multiplications , the total number of long operations is :

n divisions

$n^3/3 + n^2 - n/3$ multiplications.

Note that the divisions involved in the solution of the triangular system have been performed in the reduction step.

3.3 Pivoting for Accuracy

In our derivation of GE we assumed that a_{kk} was different from 0, k=1,...,n. However, during GE we may encounter an entry a_{kk} which is zero. For example, consider the system

$x_1 + x_2 + x_3 = 1/2$
$x_1 + x_2 + 3x_3 = 2$
$x_1 + 3x_2 + 3x_3 = 1$

which has solution $x_1 = 1/4, x_2 = 1/2, x_3 = 3/4$. After the first step of GE we have

$0x_2 + 2x_3 = 3/4$
$2x_2 + 2x_3 = 1/2$

where $a_{22}^{(2)} = 0$. A way of overcoming this difficulty is to interchange rows and/or columns. In fact, if we interchange row 2 and 3, then the new system has a nonzero element where needed by GE. In the general case, when $a_{kk}^{(k)} = 0$ one might ask whether by interchanging row k with row r, r>k, we can continue GE. The answer is yes, as long as the coefficient matrix A of the original system is nonsingular. In fact, in order to proceed with the elimination at step k when $a_{kk}^{(k)} = 0$ we only need to find an element $a_{rk}^{(k)}$ different from zero and then interchange row k with row r. If we cannot find a nonzero element in column k, then we conclude that the first k columns of A are linearly dependent which contradicts the hypothesis that A is nonsingular. This interchange is sometime necessary when MNA is used. In fact, as previously pointed out, some of the entries along the main diagonal of the coefficient matrix of MNA are zeros. The interchange is mandatory when STA is used since there is no reason why a diagonal element in the Sparse Tableau should be different from zero.

Row and/or column interchange may be needed even when all the elements $a_{kk}^{(k)}, k = 1,...n$, are different from zero. For example, suppose we have to solve the following system (this example is taken from [7])

$0.000125x_1 + 1.25x_2 = 6.25$
$12.5 \quad x_1 + 12.5x_2 = 75$

Now, the solution correct to five digits is $x_1=1.0001$, $x_2=4.9999$. However suppose that the computer we use to solve (3.7) has finite precision arithmetic, say a three digit floating point arithmetic. Then, after the first step of GE, we obtain

$$0.000125x_1+1.25x_2=6.25$$
$$-1.25 \times 10^5 x_2=-6.25 \times 10^5$$

and the solution is $x_1=0$ and $x_2=5$. On the other hand, if we interchange rows, we have

$$12.5x_1+12.5x_2=75$$
$$1.25x_2=6.25$$

which yields $x_1=1$, $x_2=5$, much closer to the solution computed with five digit accuracy. When finite precision arithmetic is used, rounding errors may cause catastrophic errors. In particular, if $a_{kk}^{(k)}$ is small compared with the other elements in the coefficient matrix of the reduced system at step k, a round-off error may cause severe accuracy problems. To avoid these problems, two interchange or <u>pivoting</u> strategies are used.

   <u>Partial</u> <u>Pivoting</u>: (row interchange only) choose r as the smallest integer such that

$$|a_{rk}^{(k)}| = \max_{j=k+1,\ldots,n} |a_{jk}^{(k)}|$$

and interchange rows r and k.

   <u>Complete</u> <u>Pivoting</u>: (row and column interchange) choose r and s as the smallest integers for which

$$|a_{rs}^{(k)}| = \max_{i,j=k+1,\ldots,n} |a_{ij}^{(k)}|$$

and interchange rows r and k and columns s and k.

In partial pivoting, we choose as pivot the coefficient of largest absolute value in column k, in complete pivoting, we choose the coefficient of largest absolute value in the entire coefficient matrix of the reduced system. It is possible to compute an upper bound on the error in the final solution introduced by GE if partial or complete pivoting is used at each step.

However, as we shall see later picking up pivots for accuracy may conflict with maintaining sparsity of the coefficient matrix. Moreover, it has been observed in practice that partial or complete pivoting is needed only when the pivot $a_{kk}^{(k)}$ is quite small. In SPICE version 2F.1, a complete pivoting is In SPICE version 2F.1, a complete pivoting is performed only if a_{kk} is found to be smaller than a given threshold. This strategy is called <u>threshold</u> <u>pivoting</u>. It is important to note that, if the coefficient matrix A is diagonally dominant, then there is no need for pivoting to maintain accuracy. Therefore if nodal analysis is used to formulate the circuit equations, we do not have to worry about pivoting in many practical cases.

3.4 LU Decomposition.

In many circuit simulators, e.g. SPICE and ASTAP, a modification of GE is implemented : the LU decomposition. The idea behind this strategy is as follows. Suppose we have a way of factoring the coefficient matrix A into the product of two matrices, L and U, which are respectively upper and lower triangular. Then we have

$$Ax=LUx=b \qquad (3.8)$$

If we set $Ux=y$, then we can solve (3.8) solving the triangular system $Ly=b$ by forward substitution and then solving the triangular system $Ux=y$ by back substitution. This strategy is very advantageous if a system has to be solved for many different right hand sides. In fact, if we use GE as explained in section 3.2, we need to repeat the entire process for each different right hand side. If we use LU decomposition, only the solution of the two triangular systems is needed for each different right hand side. Thus the cost for LU decomposition is n^2 multiplications per right hand side as opposed to $n^3/3+n^2-n/3$ for GE. Now the problem is how to factor A into the product of L and U. If we set up the matrix equation

$$\begin{bmatrix} l_{11} & 0 & \ldots & 0 \\ l_{21} & l_{22} & \ldots & 0 \\ \vdots & & \ddots & \vdots \\ l_{n1} & & \ldots & l_{nn} \end{bmatrix} \begin{bmatrix} u_{11} & u_{12} & \ldots & u_{1n} \\ 0 & u_{22} & \ldots & u_{2n} \\ \vdots & & \ddots & \vdots \\ 0 & & \ldots & u_{nn} \end{bmatrix} = \begin{bmatrix} a_{11} & a_{12} & \ldots & a_{1n} \\ a_{21} & a_{22} & \ldots & a_{2n} \\ \vdots & & \ddots & \vdots \\ a_{n1} & a_{n2} & \ldots & a_{nn} \end{bmatrix}$$

then we have n^2 equations, as many as the entries of A,

in $n^2 + n$ unknowns, the entries of L and U. By inspection we see that each of the n^2 equations has form

$$l_{ip} \, u_{pj} = a_{ij}$$

where $r=\min(i,j)$. Since we have n more unknowns than equations, we can try to set n entries of L or n entries of U to a certain convenient value. It turns out that setting the diagonal elements of L or the diagonal elements of U equal to one, the computation of L and U can be carried out very efficiently. Here we focus on the Doolittle method which sets the element of the diagonal of L equal to one. The computation of the remaining elements of L and U can be performed as follows. The first row of U can be computed from

$$l_{11} \, u_{11} = a_{11}$$
$$l_{11} \, u_{12} = a_{12}$$
$$. \quad . \quad .$$
$$. \quad . \quad .$$
$$. \quad . \quad .$$
$$l_{11} \, u_{1n} = a_{1n}$$

since $l_{11}=1$. Then, the first column of L can be computed from

$$l_{21} \, u_{11} = a_{21}$$
$$l_{31} \, u_{11} = a_{31}$$
$$. \quad . \quad .$$
$$. \quad . \quad .$$
$$. \quad . \quad .$$
$$l_{n1} \, u_{11} = a_{n1}$$

Then, the second row of U can be computed from

$$l_{22} \, u_{22} + l_{21} \, u_{12} = a_{22}$$
$$. \quad . \quad . \quad . \quad .$$
$$. \quad . \quad . \quad . \quad .$$
$$. \quad . \quad . \quad . \quad .$$
$$l_{22} \, u_{2n} + l_{21} \, u_{1n} = a_{2n}$$

and so on, until all the entries of L and U are computed. In compact form, we have for $k=1,\ldots,n$,

$$u_{kj} = a_{kj} - \sum_{p=1}^{k-1} l_{kp} \, u_{pj}, \quad j = k, k+1, \ldots n.$$

$$(3.9)$$

$$l_{ik} = (a_{ik} - \sum_{p=1}^{k-1} l_{ip} \, u_{pj})/u_{kk}, \quad i = k+1, \ldots n.$$

where $\sum_{p=1}^{0}$ is defined to be zero. The above formula says that the computation of L and U can be arranged so that a row of U and a column of L are alternately computed.

Other computational schemes are possible. Each scheme is optimized to take advatange of a particular way of storing the coefficient matrix of A. The method we presented here is particularly efficient when the entries of A above the main diagonal are stored by rows and the elements of A below the main diagonal are stored by columns. We shall see that this method can also be implemented in the sparse matrix case if a bidirectional threaded list is used to store A. From (3.9), it is possible to see that the LU factorization of A requires n divisions and $n^3/3 - n/3$ multiplications. Therefore, since the solution of the two triangular systems require n^2 multiplications, the total computation involved in solving Ax=b by using LU decomposition is the same as GE. The similarity between LU decomposition and GE carries further. In fact, a little algebra shows that the entries of U are equal to the coefficients of the triangular system obtained by GE., i.e. for k=1,..n,

$$u_{kj} = a_{kj}^{(k)}, \quad j = k, k+1, \ldots, n. \qquad (3.10)$$

Moreover, the entries of L are equal to the multipliers $l_{jk} = a_{jk}^{(k)}/a_{kk}^{(k)}$ used in GE.

Because of (3.9) and (3.10), we see that the LU factorization of A is possible only if for k=1,...n, $a_{kk}^{(k)}$ is different from zero. Therefore, before applying the LU decomposition method we have to reorder the rows and columns of A to obtain a matrix A' such that $a'_{kk}^{(k)}$ is different from zero. If A is nonsingular, we have shown that such a reordering is always possible and the LU method can therefore be applied to any nonsingular matrix provided that the correct reordering of

A is performed. It is important to notice that this reordering can be performed "on the fly", i.e. while the LU decomposition is performed. In fact, if u_{kk} is found to be zero or too small if we are considering pivoting for accuracy also, then a row interchange can be performed without throwing away all the intermediate computation performed so far.

4. TECHNIQUES FOR HANDLING SPARSE MATRICES

4.1 Introduction

We have shown that, in absence of special structure of the coefficient matrix, the most efficient techniques available today for the solution of a linear system of algebraic equations require a number of multiplications which is of order n^3, where n is the dimension of the system. LSI circuits may require the solution of a system with thousands of unknowns. Assuming that n=10000, and that the available computer takes 1 us to perform a multiplication, we need computer time of the order of 10^6 sec.,i.e. about 115 days, to solve the system of equations. However, in the evaluation of the computer time needed, the sparsity of the coefficient matrix was not considered. It turns out that the solution of a system of linear algebraic equations assembled by a circuit simulator such as SPICE, requires only an average of $n^{1.5}$ multiplications to solve the system. In our previous example the computer time needed to solve the system would be only 1 sec. Also, in the previous example we considered that all the data needed to perform GE or LU decomposition fit in core. Now, if the coefficient matrix A is stored as a full matrix, then we need $(10^4)^2 = 10^8$ memory locations. However, if the zero entries of A are not stored, the memory locations needed to store A would be proportional to n for general sparse matrices.

In this Section we investigate how sparsity can be exploited to obtain that impressive reduction in complexity and how a sparse matrix can be stored efficiently.

4.2 Pivoting for Maintaining Sparsity

Consider the system whose coefficient matrix A has the following zero-nonzero pattern

$$A = \begin{bmatrix} x & 0 & 0 & x \\ 0 & x & 0 & x \\ 0 & 0 & x & x \\ x & x & x & x \end{bmatrix}$$

Assume that the right hand side b is a vector with no zero elements. If we apply GE or LU decomposition to A the number of multiplications needed to solve Ax=b is

36 according to the formula given in the previous section. However this operation count considers trivial operations of the form 0x0. If only the nontrivial operations are counted then we notice that the first step of GE on A requires 2 multiplications and 1 division, the second step 2 multiplications and 1 division, the third step 2 multiplications and 1 division as well. The modification of the right hand side requires a total of 3 multiplications and the back substitution requires 7 multiplications and 1 division only. Then the total number of long operations is 4 divisions and 16 multiplications. Therefore the presence of zero elements in A can save a considerable amount of operations if trivial operations are not performed. The general formula to compute the number of multiplications needed to solve Ax=b, A sparse, can be shown to be

$$\sum_{k=1}^{n-1} (r_k - 1) c_k \; + \; \sum_{k=1}^{n-1} (c_k - 1) \; + \; \sum_{k=1}^{n-1} (r_k - 1) \; + \; n$$

where r_k is the number of nonzero elements in the k-th row and c_k is the number of nonzero elements in the k-th column of the reduced system of dimension n-k+1 obtained after k-1 eliminations have been performed. Now suppose that the rows and columns of A have been rearranged to give

$$A' = \begin{bmatrix} x & x & x & x \\ x & x & 0 & 0 \\ x & 0 & x & 0 \\ x & 0 & 0 & x \end{bmatrix}$$

The first step of GE applied to A' requires 1 division and 12 multiplications. Moreover, the reduced system after elimination has all its coefficients different from zero. This implies that we need 36 multiplications to obtain the solution of the system A'x=b with no advantage with respect to the full matrix case. This example shows that the number of nontrivial operations required to solve Ax=b when A is sparse, depends upon the row and column permutations of the coefficient matrix.

In general, it is obvious that we should find a permutation of A for which GE or LU decomposition requires the minimum number of multiplications. Another criterion which can be followed in the selection of the permutation of A is the minimization of the number of nonzero elements created in A during GE. In fact, while a zero element of A which remain zero during GE need not be stored or operated upon, a new nonzero, also called a <u>fill-in</u>, must be stored and is used during GE. Thus the minimization of the number of fill-ins is related directly to the minimization of the storage requirements but only indirectly to the minimization of the operations needed to perform GE. Since,

$$a_{ij}^{(k+1)} = a_{ij}^{(k)} - \left(a_{ik}^{(k)}/a_{kk}^{(k)}\right)a_{kj}^{(k)}$$

a fill-in is created in position i,j when pivoting in position k,k at step k, if and only if $a_{ik}^{(k)}$ and $a_{kj}^{(k)}$ are both different from zero.

Because of the relations between GE and LU decomposition, it turns out that the zero-nonzero pattern of the matrix formed by L+U is the same of the zero-nonzero pattern of the matrix A with fill-ins added. Therefore, minimizing the fill-ins of GE also minimizes the storage requirements for LU decomposition. Moreover it is easy to see that the number of operations for GE and LU decomposition is the same even in the case of sparse matrices. Thus, from now on we shall discuss only the case for GE since LU decomposition has the same computational complexity and storage requirements.

To find a permutation of a sparse matrix A which minimizes the number of fill-ins has been proven to be a very hard problem. In fact no algorithm is available today to find such an optimal permutation with a reasonable computational complexity. The minimum fill-in problem belongs to a class of combinatorial problems called NP-complete problems. This means that the worst case complexity of an algorithm for the minimum fill-in problem is 2^q where q is the number of nonzero elements in A, and that there is almost no hope of finding an algorithm whose worst case complexity is bounded by a polynomial in q [10]. Therefore we must use some heuristics to select a permutation such that the number of fill-ins during GE is reasonably small. Among the numerous heuristic algorithms proposed, one of the first to be published, the Markowitz criterion [11], has given the best results : it is cheap and produces very good permutations. The Markowitz criterion is based on the following fact. At step k, the maximum number of fill-ins generated by choosing $a_{ij}^{(k)}$ as pivot is $(r_i-1)(c_j-1)$ where r_i-1 is the number of nonzero elements other than $a_{ij}^{(k)}$ in the i-th row of the reduced system and c_j-1 is the number of nonzero elements other than $a_{ij}^{(k)}$ in column j of the reduced system. Markowitz selects as pivot element at step k the element which minimizes $(r_i-1)(c_j-1)$. For the matrix

$$
A = \begin{bmatrix}
a_{11} & a_{12} & a_{13} & a_{14} \\
a_{21} & a_{22} & 0 & 0 \\
0 & a_{32} & a_{33} & 0 \\
0 & a_{42} & a_{43} & a_{44}
\end{bmatrix}
$$

Markowitz selects as first pivot element a_{21}. For the reduced system

$$
\begin{bmatrix}
a_{12}^{(2)} & a_{13}^{(2)} & a_{14}^{(2)} \\
a_{32}^{(2)} & a_{33}^{(2)} & 0 \\
a_{42}^{(2)} & a_{43}^{(2)} & a_{44}^{(2)}
\end{bmatrix}
$$

the elements in position $(1,4),(3,3),(3,2)$ and $(4,4)$ have the same Markowitz count. In SPICE the ties are broken according to the following rules : the element with minimum column count is chosen, if ties remain then the element with minimum row count is chosen. In our example, $(1,4)$ and $(4,4)$ are selected according to the tie break rules. The choice between the two is now purely arbitrary. After the second elimination step the reduced system is full and any further selection yields the same results as far as computation cost is concerned.

In the previous section, we discussed pivot selection to maintain accuracy in the presence of round-off errors. This selection depends on the numerical value of the element while the pivot selection for maintaining sparsity depends only on the zero-nonzero pattern of the matrix, since exact cancellations during GE are ignored. The two criteria may conflict so that if complete or partial pivoting is used to select pivots, we may have a large number of fill-ins. On the other hand, if we choose pivots only on the basis of the Markowitz criterion, we may compute a solution which is affected by large errors. Circuit simulators using nodal analysis or modified nodal analysis such as SPICE restrict the selections of pivots to the main diagonal to retain the diagonal dominance property of the coefficient matrix which ensures good accuracy. This choice also considerably simplifies the application of the Markowitz criterion since the candidates for the next pivot to be considered at each step are fewer than in the unrestricted case. However, in modified nodal analysis diagonal pivoting may not be possible in some cases.

For example, consider the MNA equations (2.12). If pivots are restricted to the main diagonal and are performed sequentially, after three steps of GE we find $a_{44}=0$. An a priori reordering of the rows and columns of A is needed so that diagonal pivoting can be carried out. Some schemes have been suggested to reorder the coefficient matrix [1,5]. The most efficient scheme is described in [12]. For the simulators based on sparse tableau such as ASTAP, restricting the choice of pivots on the main diagonal does not make sense. In SPICE 2F.2, diagonal pivoting is implemented. However, since in active circuits some diagonal elements may become very small during GE, a check is made on the size of the pivot chosen by the Markowitz algorithm. If the pivot element is smaller than a threshold, then a complete pivoting to retain accuracy is performed. If some ties occur, they are broken by picking up the element which has minimum Markowitz count.

4.3 Data Structures for Sparse Matrices

There are several ways of storing a sparse matrix in compact form. (See for example the review paper by Duff for an almost exhaustive list of the available techniques [13].) Some methods require a minimal amount of storage at the cost of a difficult data accessing scheme and cumbersome updating procedures. Others require more storage but the operation performed during pivot selection and LU decomposition are made much easier. A circuit simulator designer must carefully

trade-off these aspects when selecting a data strucuture for his(her) program. Some simulators such as
ASTAP and SPICE 2 versions up to 2E even used different
data structures in different parts of the program. For
the sake of simplicity, in this section, we focus on a
bidirectional threaded list structure proposed in [14]
and used in SPICE 2F.

We describe the data structure with the help of an
example. Consider the matrix

$$A = \begin{bmatrix} 5 & 3 & 2 & 1 \\ 7 & 2 & 0 & 0 \\ 0 & 1 & 3 & 0 \\ 0 & 7 & -13 & 1 \end{bmatrix}$$

The bidirectional threaded list for A is shown in Figure 11a. Each block in the list has three data and two
pointers. The data are the element value, its row and
column index. The first pointer points to the location
of the next element in the same column and the second
points to the locations of the next element in the same
row. A set of pointers is also provided to locate the
first element in each row and in each column. In Figure
11c, a FORTRAN oriented implementation of this data
strucuture is described. The array VALU contains the
value of the element, the arrays IROW and JCOL contain
respectively the row index and the column index of the
element, the arrays IIPT and JJPT contain the pointers
to the location of the first element in the corresponding rows and columns of A, the arrays IPT and JPT contain the pointers to the next element in the same row
and in the same column of the element in the corresponding location of VALU. If we want to scan column
1, then we read JJPT(1) to find the location of the
first element in column 1. In this example, JJPT(1)=1
and the first element value is in VALU(1)=5. In
IROW(1) we find its row index, IROW(1)=1. The next
element in column 1 is located by reading JPT(1)=5.
The value of the element is VALU(5)=7, its row index is
IROW(5)=2. The location of the next element is given
by JPT(5)=0, i.e. there is no more element in column 1
of the matrix. This data structure is particularly
efficient for sparse matrices algorithms. For example,
after an element not on the main diagonal has been
selected, a row and column interchange is required. To
swap two rows, we only need to redirect the corresponding pointers in IIPT. We also need to change the IROW

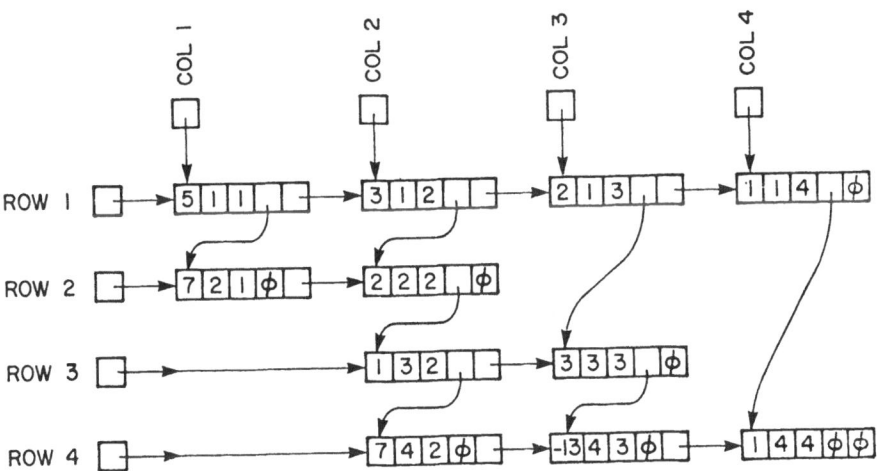

Figure 11a.
The bidirectional threaded list.

$$VALU = \begin{array}{c} (1) \\ (2) \\ (3) \\ (4) \\ (5) \\ (6) \\ (7) \\ (8) \\ (9) \\ (10) \\ (11) \end{array} \begin{bmatrix} 5 \\ 3 \\ 2 \\ 1 \\ 7 \\ 2 \\ 1 \\ 3 \\ 7 \\ -13 \\ 1 \end{bmatrix} ; IROW = \begin{bmatrix} 1 \\ 1 \\ 1 \\ 1 \\ 2 \\ 2 \\ 3 \\ 3 \\ 4 \\ 4 \\ 4 \end{bmatrix} ; JCOL = \begin{bmatrix} 1 \\ 2 \\ 3 \\ 4 \\ 1 \\ 2 \\ 2 \\ 3 \\ 2 \\ 3 \\ 4 \end{bmatrix} ; IPT = \begin{bmatrix} 2 \\ 3 \\ 4 \\ \phi \\ 6 \\ \phi \\ 8 \\ \phi \\ 10 \\ 11 \\ \phi \end{bmatrix} ; JPT = \begin{bmatrix} 5 \\ 6 \\ 8 \\ 11 \\ \phi \\ 7 \\ 9 \\ 10 \\ \phi \\ \phi \\ \phi \end{bmatrix} ; IIPT = \begin{bmatrix} 1 \\ 5 \\ 7 \\ 9 \end{bmatrix} ; JJPT = \begin{bmatrix} 1 \\ 2 \\ 3 \\ 4 \end{bmatrix}$$

Figure 11b.
The FORTRAN oriented implementation of the
bidirectional threaded list.

entries corresponding to the elements in the swapped rows. This can be done by scanning the rows with the help of IPT. Then, we need to insert fill-ins if they are generated by the elimination. Suppose that a nonzero element in position $(2,3)$ has to be added. Its value is stored in VALU(12), its row and column index are stored in IROW(12) and JCOL(12). Then, we need to redirect pointers in row 2 and column 3. To do that we scan row 2 until the element in position $(2,2)$ is reached. Then its row pointer is changed from 0 to 12 and the row pointer associated with the new element will be set to zero. For the column pointers update, we scan column 3 until the element in position $(1,3)$ is reached. Then, we change the column pointer from 8 to 12 and the column pointer associated with the new element is set equal to 8. For LU decomposition, an element of A is replaced by the element of L or U in the corresponding position as soon as it is computed. Because of this way of storing the elements of L and U, the computation of, say, u_{ij}, $j>i$, requires according to (3.9) the use of the elements of U already computed in column j and stored in the bidirectional threaded list and the elements of L already computed in row i and stored in the list. As already pointed out, the retrieval of elements in the same row and in the same column of a given element is accomplished easily in this data structure. The bidirectional threaded list requires $2n + 5l$ storage elements, where l is the number of nonzero elements of the matrix to be stored. However the information on row and column indices can be compacted to reduce the storage requirements by l. Moreover, pointers can be packed in computer words, saving further storage elements.

To speed up the execution of LU factorization, forward and back substitution, the so called compiled code approach has also been used [15]. This approach requires the implementation of a preprocessor which analyzes the structure of both the coefficient matrix and the factored one and generates a loop-free code which can rapidly solve matrices with the same structure. The LU factorization is extremely fast since no testing or branching is performed and every variable is accessed directly (Nagel has reported 70% savings in execution time [1]), but at the expenses of a tremendous increase in storage requirements (in [15]where this approach was proposed, it was reported that a code length of 1.3×10^6 bytes was needed for the LU factorization of a matrix of order 1024 with 15000 nonzeros).

The code generation approach was used in SPICE2 for the CDC version.

4.4 Comparison between Sparse Tableau and Modified Nodal Analysis

We can now compare the analysis methods introduced in Section 2 from the point of view of computational complexity by using the results obtained in Sections 3 and 4.

Assume for the sake of simplicity that the circuit to be analyzed consists of ideal two terminal resistors, ideal independent current sources and VCCS. Then modified nodal analysis coincides with nodal analysis. Let us consider the Sparse Tableau of the circuit

$$\begin{bmatrix} A & 0 & 0 \\ 0 & I & -A^t \\ I & -Y & 0 \end{bmatrix} \begin{bmatrix} i \\ v \\ e \end{bmatrix} = \begin{bmatrix} 0 \\ 0 \\ IS \end{bmatrix}$$

where I is the identity matrix, Y is the branch admittance matrix and IS is the vector of the independent current sources. Let us apply GE to the Sparse Tableau in the following order: eliminate the branch current using the branch equations to obtain

$$\begin{bmatrix} 0 & AY & 0 \\ 0 & I & -A^t \\ I & -Y & 0 \end{bmatrix} \begin{bmatrix} i \\ v \\ e \end{bmatrix} = \begin{bmatrix} -AIS \\ 0 \\ IS \end{bmatrix}$$

then eliminate the branch voltages using KVL to get

$$\begin{bmatrix} 0 & 0 & AYA^t \\ 0 & I & -A^t \\ I & -Y & 0 \end{bmatrix} \begin{bmatrix} i \\ v \\ e \end{bmatrix} = \begin{bmatrix} -AIS \\ 0 \\ IS \end{bmatrix}$$

Now the first set of equations are precisely the nodal analysis equations. In other words, NA can be obtained from the Sparse Tableau by performing GE in a precise order. If some of the elements do not have a voltage controlled characteristic, the modified nodal equations can still be obtained by choosing a precise elimination order on the Sparse Tableau. In principle, this elimination order may be far from optimal with respect to sparsity. Therefore, Sparse Tableau is theoretically more efficient than any other method of analysis if sparsity is considered as the criterion to judge the properties of an analysis method. Note that the pivot sequence we used earlier to reduce the Sparse Tableau to NA does not cost any multiplication since all the coefficients of the variables we eliminated were either +1 or -1. When using Sparse Tableau, one has to provide ways of avoiding trivial operations not only of the form 0x0 but also 1x1. When using NA or MNA the number of coefficients which are +1 or -1 is so small that one does not have to worry about this kind of trivial operations. This is to emphasize that in order to fully exploit the potential of Sparse Tableau more complicated algorithms have to be used.

5 DC ANALYSIS OF NONLINEAR CIRCUITS

5.1 Introduction

After investigating the analysis of linear resistive circuits, the next step towards the analysis of more complicated circuits consists in considering circuits containing nonlinear resistive elements. The analysis of these circuits is an important problem since bias circuits are essentially DC circuits with no capacitors or inductors and since it can be shown that, when numerical algorithms are applied to the analysis of circuit containing capacitors and inductors, we have to solve many different nonlinear resistive circuits. In this section, we first investigate how to formulate the circuit equations. Then, we study how to solve the set of equations assembled and finally we discuss how to implement the algorithms for the solution of nonlinear equations in circuit simulators.

5.2 Equation Formulation

Consider the circuit in Figure 12. KCL and KVL are the same as the circuit of Figure 5. If we append to KCL and KVL the branch equations

$$v_1 - R1\ i_1 = 0$$

$$i_2 - I2\ (\exp(v_2/V_T)) - 1) = 0$$

$$v_3 - R3\ i_3 = 0$$

$$i_4 - I4\ (\exp(v_4/V_T) - 1) = 0$$

$$i_5 = IS5$$

we have a system of 12 equations (2 KCL, 5 KVL, 5 branch equations) in 12 unknowns (5 branch currents, 5 branch voltages and 2 node voltages). These equations are the Sparse Tableau equations for the nonlinear circuit. It is obvious that this method of formulating circuit equations is general and yields a system of 2b + n equations in 2b + n unknowns for a circuit with b branches and n nodes. Circuit simulators use the Sparse Tableau either explicitly (ASTAP) or implicitly (SPICE) to formulate the circuit equations for nonlinear circuits.

5.3 Solution of the Circuit Equations

If we take the right hand side to the left of the equality sign in the circuit equations, we are now faced with the problem of solving a system of nonlinear algebraic equations of the form

$$f(x) = \begin{bmatrix} f_1(x_1, \ldots, x_N) \\ \vdots \\ f_N(x_1, \ldots, x_N) \end{bmatrix} = 0 \qquad (5.1)$$

where f is a function mapping the unknown vector x, a vector of real numbers of dimension N, into a vector of real numbers also of dimension N. The first question to ask is whether (5.1) has a solution and, if it does, whether this solution is unique. In the linear alge-braic case, the assumption that the coefficient matrix is nonsingular is necessary and sufficient to ensure the existence and uniqueness of the solution. Unfortu-nately, the situation is much more complicated in the nonlinear case. There are no necessary and sufficient conditions for the existence and the uniqueness of a solution of (5.1) in the general case. If the circuit is well modeled and well designed, it is obvious that (5.1) should have at least one solution. However the circuit may have more than one solution. For example, the DC equations of a flip-flop have three solutions. In the sequel, we shall assume that a solution of (5.1) exists but we make no assumption on its uniqueness.

Since there are no closed form solutions for the general case, we have to use numerical methods to com-pute x which solves (5.1). The numerical methods available are all iterative, i.e. they produce a se-quence of approximate solutions $x^1, \ldots, x^i, \ldots$ which hopefully converges to one of the exact solutions x^* of (5.1). In circuit simulators, the most successful tech-niques to solve equations of the form (5.1) belong to two classes:

1-Newton-Raphson methods

2-Relaxation methods.

Newton-Raphson methods have been used in almost all the simulators available today, while relaxation methods have been implemented mostly in timing simulators. In this section we shall only consider Newton-Raphson(NR) methods since relaxation methods will be discussed in the chapters dedicated to timing and hybrid simulation.

We introduce the NR methods by means of a simple case:

$$g(y)=0 \qquad (5.2)$$

where y is a scalar and g is a scalar function. If we plot $g(y)$ as a function of y, the solution of (5.2), y^*, is represented by the intersection of the curve representing $g(y)$ with the y axis, as shown in Figure 13a. The idea behind the NR techniques is simple: suppose we start with an initial approximation y^0 to y^*. Then, we approximate the nonlinear function with a linear function which is as "close" as possible to the nonlinear function in a neighborhood of the initial approximation. If we use the Taylor expansion of g, we obtain:

$$g(y)=g(y^0)+(dg(y^0)/dy)(y-y^0) + \text{higher order terms} \quad (5.3)$$

We can truncate the Taylor expansion to obtain

$$g_L(y) = g(y^0)+(dg(y^0)/dy)(y-y^0). \qquad (5.4)$$

Note that $g_L(y)$ is the equation of the tangent line of $g(y)$ at y^0, as shown in Figure 13b. Now we can find a better approximation to y^* by finding the intersection of $g_L(y)$ with the y-axis as shown in Figure 13b,i.e. we can obtain a better approximation y^1 by solving

$$0=g_L(y^1)=g(y^0)+(dg(y^0)/dy)(y^1-y^0) \qquad (5.5)$$

which yields

$$y^1=y^0-[dg(y^0)/dy]^{-1}g(y^0). \qquad (5.6)$$

If we are not satisfied with this approximation, we can repeat the linearization process at y^1. Then we compute

$$y^2=y^1-[dg(y^1)/dy]^{-1}g(y^1). \qquad (5.7)$$

In general, we can repeat this computation many times to obtain a sequence of points y^k where

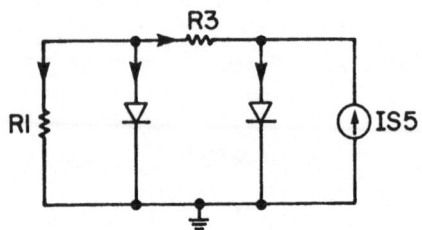

Figure 12
A nonlinear resistive circuit.

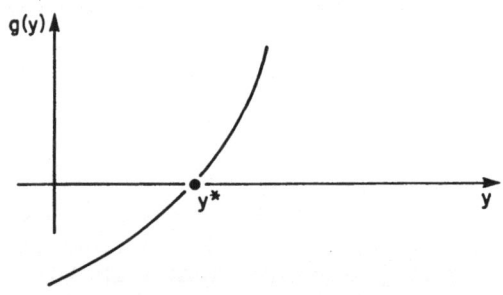

Figure 13a.
The plot of g(y) and the geometric interpretation
of the solution of g(y) = 0.

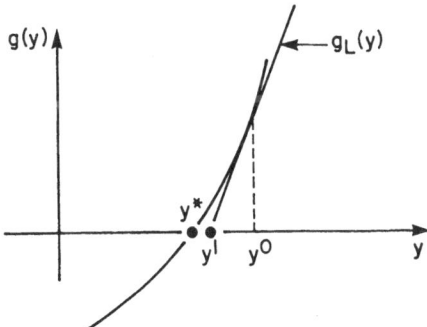

Figure 13b
The linear approximation of $g(y)$ at y^o

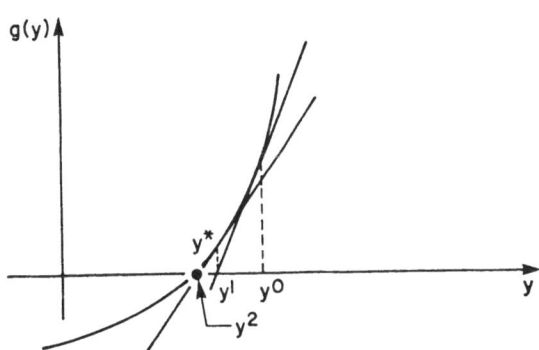

Figure 13c
Convergence of the iterations.

$$y^{k+1} = y^k - [dg(y^k)/dy]^{-1} g(y^k).$$ (5.8)

As we can see from Figure 13c, after a few iterations our approximation gets very close to y^* and in the limit we have

$$y^k \to y^*$$

where our notation means that when k goes to infinity y^k converges to y^*. It is natural to ask whether we can always ensure convergence of this linearization process to a solution of (5.2) in the general case and, if the process converges, how fast it converges to y^*. In order to apply the method we have to ask for the existence of the first derivative of g and of its inverse at the points of the sequence. Some additional conditions stated in the following theorem given without proof are sufficient to guarantee a "fast" convergence to y^*.

Theorem 5.1: Convergence of NR method in the scalar case. Let g be two times continuously differentiable (i.e. g has second derivative which is continuous) and let $dg(y^*)/dy$, where y^* is a solution of (5.2), be different from zero (i.e. y^* is a simple root). Then the NR method always converges to y^* provided that the initial guess is sufficiently close to y^*. Moreover, let

$$\epsilon_k = |y^k - y^*|$$

be the error at iteration k, then

$$\epsilon_{k+1} \le C\epsilon^2_k,$$

where C is a positive constant, i.e. NR method is quadratically convergent.

Note that we have to choose the initial approximation sufficiently close to y^* in order to have convergence. Unfortunately there is in general no way to make sure that the initial guess is sufficiently close to the solution. In Figure 14 we show cases in which the NR method does not converge to a solution due to a poor choice of the initial guess.

A very nice feature of the NR method is its quadratic convergence. To give an idea how fast quadratic convergence is, note that after k iterations we have

$$\epsilon_k \leq (1/C)(C\epsilon_0)^2$$

If $C\epsilon_0 < 1$, say $C\epsilon_0 = 0.9$ and $C=1$, then the error sequence is bounded by

$$0.81, 0.66, 0.44, 0.19, 0.036, 0.0013, 0.000016, \ldots$$

Note that for $k \geq 6$ the number of significant decimals is almost doubled at each iteration.

The generalization to the multidimensional case is rather straightforward. The NR iteration has form

$$x^{k+1} = x^k - J(x^k)^{-1} f(x^k). \qquad (5.9)$$

where $J(x^k)^{-1}$ is the inverse of the Jacobian of f computed at x^k, i.e.

$$J(x^k) = \begin{bmatrix} \partial f_1(x^k)/\partial x_1 \ldots \partial f_1(x^k)/\partial x_N \\ \cdot \quad \cdot \quad \cdot \quad \quad \cdot \quad \cdot \quad \cdot \\ \cdot \quad \cdot \quad \cdot \quad \quad \cdot \quad \cdot \quad \cdot \\ \partial f_N(x^k)/\partial x^1 \ldots \partial f_N(x^k)/\partial x_N \end{bmatrix} \qquad (5.10)$$

Under the assumptions that :

1-$J(x)$ is Lipschitz continuous, i.e. that $J(x)$ is continuous and that the rate of change of $J(x)$ is bounded;

2-$J(x^*)$ is nonsingular

it can be proven that

1- $x^k \to x^*$ provided that the initial guess x^0 is sufficiently close to x^*

2- the rate of convergence is quadratic, i.e.

$$\epsilon_{k+1} \leq C\epsilon^2_k$$

where C is a constant and $\epsilon_k = ||x^k - x^*||$.

Note that the NR iteration is expressed in terms of the inverse of $J(x^k)$. From a computational point of view, it is more convenient to write the iteration as

$$J(x^k)(x^{k+1}-x^k) = -f(x^k). \qquad (5.11)$$

Then, the NR process at each iteration requires:

1-The evaluation of $f(x^k)$;

2-The computation of $J(x^k)$;

3-The solution of a linear system of algebraic equations whose coefficient matrix is $J(x^k)$ and whose right hand side is $-f(x^k)$.

5.4 Application of the NR method to Circuit Equations

Let us consider the Sparse Tableau equations of the circuit in Figure 13. Let v^k, i^k and e^k be the vector of branch currents, the vector of branch voltages and the vector of node voltages at iteration k. Then we have $x=[i_1,\ldots,i_5,v_1,\ldots,v_5,e_1,e_2]^t$ and

$$f_1(x) = i_1 + i_2 + i_3$$
$$\cdot \quad \cdot \quad \cdot \quad \cdot \quad \cdot$$
$$\cdot \quad \cdot \quad \cdot \quad \cdot \quad \cdot$$
$$f_9(x) = i_2-I2(\exp(v_2/V_T)-1)$$

$$(5.12)$$

$$\cdot \quad \cdot \quad \cdot \quad \cdot \quad \cdot \quad \cdot \quad \cdot$$
$$\cdot \quad \cdot \quad \cdot \quad \cdot \quad \cdot \quad \cdot \quad \cdot$$
$$f_{11}(x) = i_4-I4(\exp(v_4/V_T)-1)$$
$$f_{12}(x) = i_5-IS5$$

If we apply the NR method to this system of equations, we find that at iteration k+1 all the coefficients of KCL, of KVL and of all the branch equations of the linear elements remain unchanged with respect to iteration k. For example consider the first equation. Then at iteration k+1, we have

$$\partial f_1(x)/\partial x_1 = \partial f_1(x)/\partial x_2 = \partial f_1(x)/\partial x_3 = 1$$

and

$$f_1(x^k) = i_1^k + i_2^k + i_3^k$$

Thus the NR method yields

$$1(i_1^{k+1}-i_1^{k})+ 1(i_2^{k+1}-i_2^{k+1})+1(i_3^{k+1}-i_3^{k})= \qquad (5.13)$$

$$-i_1^{k}-i_2^{k}-i_3^{k}.$$

By simplifying (5.13), we obtain

$$i_1^{k+1}+i_2^{k+1}+i_3^{k+1}=0 \qquad (5.14)$$

For branches 1 and 3, we have

$$v_1^{k+1}-R1\ i_1^{k+1}=0 \qquad (5.15)$$

$$v_3^{k+1}-R3\ i_3^{k+1}=0$$

For the diode equations we have

$$\partial f_9(x^k)/\partial x_2 = 1$$

and

$$\partial f_9(x^k)/\partial x_7 = I2/V_T \exp (v_2^k/V_T).$$

Thus, we have

$$1(i_2^{k+1}-i_2^{k})-I2/V_T\exp(v_2^k/V_T)(v_2^{k+1}-v_2^k) \qquad (5.16)$$

$$=-i_2^{k}+I2(\exp(v_2^k/V_T)-1)$$

By simplifying (5.16), we obtain

$$i_2^{k+1}-I2\ \exp\ (v_2^k/V_T)(v_2^{k+1}-v_2^k) = I2\ \exp\ (v_2^k/V_T)-1) \qquad (5.17)$$

If we rearrange (5.17), we obtain

$$i_2^{k+1}-I2/V_T\exp(v_2^k/V_T)v_2^{k+1}=$$

$$\qquad (5.18)$$

$$I2(\exp(v_2^k/V_T)-1)-I2/V_T\exp(v_2^k/V_T)v_2^k$$

Note that all the quantities which contain v^k are numbers, since v^k has been previously computed. Then, (5.18) can be written as

$$i_2^{k+1}-G2(k)v_2^{k+1}=IS2(k) \qquad (5.19)$$

where G2(k) is a conductance and IS2(k) is an independent current source. Therefore equation (5.19) can be interpreted as the branch equation of a parallel combination of a linear resistor whose conductance is G2(k) and of an independent current source whose value is IS2(k) as shown in Figure 15. The circuit of Figure 15 is called the <u>companion element</u> of branch 2. The same procedure can be followed for the diode in branch 4 to obtain

$$i_4^{k+1} - G4(k) v_4^{k+1} = IS4(k).$$ (5.20)

where

$$G4(k) = I4/V_T \exp(v_4^k/V_T)$$ (5.21)

and

$$IS4(k) = I4(\exp(v_4^k/V_T) - 1) - I4/V_T \exp(v_4^k/V_T) v_2^k.$$ (5.22)

Thus at iteration k+1 we have a linear system of equations of the form

$$i_1^{k+1} + i_2^{k+1} + i_3^{k+1} = 0$$

$$\begin{matrix} \cdot & \cdot & \cdot & \cdot \\ \cdot & \cdot & \cdot & \cdot \\ \cdot & \cdot & \cdot & \cdot \end{matrix}$$ (5.23)

$$i_4^{k+1} - G4(k) v_4^{k+1} = IS4(k)$$

Now this system of equations can be interpreted as the Sparse Tableau of a linear circuit whose elements are specified by the linearized branch equations. This linear circuit is called <u>the companion network</u>.

This procedure can be applied in the general case to yield the following result:

<u>The NR method applied to a nonlinear circuit whose equations are formulated in the Sparse Tableau form produces at each iteration the Sparse Tableau equations of a linear resistive circuit obtained by linearizing the branch equations of the nonlinear elements and leaving all the other branch equations unmodified.</u>

Therefore, a circuit simulator, when analyzing a circuit in DC, can first linearize the branch equations of the nonlinear elements and then assemble and solve the circuit equations for the companion network. After the

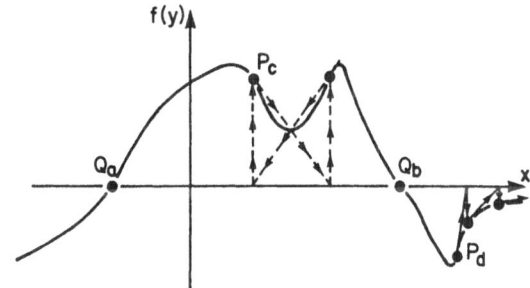

Figure 14.
An example of nonconvergence of NR method
(starting points P_c and P_d)

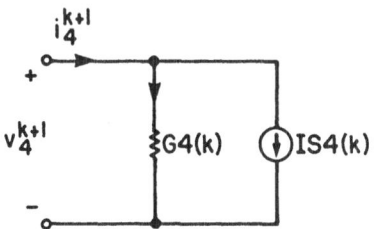

Figure 15.
The companion element of the diode in the
example circuit.

linear circuit is produced, there is no need to stick
to the Sparse Tableau equations, but other methods such
as MNA can be used to assemble the circuit equations.
Note that, since the topology of the circuit, the
branch equations of the linear elements and the structure of the branch equations of the nonlinear elements
remain unchanged from iteration to iteration, the circuit equations need to be assembled only once, provided
that a set of pointers specifies the entries of the
coefficient matrix and of the right hand side which
change from iteration to iteration. This fact implies
that the structure of the coefficient matrix does not
change from iteration to iteration and, hence, the
pivoting for sparsity needs to be done only once. However, since the value of some of the coefficients may
change drastically, it is always a good strategy to
check the numerical values of the pivots while performing the LU factorization of the coefficient matrix. If
a pivot element becomes too small at a certain iteration then a reordering which perturbs as lightly as
possible the pivoting sequence previously determined,
should be accomplished on the fly. Spice version 2G.0
checks the size of the pivot elements and performs a
complete pivoting if the size of the element is found
to be below a threshold.If the bidirectional list
structure presented in the previous section is used,
changing the pivoting order is rather easy and fast. A
difficulty encountered in this procedure is that ,
since the pivoting sequence is changed, the fill-ins
also change. It is an expensive procedure to locate the
fill-ins due to the previous pivoting sequence and
possibly remove the ones that are not generated by the
new sequence. The strategy followed in SPICE 2 is simply to add any new fill-in due to the new pivoting
sequence but to leave alone the fill-ins due to the
previous pivoting sequence. This strategy implies some
unnecessary storage and computation during the LU factorization but the change of the pivoting sequence can
be done much faster.

If Sparse Tableau is used to assemble the linear
circuit equations, then several coefficients of the
Sparse Tableau remain unchanged from iteration to iteration. This also implies that if the pivoting sequence
is chosen carefully many elements of L and U may remain
unchanged from iteration to iteration and a complete LU
factorization is not needed at each iteration. In fact,
if a_{ik}, l_{ip} and $u_{pk}, p=1,...,min(i-1,k)$ do not change
from iteration to iteration, neither does l_{ik}, if $i >$

k or u_{ik} if $i \leq k$. Therefore, the number of elements which remain constant is dependent upon the pivoting sequence. Hachtel et al. [4] have proposed an algorithm for choosing the pivot sequence which takes this effect into account.

Before discussing some of the drawbacks of the NR method, it is important to note that, since convergence to an exact solution of the circuit equation is only possible in the limit, a circuit simulator must decide when to truncate the sequence and accept the last point generated as a good approximation. In general the stopping criterion is expressed as

$$|| x^{k+1}-x^k || < \epsilon_a + \epsilon_r \min\{||x^{k+1}||, ||x^k||\}, \qquad (5.24)$$

where ϵ_a is an absolute error and ϵ_r is a relative error both specified by the user. To avoid "false convergence", the stopping criterion should be applied to both currents and voltages of the nonlinear elements. Nagel [ ] reports that if the criterion is only applied to the voltage of the junctions of bipolar devices, then one can stop quite far away from the exact solution.

5.5 Overflow and Nonconvergence Problems

Two major problems are encountered in the application of the NR method to electronic circuits:

1-Numerical overflow during the iterations.

2-Nonconvergence of the iterations.

The first problem is due to the characteristics of active devices, in particular the exponential characteristic of bipolar devices. Consider the circuit shown in Figure 16a (this example is taken from [1]). The solution v^* is shown graphically in Figure 16b. Now, if the initial guess is v^0, then the result of the first NR iteration is v^1 shown in Figure16b. To carry out the next iteration, we need to evaluate the current in the diode. However, as shown in Figure 16b, the value of the current is very large and we may have a numerical overflow(if v^1 is 10 volts we need to evaluate e^{400}).

Many methods have been proposed for avoiding overflows. All these methods limit the step $\Delta v = v^{k+1} - v^k$

which the NR method prescribes. Therefore, they are generally called limiting methods. In this section we shall only consider the methods implemented in SPICE.

The idea behind the SPICE limiting scheme is to iterate on the current of the junction of bipolar devices when the voltage v^{k+1} obtained by the NR method exceeds a heuristically determined critical junction voltage. In Figure 16c the value of v^{k+1}_{lim} determined according to the limiting scheme is shown. Basically, v^{k+1}_{lim} is the voltage corresponding to the current i^{k+1} determined by the NR iteration. Therefore, the voltage v^{k+1}_{lim} can be computed by equating the current of the linearized element and the current on the exponential characteristic, i.e.

$$i^{k+1} = I/V_T \exp(v^k/V_T)(v^{k+1}-v^k) + I\exp(v^k/V_T)$$

$$= i = I(\exp(v^{k+1}_{lim}/V_T)-1) \quad \text{(junction equation)} \quad (5.25)$$

In (5.20) the only unknown is v^{k+1}_{lim}. The solution of (5.20) yields

$$v^{k+1}_{lim} = v^k + V_T \ln((v^{k+1}-v^k)/V_T + 1) \quad (5.26)$$

This formula simply says that the step Δv should be the natural logarithm of that prescribed by the NR method. As previously pointed out, this formula applies only if v^{k+1} is larger than V_{crit} and than v^k. Otherwise the step suggested by the NR method is accepted. In SPICE2 the value of V_{crit} is given by

$$V_T \ln(V_T/I\sqrt{2}). \quad (5.27)$$

The second problem is the nonconvergence of the NR method in some circuits. The reasons for nonconvergence are usually:

1-The first derivatives of the branch equations of the nonlinear elements are not continuous.

2-The initial guess is not sufficiently close to the solution.

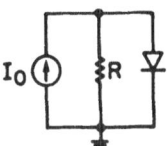

Figure 16a.
A diode circuit.

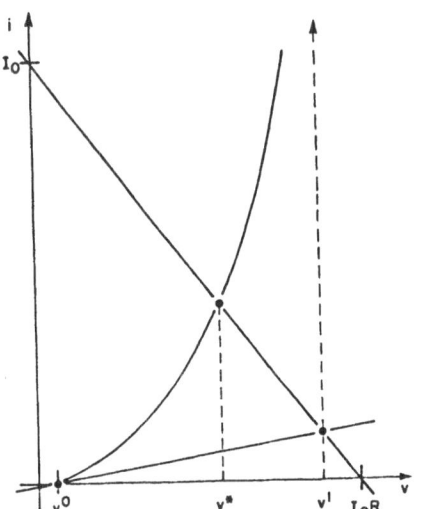

Figure 16b.
Graphical interpretation
of NR method applied to
the diode circuit

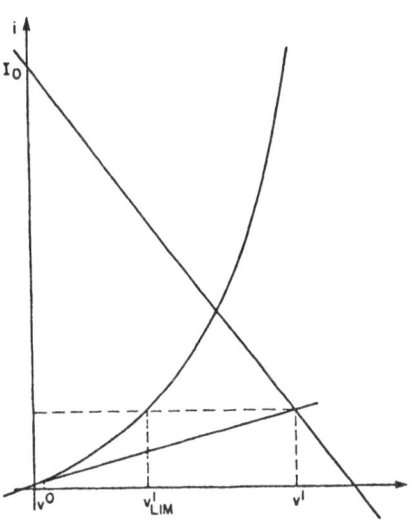

Figure 16c.
The limiting step algor-
ithm: graphical interpre-
tation

The first problem often arises when certain models of electronic devices are used. As Vladimirescu et al. point out [16] some MOSFETs models have discontinuities in their first derivatives. This problem can be fixed by modifying the model equations to have continuity in the first derivative. This requires the model designer to think carefully of the implications of his(her) decisions on the behavior of the numerical algorithms.

The second problem is much more difficult to handle. Several algorithms have been proposed to "stabilize" the NR method, i.e. to enlarge the region in which the initial guess can be made without loosing convergence. These algorithms can be roughly divided into two classes:

1-Optimization based methods.

2-Continuation methods.

The first class of methods constrains the selection of the step taken by the NR method to produce a sufficient decrease in the error, i.e. they require x^{k+1} to be such that $f(x^{k+1})$ is closer than $f(x^k)$ to zero by a prescribed amount. In other words, they solve the problem $\min||f(x)||^2$ by a descent strategy. Of course the solution of the minimization problem coincides with the solution of $f(x)=0$ (if a solution to this problem exists). It can be rigorously proven that the convergence of the NR method is greatly improved at the expense of a certain increase in complexity [17].

The second class of methods requires the solution of a sequence of parametrized problems $P(a)$ whose solution sequence approaches the solution of the original problem when the parameter a ranges from 0 to a value a^*. A method belonging to this class is the so called source stepping algorithm [18]. In this method, the vector of the independent sources values $S(a)$, is determined by $S(a)=aS$, where $0 \leq a \leq 1$ is the parameter and S is the actual value of the independent sources of the circuit to be analyzed. If we assume that:

1-the solution of the circuit is either known or can be computed efficiently when a=0,

2-the solution of the problems is continuous with respect to a, then we can find an a^1 small enough so that the solution of $P(a^1)$ is close to the solution of $P(0)$ and solve $P(a^1)$ using the solution of $P(0)$ as initial guess. This procedure can be repeated until a reaches 1.

In ASTAP, a similar strategy is followed, where the parameter a is time. The circuit is first modified by appending inductors in series to current controlled elements and independent voltage sources and capacitors in parallel to voltage controlled elements and independent current sources. The initial conditions are set to zero for all the inductors and capacitors and a transient analysis is performed. In the next section, we shall see that the transient analysis of a circuit is performed by solving a sequence of DC analysis, one at each time step. If the time step is sufficiently small, then the solution of the DC equations at time t^{k+1} is close to the solution of the DC equations at time t^k, so that the latter can be used as a guess for finding the former solution. When t increases, the circuit will eventually reach a "steady-state" where the current through the capacitors and the voltage across the inductors are both zero. But then, the steady-state solution corresponds to the solution of the original DC equations.

Nagel [1]has observed that the limiting scheme used in SPICE2 solves not only the overflow problem, but also improves the overall convergence of the NR method. Moreover, he has experimentally determined that the use of the stabilized methods increases considerably the number of iterations needed to obtain the solution, so that the price paid to obtain a more reliable algorithm was considered to be to high for these methods to be included in SPICE2.

5.6 Techniques for Reducing the Complexity of the NR method

For circuits with less than a thousand devices, because of the efficient solution of sparse systems of linear equations, the computation of $J(x)$ and the evaluation of $f(x)$ dominate the complexity of the NR method.

In circuit analysis the computation of $J(x)$ implies the computation of the derivative of the branch equations as discussed in the previous section. This computation can be done by the program or the derivatives may be provided by the user or built into the simulator. For example, ASTAP computes the derivatives of the branch equations by finite differences. This method has the advantage of being simple to implement, but it is error prone and numerically unstable. Its computational cost is of one additional function evaluation plus a division per branch equation.

If the derivatives are provided in analytic form by the user, then the cost of computing the Jacobian is equivalent to a function evaluation. This approach has the disadvantage of requiring additional work by the user and there is no guard against possible errors in inputting the derivatives.

Another possible approach could be to write a preprocessor which produces the derivatives of the branch equations in symbolic form. Then the cost of computing $J(x)$ would be the computation time required by the preprocessor and by a function evaluation.

Another strategy, followed by SPICE, is to provide built-in models and derivatives. The obvious disadvantage is that the user cannot use models of his(her) choice. However, function evaluation is much faster. In fact, for ASTAP and other programs which allow user defined models as input, the user defined functions must be parsed each time the code is run, i.e. user defined functions must be separated into symbols and operations. The symbols must be found in an internal symbol table, giving addresses. Then, the values at the addresses must have the operations performed on them. For a predefined function, such as the ones provided in a built-in model, all the steps except the operations are performed only once at the time of program compilation. Thus, only the operations need to be carried out at each time the code is run. Another advantage of built-in models is in the use of limiting step algorithms tailored for each model equation. It seems inefficient to leave the burden of finding a suitable limiting step algorithm for his(her) models to the user.

The techniques previously presented may be labelled as programming techniques. Some algorithms have also been proposed to reduce the number of computations of $J(x)$. Among the most efficient we have:

1-The Broyden rank-one update method, which uses an approximation to the Jacobian $J(x^{k+1})$ by a rank-one modification of $J(x^k)$ (for the full details of this and other related method see the excellent book [19]).

2-The Shamanskii algorithm [20] which uses the Jacobian $J(x^k)$ for 1 steps before it is updated.

Both algorithms can be proven to converge under suitable conditions. However they have rate of convergence which is smaller than the NR method (for example the Shamanskii algorithm has rate of convergence $(1+1)^{1/1+1}$. Thus, they may require more iterations to converge than the NR method and, hence, more function evaluations. This is the main reason for which circuit simulators do not use these algorithms.

To save both Jacobian and function evaluations SPICE2 uses a so-called bypass algorithm. The basic idea is that if the arguments of a nonlinear branch relation do not change "much", then the value of the function and its derivative are also close to their previous values. Therefore, if the difference between arguments of a nonlinear function at two succesive iterations is smaller than a certain value, the function and its derivative are not evaluated but their previous value is used in the next NR iterate. Nagel has shown experimentally that the use of the bypass algorithm saves an average of 4% of computation time. However for large digital circuits, where large part of the circuit is inactive at any given time, experimental studies reported by Newton et al. [30] have shown that savings up to 40% are possible.

6 TRANSIENT ANALYSIS OF DYNAMICAL CIRCUITS

6.1 Introduction

One of the most complex problem in circuit simulation is the transient analysis of dynamical circuits, i.e. of circuits containing capacitors and inductors. In general, the value of some branch voltages and branch currents must be computed for some time interval, say for $0 \leq t \leq T$, given initial conditions on capacitor voltages and inductor currents.

In this section, we discuss how to formulate the circuit equations for transient analysis, then we introduce the methods for the solution of the circuit equations, and finally we discuss their properties and implementation in circuit simulators.

6.2 Circuit Equation Formulation

Let us consider the example in Figure 17, where one of the diodes of the circuit in Figure 12, has a capacitor in parallel to model the minority carrier diffusion process of the bipolar junction. The q-v characteristic of the capacitor is

$$q_6 - q_0(\exp(v_6/V_T) - 1) = 0.$$

Since $dq_6/dt = i_6$, we have the branch equation

$$i_6 = (q_0/V_T) \exp(v_6/V_T) \, dv_6/dt$$

with initial condition $v_6(0) = V_6$.

If we modify the Sparse Tableau equations of the circuit by adding i_6 to node equation 2, a KVL for branch voltage v_6 and the branch equation of the capacitor, then we obtain a set of mixed algebraic and differential equations: 14 equations in 14 unknowns. These equations are the Sparse Tableau equations for the dynamical circuit of Figure 17.

It is trivial to generalize the above procedure for any dynamical circuit formed by interconnecting the ideal elements shown in Table 1: the system of 2b + n equations in 2b + n unknowns so obtained are the Sparse Tableau equations. It turns out that many simulators use implicitly (SPICE) or explicitly (ASTAP) the Sparse Tableau formulation.

6.3 Solution of the Circuit Equations

The Sparse Tableau equations can be written compactly as

$$F(\dot{x},x,t) = 0, x(0) = X_0 \tag{6.1}$$

where x is the vector of the circuit variables, $\dot{x}$ is the time derivative of x, t is time and F is mapping x, $\dot{x}$ and t into a vector of real numbers of dimension 2b + n.

As in the DC case, in general we do not have closed form solutions of (6.1). Thus we must resort to numerical algorithms. The first question to ask is whether a solution of (6.1) exists and is unique. It turns out that, under rather mild conditions on the continuity and differentiability of F, it can be proven that there exists a unique solution. Since in most of the practical cases, well modeled circuits satisfy these conditions, from now on we shall assume that (6.1) has unique solution.

The class of methods to be discussed in this section subdivides the interval [0,T] into a finite set of distinct points:

$$t_0 = 0, \quad t_N = T, \quad t_{n+1} = t_n + h_{n+1}, \quad n = 0,1,\ldots,N.$$

The quantities h_{n+1} are called <u>time steps</u> and their values are called <u>step size</u>. At each point t_n, the numerical methods we are investigating compute an approximation x_n of the exact solution $x(t_n)$. The values of x_n are determined from a set of algebraic equations which approximate (6.1).

Among all the methods available for computing x_n, we shall consider the ones which are mostly used in circuit simulations: the so-called <u>linear multistep methods</u>. For the sake of simplicity, we shall introduce them and discuss their properties for the case of the general first order ordinary scalar differential equation

$$\dot{y} = f(y,t). \tag{6.2}$$

A multistep method is a numerical method which computes y_{n+1} based on the values of y and $\dot{y}$ at previous $p+1$ time points, i.e.

$$y_{n+1} = \sum_{i=0}^{i=p} a_i y_{n-i} + \sum_{i=-1}^{i=p} h_{n-i} b_i \dot{y}_{n-i}$$

$$\tag{6.3}$$

$$= \sum_{i=0}^{i=p} a_i y_{n-i} + \sum_{i=-1}^{i=p} b_i f(y_{n-i}, t_{n-i}).$$

An example of multistep method is the Forward Euler (FE) method which is characterized by

$$y_{n+1} = y_n + h_{n+1} \dot{y}_n \tag{6.4}$$

Here $p=0$, $a_0=1$, $b_0=1$ and all the other coefficients are zero. This method can also be considered as a Taylor expansion of the solution $y(t)$ around t_n truncated at its first term.

Other well known multistep methods with $p=0$, also called single step methods, are:

Backward Euler (BE): $y_{n+1} = y_n + h_{n+1} \dot{y}_{n+1}$ $\hspace{1cm}$ (6.5)

TRapezoidal (TR): $y_{n+1} = y_n + (h_{n+1}/2)(\dot{y}_{n+1} + \dot{y}_n).$ $\hspace{0.5cm}$ (6.6)

TR has $a_0=1$, $b_{-1}=b_0=1/2$.

Multistep methods can be divided into two classes:

1-Explicit methods.

2-Implicit methods.

Explicit methods are characterized by $b_{-1}=0$ and implicit methods by $b_{-1} \neq 0$. For example, FE is an explicit method while both BE and TR are implicit methods. Note that an explicit method does not require any solution of equations to generate y_{n+1}, just function evaluations. An implicit method requires the solution of a nonlinear equation to obtain y_{n+1}, since y_{n+1} is an argument of f. Explicit methods are very cheap to use, but unfortunately, as we shall see later, their stability properties make them unsuitable for circuit simulation.

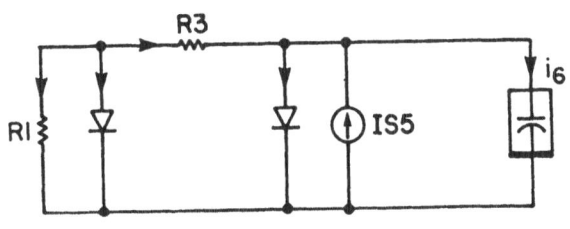

Figure 17.
A dynamical circuit.

6.4 Truncation Error of Linear Multistep Methods

At this point it is natural to ask how well y_n approximates $y(t_n)$ for the different linear multistep methods that can be generated. If we assume that all the computations are made with infinite precision, then the error, i.e. the difference between the computed and the exact solution is due to the particular integration method used. This error is called truncation error of the integration method. It is important to distinguish between local and global truncation error. The Local Truncation Error (LTE) of an integration method at t_{n+1} is the difference between the computed value y_{n+1} and the exact value of the solution $y(t_{n+1})$, assuming that no previous error has been made, i.e. that

$$y(t_i) = y_i \ . \ i=0,1,\ldots,n.$$

The Global Truncation Error (GTE) of an integration method at t_{n+1} is the difference between the computed value y_{n+1} and the exact value of the solution $y(t_{n+1})$ assuming that only the initial condition is known exactly, i.e. that $y_0=y(t_0)$.

For the sake of simplicity, we shall assume that the step size used in the method are uniform, i.e. that $h_1=h_2=\ldots=h_N$. Let us rewrite (6.3) as

$$\sum_{i=-1}^{i=p} a_i y_{n-i} + h \sum_{i=-1}^{i=p} b_i \dot{y}_{n-i} = 0 \tag{6.7}$$

where $a_{-1}=-1$. If we plug the values of the exact solution $y(t_k)$, $k=n+1,\ldots,n-p$ in (6.7), in general we will not obtain 0, but a number which turns out to be the LTE of the method described by (6.7). In particular,

$$LTE_{n+1} = \sum_{i=-1}^{i=p} a_i y(t_{n+1}-ih) + h\sum_{i=-1}^{i=p} b_i y(t_{n+1}-ih). \tag{6.8}$$

where $t_{n-i} = t_{n+1}-ih$ because of the assumption of uniform step sizes.

Now LTE_{n+1} can be seen as the result of the application of an "error operator" $E[y;h]$ to the function y and to the step size h. We are interested in the behaviour of the truncation error when h varies. Thus, let us expand the operator in power series of h to obtain

$$E[y;h] = E[y;0] + E^{(1)}[y;0]h + E^{(2)}[y;0]h^2/2$$

$$+...+E^{(k+1)}[y;0]h^{k+1}/(k+1)+O(h^{k+2}) \tag{6.9}$$

where $E^{(i)}[y;0]$ is the i-th derivative of E with respect to h computed for h=0. From (6.8), we obtain

$$E[y;0]= \sum_{i=-1}^{i=p} a_i y(t_{n+1}) \tag{6.11}$$

$$E^{(1)}[y;0] = \sum_{i=-1}^{i=p}(-i)a_i y(t_{n+1}) + \sum_{i=-1}^{i=p} b_i \dot{y}(t_{n+1}) \tag{6.12}$$

$$E^{(k+1)}[y;0] = \sum_{i=-1}^{i=p}(-i)^{k+1}a_i y^{(k+1)}(t_{n+1})+ \tag{6.13}$$

$$+ (k+1)\sum_{i=-1}^{i=p}(-i)^k y^{(k+1)}(t_{n+1})$$

In general, multistep algorithms of interest are designed so that they are <u>exact</u> for a polynomial of order k, i.e. compute the exact solution of (6.2) if this solution is a polynomial of degree k in t. These algorithms are said to be of <u>order k</u>. Now, the requirement that a linear multistep method be of order k imposes some constraints on the coefficients a_i and b_i. In fact, suppose that y(t) is a polynomial of degree k and let us apply (6.8) to it. Since the method is supposed to be exact, the LTE should be identically zero. Since $E^{(k+1)}$ and all the terms of the power series which follow it contain derivatives of y of order larger than k, they are all identically zero. Therefore, the exactness of the algorithm for polynomials of degree k implies that $E[y;0]$ and $E^{(i)}[y;0],i=1,...,k$ must be identically zero. It is also true that if $E[y;0]$ and its k-th derivatives computed at y and 0 are identically zero, then the method is of order k. Therefore we can conclude that a method is of order k if and only if its coefficient satisfy a set of linear algebraic equations of the form

$$\sum_{i=0}^{i=p} a_i = 1$$

$$\sum_{i=0}^{i=p} (-i) a_i + \sum_{i=-1}^{i=p} b_i = 1 \qquad\qquad (6.14)$$

$$\sum_{i=0}^{i=p} (-i)^k a_i + k \sum_{i=-1}^{i=p} (-i)^{k-1} b_i = 1$$

since $a_{-1} = -1$ by definition. This set of equations determines the so called <u>exactness constraints</u>. Note that we have k+1 equations in 2p+3 unknowns, the coefficients of the method and hence that $p \geq |(k-2)/2|$ where $|c|$ denotes the ceiling of c, i.e. the smallest integer larger than or equal to c.

Now, for a linear multistep algorithm of order k, the LTE can be approximated by $E[y;h]=E^{(k+1)}[y;0]$ since all the terms preceding it in the power series are identically zero. Therefore, the LTE of a multistep method of order k can be approximated by

$$LTE_{n+1} = [C_k h^{k+1}/(k+1)] y^{(k+1)} (t_{n+1}). \qquad (6.15)$$

where C_k is a constant depending on the coefficients and the order of the method. Note that the coefficients of BE and FE satisfy the first and second exactness constraints and hence they are both first order methods, while the coefficients of TR satisfy also the third exactness constraint and is therefore of second order. Applying (6.13) we see that the constant c_k of the LTE expression for FE, BE and TR are:

FE: 1/2

BE: -1/2

TR: -1/12

Note that the expression for the LTE of a k-th order method implies that the error can be driven to zero as h is made smaller and smaller. Now the question to answer is how the LTE accumulates to produce GTE.

6.5 Stability of Linear Multistep Methods

To see how errors can accumulate, let us consider the so called <u>explicit midpoint method</u>

$$y_{n+1} = y_{n-1} + 2h\dot{y}_n \qquad (6.16)$$

It is easy to see that this is an explicit method of order 2, with p=1.

Assume that we use this method to compute the solution of the linear differential equation

$$\dot{y} = -y, y(0) = 1 \qquad (6.17)$$

whose exact solution is y(t)=exp(-t). Also assume that we use uniform step size, e.g. h=1. Note that to apply the method we need to know also y(1), which we suppose is given exactly, i.e. y(1)=exp(-1).

The method computes the following sequence of points

0.264,-0.161,0.585,-1.331,3.247,-7.825,18.897,...

i.e. gives rise to growing oscillations, so that , when n goes to infinity, $|y_{n+1}|$ also goes to infinity, while the exact solution y(t) obviously goes to zero when t goes to infinity. It can be easily shown that the behavior of the method does not change qualitatively if h changes. The only difference is that if h is small then the oscillations start growing later. The oscillations are set off by the LTE or even by round-off errors in the evalution of the formula.

Now let us apply FE to the same equation. The sequence of points computed by FE with h=1 is a sequence of all zeros, i.e. the method underestimates the exact solution but does not exhibit the unstability of the previous method. However, if we choose h=3, we have the sequence

-2,4,-8,16,-32,...

which shows growing oscillations.

If we apply BE to the same problem with h=1, we obtain the sequence

1/2,1/4,1/8,1/16,....

which is obviously well behaved. Even if we increase h, the sequence computed by BE has the property that, when n goes to infinity, y_{n+1} goes to zero. As we shall see later, it can be proven that for all h greater than zero the sequence of points computed by BE stays bounded.

Unfortunately it is very difficult to obtain stability results for general differential equations of the form (6.2). Therefore, in this section we shall discuss the stability of linear multistep method with the help of a simple test problem

$$\dot{y} = -\lambda y, y(0) = 1. \qquad (6.18)$$

where λ is in general a complex constant. This test problem is simple enough to be analyzed theoretically and yet complex enough that any result obtained for it is of practical relevance.

When a linear multistep method with uniform stepsize is applied to the test equation, we obtain a difference equation of the form

$$y_{n+1} = \sum_{i=0}^{i=p} a_i y_{n-i} + h\sum_{i=-1}^{i=p} b_i (-\lambda y_{n-i}). \qquad (6.19)$$

It is convenient to write (6.19) in the following form

$$(1+\sigma b_{-1})y_{n+1} - (a_0-\sigma b_0)y_n - \ldots - (a_p-\sigma b_p)y_{n-p}=0 \qquad (6.20)$$

where $\sigma=\lambda h$. The stability properties of the multistep method can be studied by looking at the solutions of (6.20). Note that the values computed by the multistep method satisfy the difference equation. As we have seen from the examples dicussed earlier, stability is related to the "growth" of the solution of the difference equations. Now we can state precisely the definition of stability of a linear multistep method.

A linear multistep method is said to be stable if all the solutions of the associated difference equation obtained from (6.20) setting $\sigma=0$ remain bounded as n goes to infinity. The region of absolute stability of a linear multistep method is the set of complex values σ such that all the solutions of the difference equation (6.20) remain bounded as n goes to infinity. Note that a method is stable if its region of absolute stability contains the origin.

Since we can obtain the solutions of (6.20) in closed form, we can find rather easily the region of absolute stability of a given method. In fact, given a value of σ, (6.20) has a complete solution of the form

$$y_n = c_1 z_1^n + \ldots + c_{p+1} z_{p+1}^n \qquad (6.21)$$

where z_i, $i=1,\ldots,p+1$, are the roots of the polynomial equation

$$P(z) = (1+\sigma b_{-1}) z^{p+1} - (a_0 - \sigma b_0) z^p - \ldots - (a_p - \sigma b_p) = 0 \quad (6.22)$$

provided they are all distinct. If (6.20) has multiple roots, z_k of multiplicity m_k, then z_k appears in (6.21) as

$$(c_{k_1} + c_{k_2} n + \ldots + c_{k_{m_k}} n^{m_k - 1}) z_k^n. \qquad (6.23)$$

From (6.21) and (6.23), we can conclude that the region of absolute stability of a multistep method is the set σ such that all the roots of (6.22) are such that $|z_i| \le 1$, $i=1,\ldots,k_{p+1}$, where k_{p+1} is the number of distinct roots of (6.22), and that the roots of unit modulus are of multiplicity 1.

For example, consider FE and its associated difference equation

$$y_{n+1} - y_n + \sigma y_n = 0 \qquad (6.24)$$

whose solution is

$$y_n = c (1-\sigma)^n. \qquad (6.25)$$

The region of absolute stability of FE is found immediately to be described by

$$|1-\sigma| \le 1, \qquad (6.26)$$

This is a circle in the σ plane with radius 1 and center at 1, as shown in Figure 18. The region of absolute stability explains the behavior of FE in the previously discussed example In fact, h=1 is inside the region of absolute stability since $\lambda=1$, while h=3 is outside the region of absolute stability. Consider now the difference equation for BE:

$$y_{n+1}+\sigma y_{n+1}-y_n=0 \qquad (6.27)$$

The region of absolute stability of BE is given by

$$|1/(1+\sigma)|\leq1 \qquad (6.28)$$

the region shaded in Figure 19. Note that, for stable test equations, i.e. for equations with Re$\lambda \geq0$, the step size can be chosen arbitrarily large without producing any growing oscillation.

For TR the difference equation is

$$y_{n+1}+(\sigma/2)y_{n+1}-y_n+(\sigma/2)y_n=0 \qquad (6.29)$$

and the region of absolute stability is described by

$$|-1+\sigma/2|/|1+\sigma/2| \qquad (6.30)$$

which is shown in Figure 20. TR has the same properties as BE: the time step can be chosen arbitrarily large without causing numerical instability.

Note that the regions of absolute stability of BE and TR both contain the entire right half plane. Methods which have this property are called A-stable. We shall see in the next section the importance of this property for the solution of circuit equations.

Dahlquist [22] has shown that A-stability implies k ≤2, i.e. that only first and second order linear multistep methods can be A-stable. Moreover he has shown that TR is the most accurate A-stable method, i.e. the A-stable method with the smallest LTE.

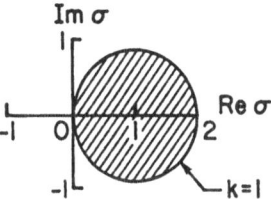

Fig. 18   Region of absolute stability of FE.

Fig. 19 Region of absolute
stability of BE.

Fig. 20 Region of absolute
stability of TR.

This property has made TR popular in circuit simulation. For example, SPICE2 is mostly run with TR as integration method.

6.6 Stiff Problems

The accurate analysis of analog circuits often requires that the parasitics in the circuit components and interconnections be accounte for. These parasitic elements are usually modeled as lumped capacitors and inductors which are in general very small. When these circuits contain bypass and/or coupling elements other capacitors and/or inductors several order of magnitude larger than the parasitic elements have to be considered in circuit simulation. This situation often implies that the solution of the dynamical circuit equations has "fast" and "slow" components, the fast components due to the parasitics and the slow components due to the bypass and coupling elements. Digital circuits also exhibit this mix of slow and fast components as well as many other engineering systems. Circuits (and in general systems) which exhibit widely separated "scale of motions" as well as their equations are called stiff.

The integration of stiff differential equations presents several difficulties. For example, consider the linear differential equation

$$\dot{y} = -\lambda_1(y-s(t)) + ds/dt, y(0) = y_0$$

with $s(t)=1-\exp(-\lambda_2 t)$. The exact solution of (6.31) is

$$y(t)=y(0)\exp(-\lambda_1 t)+(1-\exp(-\lambda_2 t))$$

and is plotted in Figure 21. Suppose that λ_1 and λ_2 are widely separated, say $\lambda_1=10^6$ and $\lambda_2=1$, so that (6.31) is stiff. Note that to have enough information about the solution, we have to integrate (6.31) for at least 5 sec., since the exponential waveform of the source has a time constant of 1 sec. Note also that the first exponential waveform of the solution dies out after about 5 μsec.

To minimize the computation needed to find the solution of (6.31), we should minimize the number of time steps needed to cover the interval [0,5] while maintaining the desired accuracy. If we use a uniform time

step say h, then h must be chosen so that the fast
transient is observed. This implies that h must be of
order of magnitude comparable to 10^{-6} sec. Since the
integration must be carried out for 5 sec., we have a
total of 5×10^6 time points needed to integrate a trivial equation such as (6.31). It would make more sense
to use an adaptive choice of step size, so that , after
the fast transient is over, the step size could be
increased taking into account only the error in computing the "slow" transient. In this example, after approximately 5 steps with $h = 10^{-6}$ sec, one would like to
increase h to about 1 sec. to observe the slow transient, for a total of about 10 time points.

Unfortunately, this strategy cannot be implemented
for all the linear multistep methods. In fact, consider
FE applied to (6.31). Let us choose a step size of
10^{-6} to start the integration. After 5 steps, let us
change the step size to 10^{-4} sec. Assume $y(0) = 2$, then
the sequence of points obtained by FE is

$$10^{-6}, 2 \times 10^{-6}, 3 \times 10^{-6}, 10^{-4}, 2 \times 10^{-4},$$

$$3 \times 10^{-4}, -4 \times 10^{-3}, .47, -46, 4613$$

and we have growing oscillations, i.e. numerical instability. It turns out that the values of h for which FE
applied to (6.31) is stable, can be obtained by looking
at the region of absolute stability of FE with $\sigma = h\lambda_1$,
i.e. the stability of the method depends on the
smallest time constant. In Figure 22 we show graphically the mechanism which causes FE to explode. Note
that y_{n+1} computed by FE is based on the first derivative of the solution passing through y_n. If y_n is not
very close to the exact solution, then FE picks up a
spurious fast transient which causes instability if h
is not chosen small enough.

This example shows that:

1-The analysis of stiff circuits requires the use of
variable step sizes;

2-Not all the linear multistep methods can be efficiently used to integrate stiff equations.

In order to be able to choose the step size based
only on the accuracy considerations, the region of
absolute stability of the integration method should
allow a large step size for large time constants, without being constrained by the small time constants.

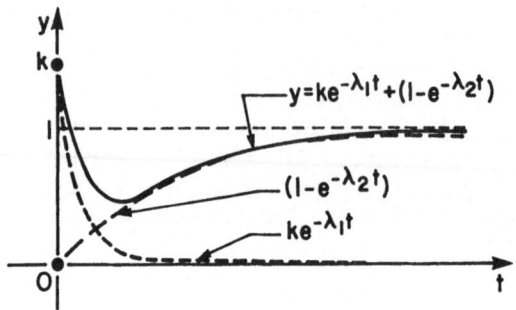

Fig. 21.   Exact solution of eq. (6.31)

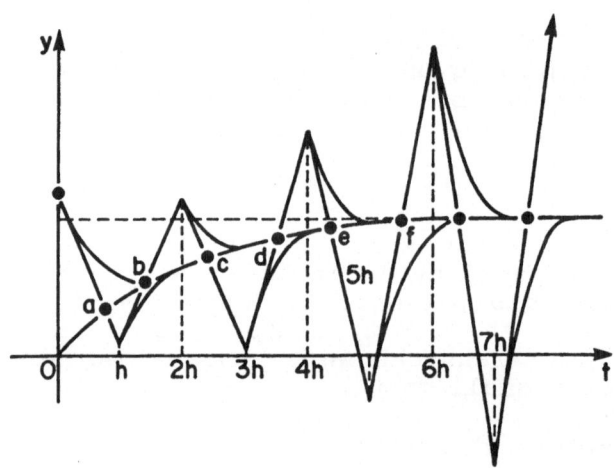

Fig. 22.   Numerical instability of FE applied to (6.31)

Clearly, A-stable methods satisfy this requirement. However it is reasonable to ask whether it is possible to devise integration methods of order larger than 2 which are suitable for stiff differential equations.

Gear [23] has discussed the characteristics of the region of absolute stability of a linear multistep method suited for the integration of stiff differential equations. In essence, Gear considerations imply that the region of absolute stability include the sector | arg (λh) |$<\alpha$ for some $\alpha>0$. A method which satisfies this requirement is said to be <u>stiffly</u> <u>stable</u> or also <u>A(α)-stable</u>. Gear has also proposed a particular class of linear multistep methods characterized by

$$y_{n+1} = \sum_{i=0}^{i=p} a_i y_{n-i} + h_{n+1} b_{-1} \dot{y}_{n+1}$$

A k-th order Gear's method has p = k-1.

The regions of absolute stability of Gear's methods for k=1,2,...,6 are shown in Figure 23. Note that Gear's first order method is BE. From Figure 23, it is immediate to see that Gear's methods up to order 6 are A(α)-stable and, hence, well suited for the analysis of stiff differential equations. However it is possible to show that Gear's methods of order higher than 6 are not stiffly stable. In the literature Gear's method are also called <u>Backward</u> <u>Differentiation</u> <u>Formulae</u>(BDF) [24].

6.7 Linear Multistep Methods with Variable Step Size

As discussed in Section 6.4, the coefficients of a linear multistep methods of order k can be computed by solving a system of linear algebraic equations derived by constraining the method to be exact for polynomials of order k. When the time step is uniform, the coefficients can be used for all problems and time steps. However, when the time step is changed during the integration, the computation of the coefficients of a multistep method becomes more complicated. For example, consider Gear's method of order 2, characterized by

$$y_{n+1} = a_0 y_n + a_1 y_{n-1} + h_{n+1} b_{-1} \dot{y}_{n+1} \qquad (6.33)$$

92

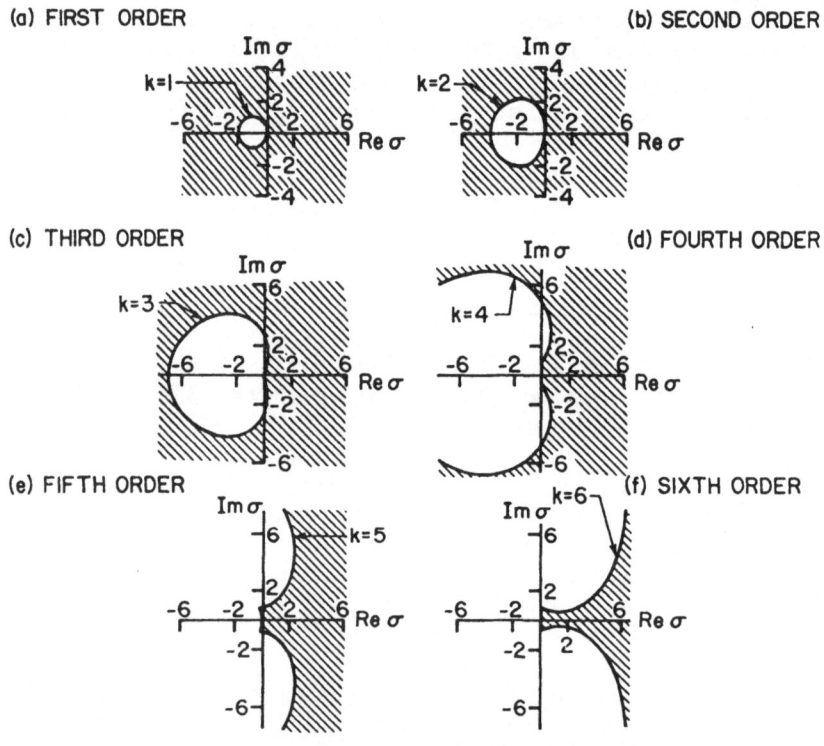

Fig. 23.   Region of absolute stability of Gear's method

where $h_{n+1} = t_{n+1} - t_n$. Let us now consider the most general polynomial of order 2

$$y(t) = c_0 + c_1 t + c_2 t^2 \qquad (6.34)$$

and plug it into (6.33) to obtain

$$y(t_{n+1}) = a_0 y(t_n) + a_1 y(t_{n-1}) + h_{n+1} b_{-1} \dot{y}(t_{n+1})$$

Since

$$y(t_i) = c_0 + c_1 t_i + c_2 t^2{}_i$$

and

$$\dot{y}(t_i) = c_1 + 2c_2 t_i$$

then, setting $t_n = t_{n+1} - h_{n+1}$, $t_{n-1} = t_{n+1} - h_{n+1} - h_n$, we have

$$c_0 + c_1 t_{n+1} + c_2 t^2{}_{n+1} \equiv$$

$$\equiv a_0 [c_0 + c_1(t_{n+1} - h_{n+1}) + c_2(t_{n+1} - h_{n+1})^2] + a_1 [c_0 + c_1$$

$$(t_{n+1} - h_{n+1} - h_n) + c_2 (t_{n+1} - h_{n+1} - h_n)^2]. \qquad (6.35)$$

From (6.35), we obtain

$$a_0 = (h_{n+1} + h_n)^2 / [h_n (2h_{n+1} + h_n)]$$

$$a_1 = -h^2{}_{n+1} / [h_n (2h_{n+1} + h_n)]$$

$$b_{-1} = (h_{n+1} + h_n) / [(2h_{n+1} + h_n)]$$

Note that the coefficients depend on both h_{n+1} and h_n, and hence they must be recomputed at each time step.

Brayton et al. [24] have proposed an efficient algorithm for computing the coefficients of Gear's methods with variable time step. They have shown that the number of operations needed to update the coefficients at each step is proportional to k.

To conclude this section, we would like to mention that Gear's methods can be made unstable by selecting a sequence of rapidly varying time steps. (See for example [24].) This seems to be in contrast with the theoretical results based on the region of absolute stability of Gear's methods. A little thought shows that there is no contradiction. In fact, the regions of absolute stability of Gear's methods have been obtained for <u>uniform step size</u>. Unfortunately, it is very difficult to discuss the stability of general linear multistep methods when the time step is dynamically changed. However, Nevannlina et al. and Liniger [26,27,28] have introduced multistep methods with proven stability properties for the case of rapidly changing time step. Even though Gear's methods can be made unstable, in most of the practical cases they are well behaved and hence useful for circuit simulation.

6.8 Time Step Control for Multistep Methods

As previously pointed out, to minimize the computation time needed to integrate differential equations, a step size must be chosen as large as possible provided that the desired accuracy is achieved. In this section, we concentrate on methods to choose the step size efficiently.

Suppose that a bound, E_{n+1}, on the absolute value of the LTE at time t_{n+1} is given. We know that, for a multistep method of order k,

$$| LTE_{n+1} | = |[c_{k+1}h_{n+1}^{k+1}/(k+1)!]y^{(k+1)}(t_{n+1})| + O(h^{k+2}) \quad (6.36)$$

where c_{k+1} can be computed from the coefficients and the order of the method. If we neglect the higher order term in (6.36), we have that

$$h_{n+1} \leq |[(k+1)! |E_{n+1}/c_{k+1}y^{(k+1)}(t_{n+1})]|^{1/k+1} \quad (6.37)$$

in order to meet the error specifications. If we knew $y^{(k+1)}(t_{n+1})$ then we could use (6.37) directly to compute the time step. Since this quantity cannot be computed exactly, we have to use some approximation.

Several approximation are available. Here we shall discuss the method used by SPICE2: Divided Differences (DD). DD are defined recursively as follows :

$$DD_1(t_{n+1}) = \frac{y_{n+1}-y_n}{h_{n+1}}$$

$$DD_2(t_{n+1}) = \frac{DD_1(t_{n+1})-DD_1(t_n)}{h_{n+1}+h_n}$$

$$\cdot \qquad \cdot$$
$$\cdot \qquad \cdot$$
$$\cdot \qquad \cdot$$

$$DD_{k+1}(t_{n+1}) = \frac{DD_k(t_{n+1})-DD_k(t_n)}{\sum\limits_{i=-1}^{k-1} h_{n-i}}$$

It is relatively easy to show (see [29] for example) that

$$y^{(k+1)}(t_{n+1}) \cong (k+1)!\,DD_{k+1}(t_{n+1}).$$

Thus the algorithm for time-step control should be as follows.

Given a step h_{n+1}, y_{n+1} is computed according to the method chosen for integration. Then, y_{n+1} and h_{n+1} are used to compute $DD_{k+1}(t_{n+1})$. and h_{n+1} is checked by using

$$h_{n+1} \leq [\frac{E_{n+1}}{c_{k+1}|DD_{k+1}(t_{n+1})|}]^{1/k+1} \qquad (6.38)$$

If h_{n+1} satisfies the test, it is accepted; otherwise it is rejected and a new h_{n+1} given by the right hand side of (6.38) is used.

After h_{n+1} has been accepted h_{n+2} has to be chosen. A commonly used strategy is to use the right hand side of (6.38) as the new step size.

An important factor in the selection of the step size is the choice of E_{n+1}. In general, a circuit de-

signer has a rough idea of the precision which (s)he
wants in his(her) simulation. However this is in general related to the GTE not to LTE. Therefore a simulation program must translate this requirement into a
bound on the LTE. In general this is done by assuming
that the LTE accumulates <u>linearly</u>, i.e. that the worst
GTE is achieved at the final time T of the simulation
and that

$$|GTE_{n+1}| \leq \sum_{i=1}^{n+1} |LTE_i|$$

(6.39)

For <u>stable</u> <u>systems</u> <u>and</u> <u>for</u> <u>A(α)-stable</u> <u>methods</u>, certain
bounds can be rigorously obtained given the integration
method, the equations of the system to be analyzed and
the LTE at previous time points. (See for example
[23,29]). These results confirm that the LTE accumulates less than linearly but also show that (6.39) is
quite conservative.

In practice, E_{n+1} is computed as follows. Let E_{max}
be the bound on the GTE given by the user, then the
quantity E_{max}/T is called the <u>maximum</u> <u>allowed</u> <u>error</u> <u>per</u>
<u>unit</u> <u>step</u> and is denoted by ϵ_u. Then E_{n+1} is obtained
by

$$E_{n+1} = \epsilon_u h_{n+1}.$$

(6.40)

Because of the approximation used in:

1-deriving (6.37),

2-replacing $|y^{(k+1)}(t_{n+1})|$ with $DD_{k+1}(t_{n+1})$,

3-defining E_{n+1},

it has been observed experimentally that the time step
computed by means of (6.38) is too conservative. Thus
in SPICE2 the right hand side of (6.38) is multiplied
by a parameter larger than 1 which can be controlled by
the user. The default value used in the program is 7.

In SPICE2, E_{n+1} is also given by means of a combination of absolute and relative errors as follows

$$E_{n+1} = \epsilon_a + \epsilon_r |y_{n+1}| . \qquad (6.41)$$

6.9 Choice of the Integration Method

As previously mentioned, to minimize the cost of computation, we should choose a step size as large as possible. At a certain time point say t_{n+1}, different integration methods would allow different step size since (6.37) obviously depends on the coefficients and on the order of the method. If we plot the maximum allowed error per unit step versus the step size for multistep methods of order 1,2,3 and 4, we have the result shown in Figure 24. By examining Figure 24, it is immediately seen that for a given error, there exists a method which gives the best (i.e. the largest) step size. Thus, it seems advantageous to implement a strategy which allows a change of method as well as of time step.

If we restrict ourselves to the use of stiffly stable algorithms, then a variable step size-variable method strategy is most efficient when Gear's methods are used, since we have a choice of methods of order from 1 to 6.

The strategy used in ASTAP and in SPICE2 when using variable order methods is essentially the following.

At time t_{n+1}, after y_{n+1} has been computed, (6.37) is evaluated for the current method, which we assume to be of order k, and for the methods of order k+1 and k-1. Then, the method which allows the largest step size h_{n+2} is chosen to integrate the circuit equations at time t_{n+2}. This implies that the necessary information to compute (6.37) for the method of order k+1 must be stored. If divided differences are used to compute (6.37), then we need to store the value of y_{n-k} in addition to $y_{n+1}, y_n, \ldots, y_{n-k+1}$ needed for the method of order k.

If this strategy is followed literally, we might end up with changing the order of the method and the step size at each time point, causing the computation time spent in overhead to grow considerably. Moreover if h_{n+2} is taken as the largest step size predicted by (6.37), we might need to reduce the step size a few times before an acceptable error is produced at time t_{n+2}. Therefore, the following heuristic rules are followed:

1-Consider a possible change of order only after m steps have been taken with the current method.

2-Change the order of the method only when the step size for the new method is at least R_0 times the step size allowed by the current method, where R_0 is a constant larger than 1.

3-Increase the step size only if the new step size is R_1 times the old step size, where $R_1 \leq R_0$ is a constant larger than or equal to 1.

4-Increase the step size at most of R_2 times the old time step.

The optimal choice of these parameters is rather difficult. In general they are tuned by tests on sample systems to be analyzed.

6.10 Application of Linear Multistep Method to Circuit Equations

We pointed out in the previous sections that variable step methods have to be used to analyze stiff circuits efficiently. Stiffly stable methods are an obvious choice for circuit simulators, however they are implicit. This fact implies that the application of stiffly stable methods is not straightforward. In this section, we shall investigate how to apply stiffly stable methods (in particular TR and Gear's methods).

The major difficulty in applying numerical integration methods to circuit equations is the fact that circuit equation are usually given at the input in implicit form. In the past, a lot of research effort has been spent in trying to devise efficient methods to formulate the circuit equations in the form

$$\dot{w} = f(w, t)$$

where w is the vector of the so called <u>state</u> <u>variables</u> of the circuit (in general, capacitor voltages (charges) and inductor currents (fluxes)). Unfortunately, these trials have been unsatisfactory.

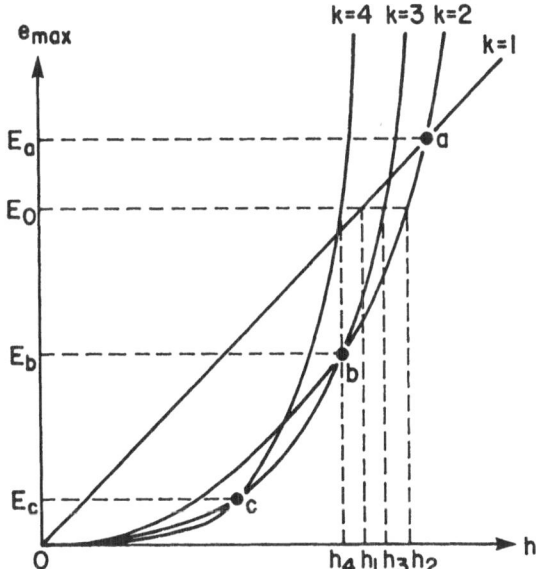

Fig. 24. Maximum error per unit step vs. step-size
for multistep methods of order 1, 2, 3 and 4.

Now if we have a set of mixed differential-algebraic equations of the form

$$F(\dot{x}, x, t) = 0$$

(recall that in our notation x is a vector), at time t_{n+1} we have to solve

$$F(\dot{x}_{n+1}, x_{n+1}, t_{n+1}) = 0 \qquad (6.42)$$

The problem with (6.42) is that $\dot{x}_{n+1}$ is not known. However, let us rewrite Gear's method equations as

$$\dot{x}_{n+1} = -\frac{1}{h_{n+1}b_{-1}} \sum_{i=-1}^{i=p} a_i x_{n-1} \qquad (6.43)$$

Now (6.43) gives $\dot{x}_{n+1}$ as a function of x_{n+1} and of x at previous time points. Since the value of x at previous time points is known, (6.43) can be written compactly as

$$\dot{x}_{n+1} = \dot{x}_{n+1}(x_{n+1}). \qquad (6.44)$$

We can use (6.44) to replace $\dot{x}_{n+1}$ in (6.42) to obtain at t_{n+1}

$$F(\dot{x}_{n+1}(x_{n+1}), x_{n+1}, t_{n+1}) = 0 \qquad (6.45)$$

Now (6.45) is a set of algebraic nonlinear equations with x_{n+1} as unknowns and the techniques discussed in Section 5 can be used to solve it.

For TR we have an analogous situation since we can write

$$\dot{x}_{n+1} = 2 \frac{x_{n+1} - x_n}{h_{n+1}} - \dot{x}_n \qquad (6.46)$$

Now we can use (6.46) to obtain an algebraic equation

with x_{n+1} as unknown.

Now let us apply this strategy to the Sparse Tableau Equations of the circuit of Figure 17. Assume that we are using BE as integration method. All KVL, KCL and branch equations of resistive elements do not contain derivatives with respect to time. Thus, the application of the BE formula does not involve these equations. The only change is in the branch equation of the capacitor where at t_{n+1} we have

$$\dot{v}_{6,n+1} = \frac{1}{h_{n+1}} (v_{6,n+1} - v_{6,n}) \qquad (6.47)$$

yielding

$$i_{6,n+1} = \frac{q_0}{V_T} \exp \ (v_{6,n+1}/V_T) [\frac{1}{h_{n+1}} (v_{6,n+1} - v_{6,n+1})]$$

Since at t_{n+1}, $v_{6,n+1}$ is known, (6.48) can be interpreted as the branch equation of a nonlinear resistor. Therefore, at time t_{n+1}, we have to solve the Sparse Tableau equations of a nonlinear resistive circuit obtained by applying the integration method to the branch equation of the capacitor.

This result is general in the sense that it applies to any dynamical circuit and to any implicit integration method. Circuit simulators such as ASTAP and SPICE2, when analyzing the transient behavior of a dynamical circuit, discretize the branch equations by replacing the derivatives with respect to time of the circuit variables with the integration method formulae. ASTAP uses Gear's variable step-variable order methods, while SPICE2 uses variable time step-TR with variable step-variable order Gear's methods as option for the user.

After the branch equations have been discretized, we have a DC problem which can be solved with the NR method. In particular, the branch equations are linearized and the linear circuit equations are assembled and solved by using the methods described in Section 2, 3 and 4. The overall structure of a circuit simulator is shown in Figure 25.

It is worth mentioning at this stage that the analysis of the nonlinear circuit produced by the integration methods applied to the branch equations of the dynamical circuit can be made faster and more reliable by using the following strategy.

As pointed out in Section 5, most of the problems associated with the use of the NR method in solving circuit equations came from the fact that the initial guess may be so far away that convergence is slow and sometime is not even achieved. Now if the time step is taken small enough, the solution of the resistive nonlinear circuit at time t_{n+1} cannot be too different from the solution of the circuit equations at time t_n. In fact in most practical cases the solution is a continuous function of time. Therefore, we can use the solution at the previous time point as educated guess. However, we can do even better to obtain a better guess.

We pointed out in the previous sections that explicit methods are cheap to apply but have poor stability properties to be used in circuit simulation. However we can still use them at time t_n just to provide an estimate of the value given by the implicit method used, i.e. they can be used as predictors. SPICE2 uses a simple first order predictor to improve convergence of the NR method.

A more complex strategy involving the use of predictors is suggested by Brayton at al. [24] The strategy calls for the use of k-th order predictors in conjunction with k-th order Gear's methods. By doing this, they showed that the computation of the LTE is greatly simplified. In fact, they proved that the local truncation error on the i-th component of the unknown vector x, is given by

$$\text{LTE}_{i,n+1} = \frac{h_{n+1}}{\sum\limits_{i=-1}^{i=p} h_i} \ (x_{i,n+1} - x_{i,n+1}^p) + O(h^{k+2})$$

where $x_{i,n+1}^p$ is the predicted value of the unknown and $x_{i,n+1}$ is the value computed after the nonlinear resistive circuit has been analyzed.

In SPICE2 another strategy for the choice of the step size is implemented. The strategy is related to the minimization of the NR iterations needed to achieve convergence in the analysis of the nonlinear resistive circuit. In particular, there are two predetermined numbers n_1 and n_2, where n_2 is less than n_1. If the NR method does not converge in n_1 iterations, the step size is reduced by a factor of 8. On the other hand if it converges in fewer than n_2 iterations the step size is doubled. Nagel [1] has compared the time step control strategy based on LTE with this strategy and has concluded that the iteration count strategy may produce a large error in the solution. However, more recent tests have shown that the iteration count strategy works well for MOS circuits. Note that the time step control based on iteration count requires much less computation time than the time step control strategy based on LTE.

104

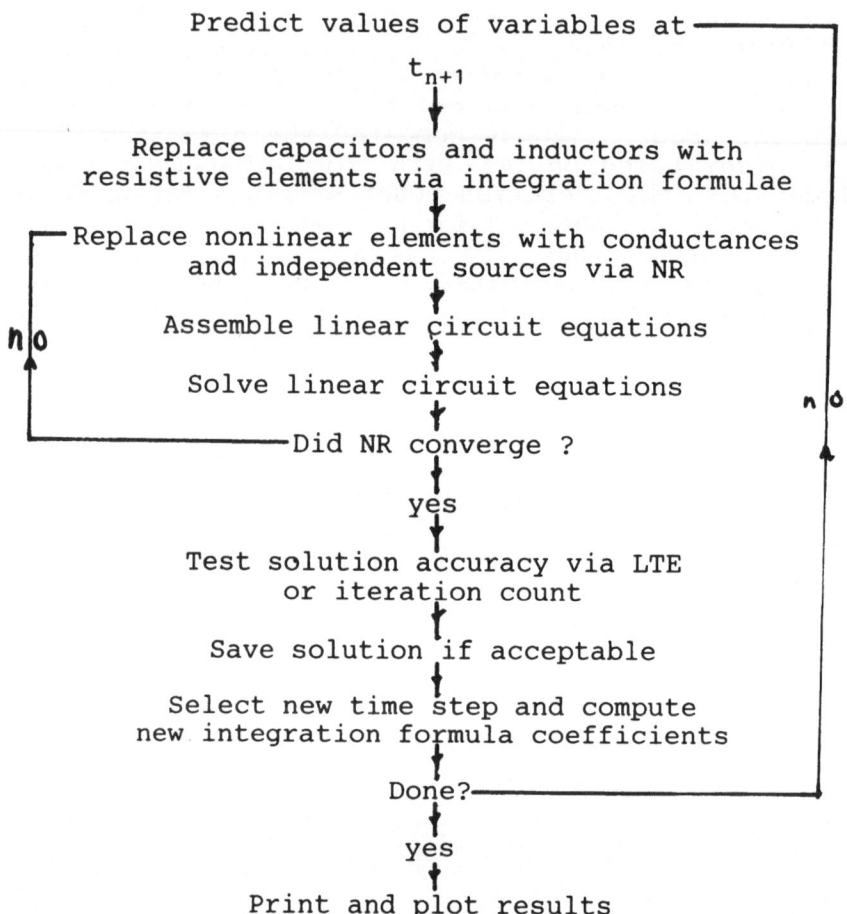

Predict values of variables at t_{n+1}

Replace capacitors and inductors with resistive elements via integration formulae

Replace nonlinear elements with conductances and independent sources via NR

Assemble linear circuit equations

Solve linear circuit equations

no

Did NR converge ?

no

yes

Test solution accuracy via LTE or iteration count

Save solution if acceptable

Select new time step and compute new integration formula coefficients

Done?

yes

Print and plot results

Fig.25. Structure of a Transient Analysis Simulator

7 CONCLUSIONS

We have reviewed the basic techniques used in circuit simulation and discussed their implementation. This section is devoted to the analysis of performance limitations of circuit simulators available today. Our discussion will be based on the performance of SPICE2. Most of the data are derived from Newton and Pederson [30].

7.1 Memory Limitations

As already mentioned in Section 5, SPICE2 has built-in models for electronic devices such as diodes, bipolar junction transistors, MOSFETs and JFETs. In Table 2 we list the memory required for storing information related to the circuit component. The data are given for the CDC6400 which has 60-bit words. Unfortunately, at this moment no structured data are available for SPICE2.F2.

Each device requires a number of computer words to store the following information

1-a model pointer,

2-device-dependent parameters (e.g. resistor values, MOSFET channel length),

3-pointers to the locations of the elements of the MNA matrix which depend on the model parameters,

4-past iteration values for capacitor charges and voltages and for inductor fluxes and currents. This information is needed to implement the time-step control based on LTE as described in Section 6.

5-branch equation derivatives used in building up the linearized circuit associated with the NR method.

The second column of Table 2 shows the number of computer words needed to store the information described in 1,2 and 3, and the third column of Table 2 the information described in 4 and 5. The data in the third column of Table 2 are related to TR. In fact, when using TR and the time-step control based on LTE, 4 past values of capacitor and inductor variables are needed to compute DD_3. This is the reason for labelling the third column of Table 2 "four state vectors".

element	device storage	four state vectors	total words
resistor	14	0	14
capacitor	12	8	28
inductor	14	8	22
voltage source	16	0	16
current source	11	0	11
BJT	35	56	91
MOSFET	45	88	133
JFET	38	52	82
DIODE	28	28	48

Outputs: words = (noutp+1)*numt tp
Codgen: words = 2.5 *iops

Table 2. Storage Requirements in SPICE2D.3.

For MOSFETs 32 words/device are required. Finally, computer memory is needed to store the outputs and the code generated for LU decomposition (recall that SPICE2F.2 does not generate machine language code for the VAX version, these data are for CDC6400 version).

Newton and Pederson [30] reported that for a binary-to-octal decoder with 7500 words for the elements, a total of 13000 words were needed for the analysis. Note that the entire data are stored in core to improve speed. When we look at VLSI circuits containing well over 10000 active devices, it is clear that other methods have to be used to store circuit information. For example, for a 256-by-1 bit RAM circuit Newton estimated that SPICE2 would have needed 220000 computer words to perform the analysis.

7.2 Computer Time Limitations

Figure 26 shows a bar chart displaying the effective time spent by SPICE2 in each subroutine. Note that the routines with name ALNLOG,SQRT MOSFET, MOSEQN and MOS-CAP are all related to the model evaluations needed to build up the discretized and linearized circuit. Therefore for a medium size circuit about 80 of the total computer time is actually spent in model evaluation. The solution of the linear equation is performed in the routine CODEXEC which accounts for only a small percentage of the total cpu time.

Many attempts have been made to reduce the cpu time spent in model evaluation. Lookup tables and simplified device models have been the most successful. These techniques will be discussed thoroughly in the following chapters.

According to Table 3, where the model of the three MOSFET models implemented in SPICE2D.8 are shown, the analysis of a circuit containing 10000 MOS transistors, for 1000 ns of simulation time, would require more than 30 hours on an IBM 370/168 computer. This shows the tremendous cost of using present circuit simulators for the analysis of LSI.

It is important to note that, even if most of the cpu time is spent in model evaluation for a medium size circuit, it has been experimentally observed that the computational complexity of model evaluation grows only linearly with circuit size, while the computational

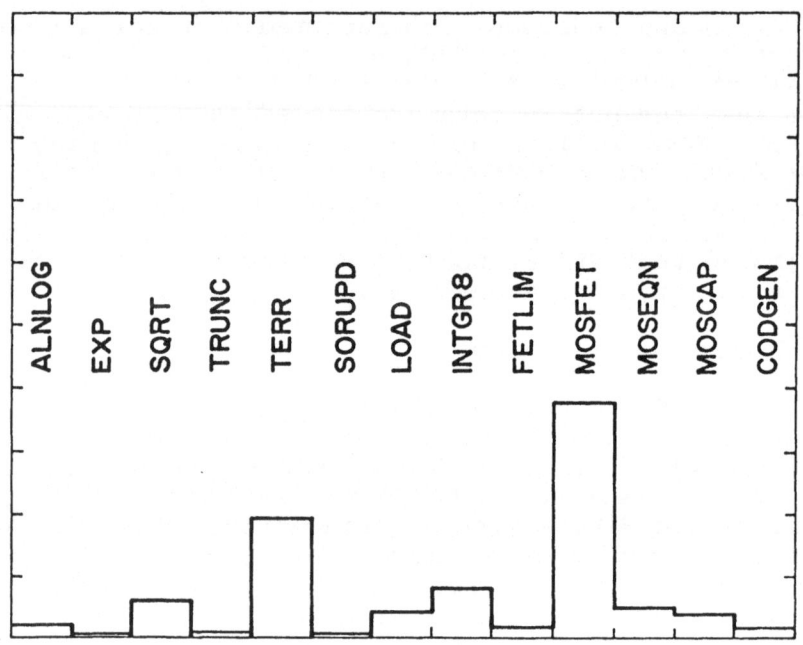

Fig. 26.   Time Statistics for Spice Runs.

model	cp/d/it (ms)	%cp load	it/tp averg	cp/pp (ms)	cp/pp norm
GP	2.9	81	4.5	2.6*	12*
EMI	2.3	77	4.7	2.3*	18*
MOS3	9.8	88	3.7	2.2	18
MOS2	4.3	79	3.1	1.3	6
MOS1	3.7	78	4.1	0.9	4
MOS0	0.24	21	4.6	0.22	1

```
 *   normalized by device count (48/88)
cp: central processor time for analysis
 d: active device
it: Newton-Raphson iteration
tp: time point
pp: user requested print point
```

Table 3. Relative model performance for
         Binary-to-Octal Decoder.

complexity of the linear equation solver grows approximately as $n^{1.4}$ where n is the "size" of the circuit. Nagel (private communication) has observed that for circuits containing approximately 3000 active devices the cpu time spent in model evaluation is equal to the time spent in the solution of the linear system of equations. Therefore, we can conclude that the most limiting factor in the application of circuit simulators to VLSI circuits is the fact that the solution of a large system of linear equation is needed at each iteration of the NR algorithm.

Sparsity has already been exploited so much that it does not seem possible to improve the sparse matrix algorithms by a factor which would make circuit simulators suitable for VLSI analysis. In our opinion the structure of large circuits has to be exploited beyond sparsity. Decomposition algorithms based on the fact that a large circuit is formed by repetitive subcircuits must be developed and tested. Decomposition must be applied at the linear equation solution level as well as at the nonlinear and differential equation level in order to cope with the LSI complexity. One of the important advantages of decomposition is the capability of using parallel computing in an effective way.

We believe that the future generation of circuit simulators will be extensively using decomposition at all levels. We also believe that there is a definite trend towards closing the gap between timing an circuit simulators performances both in accuracy and computing speed.

References.

1. L. Nagel, SPICE2: A Computer Program to Simulate Semiconductor Circuits, ERL Memo. No. UCB/ERL M75/520, May 1975.

2. Weeks, W.T. et al., Algorithms for ASTAP - A Network Analysis Program, IEEE Trans. on CT., vol. CT20, pp. 628-634, Nov. 1973.

3. C.A. Desoer and E.S. Kuh, Basic Circuit Theory, McGraw-Hill, 1969.

4. G.D. Hachtel et al., The Sparse Tableau Approach to Network Analysis and Design, IEEE Trans. on Circuit Theory, vol. CT-18, Jan. 1971, pp. 101-108.

5. C.W. Ho et al., The Modified Nodal Approach to Network Analysis, IEEE Trans. on Circuits and Systems, vol. CAS-22, No. 6, pp. 504-509, Jan. 1975.

6. T.K. Young and R.W. Dutton, MSINC - A Modular Simulator for Integrated Nonlinear Circuits with MOS Model Example, Proc. 1974 International Symposium on Circuits and Systems, April 1974, pp. 720-724.

7. L.O. Chua and P.M. Lin, Computer Aided Analysis of Electronic Circuits, Prentice-Hall, 1975.

8. V. Strassen, Gaussian Elimination is not Optimal, Numerical Math. 13, pp. 354-356, 1969.

9. R.M. Karp, Reducibility among Combinatorial Problems, in Complexity of Computer Computations, (R.E. Miller and J.W. Thatcher, Eds.) New York: Plenum, 1972.

10. R.M. Karp, On the Computational Complexity of Combinatorial Problems, Networks, vol. 5, pp. 45-68, 1975.

11. H.M. Markowitz, The Elimination Form of the Inverse and Its Application to Linear Programming, Management Sci., vol. 3, April 1957, pp. 255-269.

12. P. Yang, I.N. Hajj and T.N. Trick, On Equation Ordering in the Modified Nodal Approach, Proc. ISCAS, Nov. 1979.

13. I.S. Duff, A Survey of Sparse Matrix Research, Proc. of the IEEE, April 1977.

14. W.J. McCalla, unpublished manuscript.

15. F.G. Gustavson et al., Symbolic Generation of An Optimal Crout Algorithm for Sparse Systems of Linear Equations, JACM, vol. 17, pp. 87-109, 1970.

16. A. Vladimirescu and S. Liu, The Simulation of MOS Integrated Circuits Using SPICE2, ERL Memo. No. UCB/ERL M80/7, Feb. 1980.

17. E. Polak, On the Global Stabilization of Locally Convergent Algorithms, _Automatica_, vol. 12, pp. 337-342, 1976.

18. D.F. Davidenko, On a New Method of Numerical Solution of Systems of Nonlinear Equations, _Dokl. Akad. Nauk. SSSR_, vol. 88, 1953, pp. 601-602.

19. J.M. Ortega and W.C. Rheinboldt, _Iterative Solution of Nonlinear Equations in Several Variables_, Academic Press, 1970.

20. V. Shamanski, On A Modification of the Newton Method, _Ukrain Math. Journal_, 19, pp. 133-138.

21. E. Polak and Teodoru, Newton Derived Methods for Nonlinear Equations and Inequalities, _Nonlinear Programming_, 2, Academic Press 1975.

22. G. Dahlquist, Stability and Error Bounds in the Numerical Integration of Ordinary Differential Equations, _Trans. of the Royal Inst. of Technology_, No. 130, pp. 1-86, 1959.

23. C.W. Gear, _Numerical Initial Value Problems in Ordinary Differential Equations_, Prentice-Hall, 1971.

24. R.K. Brayton et al., A New Efficient Algorithm for Solving Differential-Algebraic Systems Using Implicit Backward Difference Formulas, _Proc. IEEE_, vol. 60, No. 1, pp. 98-108, Jan. 1972.

25. R.K. Brayton and C.C. Conley, Some Results on the Stability and Instability methods with Non-uniform Time Steps, in Topics in Numerical Analysis, _Proc. Royal Irish Academy Conf. Numerical Analysis_. London: Academic Press, 1972, pp. 13-33.

26. O. Nevanlina and W. Linger, Contractive Methods for Stiff Differential Equations: Part I, _BIT_, 18, pp. 457-474, 1978.

27. O. Nevanlina and W. Liniger, Contractive Methods for Stiff Differential Equations: Part II, _BIT_, 19, pp. 53-72, 1979.

28. W. Liniger, Multistep and One-leg Methods for Implicit Mixed Differential Algebraic Systems, _IEEE Trans. on Circuits and Systems_, vol. CAS-26, pp. 755-760, Sept. 1979.

29. G. Dahlquist et al., _Numerical Methods_, Prentice-Hall, 1974.

30. A.R. Newton and D.O. Pederson, Analysis Time, Accuracy and Memory Requirement Tradeoffs in SPICE2, _Proc. Asilomar Conference_, 1978.

MIXED-MODE CIRCUIT SIMULATION TECHNIQUES AND THEIR IMPLEMENTATION
IN DIANA

H. De Man, G. Arnout, P. Reynaert

Catholic University of Leuven, Belgium

1. INTRODUCTION

The complexity of integrated circuits has increased from a few bipolar transistors per chip in 1960 to 80.000 MOS devices/chip in
1980.
As a result the simple design techniques whereby IC's are designed
directly at the transistor level both for simulation and manual
layout are no longer sufficient.
Such a design style would indeed give rise to unreasonably high
design costs and product development time. If no changes in design
methodology are made it has been predicted that the design of the
1982 microprocessor would take 68 manyears [1] which is clearly intolerable. Indeed it would make design cost so high that it is
hard to imagine which products would be candidates for realization
in VLSI technology [2] .
Recently formal descriptions of new design methodologies [3] have
been formulated calling for hierarchical, structured design. However important such methods may be, perhaps the biggest concern
in such a complex design is to insure design correctness during
the design process by a continuous verification at, and amongst,
all levels of design i.e. from concept to realization of masks and
test patterns. Only in this way can we prevent that a methodolo-
gical design does not become a disaster to debug after prototype
realization.
Clearly here the role of CAD tools is becoming an absolute must
instead of a luxury.
In the first contribution of this course circuit simulation has
been studied. It is a CAD tool which stands closest to technology
and layout and is invaluable for software "breadboarding" at
transistor level.

In section 2 of this contribution we will show how the hierarchical design methodology requires the use of simulation tools at
different levels of abstraction, many of which result from digital
systems design.
However, the requirement for design correctness and the nature of
the VLSI design process imposes a "unification" or integration of
the different levels of simulation such that CAD does not become
the use of loosely coupled programs each with their own language,
data structure and algorithm.
Also traditional disciplines such as register transfer simulation
[4] and logic simulation [5] need to be adapted to VLSI requirements. Therefore in section 3 the limits of circuit simulation
will be studied.
Based on these limits and the properties of VLSI design the concepts of event-driven circuit simulation, timing simulation, macromodeling and mixed-mode simulation are introduced.
The implementation of these concepts in the DIANA program will be
discussed. Section 4 is concerned with the concept of Boolean
Controlled network Elements (BCE) and their application to a new
form of logic simulation implemented in DIANA. In section 5 the
multiphase MOS and CMOS macromodels in DIANA are introduced as well
as the full mixed mode capabilities and interfacing problems.

2. VLSI DESIGN AND THE NEED FOR MIXED MODE SIMULATION

In order to cope with the enormous amount of transistors in circuitry and rectangles in layout one has to introduce a design
methodology and to insure the correctness of all steps in the design of a VLSI chip. The methodology should be such that a systematic approach from system to transistor level is possible. Therefore it is useful to consider first a few common characteristics
of VLSI design which are likely to occur and which may have a
strong influence on CAD tools to insure design correctness.

As a first characteristic one can say that VLSI IC's are
integrated systems which for over 90% will be digital and, for testability reasons [6], synchronous. A minor part will be analog
circuitry at the periphery of the same chip (A/D, D/A, filters).
These analog circuits are likely to be sampled systems [7].
This property could be called time partitioning of the circuits by
multiphase clocks.

As a second characteristic one can see a clear tendency towards the use of highly repetitive structures for at least the
following reasons :
- digital systems handle 8...60 bit data in parallel (bus-bit
    or serially (shifting networks),

- analog networks are built out of "unit cells" (opamp-switch-capacitor in ladder filters);
- memory as superregular block is very cheap ;
- reduction of data manipulation and storage;
- reduction of design effort;
- solution of the problem of interaction of the global and local parts in LSI design.

We call this property : repetitivity in space.

As a result of these two characteristics it is easy to understand that design correctness can only be insured by a so called top-down design approach whereby however at the implementation phase very often design iterations have to be made due to the strong link between technology, layout and behavior. This contrasts sharphy with PCB design unless a high degree of standardization is imposed such as standard cell design [8] [9] [10] procedures. The latter however can be area consuming and only justifiable for custom LSI.

The design levels which are likely to be considered are illustrated in Table 1.

Level	activity	CAD tool for verification
System	protocols, algorithms	behavioral languages
Register level	synchronous registers & boolean vector operations	register transfer language & simulation
Gate-bit-opamp	Boolean realization in gates or use of analog cells	logic and timing simulation verification of time partitioned blocks
Transistor	realization of cells	circuit simulation of cells verification and parametrization.

Table 1.

Register-transfer level (RTL) simulation will be covered elsewhere in this course. It exploits the fact that a finite state synchronous machine can be considered as a synchronous concurrent number of combinatorial Boolean vector operations between registers (one- or more dimensional).

Once this architecture is behaviorally verified by simulations then the combinatorial vector operations and registers are synthesized as (time partitioned) subblocks such as PLA's, ALU's, SR, MPX, RAM, ROM, I/O networks etc...

These networks are only logically characterized i.e. no timing information is present if they are not layed out yet ! (local-global problem).

Therefore logic simulation at (1,0,X) level and at most unit delay is usually performed to verify the logic behavior against the RTL behavior) not including timing yet.

Once the logic diagram is fixed it is translated into a transistor schematic including transmission gates, buffers etc... Reasonably accurate circuit simulation is only feasable after this circuit schematic has been translated into actual layout e.g. using symbolic layout (see elsewhere in the course).

This circuit simulation will most probably be used for parametrization of the space repetitive cells. Then the bottom-up implementation phase has to be started. The parametrized cells can then be used for accurate timing verification of the time partitioned blocks between registers. There accurate gate and transmission gate delay models linked to technology are necessary.

Similar arguments can be given for the design of sampled data analog circuits as will be discussed in another lecture on this particular topic.

The main point here is that the design of VLSI systems is a hierarchical process exploiting space repetitivity and time partitioning whereby :
- all levels of simulation (RTL, logic, timing, circuit) are necessary;
- simulation results of one level need to be stimuli of lower levels;
- a close link to technology and layout is necessary;
- the possibility must exist to use several levels of simulation simultaneously (e.g. analog-digital circuits, well defined logic and sensing amplifiers, bootstrapping, dynamic-logic, etc...).
- if possible data transfer and storage as well as description language should reflect hierarchy and insure design correctness.

Therefore, in contrast to a simulation system consisting of a collection of separate packages with different languages, data structures, algorithms and data-bases, VLSI design requires an integrated simulation package. Such a system should be adapted to the VLSI characteristics of time partitioning and space repetitiveness. We hereby look for a harmoneous way of combining "matrix" based analysis with "event" driven analysis into what is called mixed-mode simulation.

The ideal would indeed be to combine :
- Sampled data system simulation;
- Register transfer simulation;
- Logic simulation;
- Timing simulation;
- Circuit simulation;
driven by a :
- Common hierarchical description language.

Today several incomplete approaches exist such a SPLICE [1] combining logic simulation, timing and circuit simulation and discussed in detail in this course, DIANA [12] ,which in addition to the above also covers sampled data system simulation and ADLIB-SABLE [13] covering system down to logic simulation.

In what follows we will build up the concept of mixed-mode from bottom (circuit simulation) to top (logic-simulation) as it also occurred historically in the "design" of the DIANA and SPLICE program. In such a way both the principles of mixed-mode and the understanding of DIANA are achieved.

3. INTRODUCTION TO EVENT-DRIVEN CIRCUIT SIMULATION

3.1. The limitations to circuit simulation (CS).

There is no doubt that circuit simulation gives the highest accuracy in behavioral verification which, for digital circuits, is usually time-domain simulation.
A direct link to technology and layout exists through device models and therefore CS has been the traditional workhouse of IC designers.

We will first look at the problems of extending circuit simulation up to higher complexity. All CS programs are based on the following principles, explained in lecture 1 :
1. Implicit integration which transforms nonlinear differential equations into non-linear algebraic difference equations.
Hereby the user wants analysis results for N_{tu}(NOUT*) time steps of length h_u (DELTA).
Time step control for a given truncation error will create an internal timestep $h_i < h_u$, thereby creating a true number of time steps $N_t > N_{tu}$.
2. Within each internal time step and for every non-linearity in the network a (Newton-Raphson) iteration process takes place. Hereby the nonlinear network analysis consists of N_i analyses of the linearized network solved by :
3. Matrix inversion by L U decomposition of the sparse network-matrix of which the dimension is close to the number of nodes n in the network.
If sparse storage and sparse inversion is used then the matrix inversion time t_{INV} is proportional to n or :

$$t_{INV} = \overline{t}_i . n \qquad (1)$$

$\overline{t}_i$ is the average matrix inversion time/node.
For circuit analysis according to the above characteristics it follows :

Property 1 : for circuit simulation over a fixed number of time points N_{tu} the CPU time for large circuits increases more than proportional to the number of nodes n in the circuit i.e.

$$t_{SIM} = Cn^{\alpha} \qquad (1 < \alpha \leq 2) \qquad (2)$$

---

* Symbols used in DIANA manual.

This can be understood as follows :

$$t_{SIM} = (t_M + t_{INV}) \, N_i N_t \tag{3}$$

Hereby t_M is the time to <u>recalculate</u> (update) the costly <u>non-lineari-ties</u> in e.g. MOS transistormodels and the set up of the network matrix. This time is proportional to the number of device models m, the number γ of nonlinearities per device model and the average computation time $\bar{t}_{nl}$ per nonlinearity or :

$$t_M = \gamma \, \bar{t}_{nl} \, m$$

Since furthermore m is proportional to n :

$$t_M = k \, \gamma \, \bar{t}_{nl} \, m \tag{4}$$

whereby k is the number of devices/node.
Therefore (3) with (4) and (1) becomes :

$$t_{SIM} = Kn \, N_t N_i \tag{5}$$

with

$$K = k \, \gamma (\bar{t}_{nl} + \bar{t}_i) \tag{6}$$

Let us concentrate now on (5) and come back to (6) later. Consider the network in Fig. 1a which consists of <u>repetitive</u> circuits (here e.g. inverters) in both x and y direction. Let the total num-ber x.y = N be fixed and let there be k_N nodes per inverter. Then

$$n = k_N N = k_N xy \tag{7}$$

If a pulse is applied to the y inverters of x=1 (Fig. 1b) then the first two waveform vectors $V_{01}(t)$ and $V_{02}(t)$ are shown in Fig. 1b. Due to time step control $h_i \lessapprox h_u$ in the rising and falling edges during pulse propagation through the chain and therefore the com-<u>putational activity</u> defined as the <u>number of matrix evaluations/user defined time step</u> increases when for some network response $V_i(t)$

$$\left| \frac{1}{V_i(t)} \frac{dV_i}{dt} \right| > \varepsilon \tag{8}$$

From Fig. 1a. and 1b. it is clear that the total computational ef-fort can be expressed by $N_t > N_{tu}$ and obviously :

$$N_t = N_{tu} \, (1 + k.x) \tag{9}$$

since the longer the chain, the longer it will be <u>active</u>.
From (5), (7) and (9) :

$$t_{SIM} = k_N KN_i N_{tu} xy \, (1 + kx) \tag{10}$$

or

$$t_{SIM} = C_1 n + C_2 x^2 y \tag{11}$$

Now consider the following cases :

  a) y=1 (serial network - no parallelism) :

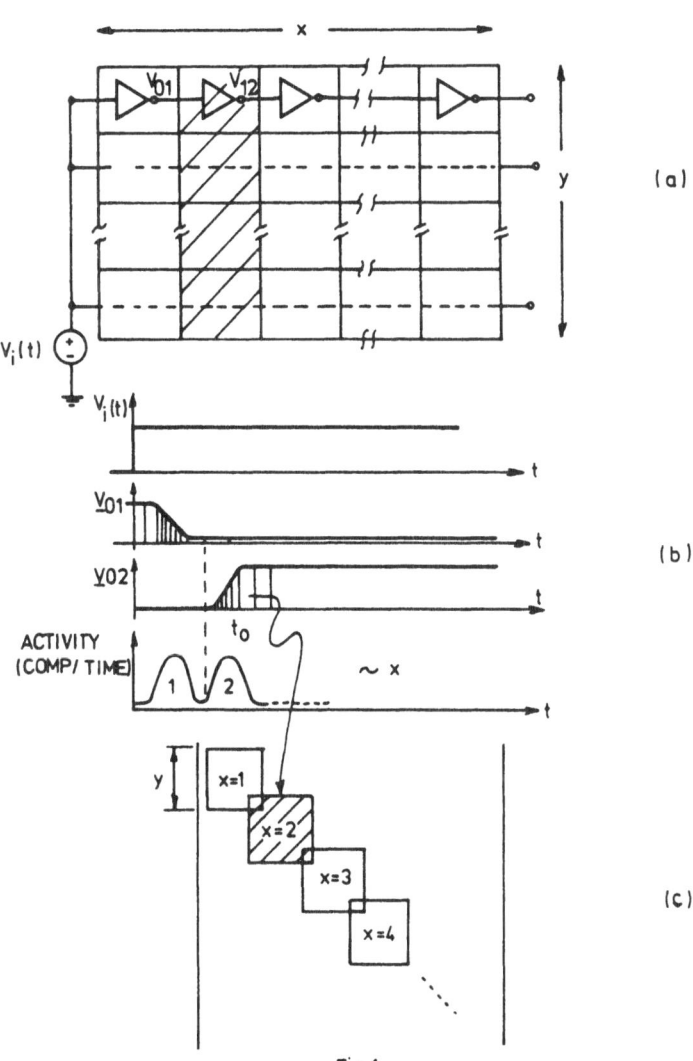

Fig.1

$$t_{SIM} = C_1 n + C'_2 n^2 \tag{12}$$

b) $x=y= k_N n^{1/2}$ (square network)

$$t_{SIM} = C_1 n + C_2'' n^{3/2} \tag{13}$$

c) $x=1$ (parallel network - serial depth = 1)

$$t_{SIM} = C'_1 n$$

i.e. for large n :

$$t_{SIM} = C n^{\alpha} \qquad (1 \leqslant \alpha \leqslant 2)$$

The above result has been confirmed experimentally [14].
Typically a 100 MOS transistor circuit over 1000 time points takes
ca. 30 minutes CPU on an IBM 378-158. From (2) and $\alpha = 1.5$ it
follows that a 1000 transistor circuit would require ca. 15 hours
of CPU time which is clearly intolerable even if the program
would run on a dedicated mini. In the following subsection we will
show a way of improving the performance of circuit analysis by exploiting characteristics of digital circuits.

3.2. Event-driven circuit simulation for digital circuits.

Figs. 1b. and 1c. suggest that at t_o only circuit block x=2 needs
to be evaluated since the rest of the circuit is in a stable state.
It is therefore a waste of effort to set-up and invert the <u>full</u>
matrix of the network provided a <u>criterion for activity of a given</u>
<u>network part exists.</u>
This leads to :
<u>Property 2 : Consider a digital network driven from an excitation</u>
<u>with pulsewidth larger than its ripple through propagation time.</u>
<u>If in such a network only the submatrices according to the active</u>
<u>networks are set-up and inverted then simulation time becomes pro-</u>
<u>portional to n</u>
i.e.

$$t_{SIM} = C n \tag{14}$$

if N_{tu} and h_u are constant.

This follows from (10) which becomes :

$$t_{SIM} \approx K y (1+kx)$$

$$\approx K y + Kkn/k_N$$

$$\sim n \quad \text{for large n}$$

Since VLSI is predominantly <u>digital</u> and since the analog sampled
MOS circuits are <u>time partitioned</u> (sequence of equilibrium states)
property 2 should be <u>exploited.</u>
The exploitation of this property requires the introduction of :

    1. <u>Events</u> : defined as occurences in the time domain when net-
       work excitations and responses fulfill the conditions to

2. Schedule or

3. Deschedule the defined subcircuits.

4. Active : a given subcircuit which is scheduled is said to be active, otherwise it is inactive (or stable).

5. Activity : is the percentage of subcircuits active at a given time t. The exact definition of event and activity will be discussed for each practical case in sections 4 and 5.
The procedure to exploit the inactivity of subnetworks is called "selective trace" [15]. It has been used intensively in logic simulators to be discussed later in this course. Sometimes the fact of inactivity in digital or sampled circuits is called "latency" [16] . Simulation according to this principle is said to be event-driven.
Since in digital networks with gates as subcircuits the average activity is generally 1...20%, considerable savings in computational effort will result.

3.3. The concepts of macromodeling, table models and piecewise linear models.

From eqs. (5) and (6) it follows :

$$t_{SIM} \sim (t_M + t_{INV}) N_i \qquad (15)$$

Using todays efficient sparse matrix inversion algorithms one generally finds [17] :

$$\frac{t_M \text{ (nonlinear, model evaluation)}}{t_{INV} \text{ (matrix inversion)}} \approx 4...8 \qquad (16)$$

This is the result of the fact that device non-linearities are usually based on costly FORTRAN fucntions evaluated sequentially[*].
Table 2 shows the relative cost of mainly used functions with respect to multiplications on an IBM 370/158.

Function	Cost
A*B	1
SQRT(A)	12
EXP(A)	20
A**B	42

Table 2.

---

[*] In principle full model evaluation could be done in parallel if a parallel computer architecture could be used !

This clearly leads to the following conclusions :
  a) Decrease or avoid non-linear function evaluation/model.
  b) Decrease or avoid nonlinear iterations.
(N_i = 1 instead of 3...5).
Requirement a) favors the introduction of piecewise linear or
table-look up models.
This will in section 4 lead to so called "Boolean controlled network elements" (BCE's) and in section 5 to table models. The number
of nonlinearities can also strongly be reduced by no longer modeling
at device level but at the level of repetitive cells. One functional nonlinearity (or BCE, or table) then substitutes for a large
number of device nonlinearities. This will of course be the subcircuit to be scheduled according to the discussion in the previous
subsection.

Requirement b) will be satisfied by the use of a single secant
iterate per time-step as will be discussed in section 5.
In the following sections we will apply the above principles in
the sequence of their implementation in the DIANA program as this
method allows for a better understanding of the underlying principles and the use of "event-driven" mixed mode simulators in general.

4. MIXED LOGIC- AND CIRCUIT SIMULATION IN DIANA USING THE MACRO-
   MODEL MØ.

4.1. Introduction

In section 1 the need for logic simulation in VLSI has been discussed.
Although logic simulation will be discussed in more detail later,
its principles are illustrated in Fig. 2.
In contrast to a circuit simulator a logic simulator operates on
signal states rather than waveforms and models are logic operators
(gates) in the algebra of signal states and have more or less
sophisticated delay models [15] [18] .
As shown in Fig. 2 signal states can be e.g. $\{0,1,X\}$ with X the
"undefined state"; $\{0,1,X,Z\}$ with Z a tri-state high impedance bus
state; $\{0,1,X,U,D,Z\}$ etc... Fig. 2. also shows that each logical
operator corresponds to a "truth table" which, for high fan-in and
many states rapidly becomes very complicated.
Most logic simulation is event-driven, as also is indicated in
Fig. 2., but scheduling becomes fairly complicated when e.g. different rise and fall delays as well as inertial delay is used [18] .
Furthermore no electrical loading effects can be taken into account
unless analytical relations with fan-out and load capacitance are
programmed in the simulator.

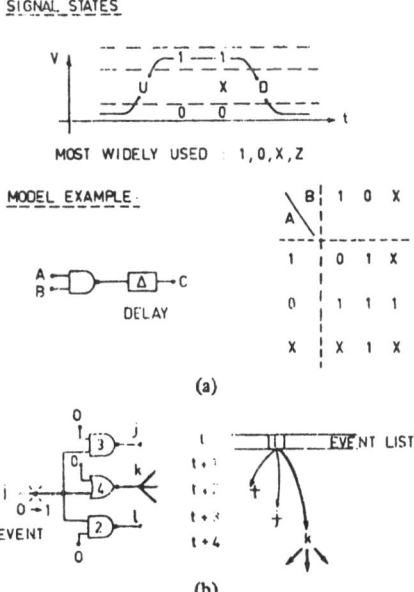

SIGNAL STATES

MOST WIDELY USED : 1,0,X,Z

MODEL EXAMPLE.

DELAY

(a)

EVENT

EVENT LIST

(b)

Fig. 2.

In DIANA we have introduced a "logic" simulator mode [19] which is based on the principles discussed in section 3 in which the simplicity of unit-delay logic simulator scheduling is combined with simple macromodeling based on piecewice linear macromodels for gates and J-K flip-flops.
A similar approach has recently been proposed by Da Costa and Nichols [20] . Externally to the user it behaves as a {1,0,X,U,D} simulator with <u>inertial</u> rise-fall delay and first order high and low impedance modeling. We will refer to the macromodel as the MØ macromodel based on the concept of Boolean-controlled network Elements (BCE's) [19] .

124

4.2. Boolean Controlled network Elements.

Consider in Fig. 3. a simple ratioed MOS inverter.

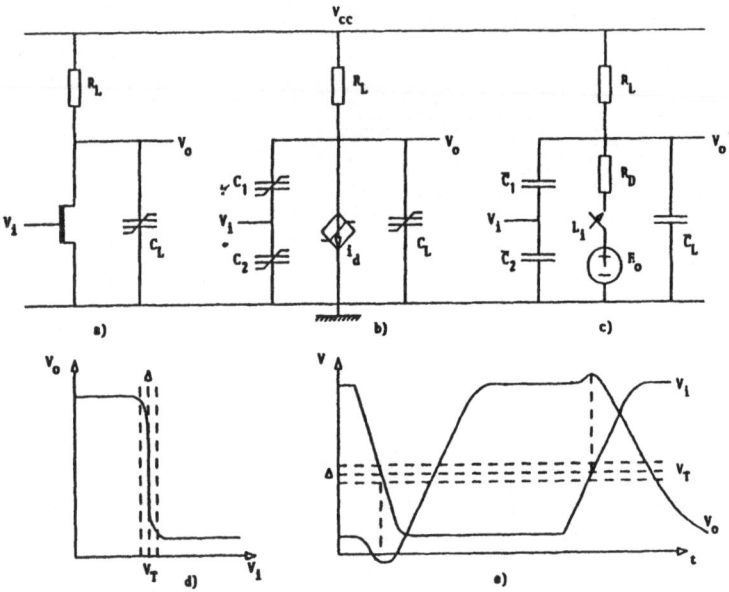

Fig. 3.   Inverter macromodel
a) Inverter   b) Equivalent circuit model
c) Macromodel   d) DC transfer curve
e) Transient response.

In a well designed inverter the circuit exhibits a high gain in a region   around the threshold V_T (VT*) of the DC transfer characteristic also shown in Fig. 3 (d). As a result, in a dynamic situation $V_i(t)$ rapidly crosses the region   (Fig. 3e) and the waveform at the output is largely defined by the impedance states before and after switching (for a more refined treatment see section 5).

---

* See DIANA manual. Not to be confused with the threshold of the MOS driven transistor.

Therefore the MOS driver transistor, to a first approximation, can be considered as a switch branch (R_D, E_o, L_i) (Fig. 3c.) which is open $(L_i=0)$ for $V_i(t) < V_T$ and closed $(L_i=1)$ for $V_i(t) \geqslant V_T$. So we can say :

$$L_i = TF \ (V_i(t) - V_T) \tag{17}$$

TF(x) is a <u>threshold function</u> [19] i.e. TF = 0 for x < 0 and TF = 1 for x $\geqslant$ 0

A threshold function TF translates <u>continuous voltages</u> (currents) <u>into Boolean variables</u> ($L_i \in \{0,1\}$).

In Fig. 3c. also the capacitors are linearized. They are substituted by linear capacitors displacing the same charge as the non-linear ones(see also section 5).

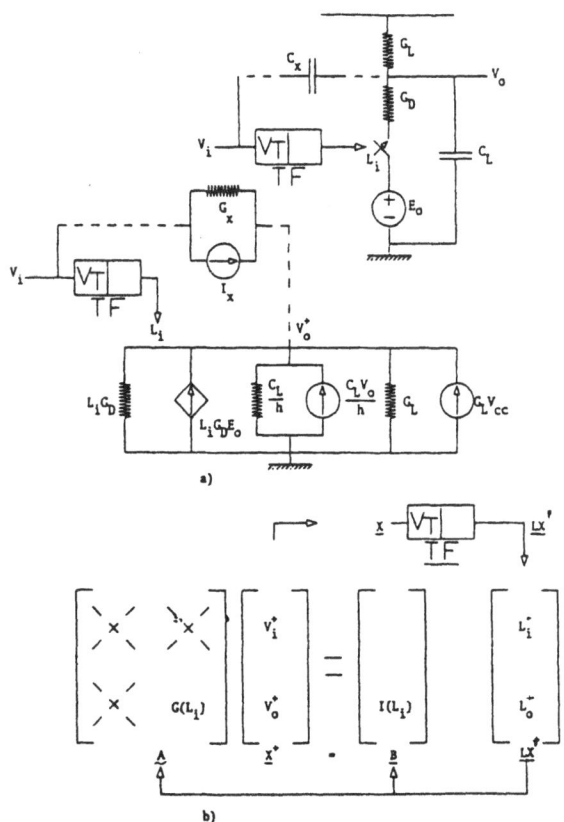

Fig. 4. Implementation of the invertor macromodel.
a) Macromodel and equivalent network
b) Matrix representation

If for each inverter we assign the input capacitance $\overline{C}_2$ to the load capacitance of the driving inverter (or gate), then Fig. 4a. represents a first order inverter model which by application of Norton's theorem and the use of the backward Euler companion model for C_L transforms into the network also shown Fig. 4a. V_o^+ is the voltage at the next timepoint t+h whereas V_o is the actual voltage at t. Now notice the following points :

a) The inverter circuit without floating capacitor C_x reduces to a linear one node circuit consisting of a conductance

$$G(L_i) = L_i G_D + G_L + \frac{C_L}{h} \tag{18}$$

in parallel with a current source :

$$I(L_i) = L_i G_D E_o + G_L V_{CC} + \frac{C_L V_o}{h} \tag{19}$$

Notice that the circuit elements $G(L_i)$ and $I(L_i)$ depend on a Boolean variable defined by a threshold function. We therefore define :

A Boolean Controlled network Element (BCE) is a linear network element of which the parameter-vector is a function of one or more boolean variables or a boolean function of such variables.

Notice that through TF function a BCE is piecewise linearly dependent on voltages and currents but this dependence can be and usually is discontinuous. This fact introduces numerical problems in finding the DC solution of networks with BCE's as well as in applying higher order integration. This will be discussed later.

b) Since the network with BCE's is linear it can be solved iterationless without nonlinear function evaluation which is in accordance with subsection 3.3.

c) Shown in Fig. 4b. is the matrix-stamp of the BCE inverter model. This scheme indicates the time domain algorithm :

1. $\underset{\sim}{A}(\underline{L}) . \underline{X}^+ = \underline{B}(\underline{L}, \underline{X})$
2. $\underline{L}^+ = TF(\underline{X}^+ - \underline{VT})$
3. $\underline{L} \leftarrow \underline{L}^+$

$\tag{20}$

4. Repeat step 1 for all time steps

d) If $C_X = 0$ or negligible i.e. if there is no network element between input or output (or its effect can be modeled by separate grounded loading of input and output) then there is no continuous coupling between input and output i.e. the TF effectively decouples the inverter models in the matrix equations.
This allows us to consider each inverter subnetwork separately as was required for event-driven circuit simulation discussed in subsection 3.2.
Before exploiting this fact let us generalize the above to more complex logic functions called the MØ macromodel in DIANA.

4.3. The MØ macromodel.

We call L_i in Fig. 3c. the <u>internal boolean variable</u> of the gate model.

Now let $L_i = f(\underline{L})$ whereby f is a boolean function of boolean variable vector $\overline{\underline{L}}$ generated from $\underline{V}(t)$ by $\underline{L} = TF(\underline{V}-\underline{VT})$ then it is clear that we have a <u>macromodel for a gate performing the logic operation</u>

$$L_o = \overline{f}(\underline{L}) \qquad (21)$$

with $\overline{f}$ the boolean complement of f.

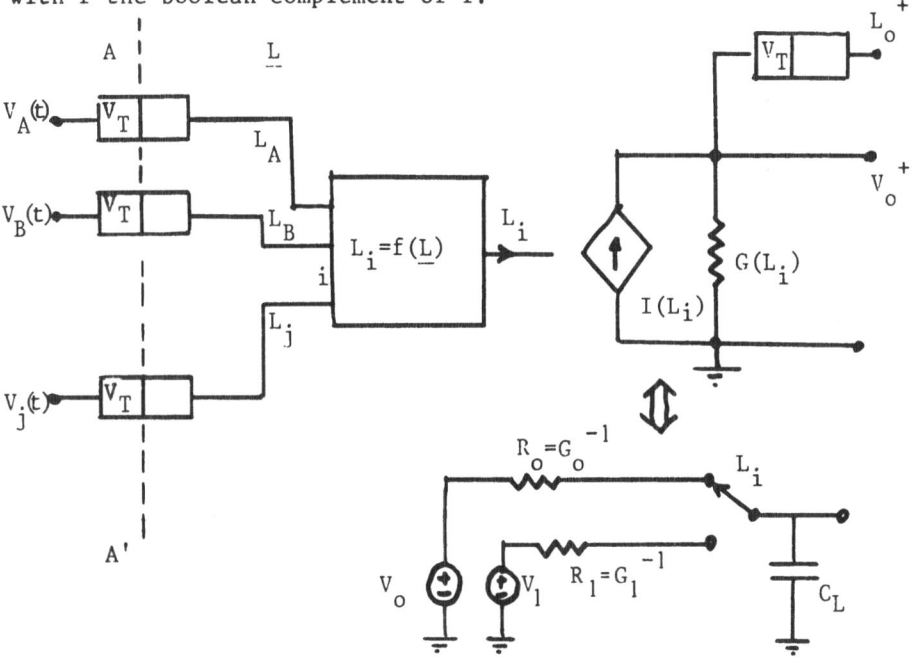

Fig. 5.

This gives rise to the model shown in Fig. 5. in which :

$$I(L_i) = V_1 G_1 \overline{L}_i + V_\emptyset G_\emptyset L_i + \frac{C_L}{h} V_o \qquad (22)$$

$$G(L_i) = G_\emptyset L_i + G_1 \overline{L}_i + \frac{C_L}{h} \qquad (23)$$

This model then has a transient response as defined in Fig. 6 (see also manual DIANA).

128

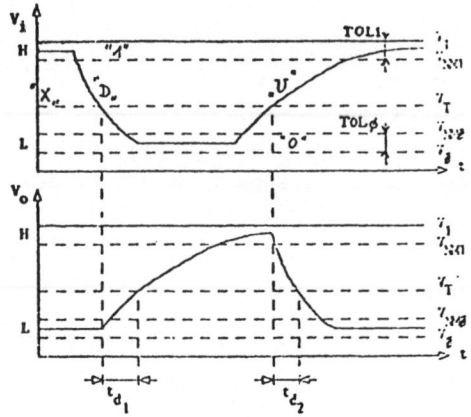

Fig. 6. Logic signals of MØ model.

We call the above model the MØ model with the following parameters

V_1 =   is the logic "one" level

$V_\emptyset$ is the logic "zero" level

t_{d1} is the trailing edge delay

t_{d2} is the leading edge delay

V_T is the gate threshold

TOL_1 latency criterion for logic "one"*

$TOL_\emptyset$ latency criterion for logic "zero"*

C_L load capacitance.

These parameters can be specified in the program. From these parameters the values of G_0 and G_1 are calculated as follows :

$$G_\emptyset = C_L \ln \left(\frac{V_1 - V_\emptyset}{V_T - V_\emptyset} \right) / t_{d2} \tag{24}$$

$$G_1 = C_L \ln \left(\frac{V_1 - V_\emptyset}{V_1 - V_1} \right) / t_{d1} \tag{25}$$

Using these relationships the user can define model parameters which give a first order modeling of the logic gates used. For example the default values :

$$V_1 = 3.5\ V \quad V_\emptyset = 0.4\ V \quad V_T = 1.4\ V$$
$$TOL_1 = 0.5V;\ TOL_\emptyset = 0.3\ V \qquad \left.\right\} I$$

---

* See next subsection.

and

$$C_L = 90 \text{ pF}; \ t_{d1} = 12 \text{ ns}; \ t_{d2} = 8 \text{ ns} \quad \Big\} \text{ II}$$

are well suited for simulating the 74 secties of TTL. The group I parameters are defined on a .LEVEL card and the group II on a .MOD (model) card in DIANA (see manual).

The following logic functions $\overline{f}$ are implemented : (N)AND; (N)OR; (N)EXOR; ANDORI; ORANDI; JK flip-flop whereby the fan-in is unlimited i.e. very complex sum of products (product of sums) can be formulated in one ANDORI resp. ORANDI function (see DIANA manual).
The output of this type of simulation can be :
- plots of wave forms
- tables of signal states whereby :

$\emptyset$: output in range $[V_\emptyset, \ V_\emptyset + TOL_\emptyset]$

U : output rising and in $[V_\emptyset + TOL_\emptyset, \ V_1 - TOL_1]$

D : output down and in $[V_\emptyset + TOL_\emptyset, \ V_1 - TOL_1]$

1 : output in range $[V_1 - TOL_1, \ V_1]$

X : output undefined in searching for a DC condition (see further)

The algorithm defined in (20) now becomes :

$$\left.
\begin{aligned}
&1. \quad \underset{\sim}{A}(\underline{L_i}) \cdot \underline{X}_d{}^+ = \underline{B}(\underline{L_i}, \underline{X}_d) \\
&2. \quad \underline{L}^+ = \underline{TF}(\underline{X}_d{}^+ - \underline{VT}) \\
&3a. \quad \underline{L}^+_i = \underline{f}\ (\underline{L}^\tau) \\
&3b. \quad \underline{L}_i \leftarrow \underline{L}^+_i \\
\end{aligned}
\right\} \quad (26)$$

4. Repeat step 1 for all time steps

Notice than $\underset{\sim}{A}$ is now a diagonal matrix i.e. a set of decoupled equations of the type :

$$a_{i,i}(L_i) \cdot x^+_{d,i} = b_i(L_i, \ x_{d,i})$$

The index d stands for "digital" and nodes belonging to M$\emptyset$ models are called digital nodes.
Algorithm (26) indicates that there is a "delay" of one time step h between L_i and the computed value $X_d{}^+$. Therefore the time step h_d (DDIG in .TRAN card in DIANA (see manual)) for logic simulation should always be smaller than MIN(t_{d1}, t_{d2}) of the gate models used.
Timing errors resulting from threshold crossings between user defined time points are minimized by interpolation techniques as will be discussed in subsection 4.4) and ref. [19] ; p. 329.
For more details on M$\emptyset$ see DIANA manual and example at the end of this section.
Before discussing mixed circuit and logic simulation with M$\emptyset$ let us first discuss the implementation of event-driven simulation with

MØ as well as some algorithmic aspects of simulation using BCE's.

4.4. Event-driven simulation with MØ.

4.4.1. Stability definitions.

In 3.2. we have shown that only active subcircuits should be "scheduled". We can now apply this to the MØ model by defining the meaning of "event" and of "activity" and "stability". For the MØ macromodel we define :

"Event" : an event for MØ gate i is defined as :

$$L_i^+ \oplus L_i = 1 \tag{27}$$

whereby L_i is an internal boolean variable or an L_i belonging to an input source driving an MØ model.

"Stability" : let V_i be the output voltage and L_i the internal boolean variable of MØ gate i, then i is stable if :

$$
\left.
\begin{array}{ll}
V_i \in [V_1 - TOL_1, V_1]. \& . \ L_i = 0 & \text{(a)} \\
V_i \in [V_0, V_0 + TOL_0]. \& . \ L_i = 1 & \text{(b)}
\end{array}
\right\} \tag{28}
$$

or

Otherwise i is active.

This explains the significance of TOL_1 and TOL_0 which can usually be defined from the noise margins of the logic family simulated. When (28a) is satisfied then $V_i^+ \triangleq V_1$, when (28b) is satisfied then $V_i^+ \triangleq V_0$.

4.4.2. Data-structure for scheduling.

The primitives for MØ logic simulation are gates, a J-K flip-flops and input sources. From the input interconnection description (see manual) the following lists are produced :

Fan-in list : $I(j) = \{$input names of element $j\}$

Fan-out list : $U(j) = \{$fan-out primitives of $j\}$

A J-K flip-flop has two outputs Q and Q' and has two-fan-out lists. If a primitive j drives the J-K then the J-K is in U(j) with the name of the Q output.

Activity lists : the model MØ can be scheduled using two activity lists as in a unit delay logic simulator [18] :

L_a : list of primitives (names) active at actual time t_n
L_b : list of primitives (names) to become active at t_{n+1}.

In reality these two lists are implemented into one single ring datastructure symbolically represented in Fig. 7.
The significance is illustrated by the following example : s is the startadress of the MØ list with i the first adress where (Fig. 7a) we find the adress j of next element (MØ gate) GN together with information of element A7 e.g. name (A7), actual and future state (L_{A7} and L_{A7}^+), $x_{d,A7}$, adress U(A7), adress I(A7) etc...

The last element of MØ points back to s. This ring structure is part of the <u>dynamic storage</u> vector used in most actual programs.

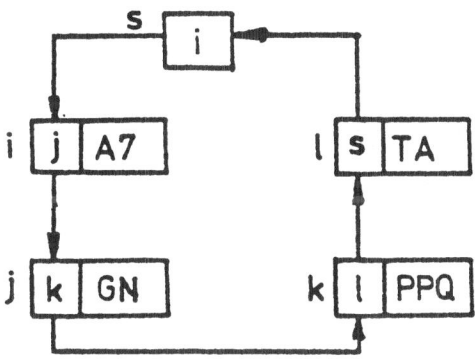

Fig. 7a. ring datastructure.

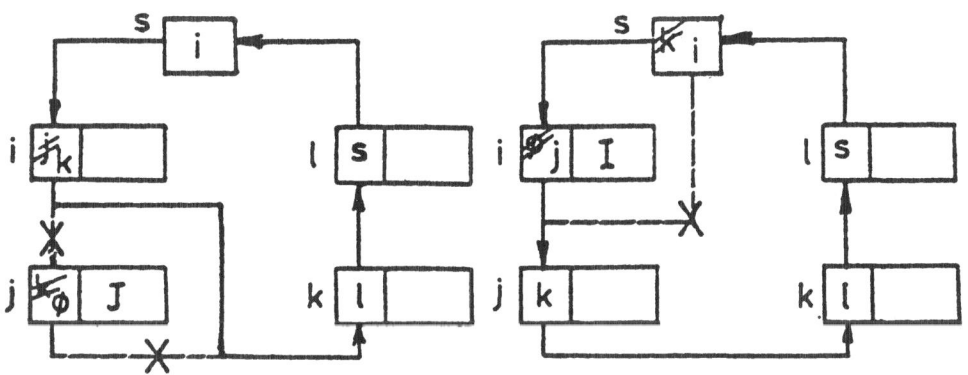

Fig. 7b. descheduling J.   Fig. 7c. scheduling I.

Fig. 7b. illustrates the "descheduling" operation of node j (DS_j) whereas Fig. 7c. shows the "scheduling" of node i (S_i) which is placed in the head of the list in order to avoid its use as a rescheduling event.

4.4.3. <u>Scheduling algorithm for MØ</u>.
The scheduling algorithm as implemented in DIANA for MØ models is shown in table 2. It is quite general and is also used for timing simulation for MOS to be discussed in section 5.
The numbers ⟨ number⟩ in the algorithm refer to notes following the algorithm :

SCHEDULING ALGORITHM FOR MØ

INITIALIZATION : get first input event and load into L_a.
$$L_b \longleftarrow \{\emptyset\}$$

STEP 1 : IF $L_a = \{\emptyset\}$

        THEN : set $t = t_n$ of <u>next</u> input <u>event</u> <1> and load
               into L_a. If <u>none is found</u> : STOP, else continue

        ELSE : $L_b \longleftarrow L_a$ <2>

STEP 2 : DO FOR $\forall j \in L_a$:
        get next input $x_{d,j}^+$ or calculate $x_{d,j}^+$ from
        $g_j \cdot x_{d,j}^+ = b_j$. Result : L_j^+
      END

        DO FOR $\forall j \in L_a$
          IF $L_j^+ \oplus L_j = 1$
               THEN : DO FOR $\forall i \in U(j)$
                      Logic evaluation of L_i^+ <3>
                      IF $L_i^+ \oplus L_i = 1$
                      THEN <u>Schedule</u> $L_b \longleftarrow i$ (S_i)
                      IF already scheduled
                        THEN SPIKE* <4>
                      ELSE next i

                    END

               ELSE :
                  IF j STABLE
                  THEN <u>deschedule</u> j in L_b (DS_j)
                  ELSE <u>next</u> $j \in L_a$

        END

STEP 3 : $L_a \longleftarrow L_b$; $L_b \longleftarrow \{\emptyset\}$; GO TO STEP 1.

Table 2.

Notice the following :

<1> an input event occurs when an input driving an MØ crosses V_T. These events can be predicted before analysis when the inputs to MØ are voltage sources. It is clear that stepping to a <u>next input event</u> can save a lot of computertime (e.g. $h_d = 20$ <u>ns</u> and <u>next input event</u> is 1 ms from now !). This principle of <u>NEXT EVENT</u> is used in all logic simulators.
If however the logic network in DIANA is analyzed in mixed-mode together with an "analog" circuit(or circuit blocks) then analysis of the analog circuit is performed first in order to detect wether circuit responses used as input to MØ simulation are an input to MØ.

Only if this happens or a regular logic input event occurs then $L_a \neq \{\emptyset\}$ and the logic network needs to be analyzed.

⟨2⟩ $L_b \leftarrow L_a$ because an M∅ element is most probably not stable in the next time step. If it is, it will be descheduled in STEP 2.

⟨3⟩ At this point the fan-out list is used to evaluate all possible next candidates for activity (selective trace). The fan-in list of i is used for <u>logic</u> evaluations to verify a possible event to be scheduled. Only if scheduled, <u>arithmetic</u> evaluation occurs (begin of STEP 2 next time).

⟨4⟩ is discussed in next subsection.

4.4.4. <u>Spike and hazard detection using M∅.</u>
One of the features of a logic simulator is the detection of "hazards" and "spikes".

<u>Hazard</u> : a hazard is defined as the occurance of two input events, which would cause an opposite output result, within a time interval smaller than rise or fall time caused by the first event.

In such case the output of the gate will become undefined and may cause ambiguities in an asynchronous sequential circuit when a spike exceeds the threshold voltage (see Fig. 8).

<u>Spike</u> : output change of a gate at the inputs of which a hazard occurs.
Detection of such spikes is an important feature but most logic simulators, since they only treat signal states, cannot judge the <u>value</u> of the spike. In the M∅ simulation Fig. 8b. and table 2 ⟨4⟩ show how easy it is to detect spikes and to find the voltage value of the spike.
Spike detection is an .OPTION SPIKE in DIANA. The time, gate name and spike value are printed out when they occur.

4.4.5. <u>Some properties of M∅ logic simulation.</u>
In contrast to a multistate, assignable delay logic simulator, the M∅ macromodel can be implemented using a very simple datastructure with two acitivity lists similar to a <u>unit-delay</u> logic simulator. Logic evaluation is extremely simple since it is limited to pure boolean $\{1,0\}$ operations. Nevertheless it provides in its present form the equivalence of a 5 value (0,1,X,U,D) simulator with as-signable t_{d1} and t_{d2} which are by nature <u>inertial</u> with spike detection and spike estimation facility.
In contrast to a pure logic simulator it provides <u>first order im-pedance states</u> as well as <u>first order capacitive</u> loading rules as a natural feature.
The output can be either <u>waveform</u> plots or state tables.
The price paid for it is that a gate is <u>scheduled and calculated</u> (one division) during its full U and D state time period and there-fore it is somewhat slower than an assignable delay simulator.
Although not yet implemented, it is easy to see that the Z state can be obtained from a floating (and possibly leaky) backward

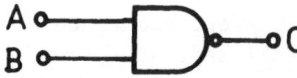

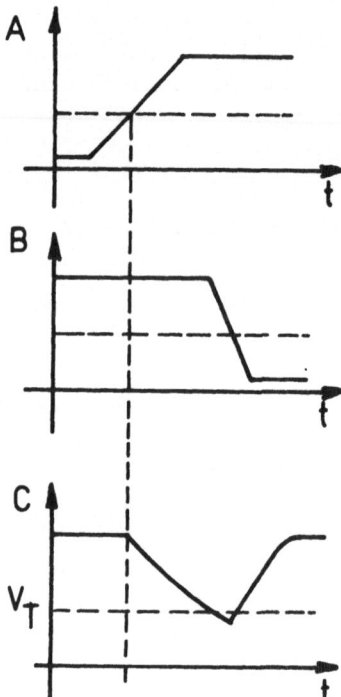

Fig. 8a.

Definition of a spike.

Fig. 8b.

Detection of a spike
in MØ scheduling
algorithm.

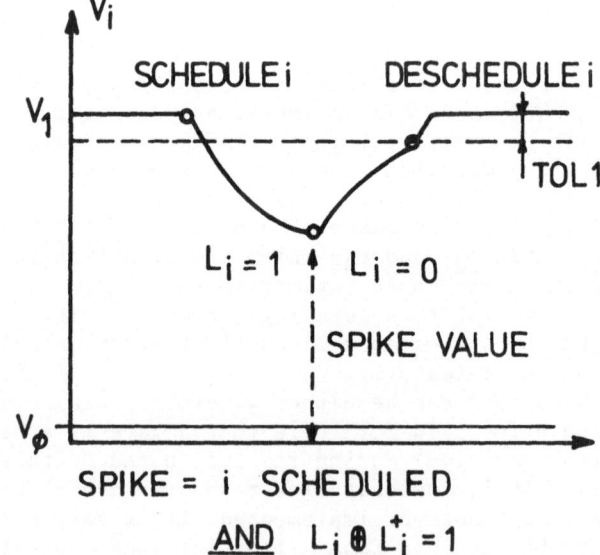

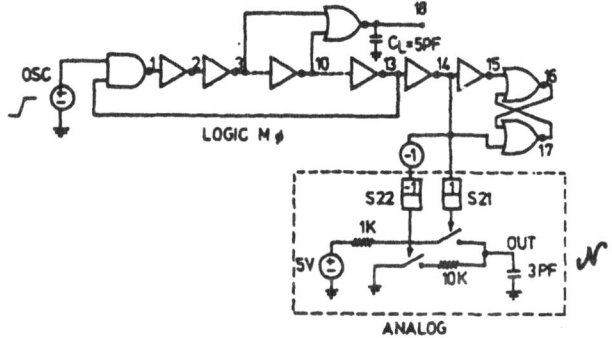

Fig. 9a. Ringoscillator and analog network

```
*****************************************************************
*
*        EXAMPLE RINGOSCILLATOR
*
*****************************************************************


        ; CONTROLCARDS

        .OPTIONS STORE LISTS SPIKE PAGE
        .TRAN AOUT=400 DELTA=3N
        .LEVEL VT=1.4 VO=0.2 V1=3.5

        ; DIGITAL PART

        .LOGIC ON
        NAND 1 IN=OSC 13
        NOR 2 IN=1
        NOR 3 IN=2
        NOR 4 IN=3
        NOR 5 IN=4
        NOR 6 IN=5
        NOR 7 IN=6
        NOR 8 IN=7
        NOR 9 IN=8
        NOR 10 IN=9
        NOR 11 IN=10
        NOR 12 IN=11
        NOR 13 IN=12
        NOR 14 IN=13
        NOR 15 IN=14
        NOR 16 IN=15 17
        NOR 17 IN=14 16
        NOR 18 IN=3 10 COUT=5P

        ; MODELKAARTEN

        .MOD O TD1=10N TD2=10N CL=1P
        .LOGIC OFF
        ;
        ; INPUTS

        INPUT OSC V-T=0 0 5 10N

        ; OUTPUT

        PRINT B: OSC B: 1 2 3 4 5 6 7 8 9 10 11 12 13 B: &
             14 15 B: 16 17 18  B: OUT

        ; ANALOG PART

        S21 OUT 0 14 0 VT=1 E=5 R=1K
        S22 OUT 0 0 14 VT=-1 R=10K
        CO OUT 0 3P

        .END
```

9b. Input to DIANA. Notice capacitive load of NOR 18.

TRANSIENT - SOLUTION
==========================

PRINT	O S C	1 2 3 4 5 6 7 8 9 1 1 1 1 /0 1 2 3	1 1 /4 5	1 1 1 /6 7 8	O U T

0.000E+00	0	1 0 1 0 1 0 1 0 1 0 1 0 1	0 :	0 1 0	0
3.000E-09	U	D 0 1 0 1 0 1 0 1 0 1 0 1	0 :	0 1 0	0
1.200E-08	1	D 0 1 0 1 0 1 0 1 0 1 0 1	0 :	0 1 0	0
1.500E-08	1	D U 1 0 1 0 1 0 1 0 1 0 1	0 :	0 1 0	0
3.300E-08	1	D U D 0 1 0 1 0 1 0 1 0 1	0 :	0 : 0	0
3.900E-08	1	0 U D 0 1 0 1 0 1 0 1 0 1	0 1	0 1 0	0
4.500E-08	1	0 U D U 1 0 1 0 1 0 1 0 1	0 :	0 1 U	0
6.300E-03	1	0 U D U D 0 1 0 1 0 1 0 1	0 1	0 1 U	0
6.900E-08	1	0 U 0 U D 0 1 0 1 0 1 0 1	0 1	0 : U	0
7.500E-08	1	0 U 0 U D U 1 0 1 0 1 0 1	0 1	0 1 U	0
9.600E-08	1	0 U 0 U D U D 0 1 0 1 0 1	0 1	0 : U	0
9.900E-08	1	0 U 0 U 0 U D 0 1 0 1 0 1	0 1	0 : U	0
1.050E-07	1	0 U 0 U 0 U D U 1 0 1 0 1	0 1	0 1 U	0
1.170E-07	1	0 1 0 U 0 U D U 1 0 1 0 1	0 1	0 1 U	0
1.260E-07	1	0 1 0 U 0 U D U 0 1 0 1	0 :	0 1 U	0
1.290E-07	1	0 1 0 U 0 U 0 U D 0 1 0 1	0 1	0 1 U	0
1.350E-07	1	0 1 0 U 0 U 0 U D U 1 0 1	0 1	0 1 U	0
1.470E-07	1	0 1 0 1 0 U 0 U D U 1 0 1	0 :	0 : U	0

=WARN=   TIME=1.5338E-07   NODE 18   SPIKE= 1.474E+00

1.560E-07	1	0 1 0 1 0 U 0 U D U D 0 1	0 1	0 : D	0
1.590E-07	1	0 1 0 1 0 U 0 U 0 U D 0 1	0 :	0 : D	0
1.650E-07	1	0 1 0 1 0 U 0 U 0 U D U 1	0 :	0 : D	0
1.770E-07	1	0 1 0 1 0 1 0 U 0 U D U 1	0 1	0 : D	0
1.860E-07	1	0 1 0 1 0 1 0 U 0 U D U D	0 :	0 : D	0
1.890E-07	1	0 1 0 1 0 1 0 U 0 U 0 U D	0 1	0 : D	0
1.950E-07	1	U 1 0 1 0 1 0 U 0 U 0 U D	U 1	0 : D	0
2.070E-07	1	U 1 0 1 0 1 0 1 0 U 0 U D	U :	0 : D	U

Fig. 9c.

Example of table output.
Logic states versus time.
Notice spike detection on
node 18.

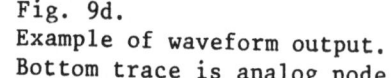

Fig. 9d.
Example of waveform output.
Bottom trace is analog node.

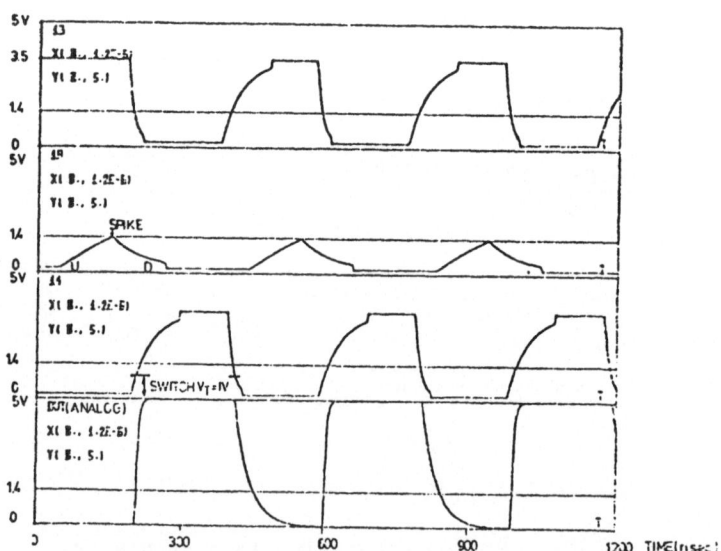

Euler capacitor connected to gates by switches (transmission gates).   Fig. 9. shows an example of the simulation of a ring-oscillator. The circuit schematic is shown in Fig. 9a.   Notice that the oscillator drives a small analog network through switches S$_{22}$ and S$_{21}$ with threshold -1V and +1V respectively. This is a simple example of <u>mixed mode simulation</u>.
Fig. 9b. shows the input language for this circuit. Notice the gate model with delay specifications and nominal load capacitance of 1 pF. Gate 18 has an external load.capacitance of 5 pF which slows this gate down six times as is clear from the waveform output plot shown in Fig. 9d. This gate clearly exhibits a <u>spike</u> since its input nodes (3,10) change too fast (hazard).
The output waveform of the analog node is also shown.
Fig. 9c. shows the usual table output whereby gate states are printed out at each change in state.
Notice the behavior as a $\{1,X,0,U,D\}$ simulator. Notice also the printout of SPIKE warnings as discussed earlier. The example took 5 sec. CPU on the VAX 11/780 under VMS for 400 time points of 3 ns. Average activity is 48%.

4.5. Mixed-mode circuit and MØ logic simulation.

   4.5.1. <u>Concept</u>.
Fig. 10. shows the general principle of mixed-mode simulation in DIANA.

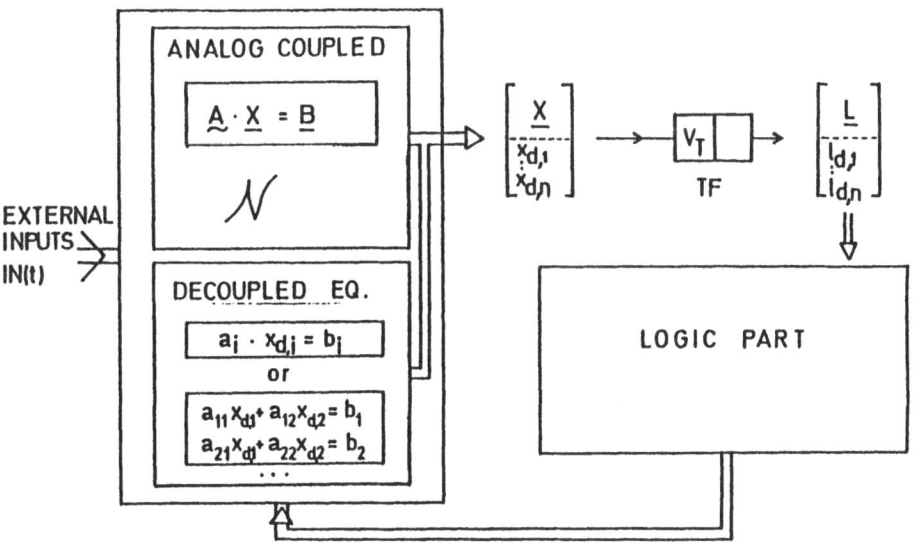

Fig. 10.

In addition to the decoupled equations

$$a_i . x_{d,i} = b_i$$

of the MØ gates also a "normal" circuit simulation part in the time domain is present which can analyse a network $\mathcal{N}$ by pure matrix formalism

$$\underset{\sim}{A} \cdot \underline{X} = \underline{B}$$

The solution vector $\underline{X}(t)$ is called the analog vector whereas $\underline{x}_d$ is the digital solution vector. The $(\underline{X}, \underline{x}_d)^T$ vector is thresholded into a logical vector, which, through the boolean functions of the gates and using selective trace for the MØ models, creates $l_{d,i}$ for the MØ models, and for switch states in $\mathcal{N}$.
In addition to the usual elements (R,L,C,VDI,IDV,VDV,IDI,p- and n-MOS transistors) the elements shown in Fig. 11 are implemented.

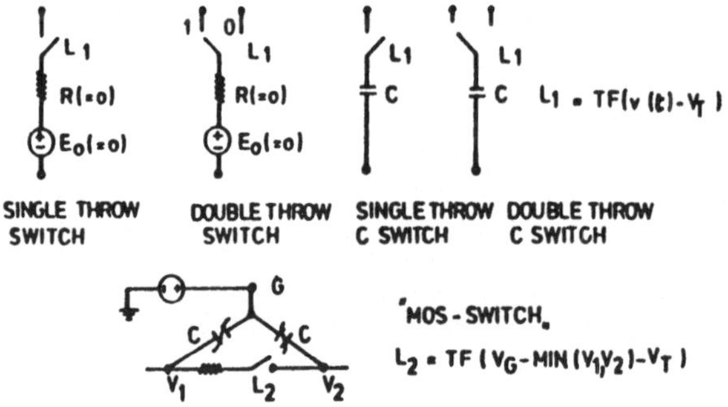

Fig. 11. New boolean controlled branches in DIANA.

For more details we refer to the DIANA manual but the switches, which are typical for this program, need some more comments.

In DIANA all switch branches are controlled by a threshold function i.e. their logical value L obeys :

$$L = TF\ (V_i - V_j - V_T) \tag{28}$$

Switches can easily be implemented into the MNA matrix. For example an ideal switch between nodes k and l and with offset E_o becomes :

$$\begin{array}{c} \\ k \\ l \\ S \end{array} \begin{array}{c} \begin{array}{ccc} k & l & S \end{array} \\ \begin{bmatrix} & & L \\ & & -L \\ L & -L & \underline{L} \end{bmatrix} \end{array} \begin{bmatrix} V_k^+ \\ V_l^+ \\ I_s^+ \end{bmatrix} = \begin{bmatrix} \\ \\ LE_o \end{bmatrix} \tag{29}$$

whereby $L \epsilon \{0, 1\}$ is its logical state and $\overline{L}$ its complement.
A switch with an offset E_0 and nonzero series resistance $R=G^{-1}$
can be implemented in a nodal matrix as follows :

$$\begin{matrix} k \\ 1 \end{matrix} \begin{bmatrix} LG & -LG \\ -LG & LG \end{bmatrix} \begin{bmatrix} V_k^+ \\ V_1^+ \end{bmatrix} = \begin{bmatrix} LGE_0 \\ -LGE_0 \end{bmatrix} \tag{30}$$

The reader can now find matrix stamps for double throw switches
etc...
of particular importance is the so called MOS switch in Fig. 11.
This switch is a first order clocked MOS switch, which can be
used to study clock feedthrough effects e.g. in A/D, D/A and MOS
sampled data filters.
Its switch state depends on :

$$L = TF(V_G - MIN(V_1, V_2) - V_T) \tag{31}$$

And its $C_{gs(d)}$ value is given by :

$$C = C_{ov} + \frac{1}{2} L C_{ox} \tag{32}$$

V_T, C_{ov} and C_{ox} can be specified per MOS switch.
C_{ov} is the gate-source (drain) physical overlap capacitance and
C_{ox} is the channel oxide capacitance of the MOS transistor. It is
assumed that the transistor will be in the triode region during
switching (as it usually is).
The on-resistance R can be estimated from clock-amplitude and,
average signal level and transistor dimensions. Notice that all
these switches are BCE's which can be driven from all logic varia-
bles $\underline{L}$ and $\underline{l}_d$ (Fig. 10) i.e. they provide a possible link or interface between circuit simulation and logic simulation. This pro-
blem is discussed in next subsection.

4.5.2. Interfacing circuit and MØ logic simulation.
As an MØ model is computed outside the network matrix $\underline{A}$, it is
not possible to have a direct coupling between an MØ model and the
analog circuit.
Fig. 12. shows the possible interfacing possibilities.

a) $\mathcal{N} \to M\emptyset$

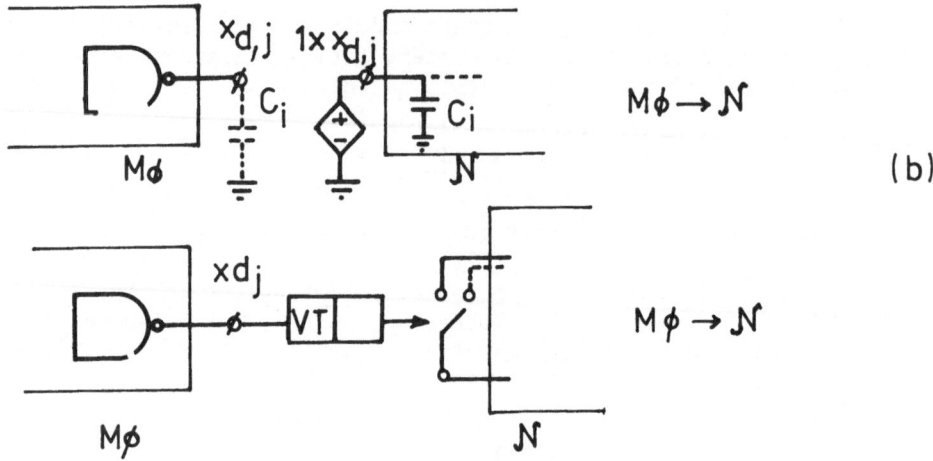

(b)

Fig. 12b. Interfacing MØ with $\mathcal{N}$ is possible through dependent sources and switches.

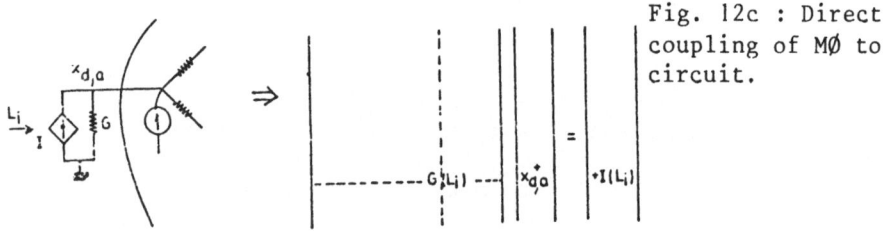

Fig. 12c : Direct coupling of MØ to circuit.

Fig. 12a. shows that an analog node X_i can drive any input in MØ. When going from MØ to $\mathcal{N}$ it is of course possible to use dependent sources or single or double throw switches.

As an example Fig. 12b. shows how a VDV with gain equal unity drives a capacitive input with value C_i in $\mathcal{N}$. If the capacitance C_i is used as load capacitance of MØ, the correct waveform is obtained. This system however does not operate correctly for resistive loads on MØ.

Fig. 12c shows how in DIANA, version 7, direct interfacing between MØ and analog circuitry becomes possible by introducing the network elements of MØ into the network matrix and excitation vector. The user should then of course be aware of the electrical interaction between the MØ gate model and the analog circuit to which it is coupled.

    4.5.3. <u>Example.</u>

Fig. 13. represents a MOS sampled data filter based on switched capacitors C_A and C_B transferring charge through M1,M2 respectively M3,M4 driven by a 100 kH$_z$ clock CL, $\overline{CL}$. It can be shown that C_A, M1,M2 and C_B,M3,M4 perform the function of a large resistor. The filter function is the treble tone control of an audio amplifier and can be digitally controlled by an up-down counter and decoding logic (27 gates, 4 flip-flops) driving the capacitors

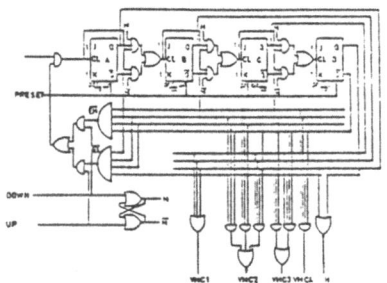

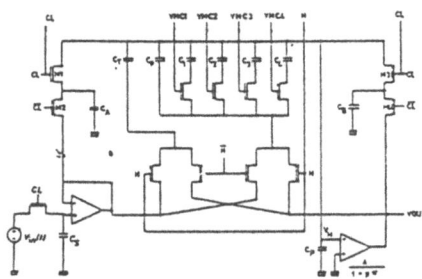

Fig. 13a. Mixed analog-digital circuit.

```
••••••••••••••••••••••••••••••••••••••••••••••••••••••••••••••••
•   SWITCHED TONE FILTER & DRIVING LOGIC
••••••••••••••••••••••••••••••••••••••••••••••••••••••••••••••••

.OPTIONS LISTS SPIKES OPTS MODE STORE
.TRAN MOUT=1000 DELTA=1U SE
:PRINT FFCL DOWN UP PRESET A0 B0 C0 D0 VMC1 VMC2 VMC3 VMC4 M LOG1
PRINT FFCL DOWN UP PRESET VIN VINS CL VM VOUT LOG1

! ••••••••••••••••••••••••••••••
! [ DRIVING LOGIC (DIGITAL PART) ]
! ••••••••••••••••••••••••••••••

.LEVEL VB=0,3 VTL=0 VCC=5 NM0=0,001 NM1=0,001
.MOS 0 TO1=10N TO2=20N

INPUT FFCL V=T=0 0 5 1U 5 9U 0 10U 0 200U CYC
INPUT DOWN V=T=0 0 0 450U 5 451U
INPUT UP V=T=0 0 5 1U 5 900U 0 901U
INPUT PRESET V=T=0 0 5 1U 5 4U 0 5U

JK QA=B0 0'=AB0' CL=ACLOCK EN'
AND A1 IN=UP EN'
AND A2 IN=DOWN CL'
OR A3 IN=A1 A2
AND ACLOCK IN=FFCL A3
JK Q=B0 2'=B0B0' CL=BCLOCK C'R=PRESET
AND B1 IN=M ACLOCK A0
AND B2 IN=M' ACLOCK A0'
OR BCLOCK IN=B1 A2
JK Q=C0 0'=C0C0' CL=CCLOCK P=PRESET
AND C1 IN=M SCLOCK B0
AND C2 IN=M' BCLOCK B0'
OR CCLOCK IN=C1 C2
JK Q=D0 0'=D0D0' CL=DCLOCK CLR=PRESET
AND D1 IN=M CCLOCK C0
AND D2 IN=M' CCLOCK C0'
OR DCLOCK IN=D1 D2
NOR M IN=DOWN M'
NOR M' IN=UP M
NAND EN' IN=A0' B0' C0' D0
NAND EL' IN= A0' B0' C0' D0'

OR VMC1 IN=A0 B0 C0
AND VC21 IN=B0' C0' D0'
AND VC22 IN=B0' C0'
AND VC23 IN=A0 B0 C0 D0'
NOR VMC2 IN=VC21 VC22 VC23
AND VC31 IN= B0' C0 D0'
AND VC32 IN= A0 B0 C0' D0'
OR VMC3 IN= VC31 VC32
AND VMC4 IN=A0' B0' C0 D0'
NOR M IN=C0 D0
NOR M1 IN=M

! ••••••••••••••••••••••••••••••
! [ SWITCHED TONE FILTER (ANALOG PART) ]
! ••••••••••••••••••••••••••••••

INP_IT IN SINE AC=0.1 FREQ=10R
RVIN VIN 0 IR 0 1,
INPUT CL TIME=9U 9U TR=1U CYC=9 VM=5
INPUT CLI V=T=0 0 5 1U 5 4U 0 5U 0 10U CYC

! CONTROLLED CAPACITOR ARRAY
C0 VM VC0 10P
C1 VM VC1 5P
C2 VM VC2 5P
C3 VM VC3 5P
C4 VM VC4 5P
S2C1 VC1 VC0 VMC1 0 R=5K VT=1,4 COV=0,005P COX=0,01P PASS
S2C2 VC2 VC0 VMC2 0 R=5K VT=1,4 COV=0,005P COX=0,01P PASS
S2C3 VC3 VC0 VMC3 0 R=5K VT=1,4 COV=0,005P COX=0,01P PASS
S2C4 VC4 VC0 VMC4 0 R=5K VT=1,4 COV=0,005P COX=0,01P PASS

! ONE POLE OPAMP
RGOPAMP VOUT 0 VM 0 1,
ROPAMP VOUT 0 1K
COPAMP VOUT 0 0,14U

! SAMPLE INPUT
C8 VIN8 0 1P
S28 VIN8 VIN CL 0 R=5K VT=1,4 COV=0,01P COX=0,02P PASS
SO02 0 VOP2 VIN8 VOP2 1,
ROP2 VOP2 0 1K
COP2 VOP2 0 0,14U

! CAPACITORS "P1" AND "P2"
S21A VCA VOP2 CL 0 R=5K VT=1,4 COV=0,01P COX=0,02P PASS
S21B VCA VM   CLI 0 R=5K VT=1,4 COV=0,01P COX=0,02P PASS
S22A VCB VOUT CL 0 R=5K VT=1,4 COV=0,01P COX=0,02P PASS
S22B VCB VM   CLI 0 R=5K VT=1,4 COV=0,01P COX=0,02P PASS
CA VCA 0 4P
CA VCB 0 4P
CP VM 0 13P

CT VM VCT 30P
S2MA VCT VOUT M 0 R=5K VT=1,4 COV=0,01P COX=0,02P PASS
S2MB VCT VOP2 M1 0 R=5K VT=1,4 COV=0,01P COX=0,02P PASS
S2MC VCB VOP2 M 0 R=5K VT=1,4 COV=0,01P COX=0,02P PASS
S2MD VCB VOUT M1 0 R=5K VT=1,4 COV=0,01P COX=0,02P PASS

.END
```

Fig. 13b. DIANA input containing MØ logic and analog circuit part.

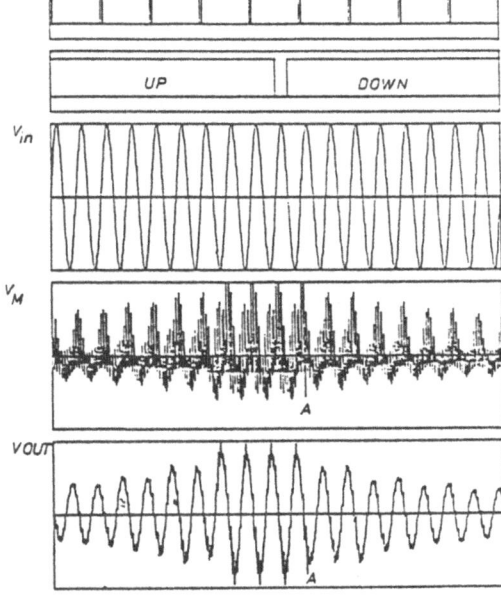

Fig. 13c. Output of DIANA. 1800 points; ½μs. A is spike in analog output introduced by spike in logic.

142

C_1 to C_4.

The opamp A has 60 dB gain and a unity gain bandwidth of 1MHz.
It is modeled using a VDI and load RC network. Fig. 14. shows a
plot made by the postprocessor PPR   on a hardware plotter
HP 7203 A of a simulation with DIANA of the complete system.
1800 time points of 1 μsec are used.
The simulation time is 120 sec. on a IBM 370/158 computer. The results shown are the digital control inputs UP,DOWN,CLOCK (up-down
rate), as well as the analog input V_{in} (10 kHz 0.1 V sine) the output V_{OUT} and the voltage V_m at the input of the op-amp.
Notice the digital control of the filter characteristic and the
spikes in V_m due to charge dumping into the band limited op-amp.
This example clearly shows the efficiency and potential of the
mixed analog/digital simulation made possible by the use of TF
(threshold functions) and BCE (Boolean controlled elements).

4.5.4. Some algorithmic problems with BCE's.
BCE's are driven by threshold functions and therefore exhibit
discontinuities as a function of other voltages or currents in the
circuit. The branch relationships of BCE's therefore cannot be approximated by polynomials which causes problems in DC algorithms
based on traditional Newton-Raphson iterations and in implicit
numerical integration routines. Moreover, when the integration step
h is too large a TF change is detected too late (see also 4.3.)
which can cause problems if this event is the possible origin of
other events in the same time slot.
These three problems however can be overcome. The reader is referred to ref. 19 for a more detailed treatment but the following
intuitively explains the solutions :

a) DC problem : if only MØ logic models are present (no analog circuit part), the initial state of the logic network can be
found from a three value {1,0,X} simulation* using a slightly modified form of the algorithm in table 2 with the excitation being the
excitation at t=0 and all other logic states being set to X at t=0.
Then the {1,0,X} simulation continues until a stable solution is
found. If some variables remain X then the network may be impossible to initialize [5]. This technique can only be applied when
no analog network is used. It is an option (DC3) in DIANA and has
also been used in MOS timing simulators [21].

If also an analog network is present the DC3 option is overwritten since it cannot take interactions with analog elements
(e.g. switches) into account. In that case in DIANA always a so
called pseudotransient analysis is used. In that case all initial
conditions are zero and all sources (also supply voltages) are
switched from zero to their value at t=0. This means that the
state of all BCE's is well defined at t=0 and then, using Backward
Euler with a time step h_{DC} (DELDC), the network as well as the MØ
models are integrated with fixed excitation and supply voltages

* See lecture an logic simulation by R. Newton or ref. [5].

untill all node voltages ·converge.
In this way the network always settles into an initial state except when undefined perfectly symmetric cross–coupled flip-flops
are present. If these flip-flops are unsymmetric they will settle
into their preferred state without warning.

b) Integration : since a BCE with a discontinuous change at
time t_o in reality corresponds to a change in topology of the network at t_o, one has to be extremely careful with multistep integration rules. Each time on a BCE event (switching) occurs one can
prove [19] that it is necessary to restart integration from order
1 (backward Euler).
Exclusive use of backward Euler never creates problems. If trape-
zoïdal integration is used, it must be changed to backward Euler
for at least one time step after the BCE change.

c) Interpolation : since a BCE event seldom coïncides with
a user defined time point, a linear interpolation is done in
DIANA to estimate the time t_e of the event. Analyses returns to
t_e, the event is executed and backward Euler step (see above) is
used for the remainder of the time slot $t_{m+1} - t_e$.
In this way timing errors are minimized. Interpolation can be
overwritten by .OPTION -INT.

5. MIXED-MODE CIRCUIT-TIMING-LOGIC SIMULATION.

5.1. Introduction to timing simulation.

The simple MØ macromodel can be used for first order timing verification taking impedance and capacitive load effects into account.
However for many logic circuits a more detailed analysis may be
necessary since e.g. for MOS logic the fall time often depends on
the rate of change of the input voltage since the driver transistor
does not change its impedance level instantaneously.
Also in dynamic logic bidirectional transmission gates are used and
effects such as threshold voltage losses, clockfeedthrough etc...
may greatly affect the timing behavior of such circuits.
Since, especially for such cases, logic simulators do a very poor
job, also since the link between behavior technology and geometry
is lost, very efficient simulation is necessary.
Historically techniques to exploit latency and macromodeling for
this purpose has first been proposed in 1975 by Ryan and Rabbat [22]
and later by Chawla-Gummel and Kozak [23] in the MOTIS program.
Here, for the first time, the term "timing simulation" has been
used. Based on MOTIS, Fan, Hsueh, Newton and Pederson have written
MOTIS-C [24] and then timing simulation along these lines has been
built into mixed-mode simulation programs SPLICE [11] and DIANA [12].
The theory of macromodeling and latency exploitation has since grown
considerably [25] [26] [27] .
In what follows we will discuss the basic underlying principles as

they are used in the DIANA program. These principles are along the lines of those used in MOTIS [23] although in DIANA use is made of mixed table and analytical macromodels rather than table models for all individual devices. Bootstrapped buffers as well as pass-transistor trees are possible.
The possibility for scheduling non-timing subcircuits in mixed mode is also provided.

5.2. Definition of unilateral networks for timing simulation.

Consider a network $\mathcal{N}$ partitioned in n subnetworks $\mathcal{N}_s$:

$$\mathcal{N} = \{\mathcal{N}_{s1}, \mathcal{N}_{s2} \cdots \mathcal{N}_{sj} \cdots \mathcal{N}_{sn}\}$$

Each $\mathcal{N}_{sj}$ can be a network described by a matrix equation :

$$\underset{\sim}{A}_j \underline{X}_j = \underline{B}_j$$

or an external input source $B_j(t)$. The $\mathcal{N}_s$ are defined to be unilateral (one-way [27]) i.e. input nodes exist each of which is characterized by an input admittance $(G_I, C_I) = Y_I$ (Fig. 15b.) and coupled to the rest of the circuit exclusively by a (nonlinear) controlled source (VDV, VDI) or a controlled conductance i.e. a relationship of the form :

$$i = f(v_2, v_i) \quad \text{(Fig. 15b.)} \tag{31}$$

Notice that a "floating" capacitor C_M is not allowed. The problem of floating capacitors will be discussed later by A.R. Newton and can also be found in [28] .
The set of input nodes per $\mathcal{N}_{sj}$ is defined at the fan-in list (Fig. 15a.) :

$$I(j) = \{i_{j1}, i_{j2} \cdots\} \tag{32}$$

The set of other external nodes per network are output-nodes $O(j)$ (Fig. 15a.) :

$$O(j) = \{o_{j1}, o_{j2} \cdots\} \tag{33}$$

The other nodes of $\mathcal{N}_{sj}$ are internal-nodes :

$$IN(j) = \{in_{j1}, in_{j2} \cdots\} \tag{34}$$

Every output node $O_k \in O(j)$ has a fan-out list

$$U(O_k) = \{\mathcal{N}_{si} : O_k \in I(i)\} \tag{35}$$

whereby N_{si} is the name (adress) of $\mathcal{N}_{si}$.
For some macromodels $\mathcal{N}_j$ for dynamic MOS and for reasons to be discussed later, inputs $i \in I(j)$ exist which will be called "clock-inputs".
We call $CI(j) \subseteq I(j)$ the set of clock-inputs for model $\mathcal{N}_{sj}$ and associate a separate clock-fan-out list $CU(k)$ with each $k \in \{0:0 = O(1) \cup O(2) \cdots \cup O(n)\}$ which is a clock input to at least one $\mathcal{N}sj$.

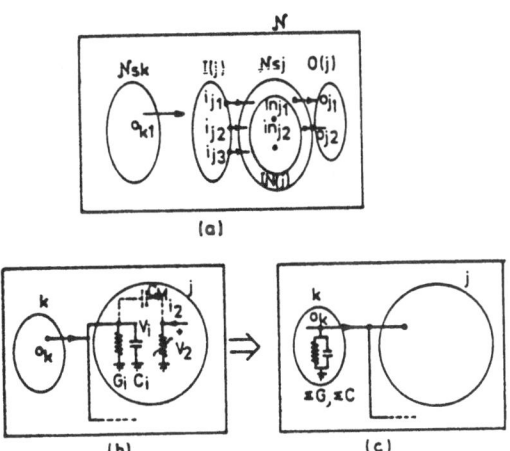

Fig. 15.

a. Note partitioning
b. Unilaterality : $C_M = 0$; $i_2 = f(v_2, v_i)$
c. Move all input impedances to driving network by
   fan-out list $U(o_k)$

We also assume that no two or more output nodes are shorted (e.g.
no "wired or" of two macromodels).
If they are, then the two (or more) subnetworks have to be merged
together into one new subnetwork (see also section 5.4.5.).
From the above we can then assume, that as a result of a preprocessing step in the algorithm all input admittances $Y_I \in U(O_k)$
can and have been associated with the output node O_k instead of
the individual inputs of the driven subnetworks (Fig. 15c.).
For each subnetwork one can then define a stability (latency)
criterion which in its simplest form could be :

$\mathcal{N}_{sj}$ is stable if for $\forall i \in \left\{ I(j), IN(j), O(j) \right\}$:

$\left| \dfrac{dV_i}{dt} \right| < \varepsilon$ with V_i the node voltage of i.

ε in DIANA is specified as the parameter TOL on the .CONVER card and is 10^6 V/sec by default (1V/μs). For timing macromodels this criterion is too severe as will be discussed in section 5.4.4.

5.3. A timing simulation algorithm according to the secant method.

The technique presented here and used in DIANA for timing simulation is based on the use of regula falsi iteration rather than the single one step Gauss-Seidel-Newton iteration method used in SPLICE and discussed further by A.R. Newton.

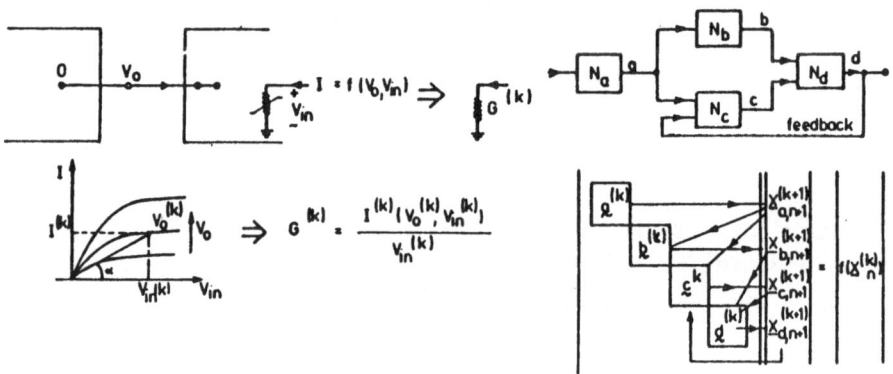

Fig. 16. The secant method applied   Fig. 17. Secant iteration ma-
          to unilateral networks.   trix for a network of
                                                    unilateral networks.

Let $\underline{X}_n$ by the actual solution at time t_n and $\underline{X}_{n+1}$ the solution at $t_{n+1} = t_n + h$ and let $\underline{X}_{n+1}^{(k)}$ denote the k-th secant iterate whereby every non-linearity has been substituted by a linear resistor as shown in Fig. 16. From this figure it is clear that such procedure always results in a fully <u>passive</u> network whereby, in contrast the Newton-Raphson, no-off diagonal elements appear outside the sub-matrices of the subnetworks. The coupling between networks is "delayed" by one secant iterate as illustrated in Fig. 17. Therefore it is, in principle, possible to use a modified version of the scheduling algorithm in Table 2 to solve the matrix by <u>scheduling per iteration per timepoint per subnetwork.</u>

Such a technique however has not been implemented in DIANA since it was found that in general for accurate circuit simulation, including analog circuits such as opamps too many iterations are necessary which takes away the advantage of scheduling. However, other techniques, still using Newton-Raphson techniques but exploiting latency and detection of strongly coupled blocks such as the feedback loops between $\mathcal{N}_c$ and $\mathcal{N}_d$ in Fig. 16. are being implemented at present. These techniques are along the lines discussed in [27] and [21] .

In this subsection we restrict ourselves to <u>timing simulation</u> of unilateral subnetworks which is generalized, along the lines in [23] , as follows :

- use exclusively unilateral digital macromodels;
- introduce latency criteria;
- use a time step h :

$$h < \min (t_{dij}) \tag{36}$$

    t_{dij} is the propagation delay from an arbitrary input i to an arbitrary output j and in principle h can be different for every macromodel;

- use only one <u>secant-iterate</u> per time point;
- use, if possible, one dimensional tables for <u>scalable</u> non-linear transcendental functions.

When applying this technique Fig. 17 becomes :

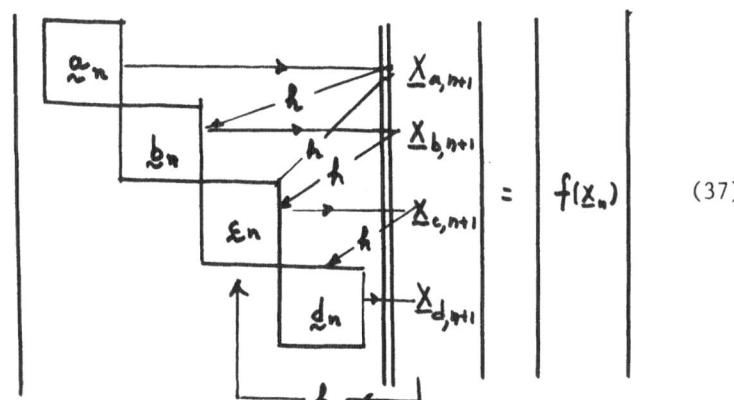

$$ = \left| f(\underline{x}_n) \right| \tag{37}$$

From (37) we see that now, just as in the case of the MØ macromodel, a time delay equal to h exists between coupled blocks.
This may cause problems in tightly coupled feedback paths and
therefore can only work properly if indeed in every subcircuit
condition (36) is satified. This problem of accuracy will be discussed in more detail by A.R. Newton. Potential measures to overcome this problem are the use of local time step control as in
SPLICE or the detection of strongly interconnected networks by
graph techniques [27] whereby, when scheduled, such blocks are
also dynamically coupled in the network matrix.
The latter technique is being implemented in DIANA at the time of
this writing.

A second error results from using only one secant iterate.
Fig. 18. illustrates the types of errors occuring in drivers,
depletion and saturated loads as well as transmission gates clocked
in saturated mode. One can clearly see that in most practical
cases the timing result is too pessimistic by underestimation of
the current which is better than optimistic estimates of timing.

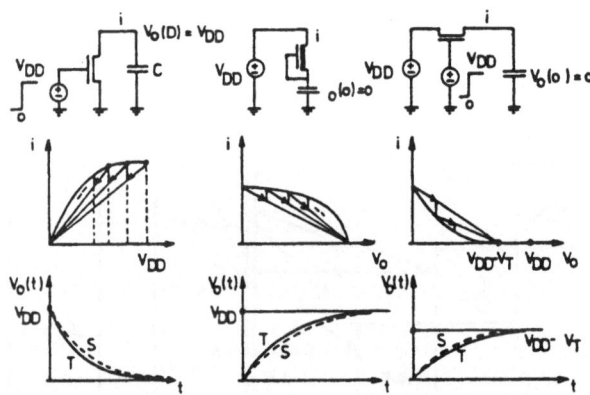

Fig. 18.   Errors in one-step secant iterates for different
circuit configuration.

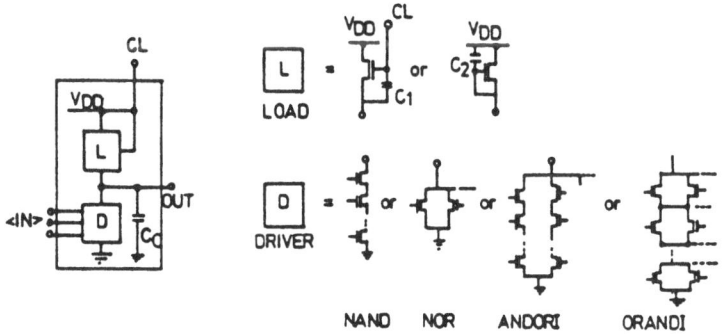

Fig.19a : basic gate macromodel

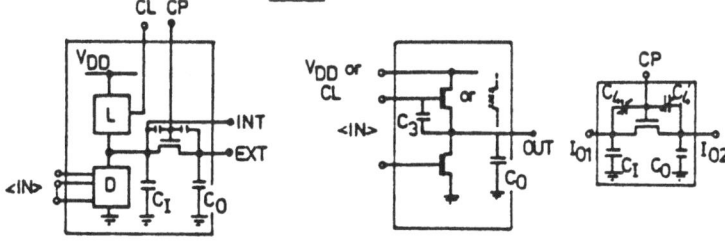

Fig.19b clocked gate
macromodel CL and CP
clock-inputs(only digital nodes)

Fig.19c
bootstrapped buffer
macromodel-clocked
or unclocked

Fig.19d
Transmission gate
or pass-transistor

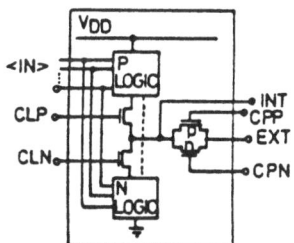

Fig.20   CMOS macromodel.CLP, CPP, CPN optional clock inputs. P-logic tree is
automatically transformed into the
complement of the n-LOGIC which can
have the same NAND,NOR,ANDORI en
ORANDI configuration as for the
basic MOS gate.

The above discussed timing simulation technique can be implemented using the same scheduling algorithm as discussed in table 2 and satisfies all properties for improved computational efficiency discussed in section 3 i.e.
- use of event-driven circuit simulation;
- use of macromodels;
- no iterations per time point;
- use of tables wherever possible.

The improvements in simulation speed obtained are between 20...500 as will be discussed in section 5.4.6.

5.4. Unilateral MOS timing macromodels in DIANA. The MOS timing models presently available in the basic DIANA version are shown in Fig. 19. Fig. 20. shows a CMOS macromodel which is being implemented.

A detailed treatment of these models falls outside the scope of this paper but the interested reader is referred to the users manual of DIANA and to ref. [29].

The following is a short summary of some characteristics of the models :

5.4.1. Geometry and technology link.

All models are computed from technological parameters such as substrate doping, oxide thickness and mobility specified on a .MOD card. The same model card contains the basic dimensions (channel length, width, overlap width etc...) of the transistors of a basic inverter. When a gate is referred to this .MOD card, the transistor geometries, inputparasitic capacitance etc... are all scaled automatically according to commonly used design rules specified in the users guide and [29] (e.g. the width of an n-input NAND driver is n times the width of the basic inverter driver etc...).

Basic is that in contrast to a logic simulator all data are related to geometry and technology.

5.4.2. Transistor and capacitance models.

A mixture of table and analytical models are used as follows [29] :

* Enhancement load transistor :
- Threshold table (default 65 points)
- Analytical Shichman-Hodges [30] secant model including mobility dependence on V_{GS}
- Of particular importance is capacitor C_1 which causes clock-feedthrough effects. It is modeled by a Boolean Controlled Capacitor the value of which is based on the state of the transistor as follows :

$$C_L = C_{ov} + L_{on} \frac{1}{2} C_{ox} + L_{sat} \frac{1}{6} C_{ox} \tag{38}$$

$L_{in} = 1$ if the load is on otherwise zero
$L_{sat} = 1$ if the load is saturated otherwise zero.

This model provides reasonably correct clockfeedthrough (see also 5.4.3.).

* Depletion load transistors
  - Threshold table (65 points)
  - One dimensional secant conductance table generated from full Meyer [31] DC model.
  - Capacitor C_2 forms a capacitive load to the inverter and given by :
  $$C_2 = C_{ov} + L_{on} \frac{1}{2} C_{ox} + L_{sat} \frac{1}{3} C_{ox}$$

* Pass-transistor (transmission gate)
  - Threshold table
  - Analytical Shichman-Hodges [30] secant.
  - Boolean controlled feedthrough
    Capacitances given by :

$$\hspace{10cm} (39)$$

$$C_{GS(D)} = C_{ov} + L_{on} \frac{1}{2} C_{ox} \pm L_{sat} \frac{1}{6} C_{ox} \quad (\text{+ source, - drain}) \quad (40)$$

  - Detection algorithm for saturation state cutting off the transistor at :

$$V_S = V_G - V_{TEeff} \, (V_{GB}) \hspace{5cm} (41)$$

In this way saturated transmission gates are possible.
The pass transistor model can either be automatically included in a clocked macromodel or it can be defined as a separate model (Fig. 19d). This model can be connected to all digital nodes except the internal node INT of the clocked model (Fig.19b).

* The treatment of pass transistor tree will be discussed in 5.4.5.

* Driver structures.

The geometry of the driver structure is derived from the basic inverter and the function it performs.

Depending on the driver structure, the total current I_D flowing into it, is calculated and the driver structure is replaced by a conductance G_D given by :

$$G_D = L_{on} \quad I_D/V_{DS}$$

The capacitance C_{FI} of the driver model (Fig. 21.) is derived later. The boolean variable L_{on}, indicates that there is a current flowing ($I_D \neq 0$) and depends on the driver structure or gate type. It is derived from the boolean variables of the gate inputs.
The calculation of the current I_D is first explained for the two basic functions NOR and NAND and then generalized to the AND-OR-INVERT and the OR-AND-INVERT gates.

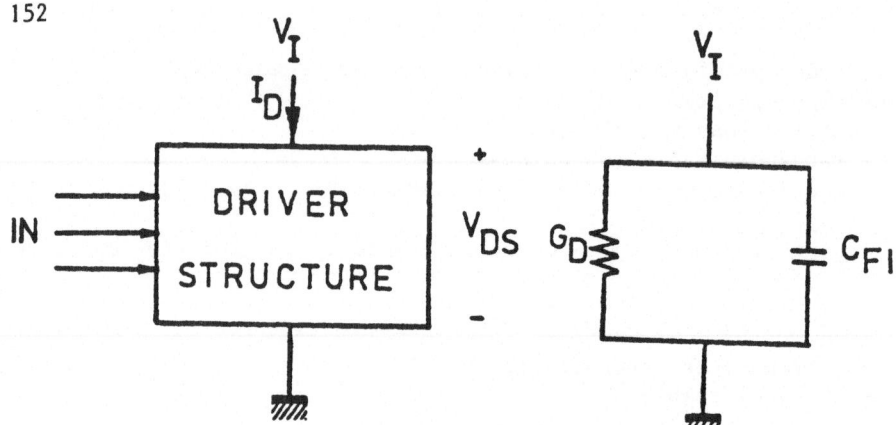

Fig. 21.

For an n-input NOR gate, the current I_D can be calculated since the gate-, source- and drain voltage of each parallel transistor are correctly known. After determining the current I_{Di} of each transistor, the total driver current is defined as :

$$I_D = \sum_{i=1}^{n} I_{Di}$$

For an n-input NAND gate consisting of n transistors in series, the channel width of each driver transistor M_i is automatically assumed to be n times the value of the channel width of the driver transistor of an invertor : $W = n \, WD$.
This fact is used to determine I_D assuming that only one gate voltage variable V_{Gi} is responsable for the total current I_D.

When, for example, in a three input NAND gate, only one gate voltage V_{G2} is changing and the other gate voltages V_{G1} and V_{G3} are kept constant at the supply voltage V_{DD}, the total current I_D only depends on V_{G2} and V_{DS2} whereby V_{DS2} represents the drain-source voltage of the corresponding transistor. Hereby we assume that $V_{DS1} = V_{DS2} = 0$.
In that case we can write :

$$I_D = L_{triode} \, \beta \, (V_{GST} + V_{GST} - V_{DS}) \, V_{DS} + L_{sat} \, \beta V_{GST}^2$$

with

$$V_{GST} = V_{G2} - V_{TE}$$

whereby β is defined from the driver dimensions of the basic inverter and not of the actual NAND transistor.
Using these approximations, the values of V_{xi} need not to be calculated and stored.

<u>Procedure</u> : When the driver structure is conducting ($L_{on}=1$), the minimum value V_G of all gate voltages V_{Gi} is searched. Then the current I_D is calculated depending on the operation region assuming that $V_{DS} = V_I$.

For an AND-OR-INVERT gate, of n AND branches each having n_i inputs, the total current I_D is determined by adding the currents I_{Di} of all AND branches. (OR of the AND sections).

For an OR-AND-INVERT gate, we assume that the total output voltage is present as V_{DS} over each of the n OR sections. The current I_D is then defined as the minimum current of all these OR sections.

* The capacitances C_I and C_0 (Fig. 19a, b). The total capacitance C_I and C_0 at the internal node I and output node 0 are given by :

$$C_I = \overline{C_{JI}} + C_{MI} + C_{FI} + C_{FO_I}$$

$$C_0 = \overline{C_{JO}} + C_{MO} + C_{FO_0}$$

$\overline{C_{JI}}$ and $\overline{C_{JO}}$ are respectively the average junction capacitances at the internal node I and the output node 0 calculated from the zero voltage junction capacitances $C_{JI}(\emptyset)$ and $C_{JO}(\emptyset)$.
C_{MI} and C_{MO} are respectively the capacitance of the metallic interconnections between nodes I (respectively 0) and other nodes of the circuit. The capacitances C_{FI}, C_{FO_I} and C_{FO_0} are respectively the fan in (FI) capacitance at node I and fan out (FO) capacitances at the nodes I and 0. Junction capacitances, fan in and fan out capacitances are now discussed more in detail.

For a given zero voltage junction capacitance $C_J(\emptyset)$, the voltage dependent capacitance is given by

$$C_J(V) = \frac{C_J(\emptyset)}{\sqrt{1+ \dfrac{V}{2\varphi_F}}} \qquad \text{with } 2\varphi_F = \frac{kT}{q} \ln \frac{N}{N_i}$$

To reduce the computational effort during transient analysis, the capacitance $C_J(V)$ is averaged over the expected voltage swing of the nodes I and 0 which is from $\emptyset$ to V_{DD} or :

$$\overline{C_J} = \frac{C_J(\emptyset)}{V_{DD}} \, 4\varphi_F \quad \left(\sqrt{1 + \frac{V_{DD}}{2\varphi_F}} -1 \right)$$

The value $C_j(o)$ are user defined per gate and in a preprocessor $\overline{C_j}$ is calculated and added to the gate model at the appropriate node.

In as far as the fan-in (C_{FI}) and fan-out capacitance (C_{FO}) are concerned, the following rules hold :

The total fan-in capacitance C_{FI} of a gate is defined as the sum of all gate drain capacitances C_{GD} of its driver transistors

154

connected to the internal node I. This capacitance is Boolean
controlled by the operation region of its driver transistors or
its driver branches (see ANDORI, ORANDI).
The constant contribution to C_{FI} is directly added to C_I during
preprocessing while the Boolean controlled part, depending on the
operation region, is added to C_I at each gate evaluation during
transient analysis.

The total fan-out capacitance C_{FO} of a node i is defined as
the sum of all gate capacitances connected to that node. This capacitance is either C_{FOI} (added to C_I) or C_{FOO} (added to C_O)
whether the node i is the internal node or the output node.
After the fan-out list of node i is set up, the contribution to
the fan-out capacitance of each element of the fan-out list is calculated from its model parameters and driver configuration.
For nodes controlling load or pass transistors a separate fan-out
list is set up as is explained in section 5.4.4.
Based on the basic gate inverter data and the specified zero bias
junction as well as interconnection parasitics and using the scaling from basic inverter into gate models, all parasitic capacitances are preprocessed and added to the correct node before the
actual timing simulation starts.
Furthermore, wherever necessary, the gate capacitance is made dependent on the state of the transistor which is very important
for accurate clock-feedthrough prediction, to be discussed next.

5.4.3. Clock-feedthrough modeling.
Clock feedthrough effects occur in all clocked dynamic MOS circuits.
This effect can have a strong influence on the value of logic levels and therefore for example on the fall time of gate driven by
such a logic level. These effects cannot be modeled by any logic
simulator.
Referring to Fig. 19. clock feedthrough is modeled for all clock-
inputs (indicated by either CL or CP). The clock-inputs are considered to be digital nodes and/or voltage sources (or low impedance points). This is generally true because clock-inputs are
either external clockpulse voltage generators or low impedance
driving gates.

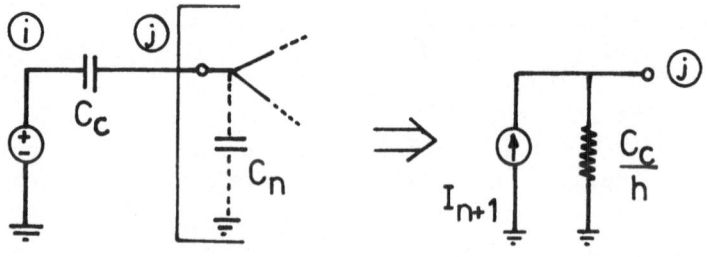

Fig. 22. Clockfeedthrough modeling.

Fig. 22. shows how clock-feedthrough is modeled when using a backward Euler companion for C_C.

The value of current source I_{n+1} at time t_{n+1} can easily be shown to be :

$$i_{n+1} = \frac{C}{h} (V_{i,n+1} - V_{i,n} + V_{j,n})$$ (42)

Notice that if C_C is loaded by a capacitor C_n also represented by a backward Euler companion model :

$$V_{j,n+1} = \frac{\frac{C_c}{h} (V_{i,n+1} - V_{i,n} + V_{j,n})}{\frac{C_c}{h} + \frac{C_n}{h}}$$

or

$$V_{j,n+1} = V_{j,n} + \frac{C_c}{C_c + C_n} (V_{i,n+1} - V_{i,n})$$ (43)

The result given by eq. (43) equals the exact value and illustrates how the backward-Euler integration models clockfeedthrough correctly in a pure capacitive divider, regardless of the step-size. Notice also that, by using boolean controlled capacitors depending on the transistor state (egs. (38)-40)), accurate clockfeedthrough modeling is obtained in spite of the use of large step size.

5.4.4. Hierarchical scheduling.

In order to be able to exploit the two-list activity scheduling algorithm of table 2, the latency criteria have to be defined as was also discussed in 5.2.

The following procedures lead to accurate results for dynamic MOS circuits :

- the fan-out list U(o) with o a digital node of a timing macromodel is scheduled when its voltage V_O crosses the threshold voltages V_{TE} of the driver section transistors or when :

$$V_O \in [V_{TE}, V_{DD} - TOL1]$$ (44)

- As discussed in 5.2. a clock-input V_{CL} belongs to a clock-fanout list and since a clock-input creates clockfeedthrough (see eq. (43)), it has to be scheduled when its node voltage derivative exceeds a tolerance value TOL or :

$$\left| \frac{V_{CL,n+1} - V_{CL,n}}{h} \right| > TOL \quad \text{(default value : 1V/}\mu\text{s)}$$ (45)

- A timing macromodel is descheduled when (45) is no longer satisfied and for the internal (if present) and external node $|dv/dt| <$ TOL except when the internal $v > V_{TE}$ and $L_{on} = 1$.

Notice that, since a functional coupling between input and output exists through the driver structure, the scheduling mechanism is based on voltage derivatives and the on state of the driver, rather than on the logic state of the model.

156

Due to the fact that clock-inputs can be derived from other digital nodes, it is clear that a hierarchy in clock evaluation exists which imposes an ordering in the future activity list L_b (defined in 5.2.).
This is illustrated in Fig. 22.
Primary or first level clocks are either external inputs (I1) or gates without pass transistor and with a clocked load (C). Clocks of level n are formed by gates clocked by clocks of level n-1. The number of clock levels is unlimited.

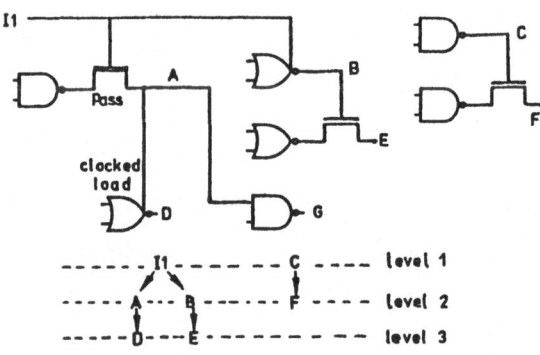

Fig. 22. Hierarchy in clock evaluation.

As for example the clock variation $V_A = V_A(t_{n+1}) - V_A(t_n)$ is needed to calculate the clockfeedthrough in gate D, gate A has to be evaluated before evaluating gate D. In general all clockgates (gates acting as clock) have to be calculated in a certain hierarchical order (from level 1 to level n) before evaluating all non-clock gates.

For the implementation of the latency principle, use is made of clock fan-out lists. They are also used to calculate the contribution of load- and pass transistors to the fan-out capacitance C_{FO} of each clock gate.
From these clock fan-out lists, the clock hierarchy is derived and unresolvable clock references or feedback loops in the clock hierarchy tree are detected (Fig. 23).

Before each time step during transient analysis, all clock gates in the active string are reordered according to the clock hierarchy. This is necessary because due to the scheduling mechanism, the order in the active list is always changing.

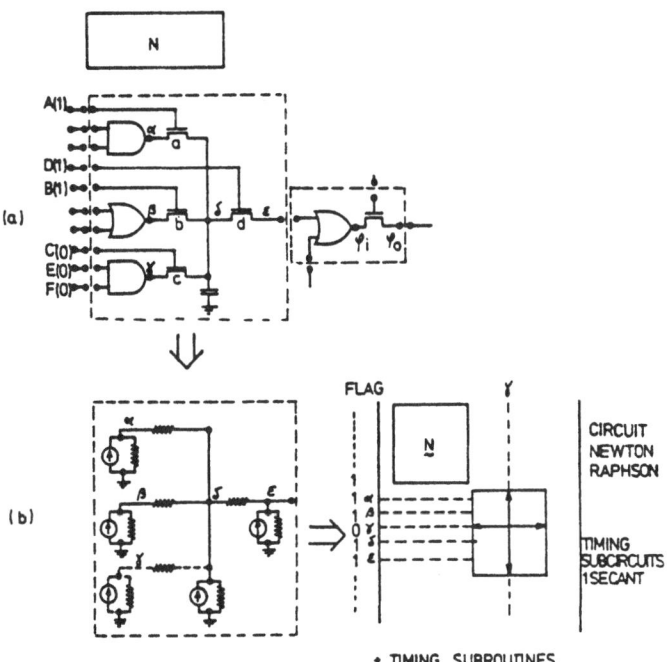

Fig. 24. Pass transistor trees, together with their driving macromodels are treated as scheduled submatrices.
Equations within submatrices can also be scheduled (e.g. γ)

158

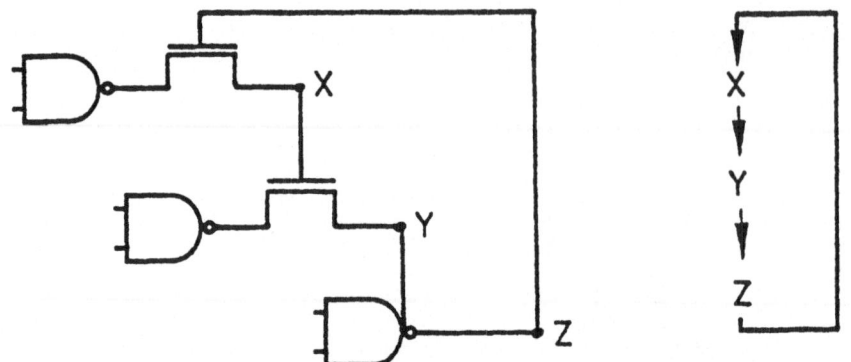

Fig. 23. Onresolvable clock references.

5.4.5. Pass transistor trees-submatrix scheduling.

In many dynamic circuits an interconnection of more than one pass-
transistor is used such as shown in Fig. 24a. We call such a network a pass transistor tree. In such case, when several of these
pass transistors are conducting at the same time, a tightly coupled
network is created which no longer allows for node decoupling.
For example when A,B,D are high and C low in Fig. 24a. then the
tightly coupled timing network shown in Fig. 24b is created.
Since D is low, node γ is latent and does not need to be evaluated.

In order to solve such problems DIANA makes use of following
procedure :

1. in input processing all pass-transistor trees are detected together and a cut-set is made through the inputs of the connected
timing models. This creates unilateral subnetworks as the one
shown in Fig. 24a.
2. all nodes within the cut-set are introduced in the network matrix as a submatrix. All rows and columns are provided with a
"bypassing" bit in a flag-vector as shown in Fig. 24c.
3. The "bypass" bit is used to "schedule" equations in the matrix.
For timing submatrices the flag vector bits are only one at "timing"
time steps for scheduled nodes. For example row and column γ are
bypassed since gate γ and pass-transistor C are not scheduled.
4. All other uncoupled macromodels for timing or logic MØ simulation are treated as single- or sets of two equations in macromodel
subroutines.
5. Tightly coupled MOS circuits are present in a circuit matrix
with normal Newton-Raphson iteration. At present techniques for
scheduling subnetworks in this main matrix $\mathscr{N}$ (Fig. 24c) also
based on the flag-vector are being studied.

5.4.6. Example of pure MOS Timing simulation.

Fig. 25 shows the block diagram of an 8 bit serial parallel multiplier in 2 phase P MOS. The circuit performs the multiplication of
a coefficient B (8 bits, entered in parallel) with a word A (entered serially one bit per clock cycle).

The output consists of the truncated product delayed by 16 clock cycles. The circuit simulated contains 8 cells each composed of 79 transistors. Fig. 26. shows a circuit schematic. V.O. is a full-adder.

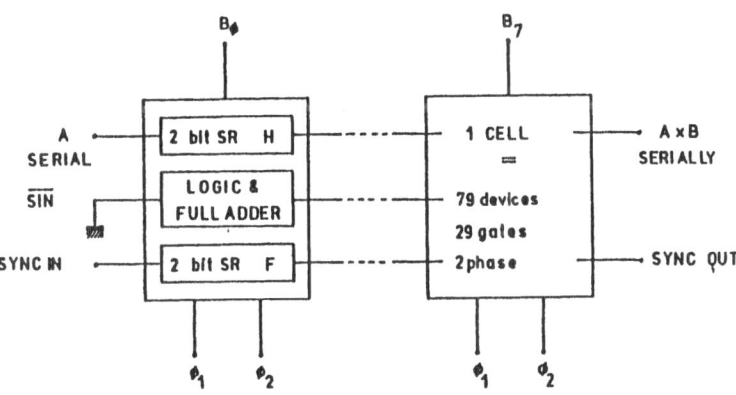

Fig. 25. An 8 bit serial parallel multiplier in 2-phase dynamic MOS.

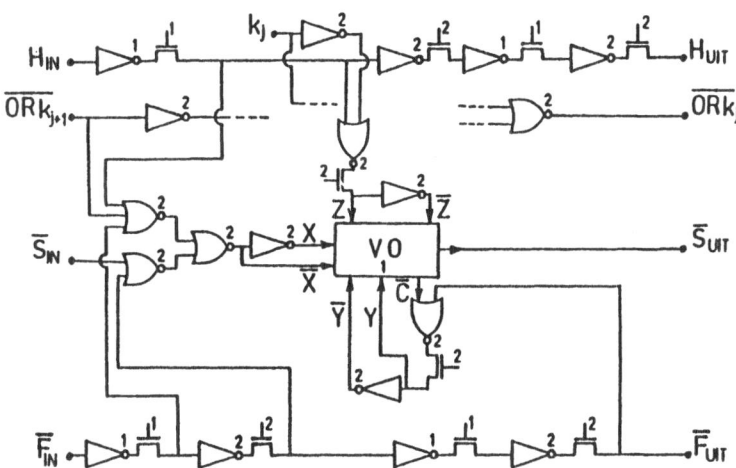

Fig. 26. Circuit schematic of one cell. V.O. is a full adder circuit.

160

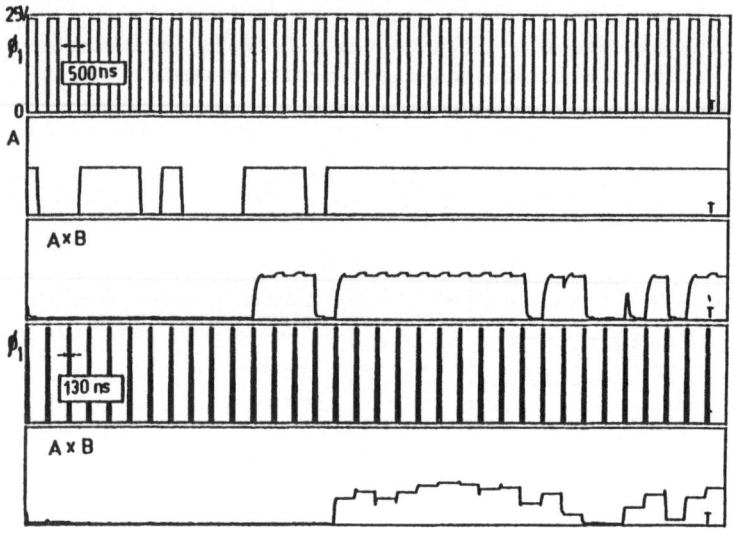

Fig. 27.   Results of simulation 1700 time steps, 20 ns; CPU time :
85 sec. IBM 370-158.

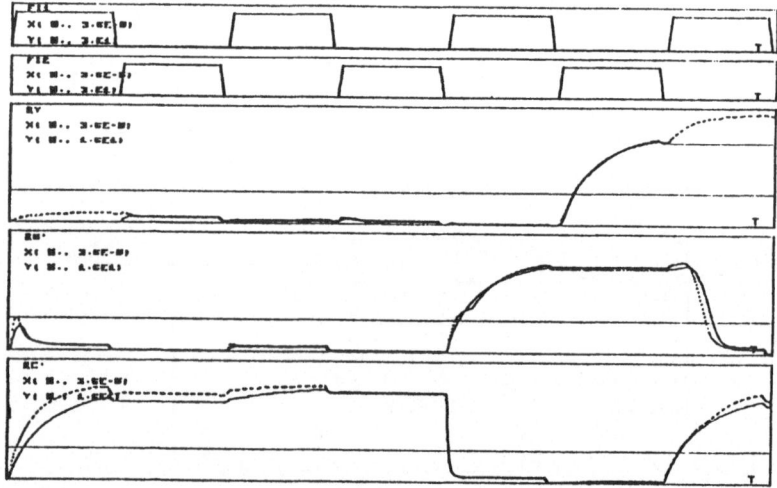

Fig. 28.   Simulation of one-cell (Fig. 26) using timing simulation (-) takes 3.5 sec.   Same circuit in circuit simulation (---) takes 28 minutes.
Notice bootstrapping problem in signal **y**.

Fig. 27. shows a simulation of the circuit from layout data of the actual circuit.
The plots in Fig. 27. are respectively the serial input word, the 1MHz clock with pulse width t_w=500 nsec. and the resulting product as simulated by DIANA. A total of 1700 timesteps of 20 nsec are used.
Also shown is a simulation with a narrow clock pulse width tw=130 nsec. which just causes wrong operation.
The effects of charge redistribution and clockfeedthrough on the signals are clearly visible. These effects cannot be obtained from pure logic simulation.
The total CPU time for this simulation is 85 sec. on an IBM 370/158 and is estimated to be about 400 times faster than circuit simulation (if the latter could cope with the 616 transistor complexity).
Fig. 28. shows a comparison of timing simulation of one cell (79 transistors) and circuit simulation (dotted line) using the circuit simulation part of DIANA with the same time step (700 time steps, 5 ns). Notice that the agreement is reasonable with exception of a "bootstrapping" effect on node Y which results from a gate-drain coupling of an internal node with a high impedance driving node. This effect is not visible in the timing simulator part as no floating capacitors are present. Excecution time in timing is 3.5 sec. against 28 minutes for circuit simulation.

    5.4.7. Example of mixed mode analog-timing simulation of a full LSI chip. Hierarchical input language MDL.

One interesting application area for mixed-mode simulation is the simulation of LSI chips with analog and digital parts.
Fig. 29a. shows a block-diagram of a complex 8 channel, 8 bit D/A convertor with software programmable scales [32].
This chip is microprocessor controlled and has ca. 1793 transistors
Under control of the handshake logic, 2 halfbytes of the microprocessor set the analog output offset and range, and 1 byte the linear interpolation for the D/A 8 bit ratioed capacitor array.
Fig. 29b. shows the capacitor array with 8 output sample and hold buffers. The adress counter scans the 16 x 8 bit RAM for reading the microprocessor data and for setting the capacitor array switches. It also multiplexes the analog output sample and hold circuits which are, each on its turn, switched into the feedback loop of an op-amp.
All sequencing of the switches for the analog circuit and the adadress counter are generated from a control PLA with built in sequencer. This control PLA has 32 minterms, 10 inputs and 31 outputs.
This example is representative for an LSI chip or part of a VLSI chip. This chip has been fully simulated over the conversion cycles for 3 output channels. One conversion cycle consists of several subconversion cycles :
    - an offset compensation step for the op-amp stored on VBB by VOFSAM (Fig. 23b).

162

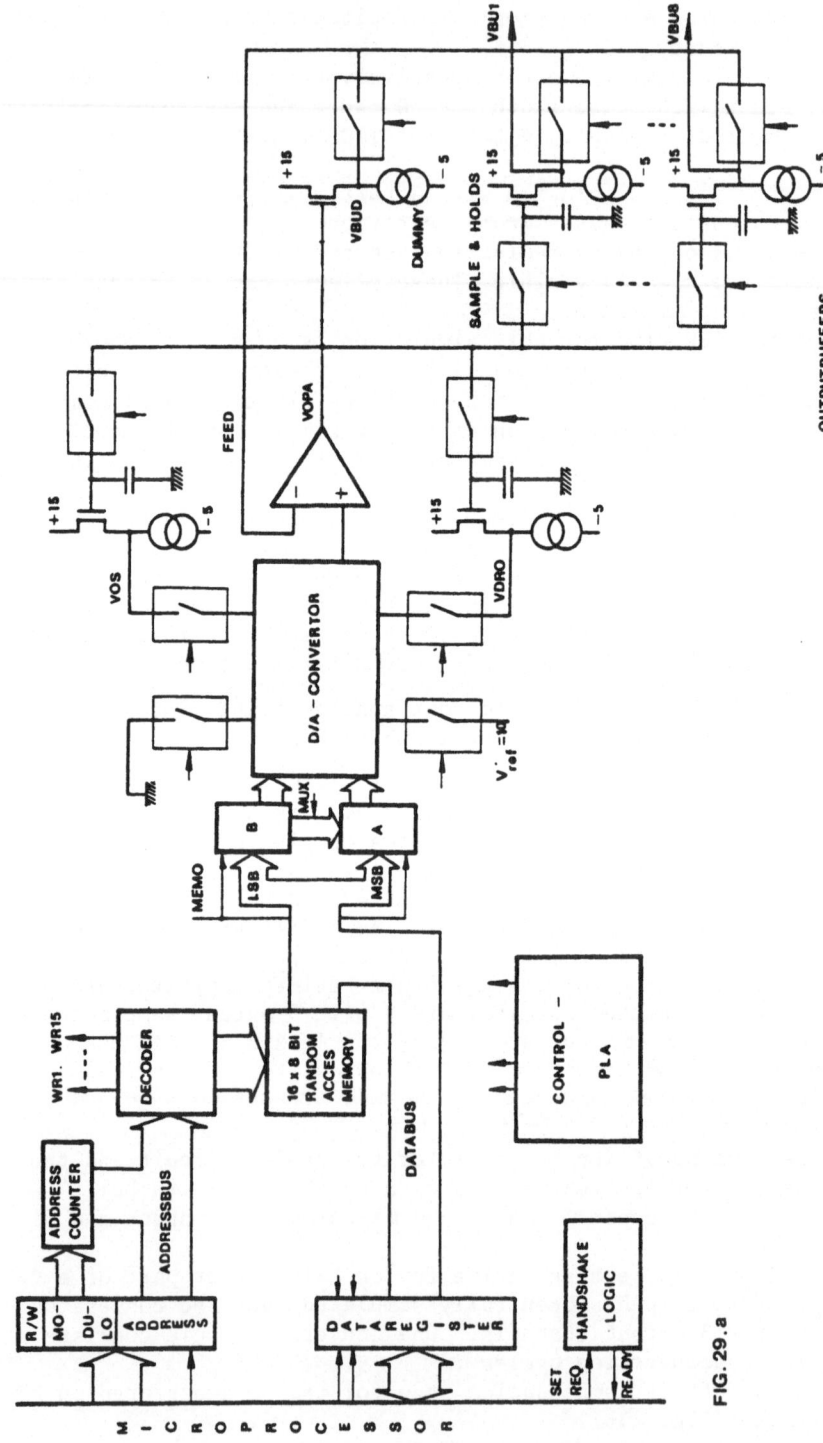

FIG. 29.a

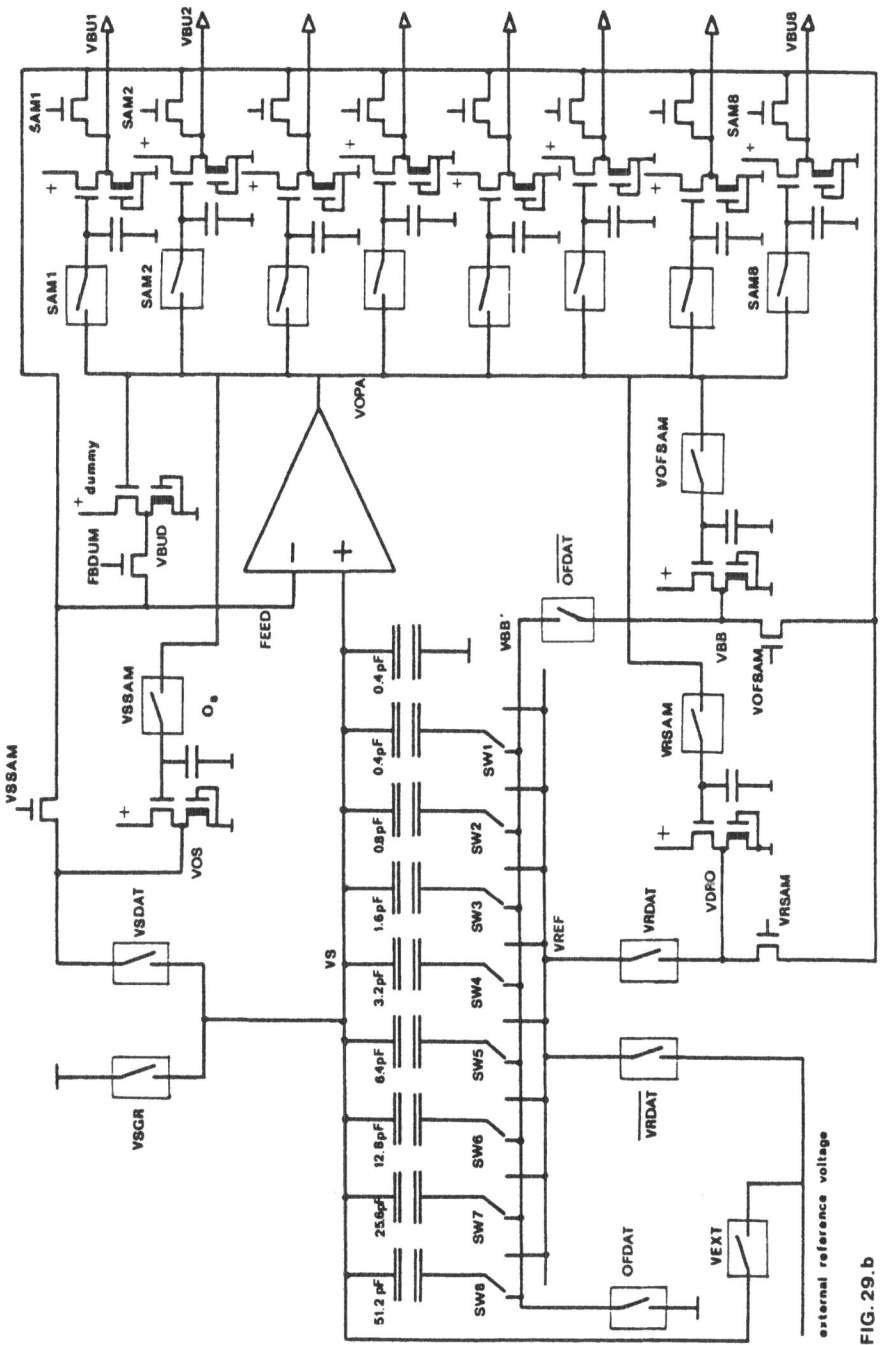

FIG. 29.b

- a scale offset setting to be stored on VOS by VSSAM
- a reference voltage generation and storage on VDRO by VRSAM
- the final conversion of the 8 bit word driven into the
  appropriate output channel.

Simulation of such a chip would be very difficult if no <u>language</u>
is used which allows a <u>hierarchical</u> description of such a chip.
Such a language preprocessor called <u>M</u>acro <u>D</u>escription <u>L</u>anguage
(MDL) has been written and allows for a hierarchical decomposition
of large circuit blocks into smaller tested subcircuits. As an
example Fig. 30. shows the full highest hierarchy of description
in the form of CALL's to the constituent blocks easily to be recognized from the block diagram Fig. 29a.

```
;
;    MIXED MODE SIMULATION OF    D A P I A
;    A DIGITAL ANALOG PROGRAMMABLE INTERFACE ADAPTER
;
;
;
CALL CONTROL(.)
CALL MODELS(.)
CALL INPUTS(2)
CALL RAM(.)
CALL COUNTER(.)
CALL PLA(.)
CALL ARRAY(.)
CALL OPAMP(.)
CALL BUFFERS(.)
;
;
```

Fig. 30. Hierarchical decomposition .

As an example of the hierarchical decomposition Fig. 31. shows
the $ MACRO BUFFERS called in Fig. 30.   These buffers represent
all the sample and hold parts and buffers shown in Fig. 29b.
Notice that $ MACRO BUFFERS <u>contains</u> two more MACROS i.e. $ MACRO
SAH and $ MACRO SAHOUT.
Notice :
    * the exploitation of repetitivity to get a very compact
description;
    * since the $ MACRO BUFFERS <u>contains</u> its own $MACRO's it can
be used a <u>self contained</u> subcircuit for separate simulation, before being taken into the full simulation.
As such every subcircuit can be fully separately tested.

Fig. 32. shows another such example for the description of the 8
bit capacitor array including switches.
Notice that the notation of MDL allows the insertion of the different capacitor values into the array (ranging from 0.4 pF to
51.2 pF). Fig. 33. shows part of the expansion of $ MACRO BUFFERS
which is used as the input to DIANA itself.

```
$MACRO BUFFERS( )


$MACRO SAH(#,@,%)
XGD@ O # NI@ O 2M
CB@ NI@ O VAL=2P
RB@ # O 500
S2B@ VOPA NI@ % O R=7K VT=1.5
S2F@ # FEED % O VT=1.5 R=150
$END SAH


$MACRO SAHOUT(#,@)
;
;   OUTPUT SAMPLE & HOLD NR @
;
CALL SAH(VBU@,@,SAM@)
NAND SHI@ MOD=PLA IN=WR# SAMI
NOR SAM@ MOD=PLA IN=SHI@
$END SAHOUT


;
;   DUMMY SAMPLE & HOLD
;
XGBD O VBUD NID O VAL=2M
RBD VBUD O VAL=500
RBDS NID VOPA VAL=7K
CBD NID O VAL=2P
S2DU VBUD FEED FBDUM O VT=1.5 R=7K
;
;   VS SAMPLE & HOLD
;
CALL SAH(VOS,S,VSSAM)
;
;   VREF SAMPLE & HOLD
;
CALL SAH(VDRO,R,VRSAM)
;
;   VBD SAMPLE & HOLD
;
CALL SAH(VBB,B,VOFSAM)
CALL SAHOUT(1,1)
CALL SAHOUT(3,2)
CALL SAHOUT(5,3)
CALL SAHOUT(7,4)
CALL SAHOUT(9,5)
CALL SAHOUT(11,6)
CALL SAHOUT(13,7)
CALL SAHOUT(15,8)
$END BUFFERS
```

Fig. 31.   MDL description of $MACRO BUFFERS

166

```
$MACRO ARRAY(.)
;
;     SWITCHED CAPACITOR ARRAY
;
CO VS 0 0.3P
$MACRO SWITCH(#,%,@)
S2R# VS@ VREF SW# 0 R=150 VT=2
S2Y# VS# VBB' SWI# 0 R=150 VT=2
NOR SWI# MOD=PLA IN=SW# CMO=0.1P
CAR# VS VS@ VAL=%P
NOR SW#  MOD=PLA IN=IN@I DI# CMO=1P
;
$END SWITCH
;
CALL SWITCH(1,0.4,B)
CALL SWITCH(2,0.8,B)
CALL SWITCH(3,1.8,B)
CALL SWITCH(4,3.2,B)
CALL SWITCH(5,6.4,A)
CALL SWITCH(6,12.8,A)
CALL SWITCH(7,25.6,A)
CALL SWITCH(8,51.2,A)
S3VR VREF VDRC EXT VRDAT 0 R=150 VT=2
S2VG VS 0 VSGR 0 R=150 VT=2
S2VD VS VGS VSDAT 0 R=150 VT=2
S2EX EXT VS VEXT 0 R=150 VT=2
S2GR VBB' 0 OFDAT 0 R=150 VT=2
S2OF VS3 VBB' OFIDAT 0 R=150 VT=2
VE EXT 0 VAL=10
;
$END ARRAY
```

Fig. 32. Description of 8 bit capacitor array.

```
        •
        •
        •

;
;    OUTPUT SAMPLE & HOLD NR 4
;
XGB4 0 VBU4 NI4 0 2M
CB4 NI4 0 VAL=2P
RB4 VBU4 0 500
S2B4 VOPA NI4 SAM4 0 R=7K VT=1.5
S2F4 VBU4 FEED SAM4 0 VT=1.5 R=150
NAND SHI4 MOD=PLA IN=WR7 SAMI
NOR SAM4 MOD=PLA IN=SHI4
;
;    OUTPUT SAMPLE & HOLD NR 5
;
XGB5 0 VBU5 NI5 0 2M
CB5 NI5 0 VAL=2P
RB5 VBU5 0 500
S2B5 VOPA NI5 SAM5 0 R=7K VT=1.5
S2F5 VBU5 FEED SAM5 0 VT=1.5 R=150
NAND SHI5 MOD=PLA IN=WR9 SAMI
NOR SAM5 MOD=PLA IN=SHI5
;
;    OUTPUT SAMPLE & HOLD NR 6
;
XGB6 0 VBU6 NI6 0 2M
CB6 NI6 0 VAL=2P
RB6 VBU6 0 500
S2B6 VOPA NI6 SAM6 0 R=7K VT=1.5
S2F6 VBU6 FEED SAM6 0 VT=1.5 R=150
NAND SHI6 MOD=PLA IN=WR11 SAMI
NOR SAM6 MOD=PLA IN=SHI6
```

        •
        •
        •

Fig. 33. Partial expansion of $MACRO BUFFERS.

The simulation of this circuit thus involves analog as well as
timing models. The following table represents a few run statistics on VAX 11/780.

Nr. of timing nodes used :	457
Nr. of transmission gates :	200
Nr. of bootstrapped buffers :	32
Nr. of analog nodes :	88
Nr. of transistors:	1793

Elements used.

Nr. of time steps	: 4800
Time step :	12.5 nsec.
CPU time :	1 h 26 min.
Dynamic memory allocation :	104.2K bytes
Average gate acitivty :	37%

Run statistics.

168

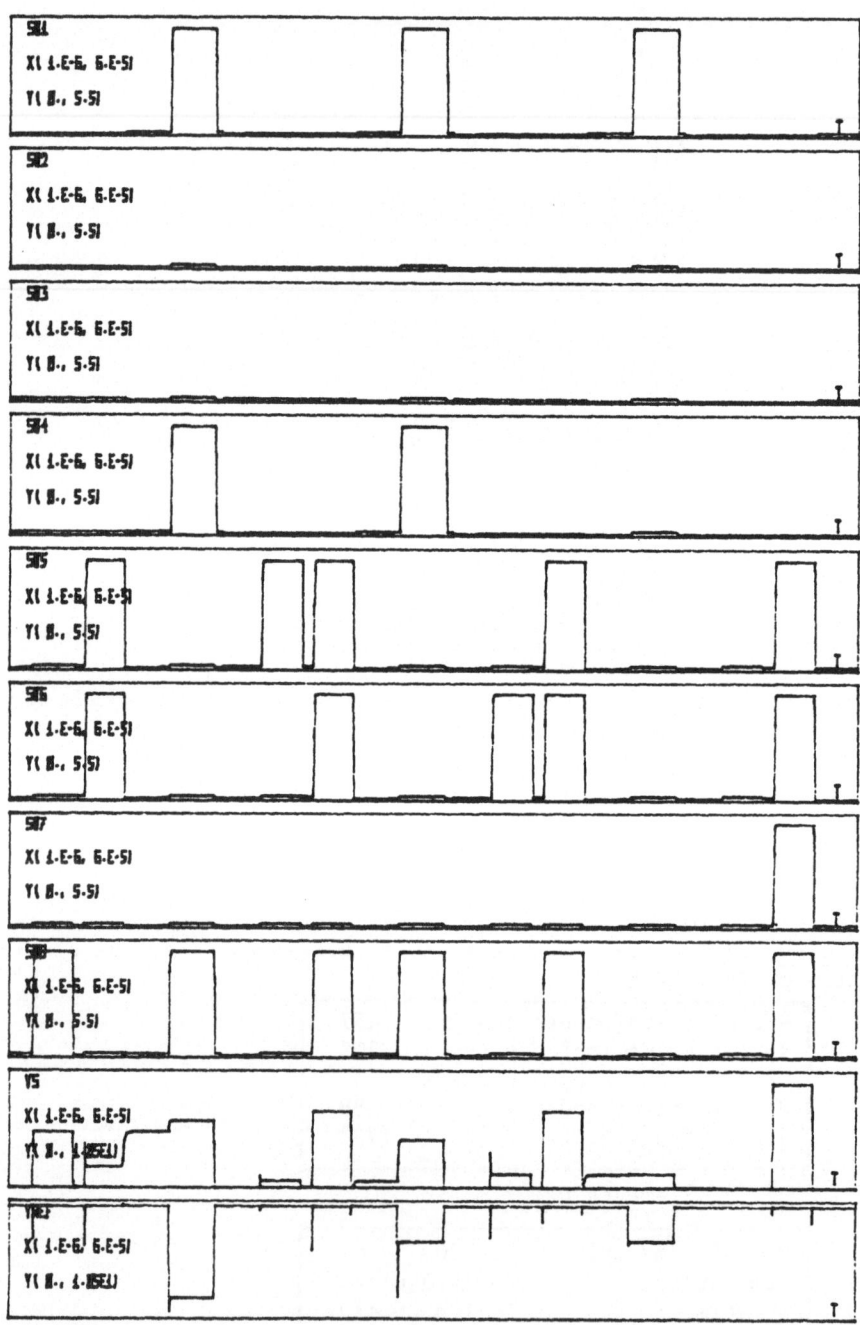

FIG.34. a

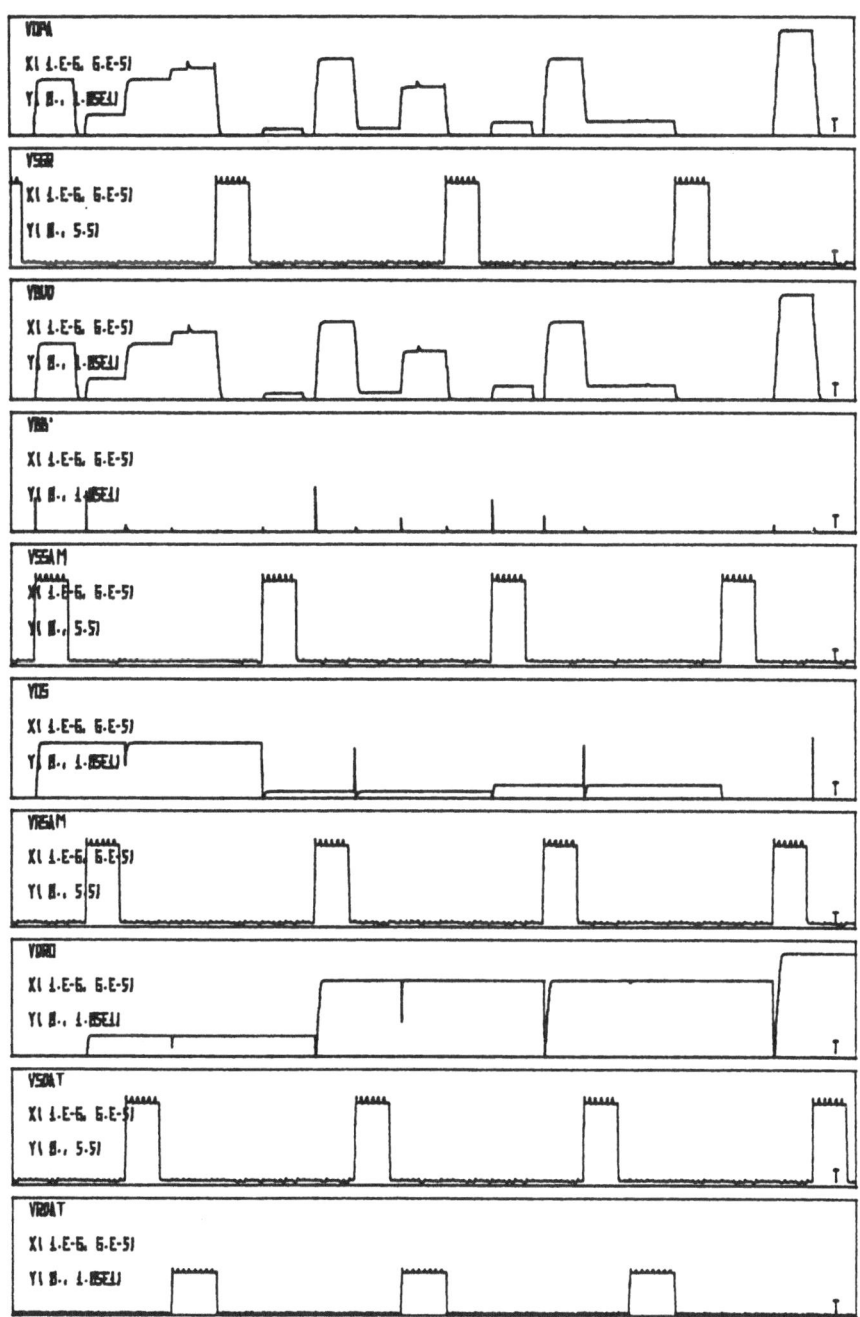

FIG.34.b

170

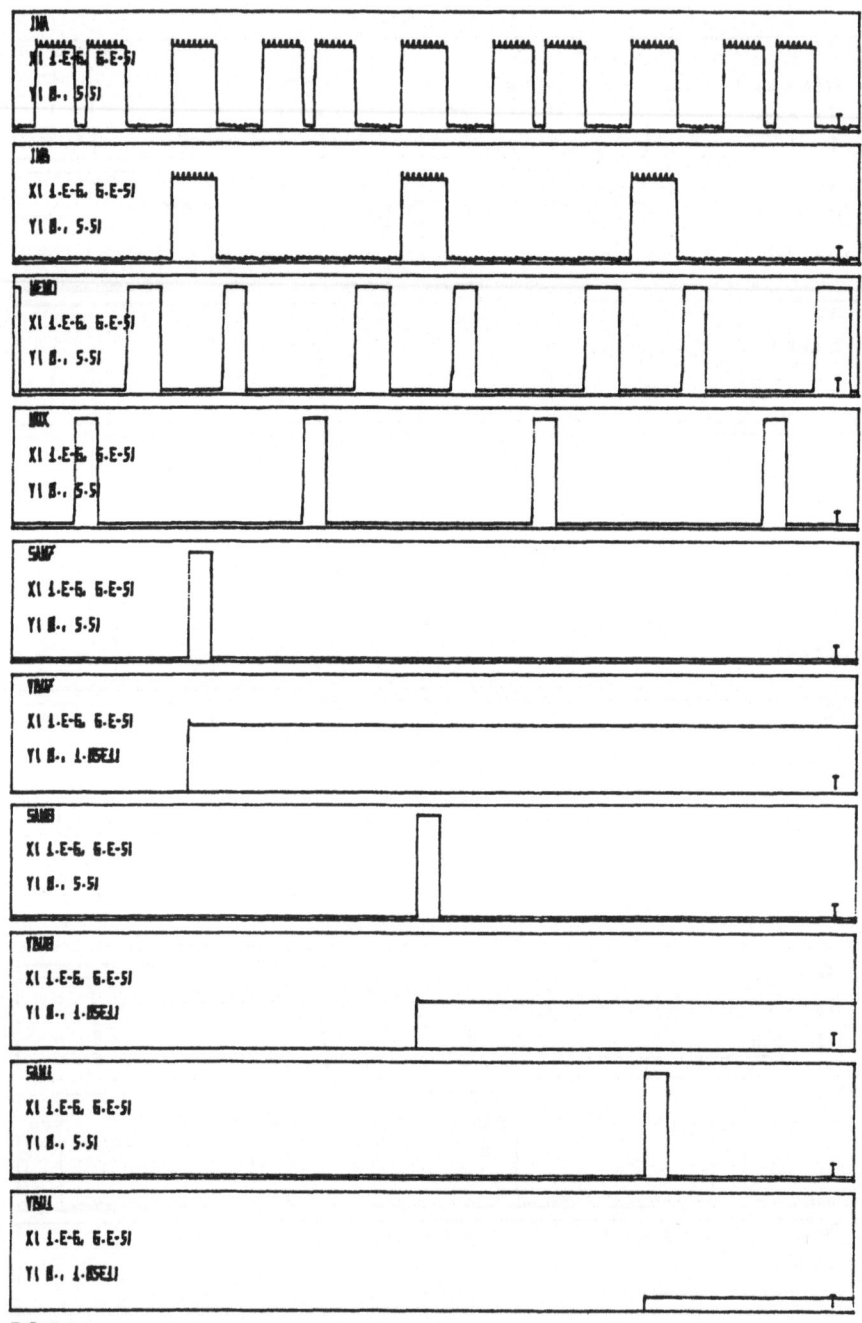

FIG.34.c

Notice that no circuit simulator can cope with such complexity within this CPU time and memory requirement. Yet all circuit interactions are reasonably predicted and thanks to the MDL languages all parts can be hierarchically built up and tested before being made part of the full circuit.

Finally Fig. 34. shows plots of a few relevant outputs. Fig. 34a. represents the signals SW1...SW8 for driving the switches of the array on Fig. 29b. and the analog voltages VS and VREF. The analog output voltage of the op-amp (VOPA) is shown in Fig. 34b. VSGR initiates an output conversion cycle. After its generation, the scale offset voltage is stored on the sample and hold VOS by the sample signal VSSAM. Hereafter the range voltage is being converted and sampled on VDRO by VRSAM. VSDAT and VRDAT are switching the scale offset VOS and the range VDRO to the capacitor array for the conversion of an output voltage. The ripple on several control signals is caused by clockfeedthrough of transmission gates. The digital information from the RAM transfers to the registers A and B (Fig. 29a) by MEMO. INA and INB (Fig. 34.c) allow registers A and B to SW8...SW5 and SW4...SW1 resp. SAM7, SAM8 and SAM1 show the sample signals of the output buffers and their resp. analog output values : VBU7, VBU8 and VBU1.

6. CONCLUSIONS

VLSI design requires simulation tools that can cope with the full spectrum of top-down design procedures from register transfer down to circuit level.

Instead of using all simulation levels in separate programs with different languages, data-structures and algorithms this contribution shows several techniques to combine logic, timing and circuit simulation into one program called a mixed-mode simulator.

The techniques have been implemented in the computer program DIANA.

Future work is necessary whereby such programs can be extended towards register transfer level simulation, provided interconnection structure is introduced in the latter.

Also further development work on better interfacing and more reliable subcircuit scheduling is necessary. A lot of attention should be paid to providing better interface also between layout and macromodelparameters.

Work on all these points as well on its implementation in DIANA is presently underway.

172

REFERENCES

[1] B. Lattin : "Computer aided design : the problem of the 80's, microprocessor design", Proc. EEC symp. on CAD of digital electronics, Brussels, pp. 239-241, Nov. 1978.

[2] G.E. Moore : "Are we Really Ready for VLSI ?", IEEE Digest of Techn. Papers of 1979 Int. Solid State Circuits Conf., Philadelphia, 54-55, Febr. 1979.

[3] C. Mead, L. Conway : "Introduction to VLSI Systems", Addison-Wesley Publ. Co., Mass. 1979.

[4] M.A. Breuer : "Digital System Design Automation : Languages, Simulation & Data Base", Ch.2-3, Pitman, Woodland Hills, Calif.'75.

[5] S.H. Szygenda, E.W. Thomson : "Digital Logic Simulation in a Time Based, Table Driven Environment - Part 1", Computer, pp. 23-36, March 1975.

[6] E.B. Eichelberger, T.W. Williams : "A Logic Design Structure for LSI Testability", Design Automation & Fault Tolerant Computing, Comp. Science Press., 1978, pp. 165-178.

[7] D.A. Hodges, P.R. Gray : "Potential of MOS Technologies for Analog Integrated Circuits", IEEE Journal of Solid State Circuits, Vol. SC-13, No. 3, pp. 285-293, June 1978.

[8] G. Perskey, D.N. Deutsch, D. Schweikert : "LTX - A Minicomputer Based System for Automatic LSI layout", Journ. of Design Automation and Fault Tolerant Computing, Vol. 1, no. 3, May 1977.

[9] H. Beke, W. Sansen : "CALMOS - A portable software system for the automatic layout of MOSLSI", Proc. 16th Design Automation Conf., June 1979, pp. 102-108.

[10] G.M. Klomp : "CAD for LSI-production's interest is in its Economics", SIGDA Newsletter, Vol. 6, no. 3, Sept. 1976, pp. 11-15.

[11] A.R. Newton, D.O. Pederson : "A Simulation Program with Large Scale Integrated Circuit Emphasis", Proc. 1978 Int. Symp. on Circuits & Systems, N.Y., pp. 1-4, June 1978.

[12] G. Arnout, H. De Man : "The use of Boolean Controlled Elements for Macromodeling of Digital Circuits", Proc. 1978 Int. Symp. on Circuits & Systems, N.Y., pp. 522-526, June 1978.

[13] D.D. Hill, W.M. Van Cleemput : "SABLE : Multilevel Simulation for Hierarchical Design", Proc. 1980 Int. Symp. on Circuits & Systems, Houston, pp. 431-434, April 1980.

[14] -- : "A Survey of computer aided design and analysis programs", Electr. Syst. Dept. Univ. of South-Florida, Tanga.

[15] E.G. Ulrich : "Exclusive simulation of activity in digital networks", Commun. Ass. Comput. Mach., Vol. 13, pp. 102-110, Febr. '69.

[16] N.B. Rabbat et al. : "A latent macromodular approach to large scale sparse networks", IEEE Trans. Circuits & Systems, Vol. CAS-23, pp. 745-752, Dec. 1976.

[17] A.R. Newton, D.O. Pederson : "Analysis Time, Accuracy and Memory Requirement Trade-offs in SPICE2", 11th Asilomar Conf. on Circuits, Systems & Computers, pp.6-9, Asilomar, Nov. 1977.

[18] M.A. Breuer, A.D. Friedman : "Diagnosis and Reliable Design of Digital Systems", Pitman Publishing Ltd. London 1976.

[19] G. Arnout, H. De Man : "The Use of Threshold Functions and Boolean-Controlled Network Elements for Macromodeling of LSI Circuits", IEEE Journ. of Solid State Circuits, Vol. SC-13, No. 3, pp. 326-332, June 1978.

[20] E.M. DaCosta, K.G. Nichols : "TWS - timing and waveform simulator", Proc. IEE Conf. on CAD Manuf. Electr. Comp. Circ. Syst., pp. 189-193, 1979.

[21] N. Tanabe, H. Nakamura, K. Kawakita : "MOSTAP : An MOS Circuit Simulator for LSI circuits", Proc. 1980 IEEE Circ. Syst. Symp. Houston, Texas, pp. 1035-1038.

[22] W.B. Rabbat and Ryan : "A computer modeling approach for LSI digital structures", IEEE Trans. Electron Devices, Vol. ED-22, Aug. 1975.

[23] B.R. Chawla, H.K. Gummel, P. Kozak : "MOTIS - An MOS Timing Simulator", IEEE Trans. Circuits & Systems, Vol. CAS-22, pp. 301-310, Dec. 1975.

[24] S.P. Fan, M.Y. Hsueh, A.R. Newton, D.O. Pederson : "MOTIS-C : A new circuit simulator for MOSLSI circuits", IEEE Proc. Int. Symp. Circ. Syst., pp.700-703, April 1977.

[25] N.B. Rabbat, H.Y. Hsieh: "A latent macromodular approach to large scale sparse networks", IEEE Trans. Circuits & Systems, Vol. CAS-22, pp. 745-752, Dec. 1976.

[26] H.Y. Hsieh, N.B. Rabbat : "Macrosimulation with quasi-general symbolic FET macromodel and functional latency", Proc. 16th design automation conf. 1979, pp. 229-234, San Diego.

[27] A.E. Ruehli, A.L. Sangiovanni - Vincentelli, N.B. Rabbat : "Time Analysis of Large Scales Circuits containing one-way macromodels", Proc. 1980 IEEE Int. Symp. Circuits & Systems, Houston, pp. 766-770, June 1980.

[28] G. De Micheli, H. Sangiovanni-Vincentelli, A.R. Newton : "New Algorithms of Timing Analysis of Large Circuits", Proc. 1980 IEEE Symp. Circuits & Systems, pp.439-443, June 1980.

[29] H. De Man, G. Arnout, P. Reynaert : "Status of macromodels in DIANA", ESAT-report available from ESAT, K.U.Leuven, Kard. Mercierlaan 94, 3030 Heverlee, Belgium.

174

[30] H. Shichman, D.A. Hodges : "Modeling and Simulation of IGFET Switching Circuits", IEEE J. of Solid State Circuits, Vol. SC-3, No. 3, Sept. 1968, pp. 285-289.

[31] I. Meyer : "MOS Models and Circuit Simulation", RCA Review, Vol. 32, March 1971, pp. 42-63.

[32] L. Bienstman, H. De Man : "An 8 channel, 8 bit μP compatible NMOS convertor with programmable ranges", Digest of Technical Papers of 1980 IEEE Int. Solid State Circuits Conference, pp. 16-17, Febr. 1980.

TIMING, LOGIC AND MIXED-MODE SIMULATION FOR LARGE MOS INTEGRATED CIRCUITS

A. R. Newton

Department of Electrical Engineering and Comp·.ler Sciences, University of California, Berkeley, Ca. USA.

1. INTRODUCTION

Many different forms of simulation can be used for the analysis of large digital integrated circuit designs at the various stages of the design process. They may be classified as *Behavioral* (also called algorithmic or functional) simulators, *Register Transfer Level (RTL)* simulators, *Gate Level Logic* simulators, *timing* simulators, and *circuit simulators*, as illustrated in Fig.1.1.

Behavioral simulators [51] are used at the initial design phase to verify the *algorithms* of the digital system to be implemented. Not even a general structure of the design implementation is necessary at this stage. Behavioral simulation might be used to verify the hand-shaking protocol on a logic buss, for example.

Once the algorithms have been verified, a potential implementation *structure* is chosen. For example, a microprocessor, some memory, and a special-purpose input-output module may be chosen to implement the handshaking protocol mentioned above. An RTL simulator can be used to verify the design at this level. Only crude timing models may be available, since the exact circuit parasitics and other implementation details are not yet known. Useful information relating to congestion and harware/firmware tradeoffs can be obtained from this level of analysis. A variety of RTL languages and associated simulators have been described in the literature [29].

Depending on the design methodology and certain technology issues, a gate-level design may be undertaken, where each of the RTL modules is further partitioned into low-level logic building blocks, or gates. A logic simulator may then be used to verify the design at this level. Sophisticated delay models may be introduced and testability analyses performed.

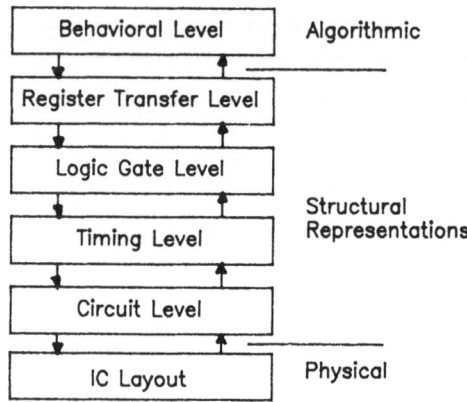

Fig. 1.1. The integrated circuit simulation hierarchy

From the gate level, transistors and associated interconnections are generated to implement the design as an integrated circuit. Accurate electrical analysis can be performed for small parts of this design using a circuit analysis program [1],[54] or larger blocks may be analyzed in less detail using a timing simulator [9-10],[13]. Once the integrated circuit layout is complete, accurate circuit parameters, such as parasitic capacitance values and transistor characteristics, may be extracted and used at the electrical level. These analyses may then be used to improve the delay models at both the gate and RTL levels to verify the circuit design using accurate timing data.

A number of simulators have been developed recently which span a range of these levels in the simulation hierarchy. These simulators are called *mixed mode* simulators [40-42] and allow different parts of a circuit to be described at different levels of abstraction. Not only does this approach permit a smooth transition between different levels of simulation (since the same simulator and associated input processor is used) but it allows the designer to take advantage of the time and memory savings available from higher-level descriptions of parts of the circuit.

This chapter addresses some of the "low end" simulation techniques described above for the analysis of large MOS integrated circuits. Section 2

describes timing simulation and is followed in Section 3 by a description of
table-driven logic simulation appropriate for MOS circuits. Finally, in Section 4 mixed-mode simulation which combines circuit, timing and logic analyses is described, with a detailed description of the SPLICE mixed-mode
simulator [41].

2. TIMING SIMULATION

2.1. Introduction

For small circuit blocks where analog voltage levels are critical to circuit
performance, or where tightly coupled feedback loops are present, a circuit
simulator such as SPICE[1] can accurately predict circuit performance. As
the size of the circuit increases, the cost and memory requirements of such
an analysis become prohibitive. This problem is illustrated in Fig.2.1 where
the computer time required per unit of simulated time is shown as a function of circuit size. As the number of circuit elements in the analysis grows,
the time required to formulate the modified nodal circuit equations grows
almost linearly with circuit size. However, the time required to solve these
sparse circuit equations $Tsol$ increases at a faster rate and rapidly becomes

the

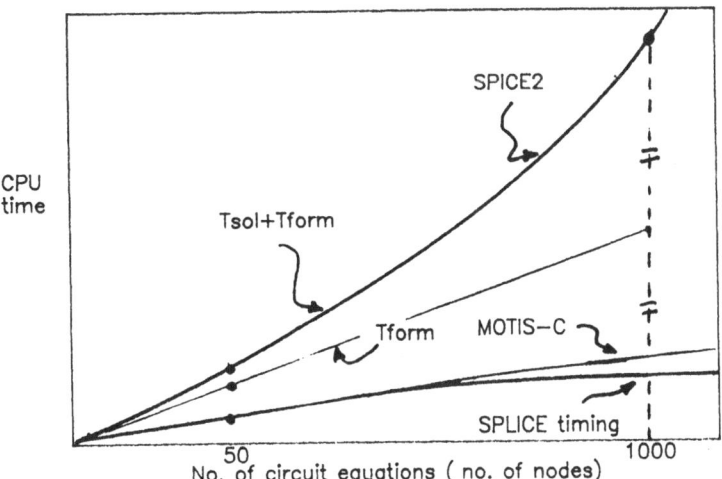

Fig.2.1 Computer time per simulated timepoint vs. circuit
size for electrical analysis.

dominant cost of the analysis. For the circuit simulator SPICE2, *Tsol* is less than 10% of the total time for a circuit of less than 30 equations but reaches half the total central processor time for a circuit containing 1000 equations or roughly 3000 MOS transistors. While Fig.2.1 shows the time required per unit of simulated time, the problem is further aggravated by the fact that for larger circuits, more information is generally required to verify the circuit performance and hence longer simulation times are required.

The majority of large integrated circuits are digital in nature and for most of these circuits only a small fraction of the circuit variables (node voltages or branch currents) are actively changing state at any time. For circuits containing over 1000 transistors, or 300 logic gates, typically more than 80% of the circuit variables are stable (unchanging) at any time. As the size of the circuit increases, the fraction of circuit variables changing at a given time tends to fall.

This inactivity, or *latency* in a large digital network can be exploited by a circuit simulator in a number of ways. At the transistor level, transistors whose controlling node voltages do not change between successive timepoints can be *bypassed* in the formulation phase and their previous circuit equation entries can be used[1]. This approach reduces the formulation time but does not help equation solution time. For digital circuits with as few as 50 transistors, *bypass* can reduce *Tform* by as much as 50%[2].

Other approaches have been proposed which reduce *Tsol* as well as *Tform*[3]-[5]. Most of these techniques rely on the identification of subnetworks of the circuit which can be processed independently. The solutions of these blocks are then factored back into the overall circuit analysis. Appropriate subnetworks may be identified directly from the circuit matrix structure[6] or from the user's input circuit description[5]. In this way, entire subnetworks which are unchanging may be bypassed and only a representation of their interface to the remainder of the network need be included in the overall analysis. This interface may be determined by the use of *tearing* techniques; both branch tearing[7] or node tearing[8] approaches have been used.

The success of these approaches on a serial computer relies on the fact that the submatricies to be bypassed are large with respect to their interconnection network. If not, the additional saving of *Tsol* is small compared to the evaluation time *Tform* and the overhead incurred in setting up the submaticies, and monitoring their activity, may reduce the overall computational savings. In many large integrated circuits, the probability that large, highly interconnected blocks of circuitry are inactive is relatively low, which may reduce the effectiveness of the circuit tearing approach.

Other approaches may be used for the solution of the nonlinear algebraic difference equations at a timepoint. Rather than formulating the problem as repeated Newton-Raphson iterations requiring direct solution of a set of linear equations at each iteration, a nonlinear displacement technique may be used to approximate the solution of the nonlinear equations directly. Timing simulators use an approach of this type, which will be described in the following section. Such techniques provide a solution time *Tsol* which grows at the same rate as the number of equations and thus the overall analysis time grows linearly with the number of circuit equations, as illustrated for the MOTIS-C timing simulator in Fig.2.1. Since the equation decoupling provided by this indirect solution approach allows the separate evaluation of circuit node voltages, the circuit is effectively torn at the individual node level. As a result, each node may be treated independently and may have its own analysis timestep, commensurate with its own electrical activity, or may be bypassed if the voltage at the node is unchanging. *Tform* would now dominate the total analysis time in a timing simulator even for large circuits. For this reason, the formulation time is often reduced in a timing simulator by the use of table look-up models. Models of this type are described in Section 2.4 and help to reduce the slope of the MOTIS-C characteristics in Fig.2.1.

For a circuit where a large fraction of the nodes are inactive, the overhead involved with checking each node and subsequently bypassing its evaluation can become a substantial fraction of the total analysis time. It is possible to generate a directed graph as illustrated in Fig. 2.2, where the branches of the graph indicate the circuit elements which can affect the voltage at a node (*fanins*) and the circuit elements whose operating conditions, and hence outputs, may be effected by a change in the voltage at a node (*fanouts*). When an input to the network changes, the graph may be traced and only those elements which may be affected by the change need be processed. Once the effect of the input change disappears along any given path in the graph that path need not be traced further. This technique is called *selective trace* and is described for timing analysis in Section 2.4 and in more detail for logic simulation in Section 3. When selective trace is applied to a timing simulation the cost of the analysis is determined primarily by the activity of the network and the time required to evaluate a single circuit node. As the size of the circuit increases, the circuit activity tends to decrease and as a result the nett computer time required per timepoint grows at a less than linear rate. The SPLICE program implements the selective trace algorithm for timing analysis. Its relative performance is illustrated in Fig.2.1.

2.2. Equation Decoupling

In the case of nodal circuit formulation for nonlinear time-domain analysis,

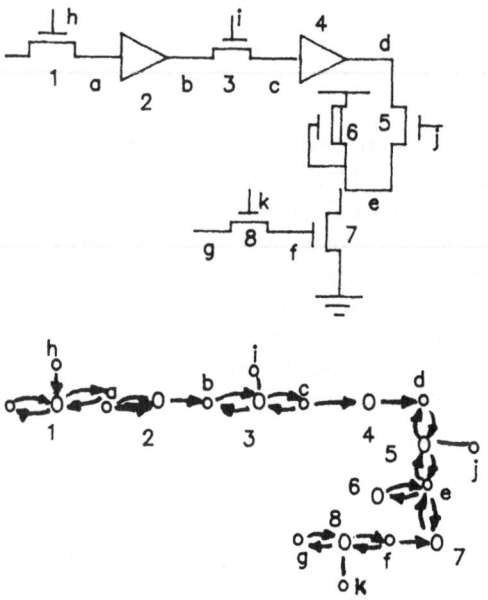

Fig.2.2(a) Circuit fragment and (b) associated directed signal
flow graph showing fanin and fanout connections.

the node voltages may be expressed in compact vector form as:

$$\dot{\mathbf{v}} = \mathbf{f}(\mathbf{v}, \mathbf{u}, t)$$
(2.1)

where $\mathbf{v} = [v_1, \cdots, v_N]^T$ are the dependent node voltages and
$\mathbf{u} = [u_1, \cdots, u_N]^T$ are the independent node voltages. To solve this initial
value problem the solution interval T is divided into small time steps and an
implicit numerical integration scheme is used to convert (2.1) into a set of
nonlinear difference equations of the form:

$$\mathbf{v}_{n+1} = \mathbf{V}(\mathbf{v}_i, \; i = n+1, \; n, \; n-1, \; \cdots)$$
(2.2)

where $\mathbf{v}_{n+1}$ is the vector of node voltages at t_{n+1}. Below, the subscript $n+1$
is assumed and has been dropped for clarity. In a timing simulator the
solution of (2.2) is obtained using a nonlinear simultaneous displacement [9]
or successive displacement [10],[13] method [14].

If the mth iterate $\mathbf{v}^m$ and the first $i-1$ components of the $(m+1)$th iterate
in the solution of (2.2) have been determined, then to obtain the ith component v_i^{m+1} using a successive displacement approach, the ith equation of

(2.2) is solved for v_i with the other $N-1$ variables held fixed:

$$j_i(v_1^{m+1}, \cdots, v_{i-1}^{m+1}, v_i, v_{i+1}^m, \cdots, v_N^m) = 0 \tag{2.3}$$

A single one-dimensional Newton-Raphson step from v_i^m is used to yield v_i, an approximate solution to (2.3). Then v_i^{m+1} is obtained from:

$$v_i^{m+1} = v_i^m + \omega(v_i - v_i^m) \tag{2.4}$$

where ω is a relaxation parameter. This results in the *nonlinear one-step SOR (successive overrelaxation)-Newton iteration* [14] and the explicit form of the iteration is:

$$v_i^{m+1} = v_i^m - \frac{\omega\ j_i(\mathbf{v}^{m,i})}{\partial j_i(\mathbf{v}^{m,i})/\partial v_i} \tag{2.5}$$

where $\mathbf{v}^{m,i} = [v_1^{m+1}, \cdots, v_{i-1}^{m+1}, v_i^m, \cdots, v_N^m]^T$ If $\omega = 1$ (2.5) reduces to a one-step Gauss-Seidel-Newton iteration [10],[13]. It is clear that (2.5) requires $\partial j_i(\mathbf{v}^{m,i})/\partial v_i$ be non-zero. This is assured in the simulator by requiring that some capacitance be included between each circuit node and the reference node. It can be shown that it does not enhance the asymptotic rate of convergence of this method to take more than one Newton-Raphson step per nonlinear SOR iteration [14].

In a timing simulator, a single one-step Gauss-Seidel-Newton iteration is used to approximate the node voltage at each timepoint. If the starting point for this iteration is $\mathbf{v}_n$ at time t_n then the ith node voltage at time t_{n+1} is $v_{n+1,i}$ and given by:

$$v_{n+1,i} = v_{n,i} - \frac{j_i(\mathbf{v}^{n,i})}{\partial j_i(\mathbf{v}^{n,i})/\partial v_i} \tag{2.6}$$

where $\mathbf{v}^{n,i} = [v_{n+1,1}, \cdots, v_{n+1,i-1}, v_{n,i}, \cdots, v_{n,N}]^T$. If circuit branch currents change significantly between timepoints the single iteration may not provide an accurate analysis. This is particularly true if strong bilateral coupling is present between nodes, as described in Section 2.5.

A major advantage of the node decoupling provided by this approach is that it is relatively easy to exploit the inactivity of the circuit under analysis, as described in the next section.

182

Since a single SOR-Newton iteration is used at each timepoint it is impor-
tant that the values of $j_i(\mathbf{v}^{n,i})$ and $\partial j_i(\mathbf{v}^{n,i})/\partial v_i$ in (2.6) accurately model
the nonlinear device characteristic between timepoints. The most satisfac-
tory timestep control algorithm for this purpose is one which produces a
small timestep when transistor nonlinearities are playing an important part
in determining the node voltage waveform and allows a larger timestep to
be used when they are not. If the change of current flowing in the nonlinear
branches is used as the metric for timestep control, small timesteps are
used when the voltage waveform has a large curvature and larger timesteps
are used when the curvature is small. This corresponds to large timesteps
where the rate of change of node voltage is zero or constant. The use of
such an algorithm is illustrated in Fig.2.3.

Rather than monitor the branch current of each active device connected at

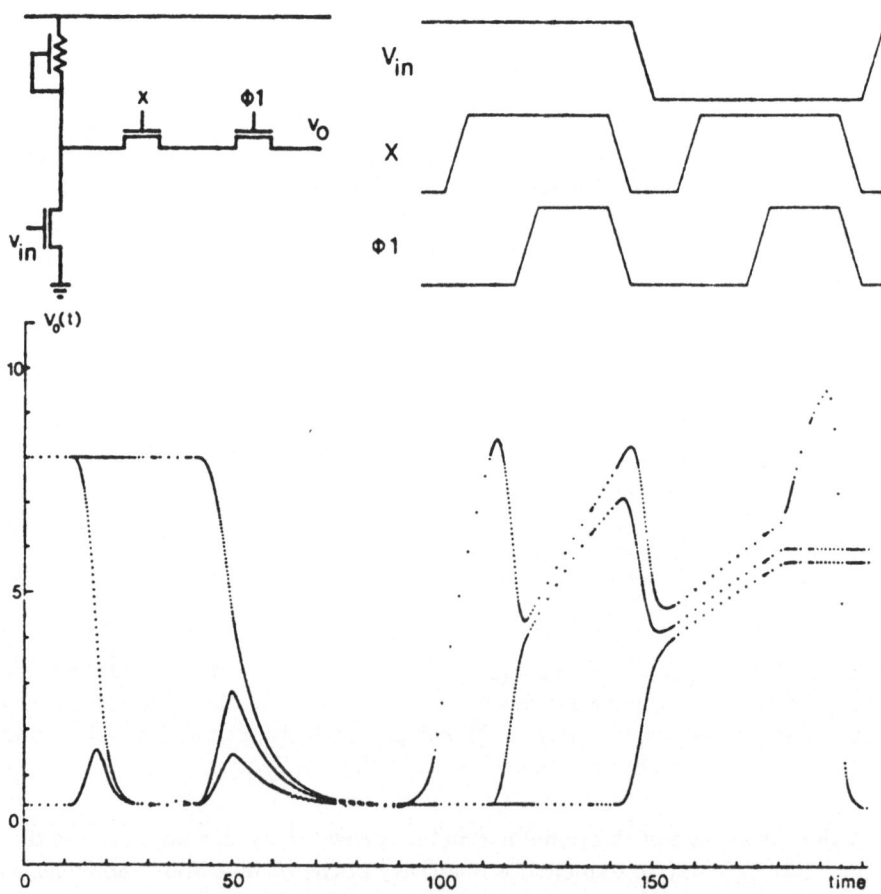

Fig.2.3(a) MOS circuit and (b) output waveforms obtained using
the stepsize control algorithm as described.

a node, the sum of those currents, as indicated by the grounded node capacitor charging current, may be used. This is a relatively accurate measure of the active device currents except in a few unlikely circumstances. With this in mind, the analysis of node i at time t_{n+1} may be summarized as:

$h_{n+1} = h_n$
loop : $Cnet_{n+1} = Inet_{n+1} = 0$
    for (each fanin element at node i) {
        compute its equivalent conductance g_{eq};
        compute its driving point current i_{eq};
        $Cnet_{n+1} = Cnet_{n+1} + g_{eq}$;
        $Inet_{n+1} = Inet_{n+1} + i_{eq}$;
        }
    if (ABS($Inet_{n+1} - Inet_n$) > DELTAI) {
        if (h_{n+1} > HMIN) {
            $h_{n+1} = \text{MAX}(h_{n+1}\text{*RFACT, HMIN})$;
            goto loop;
            }
        else {
            if (h_{n+1} = HMIN) {
                print error message, internal stepsize too small;
                stop;
                }
            }
        }
    else {
        if (h_{n+1} < HMAX) h_{n+1} = MIN (h_{n+1}*IFACT, HMAX);
    }

$$v_{n+1} = v_n + \frac{Inet_{n+1}}{Cnet_{n+1} + \dfrac{C}{h_{n+1}}}$$

where C is the grounded node capacitor value and Backward Euler integration is used. Here DELTAI, HMIN, HMAX, RFACT, and IFACT are constants which depend on the integrated circuit technology. Typical values might be 100uA, 10ps, 2ns, 4, and 2 respectively.

2.3. Exploiting Latency

Large digital circuits are often relatively inactive. A number of schemes can be used to avoid the unnecessary computation involved in the re-evaluation of the voltage at nodes which are inactive or *latent*.

A scheme used in many electrical simulators is the "bypass" scheme, described above for circuit simulators. This scheme has been employed in a number of timing simulators including the original MOTIS program, MOTIS-C, and SIMPIL[24]. In a timing simulator which uses only one iteration per timepoint, the bypass technique can introduce significant timing errors. In general, when a circuit description is read by a timing simulator and translated into data structures in memory, the elements may be read in any order. Unless some form of connection graph is used to establish a precedence order for signal flow, the new node voltages will be computed in an arbitrary order.

In a simulator such as MOTIS, where only node voltages at t_n are used to evaluate the node voltage at t_{n+1} (a form of Jacobi-Newton iteration [14]), the order of processing elements will not affect the results of the analysis. However, substantial timing errors may occur. For example, consider the inverter chain of Fig.2.5. If the input to inverter I_1 changes at time t_n, that change cannot appear at node (1) before time t_{n+1}. For a chain of N inverters, the change will not appear at output I_N before N timesteps have elapsed. If the timestep is very small with respect to the response time of any one inverter, this error may not be significant.

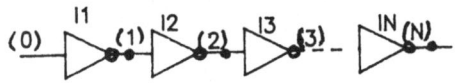

Fig.2.5 Inverter chain circuit

In a simulator such as MOTIS-C or SIMPIL, where node voltages already computed at t_{n+1} are made available for the evaluation of other node voltages at t_{n+1} (Gauss-Seidel-Newton iteration [14]), the order of processing elements can substantially affect the simulator performance. In the example above, if the inverters were processed in the order 1, 2, 3, $\cdots$, N then v_{n+1}^1 can be determined from v_{n+1}^0, v_{n+1}^2 from v_{n+1}^1, and so on. The result will be zero accumulated timing error. Should the nodes happen to be processed in the reverse order, N, $N-1$, $\cdots$, 1, then a timing error of N timesteps will occur, the same as in the Jacobi-Newton iteration.

In both of these approaches, the bypass scheme may be used to exploit latency, as described for circuit simulation. If it were possible to order the processing of nodes in the Gauss-Seidel-Newton iteration so as to follow the flow of the signal through the circuit, the timing error would be kept small. A signal flow graph would provide this information.

One way to generate this graph is to build two tables for each node. First, all circuit elements must be classified as fanin and/or fanout elements of the nodes to which they are connected. The fanin elements are defined as those which play some part in determining the voltage at the node. In an MOS circuit, any MOS transistor whose drain or source is connected to the node would be classified as a fanin since drain or source current may contribute to the determination of the node voltage.

A fanout element is one whose operating conditions are directly influenced by the voltage at the node. For MOS transistors, connection to any of the three independent ports (drain, gate, or source) would cause that MOS transistor to be included in the fanout table at the node. It is therefore possible for an element to appear as both a fanin and a fanout at the node.

The two tables constructed for each node contain the names of the fanin and fanout elements at the node. These tables may be generated as the elements are read into memory. An example of a circuit fragment and associated fanin and fanout tables is shown in Fig.2.6 When the voltage at a node changes, its fanouts are marked to be processed next. In this way the effect of a change at the input to a circuit may be *traced* as it propagates to other circuit nodes via the fanout tables, and thus via the circuit elements which are connected to them. Since the only nodes processed are those which are affected directly by the change, this technique is *selective* and hence its name: *selective trace*.

The selective trace technique has provided another major saving. By constructing a list of all nodes which are active at a timepoint and excluding those which are not, selective trace allows circuit latency to be exploited without the need to check each node for activity. The elimination of this checking process, used in the bypass approach, can save a significant amount of computer time for large circuits at the cost of some extra storage for the fanin and fanout tables at each node. For random logic circuits these tables typically require about 10 extra integers per node. The SPLICE program, described in Section 4, uses this selective trace approach for timing analysis.

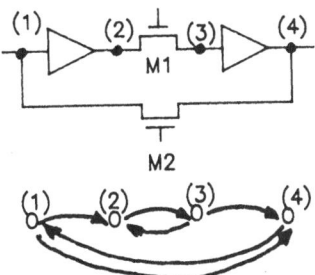

Fig.2.6 Illustration of fanin and fanout tables for selective trace analysis.

186

Even with selective trace some timing errors can occur. For example, wherever feedback paths exist, one timestep of error may be introduced. Consider the circuit fragment and its associated signal flow graph shown in Fig.2.6(b). Assume v_1 and v_2 are such that both M_1 and M_2 are conducting. If a large input appears at node (1) at time t_{n+1} it will be traced through nodes (1), (2), (3), and (4) respectively. But now node (1) is marked to be processed again. In the SPLICE program, stored with each node is the time at which it was last processed, t_s^*. If $t_s^* = t_{n+1}$, rather than process it again at t_{n+1} it is marked to be processed at t_{n+2}. This process may be summarized as:

```
while ( any nodes are marked at tₙ₊₁ ) {
    get the next marked node, nᵢ;
    if ( tₛ˙ = tₙ₊₁ ) {
        mark nᵢ for tₙ₊₂;
        next ;
    }
    else {
        process nᵢ at tₙ₊₁;
        tₛ˙ = tₙ₊₁;
        if ( vₙᵢ changed ) {
            mark nᵢ at tₙ₊₂;
            mark all fanouts at tₙ₊₂;
        }
    }
}
```

In an efficient implementation of the selective trace technique, the fanin and fanout tables would not contain the "names" of fanin and fanout elements respectively but rather a *pointer* to the actual location in memory where the data for each element is stored. The implementation of the selective trace technique will be described in more detail in Section 3.

2.4. MOS Models for Timing Analysis

To solve for the voltage at each circuit node, the contribution of each of the nonlinear circuit elements, such as transistors and diodes, as well as the contribution of linear and charge storage elements, such as resistors and capacitors, must be accounted for.

For a Newton-Raphson step, these elements are generally represented by a *companion model* which describes their linearized terminal characteristics as a function of controlling voltages and currents. In a circuit analysis program, the companion model of a nonlinear element is generally represented by a set of equations and for elements such as the MOS or Bipolar transistor, these equations can be quite complex and expensive to evaluate. In fact, the time required to evaluate the linear companion model contributions often accounts for over 80% of the total analysis time in a circuit simulation program. This is illustrated by Fig.2.4 where the fraction of total analysis time spent on a per-subroutine basis is plotted for the analysis of a 50 MOS transistor circuits using SPICE2. The shaded area indicates the time required for the evaluation of the linear companion model entries and the formulation of the sparse matrix equations.

Fortunately, for timing analysis the companion model is simplified substantially since the fixed-point iteration eliminates the need for the evaluation of most of the Jacobian matrix entries. Figure 2.7 shows the components required for the companion models of an MOS transistor for circuit analysis and timing analysis. The contributions of charge storage effects to the companion model are not shown here.

The evaluation of the timing analysis model for an MOS transistor can still be an expensive process, particularly for short-channel MOS devices where complex electronic and geometrical effects can influence the device behavior. For this reason, it is often advantageous to generate tables of device characteristics prior to the analysis and replace the expensive analytic evaluations with simple table lookups. Although these tables occupy more memory space than the equivalent code for analytic evaluation, this penalty can be justified in terms of the substantial speedup advantages achieved

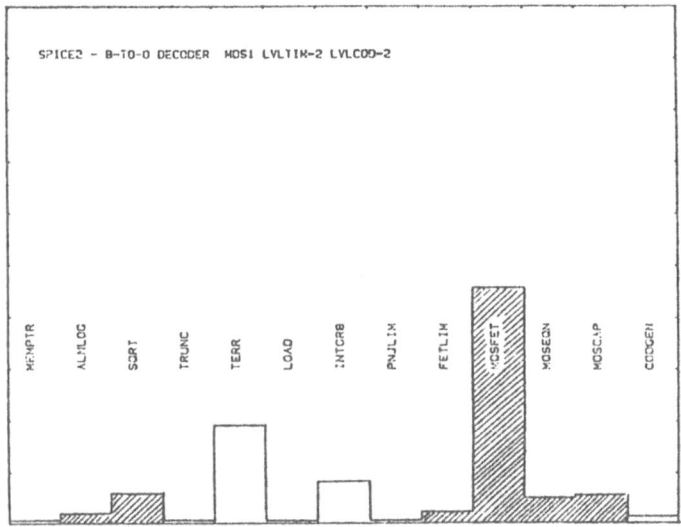

Fig.2.4 Analysis of percentage central processor time on a per subroutine basis for the analysis of a 50 MOS transistor circuit using SPICE2.

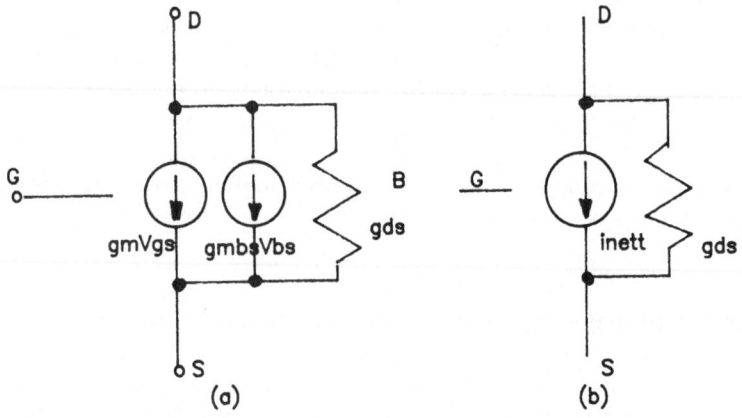

Fig.2.7 Components of companion models for static MOS transistor
analysis (a) circuit analysis (b) timing analysis

Fig.2.8 shows the controlling voltages of the MOS transistor and the companion model for timing analysis. A straightforward approach would require two tables:

$$i_{ds}(v_{gs}, v_{ds}, v_{bs}) = \mathbf{T_d}(v_{gs}, v_{ds}, v_{bs}) \tag{2.7}$$

$$g_{ds}(v_{gs}, v_{ds}, v_{bs}) = \mathbf{T_g}(v_{gs}, v_{ds}, v_{bs}) \tag{2.8}$$

where $\mathbf{T_d}$ and $\mathbf{T_g}$ are 3-dimensional tables. If N entries are required per controlling variables to maintain accuracy, this would require $2N^3$ entries per device model[1]. Since it has been observed empirically that $N>50$ for accurate results, large amounts of memory would be required for this approach. It is possible to approximate g_{ds} from $\mathbf{T_d}$ by computing gradients however this alone is not sufficient reduction in memory requirements.

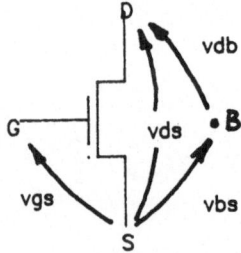

Fig.2.8 Controlling voltages of an MOS transistor

---

[1]. Note that for long-channel MOS transistors, a single model can be used for all devices and can be scaled by the W/L ratio. For short-channel devices, a separate model is required for each different MOSFET geometry. However, the number of models required is still small (10-30) for typical VLSI circuits compared with the total number of MOSFETs in the circuit.

In the MOTIS[9] program, the effect of the potential difference between the source and substrate, or "bulk", terminals, vsb, is modeled as a change in threshold voltage. A backgate bias table of N entries can be used to represent this effect:

$$v_t(v_{bs}) = T_b(v_{bs})$$ (2.9)

and combined with a two-dimensional table to obtain the drain current:

$$v_{gse} = v_{gs} - v_t$$ (2.10)

$$i_{ds}(v_{gs}, v_{ds}, v_{bs}) = T_d(v_{gse}, v_{ds})$$ (2.11)

In the MOTIS program , g_{ds} was approximated from the drain characteristic table T_d and an associated load characteristic using a secant approach. Another technique that may be used is to simply compute g_{ds} from an adjacent set of entries in T_d using a polynomial approximation. For example:

$$g_{ds}(v_{gs}, v_{ds}, v_{bs}) = \frac{T_d(v_{gse}, v_{ds} + \Delta) - T_d(v_{gse}, v_{ds} - \Delta)}{2\Delta}$$ (2.12)

where $\Delta = \dfrac{v_{ds}^{max}}{N}$ and v_{ds}^{max} is the largest drain-to-source voltage permitted in the table. This approach is used in the MOTIS-C[10] program and provides accurate results for large N ($N>100$) provided care is taken near $v_{ds} = 0$ to prevent possible jitter.[2]

The table scheme described above requires $N^2 + N$ entries per MOS model. Note that this approach is a compromise between memory requirements and model evaluation time. Some memory savings have been achieved at the expense of extra table lookups and some simple arithmetic operations. This compromise may be extended further so that only one-dimensional tables are required.

One such approach[11] uses three one-dimensional tables in the form:

$$i_{ds}(v_{gs}, v_{ds}, v_{bs}) = \frac{W}{L} T_h\left(v_{gse} - \frac{v_{ds}}{2}\right)\left(T_f(v_{ds}) - T_f(v_{db}) + v_{ds}v_{db}\right)$$ (2.13)

where T_f and T_h are referred to as the current and mobility tables respec-

---

[2]. Table entries may be computed prior to the analysis from model equations, or curve-tracer measurements of actual device characteristics may be inserted directly into the table.

tively and v_{gse} is obtained from the usual backgate bias table.

Another approach, used originally in the MOTIS-C program, is based on a transformation which will be illustrated using the Shichman-Hodges[12] MOS model where:

$$i_{ds}(v_{gs}, v_{ds}, v_{bs}) = K_p(1 + \lambda v_{ds})(v_{gse} - \frac{v_{ds}}{2})v_{ds} \qquad (2.14)$$

for operation in the linear region ($v_{ds} < v_{sat} = v_{gse}$). K_p and λ are constant for a given MOSFET and v_{sat} is the saturation voltage at v_{gs}. A typical set of drain characteristics for this model is shown in Fig.2.9. For this model, the drain current as a function of drain-source voltage for any gate voltage v_{gs} below the maximum allowed value v_{gs}^{max} can be obtained exactly from the single characteristics at v_{gs}^{max} as follows:

$$i_{ds}(v_{gs}, v_{ds}, v_{bs}) = i_{ds}(v_{gs}^{max}, v_{ds} + \Delta v, v_{bs}) - i_{ds}(v_{gs}^{max}, \Delta v, v_{bs}) \qquad (2.15)$$

where $\Delta v = v_{sat}^{max} - v_{gse}$ and v_{sat}^{max} is the saturation voltage for $v_{gs} = v_{gs}^{max}$. In the saturation region, the output conductance varies with v_{gs} and a separate table is required to model that variation. The result is that a transistor with drain characteristics of the form illustrated in Fig.2.9 can be modeled using:

$$\Delta v = v_{sat}^{max} - v_{gse}$$

$$i_{ds}(v_{gs}, v_{ds}, v_{bs}) = T_g(v_{gs})v_{ds} + (T_d(\min(v_{ds}, v_{gse}) + \Delta v) - T_d(\Delta v)) \qquad (2.16)$$

Unfortunately, the Shichman-Hodges model is not accurate for MOSFETs operating with large drain-source electric field or for small geometry MOS-FETs. However, the above model can be extended to these cases using one additional table. By redefining Δv in the form $\Delta v = v_{sat}^{max} - T_s(v_{gs})$, where T_s is a table of saturation voltages, a good fit may be obtained for more accurate models. A comparison between the extended table model and some measured small geometry NMOS characteristics are shown in Fig.2.9(b) and the model equations are summarized below:

$$v_{gse} = v_{gs} - T_b(v_{bs})$$

$$v_{dse} = \min(v_{ds}, T_s(v_{gse}))$$

$$\Delta v = T_s(v_{sat}^{max}) - T_s(v_{gse}) \qquad (2.17)$$

$$i_{ds}(v_{gs}, v_{ds}, v_{bs}) = T_g(v_{gse})v_{ds} + (T_d(v_{dse} + \Delta v) - T_d(\Delta v))$$

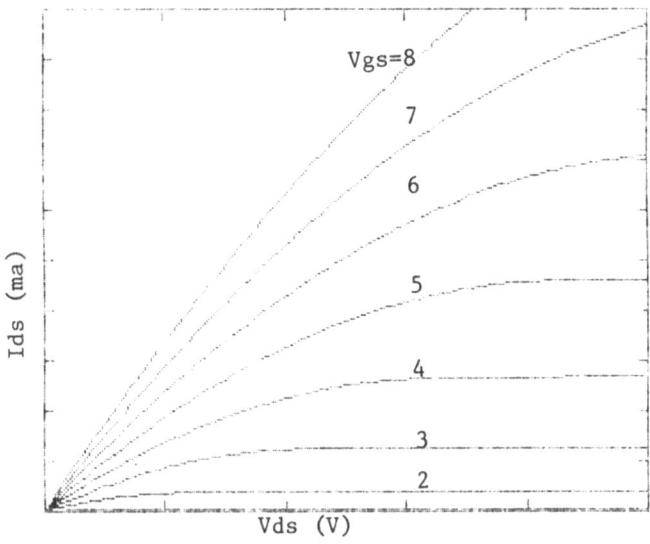

Fig.2.9 (a) Drain characteristic for the Shichman-Hodges MOS model.
Note quadratic spacing of curves in saturation.

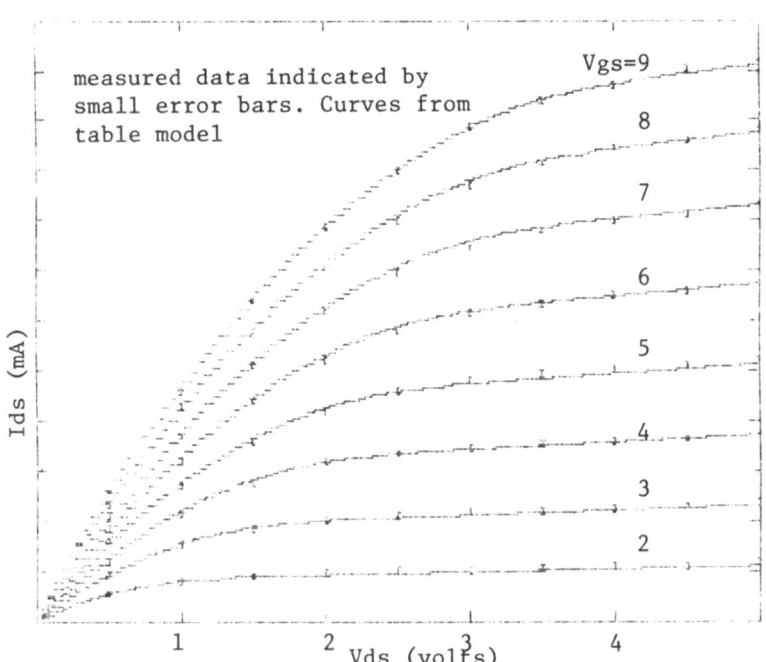

Fig.2.9 (b) Table model and measured data for MOSFET
with l_{eff} =1.75u, w=50u, t_{ox} =650angstroms
and effective junction depth of 0.5u.

A procedure which may be used to fit this model to measured characteristics is summarized below:

(1) For each measured value of v_{gs}, plot $g_{ds} = \dfrac{di_{ds}}{dv_{ds}}$ versus v_{ds}. Then

$$\mathbf{T_g}(v_{gs}) = g_{ds} \,\Big|_{\frac{dg_{ds}}{dv_{ds}} = 0}$$

$$\mathbf{T_s}(v_{gs}) = v_{ds} \,\Big|_{\frac{dg_{ds}}{dv_{ds}} = 0}$$

If more than one point of zero slope is obtained, the minimum value of v_{ds} should be used.

(2) For each data point $i_{ds}(v_{gs}^{\max}, vds, v_{BS})$, compute

$$\mathbf{T_d}(v_{ds}) = i_{ds}(v_{gs}^{\max}, v_{ds}, v_{BS}) - \mathbf{T_g}(v_{gs}^{\max})v_{ds}$$

(3) . Compute the backgate table

$$\mathbf{T_b}(v_{bs}) = v_t(v_{bs}) - v_t(v_{BS})$$

where v_{BS} id the v.due of v_{bs} at which the drain characteristic was measured.

This procedure will give initial estimates for each table. Some heuristics and an optimization program can be used to "tune" the tables to the data. Typically 2-3 iterations are required.

For timing analysis, it is generally not necessary to interpolate the tables. Rather, the nearest table entry to the present branch voltage value is used. This is not satisfactory for circuit analysis, where convergence of an iteration scheme is required at each timepoint. For table models used in circuit analysis, quadratic interpolation has provided good convergence results.

For a load device operating with a known, constant supply voltage, a one dimensional table may be used which relates load current to the source voltage vs of the load as shown in Fig.2.10. Note that the presence of backgate effects precludes the use of a table indexed by $v_{dd} - v_s$ alone. The load table characteristics of typical enhancement and depletion devices are shown in Fig.2.10. Some of the advantage of the table model can be lost in the combutation of table index. For example, if $v^{\max}$ is the maximum voltage in a given table and v is the raw index voltage, a computation of the form

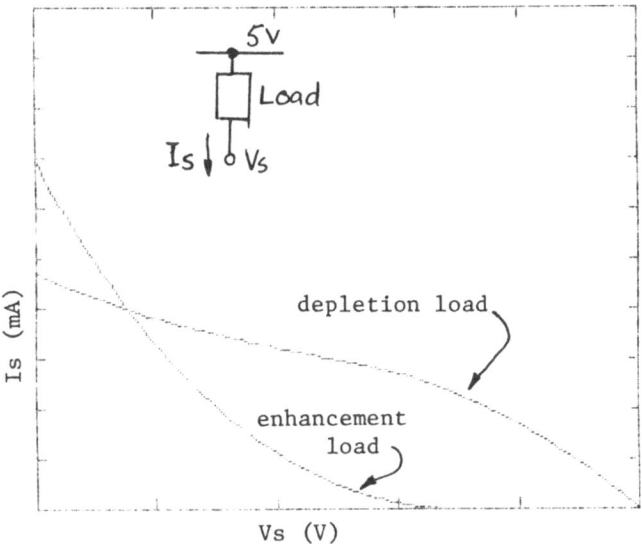

Fig.2.10 Depletion and enhancement load current vs. source voltage.

$index = round(N\dfrac{v}{v^{max}})$ must be performed for each table reference. This computation can be avoided by simply scaling the analysis so that $v^{max} = N$. In this way, a simple rounding operation is all that is required to compute the table index.

Using the above techniques, the computer time required to evaluate an MOS device model can be reduced by over an order of magnitude.

2.5. Floating Capacitors in Timing Analysis

A major drawback with the use of timing analysis is that tightly-coupled feedback loops, or bidirectional circuit elements, can cause severe inaccuracies and even instability during the analysis. For this reason, special techniques must be used to process such elements. One such element that has limited the application of timing analysis is the floating capacitor[15],[16].

The floating capacitor is often an important element in the design of integrated circuits. In Fig.2.11(a) the value of the Bootstrap capacitor C_b is generally large compared with the values of the associated parasitic grounded capacitors C_1 and C_2. The value of the intrinsic gate-drain

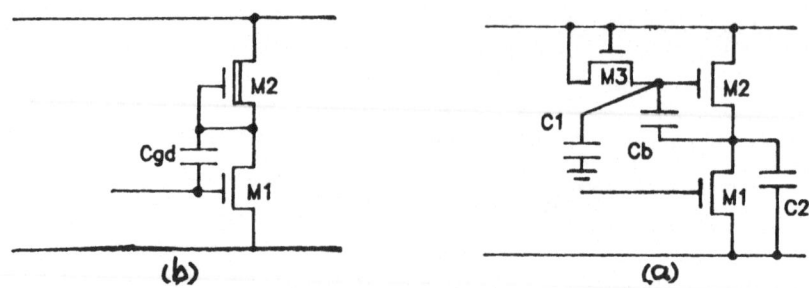

Fig.2.11 (**b**) Depletion load inverter circuit and (**a**) bootstrapped inverter circuit showing important floating capacitors.

feedthrough capacitance C_{gd} in Fig.2.11(b) is often small compared with other circuit parasitics at the gate and drain nodes, however the effect of C_{gd} on circuit performance is significant due to the large voltage gain of the stage.

Early timing simulators avoided the problem of analyzing floating capacitors by not allowing the user to include them in the circuit description. The effect of a floating capacitor may then be approximated by altering the values of the grounded capacitors at appropriate nodes in the circuit. If the operation of a circuit depends on a floating capacitor, a functional macro-model may be used [17].

In the MOTIS-C program, isolated floating capacitors are processed by maintaining the node coupling across the floating branch and solving the resulting 2×2 nodal circuit matrix at each timepoint. This approach could be extended to deal with arbitrary connections of N floating capacitors but this would require the solution of $N+1$ coupled equations at each timepoint and hence reduce the advantages of the node decoupling approach.

Another approach is to use an explicit integration method to evaluate the floating capacitor current[18],[19]. Although this technique maintains the decoupling of the two nodes, the method is unstable when, for the circuit of Fig.2.12:

$$\frac{1}{C_f} < \frac{1}{C_1} + \frac{1}{C_2} \qquad (2.18)$$

Two classes of algorithms are described here[15],[16] and referred to as "variable-in-space" and "variable-in-time" algorithms. An algorithm is defined as "variable-in-space" if different types of integration (implicit or explicit) are used for different components of the circuit at the same timepoint. An algorithm is defined as "variable-in-time" if different types of integration are used for the whole circuit at different timepoints. In the general case an algorithm can be both variable in space and time and different timesteps may be used for implicit and explicit integration.

In the circuit used to test the algorithm suppose that capacitors and conductances are linear for ease of proving algorithmic properties. For MOS circuits, where each node is assumed to have a capacitance to ground, the node equations are of the form:

$$C\dot{x} = Gx \quad x(0) = x_0 \quad x \varepsilon R^n \quad C, G \varepsilon R^{(n \times n)} \tag{2.19}$$

where x is the vector of node voltages. (2.19) can be rewritten in normal form[20]:

$$\dot{x} = Ax \quad A \varepsilon R^{(n \times n)} \tag{2.20}$$

with $A = C^{-1}G$. In order to study stability, each algorithm is associated with a companion matrix $M \varepsilon R^{(n \times n)}$ such that:

$$x_{k+p} = M_p x_k \tag{2.21}$$

Then to ensure stability of the method it is required that the spectral radius [3] $\rho(M_p) \leq 1$ being all eigenvalues of modulus unity simple [21].

There are two basic approaches to "variable-in-space" algorithms: "partitioning of variables" or "guess of variables". The former requires an implicit integration on a subset of variables and an explicit integration for the remaining ones [19]. The latter is based on the guess at step $k+1$ of some quantities which are related to the values they had at the previous steps $k, k-1, ..., k-q$. For the "guess of variables approach," suppose we are integrating the set of equations defined by (2.20) using an implicit method such as Backward Euler (B.E.). At each step we have to solve the linear system:

$$[I - hA]x_{k+1} = x_k \tag{2.22}$$

To solve (2.22) an LU factorization of $[I - hA]$ is required, which is very time consuming if a large system is involved. If A is decomposed into the sum of

---

[3] The spectral radius of a matrix M is defined to be $\rho(M) = \max\{|\lambda| \mid \lambda \varepsilon \sigma(M)\}$ where σ is the set of all the eigenvalues (spectrum) of M

two matrices:

$$A = A_t + A_g \qquad (2.23)$$

being A_t either lower or upper triangular. (2.22) is equivalent to:

$$[I - hA_t] x_{k+1} = x_k + hA_g x_{k+1} \qquad (2.24)$$

If the "guess" $A_g x_{k+1} = A_g x_k$ is made then (2.24) becomes:

$$[I - hA_t] x_{k+1} = [I + hA_g] x_k \qquad (2.25)$$

where no factorization is needed since $[I - hA_t]$ is triangular. The companion matrix is $M_1 = [I - hA_t]^{-1}[I + hA_g]$.

Two promising "guess of variables" approaches for MOS circuits are the Implicit-Implicit-Explicit (*IIE*) Method [15] and the Double Cross Voltage Guess Method [16]. Both of these methods are described below.

To illustrate the *IIE* method, consider B.E. integration with the companion model substituted for the capacitors and nodal analysis applied to the circuit of Fig.2.12. By taking $A = C^{-1}G$ in (2.22) and multiplying both members by $\dfrac{C}{h}$ we obtain:

$$[\frac{C}{h} - G] x_{k+1} = \frac{C}{h} x_k \qquad (2.26)$$

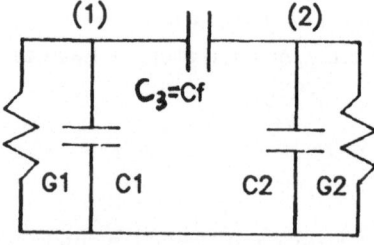

Fig.2.12 Test circuit used for the analysis of timing algorithms.

or more explicitly:

$$
\left[
\begin{array}{cc}
C_1 + C_3 + \dfrac{(C_1 + C_3)}{h} & -C_3 - \dfrac{C_3}{h} \\[2mm]
-C_3 - \dfrac{C_3}{h} & C_2 + C_3 + \dfrac{(C_2 + C_3)}{h}
\end{array}
\right]
\left[
\begin{array}{c}
v^1 \\[1mm] v^2
\end{array}
\right]_{k+1}
$$

$$
=
\left[
\begin{array}{cc}
\dfrac{(C_1 + C_3)}{h} & -\dfrac{C_3}{h} \\[2mm]
-\dfrac{C_3}{h} & \dfrac{(C_2 + C_3)}{h}
\end{array}
\right]
\left[
\begin{array}{c}
v^1 \\[1mm] v^2
\end{array}
\right]_{k}
\tag{2.27}
$$

(2.27) can be decoupled by taking the voltage at node (2) one step back in time to solve the first equation. This corresponds to guess $C_3 v_{k+1}^2 = C_3 v_k^2$ and considering the current i through the floating capacitor as given by $\dfrac{C_3}{h} [(v_{k+1}^1 - v_k^1) - (v_k^2 - v_{k-1}^2)]$. This method is stable for the test circuit but the response rings for $C3 = 0$ and stepsize greater than a critical value h_{crit} as illustrated in Fig.2.13(a).

As can be seen from the root locus plot of Fig.2.14, there exists a critical maximum timestep at which the roots of the characteristic polynomial are real. The value of the timestep at this point can be obtained explicitly from the characteristic polynomial as shown below.

For Gauss-Seidel-Newton iteration and Backward-Euler integration applied to branch i of Fig.2.15(a), the following difference equations are obtained:

$$
I_n^+ = G_{si}(V_n^+ - V_{n-1}^-) - G_{ci}(V_{n-1}^+ - V_{n-2}^-) \tag{2.28}
$$

$$
I_n^- = G_{si}(V_n^+ - V_n^-) - G_{ci}(V_{n-1}^+ - V_{n-1}^-) \tag{2.29}
$$

where $G_{ci} = \dfrac{C_i}{h}$, $G_{si} = G_{ci} + G_i$ and h is the integration timestep at the branch.

The usual test problem [22] cannot be used to explore the stability of this method since at least two dynamics are required. The circuit of Fig.2.15(b) will be used as the test circuit where branches a, b, and c are of the type of Fig.2.15(a). From KCL with node zero as the reference:

$$
G_{sa} V_n^1 - G_{ca} V_{n-1}^1 + G_{sc}(V_n^1 - V_{n-1}^2) - G_{cc}(V_{n-1}^1 - V_{n-2}^2) = 0 \tag{2.30}
$$

$$
G_{sb} V_n^2 - G_{cb} V_{n-1}^2 + G_{sc}(V_n^1 - V_n^2) - G_{cc}(V_{n-1}^1 - V_{n-1}^2) = 0 \tag{2.31}
$$

198

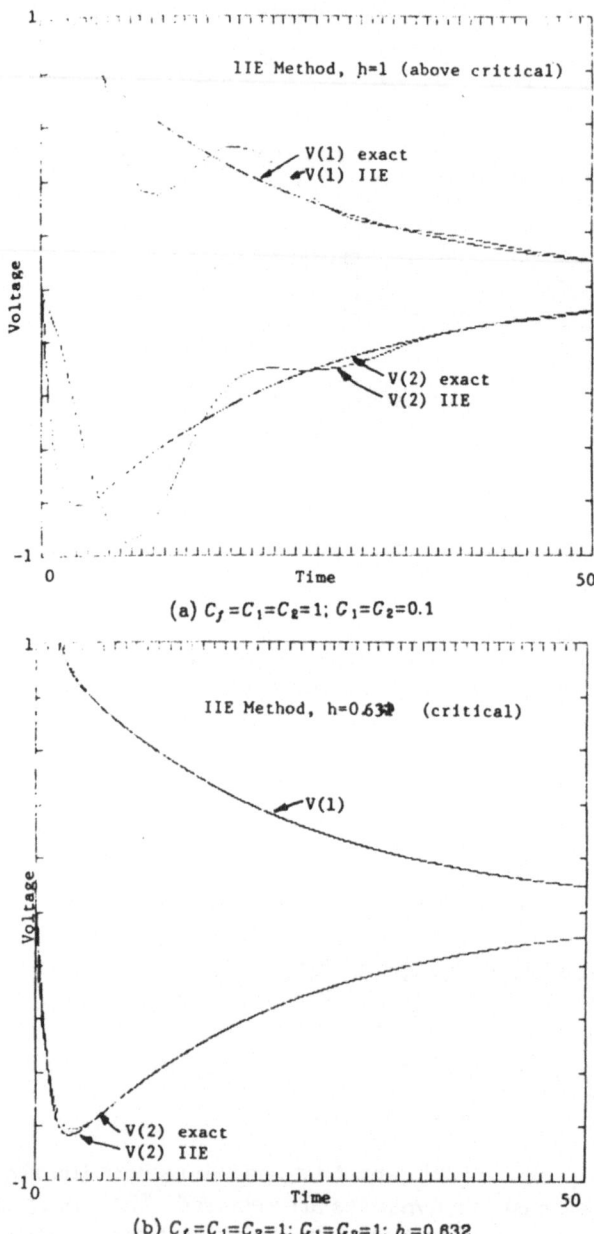

(a) $C_f = C_1 = C_2 = 1$; $C_1 = C_2 = 0.1$

(b) $C_f = C_1 = C_2 = 1$; $C_1 = C_2 = 1$; $h = 0.832$

Fig.2.13 (a) Response of circuit of Fig.2.12 using conventional
analysis and (b) response using the IIE hybrid technique.

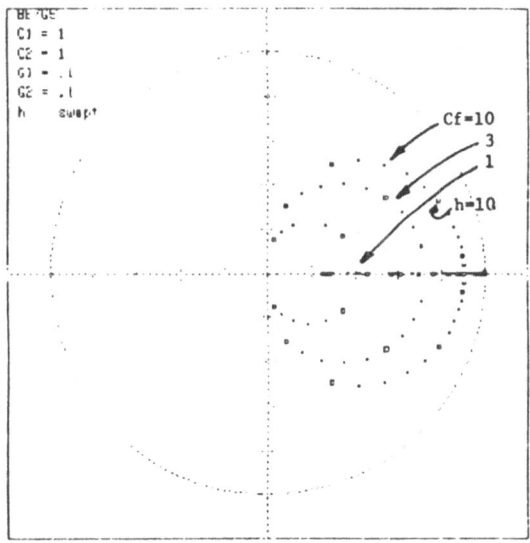

Fig.2.14 Root locus for the IIE method illustrating critical
timestep.

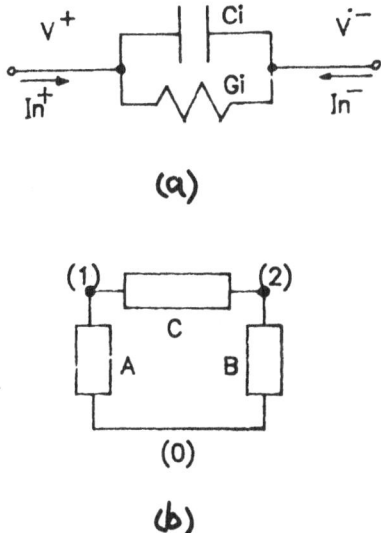

(a)

(b)

Fig.2.15 (a) Composite branch and (b) configuration
for critical timestep analysis.

$$G_{sa}V_n^1 - G_{ca}V_{n-1}^1 + G_{sb}V_n^2 - G_{cb}V_{n-1}^2 = 0 \qquad (2.32)$$

Apply the Z-transform and assume a solution of the form $V_n^1 = A_1 z^n$ and $V_n^2 = A_2 z^n$ and eliminate to obtain the characteristic polynomial:

$$Pz^3 + Qz^2 + Rz + S = 0 \qquad (2.33)$$

where:

$$P = (G_{sc} + G_{sb})(G_{sc} + G_{sa})$$

$$Q = -[(G_{cc} + G_{cb})(G_{sc} + G_{sa}) + (G_{cc} + G_{ca})(G_{sc} + G_{sb}) + G_{sc}^2]$$

$$R = [(G_{cc} + G_{cb})(G_{cc} + G_{ca}) + 2G_{sc}G_{cc}]$$

$$S = -G_{cc}^2$$

For the maximum timestep at which the roots of (2.33) are real [23]:

$$4PR^3 - Q^2R^2 + 4Q^3S - 18PQRS + 27P^2S^2 = 0 \qquad (2.34)$$

where P, Q, R, and S are as above. This equation can be further simplified and, over a range of typical parameter values, a look-up table for maximum stepsize, h_{crit}, can be used.

If the timestep associated with any floating capacitor branch is limited to its associated h_{crit}, the oscillatory error component is eliminated. An example of the output waveforms obtained for the circuit of Fig.2.12 with this h_{crit} used is shown in Fig.2.13(b). Note that for a wide range of element values, the critical timestep is still relatively large and hence does not severely degrade simulator performance.

To completely avoid the ringing in the solutions given by the above method a "double crossed voltage guess" can be performed. Refer to (2.20), (2.22), and (2.23) applied to the test circuit. The method has two steps, modeled as (2.24). In the first one, $A_l = A_l$ is lower triangular and the guess $A_{g1}v_{k+1}^2 = A_{g1}v_k^2$ is used in the first equation to solve for v_{k+1}^1 and v_{k+1}^2. In the second step $A_l = A_u$ is upper triangular and the guess $A_{g2}v_{k+2}^1 = A_{g2}v_{k+1}^1$ is used in the second equation to solve for v_{k+2}^1 and v_{k+2}^2. The companion matrix is :

$$M_2 = [I - hA_u]^{-1}[I + hA_{g2}][I - hA_l]^{-1}[I + hA_{g1}] \qquad (2.35)$$

The method is stable for the test circuit and the eigenvalues are real for all values of circuit parameters.

3. LOGIC SIMULATION

3.1. Introduction

During the initial design phase of a large integrated circuit, Behavioral Level and RTL simulation may be used to predict the functional characteristics of the design and perhaps to obtain first-order timing estimates. Once the functional modules used to describe the design at these levels have been partitioned into smaller modules, a *gate level* or *logic level* analysis may be used. A more detailed estimate of circuit timing can be introduced into the gate level description and signal interactions at circuit nodes can be modeled with more accuracy than at the higher levels of abstraction.

On the other hand, when the complexity of an integrated circuit design at the transistor level reaches the point where electrical analysis is no longer cost effective, logic simulation may also be used. Rather than dealing with voltages and currents at signal nodes discrete logic *states* are used. In fact, logic analysis may be viewed as a simplification of timing analysis where the difference equations of (2.6) are replaced by a set of discrete state equations. A fixed time step is used for the analysis and delay times are restricted to integer multiples of this timestep. Then the logic state at node i at time t_{n+1} may be written as:

$$l_{n+1,i} = L(\mathbf{l}^{n,i}) \quad i = 1, \cdots, N \tag{3.1}$$

where $\mathbf{l}^{n,i} = [l_{n+1,1}, \cdots, l_{n+1,i-1}, l_{n,i}, \cdots, l_{n,N}]^T$ and N is the number of logic nodes in the network. In this case, only simple Boolean operations are required to obtain $l_{n+1,i}$. These are generally the most efficient operations available on a digital computer. Rather than modeling the circuit at the individual transistor level, in a logic simulator transistors are grouped into logic *gates* wherever possible and a *gate-level* model is used. This form of simplification, sometimes referred to as *macromodeling* [25,26], can result in greatly enhanced execution speed by both reducing the number of models to be processed and simplifying the arithmetic required to process each transistor group. With selective trace analysis and the above simplifications, asynchronous logic simulators are typically 10 to 100 times faster than the most efficient forms of electrical analysis.

However, simulators must accurately predict the behavior, both normal and abnormal, of the physical circuits they model. It is clear that the transition from the continuous electrical domain to the discrete logic domain will result in the loss of some circuit information. It is important, therefore, that the circuit design methodology allow such an hierarchical simplification or logic simulators cannot be used effectively. In most cases, once a subcircuit of the design has been verified in detail at the electrical level, a simplified gate-level model can be used for logic simulation. However, it still may be necessary to analyze critical paths in the network at the detailed electrical level. For VLSI circuits, even a gate-level analysis may not be cost effective for the entire circuit and an RTL analysis may be necessary to verify the overall circuit performance. Timing information necessary for the RTL simulation can be obtained from the gate-level analysis of the functional modules in the circuit.

The tradeoff between the accuracy of logic simulation, and hence the amount of information it can produce about circuit operation, and the computer time required to perform the simulations, is very important. The number of logic states used in the simulator and their meaning, the logic delay models used, even the type of scheduling algorithms employed, are determined by the technology in which the circuits are to be implemented, and its associated circuit characteristics, as well as the particular design methodology being used.

It is this wide variety of factors that has resulted in the development of such a large number of logic simulators, almost every one addressing a different set of tradeoffs. In this section, some of the factors influencing the choice of logic states and delay models will be described and a number of scheduling algorithms for logic simulation will be presented. Logic simulators have been in use for the design of digital hardware since the 1950s [33], and it is impossible to address all aspects of simulator development here. Only those concepts of direct interest in the simulation of large MOS integrated circuits will be presented. A major area in which logic simulation has been used in the past is the determination of test pattern coverage and the determination of the testability of circuits. This will be covered in a later chapter by Bottdorf.

Since the major focus of this work is the simulation of large digital integrated circuits, a number of assumptions will be made which eliminate some unnecessary considerations. First, it is assumed that the circuit to be developed may not be restricted to a particular class of logic (synchronous, asynchronous, sequential, combinational) or associated form of implementation (nand, pass transistor, shift-register logic, etc.). Most circuits exhibit the characteristics of more than one kind of logic, whether intended or not. This assumption excludes some of the older *compiled simulators* [32], where the logic circuit was described in a programming language which was compiled directly to machine code and executed. Although this approach provides a very efficient mode of simulation, no compiled simulators accurately model asynchronous circuits. Compiled techniques are used, however, for higher-level Register Transfer Level (RTL) simulation and, of course, the bottom-level models in logic simulators are generally compiled.

A second assumption is that *equivalent* models for logic will not be used, where the equivalent model does not provide a reasonable approximation of the original circuit. For bipolar gate array design, where all gates available on a chip are n-input NAND gates, it would be a reasonable approximation to model a flip-flop as made of cross coupled NAND structures. However, for MOS circuits, to model 2-phase NMOS flip-flop with a NAND equivalent is unreasonable and unnecessary. This is even more important when MOS transmission gates (or *transfer* gates; sometimes called *pass transistors*) are used. For this device, its inherent bilateral nature precludes the use of composite unidirectional models. The equivalent logic approach has the advantage that it makes the implementation of a simulator simpler. However, the accuracy of the simulation suffers greatly as a result.

In this chapter, the term *logic gate* is used to refer to the intrinsic low-level models available in the simulator. They may be *simple* gates, such as AND, OR, COMPLEMENT, NAND, or more complicated gates such as multiplexors, decoders and even models with internal states, such as flip flops and registers. Simulators vary widely in the type of intrinsic models they provide.

3.2. How many states are enough?

The earliest use of logic simulation was for the verification of combinational logic. Since the logic was assumed to have zero delay and logic gates were assumed to implement ideal Boolean operations such as AND, OR and COMPLEMENT, only two states were required: a state representing *true* (logic 1) and a state representing *false* (logic 0) [34]. The zero-delay two state simulator will be used to illustrate a number of the features of more complicated logic simulators. With a two-state simulator it is not only possible to verify the logic function of a digital system (generate a *truth table*) but it is also possible to detect certain other types of potential design errors such as *hazards* and *races*. Although hazards may occur in combinational as well as sequential circuits, they are generally most important when they effect the behavior of sequential circuits. Since hazards result from paths with different delay times, any hazard *actually causing a circuit to malfunction* will be detected as a critical race or oscillation in the circuit[30]. The two-state simulator (even with random delay models) has only a limited capability for detecting races and hazards, since delay variations are not modeled.

If several inputs to a logic gate change within a relatively short period of time, it is possible that the order of occurrence of these events may change if gate delays were distributed at slightly different points within their tolerance limits. If the output state of the gate depends on the order in which the inputs change, a potential hazard exists.

It is not sufficient simply to monitor the output of a gate and look for multiple transitions during a pattern input if all *potential* hazards are to be detected. Depending on the order in which the input transitions are processed, the potential hazard may or may not be detected in the zero-delay simulator. This is illustrated in Fig.3.5 for a simple NAND gate. The potential for both static and dynamic hazards, as illustrated in Fig.3.1, can be detected. However, the errors caused by actual circuit hazards cannot be detected in a two-state simulator without the use of accurate delay models.

It should be noted that in a two-state logic system only one logic gate may drive (or fanin to) any node (often called a *net* in the context of logic design). If more than one gate did fanin to a node, there would be a potential conflict. If one gate had a logic 1 at its output and another a logic 0, it would be unclear what the signal at the node should be[4]. If it is possible for more than one output to drive a node in a particular technology, such as so called *tristate* logic (where gates may logically disconnect themselves from the node, as illustrated for MOS in Fig. 3.3), then two-state logic analysis cannot be used to verify the design.

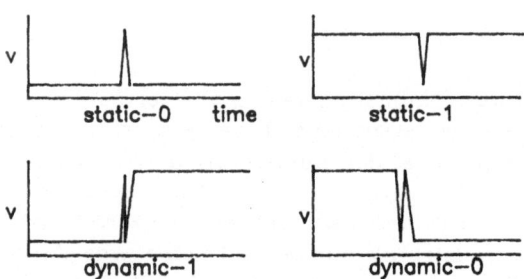

Fig.3.1 (a) Static-0 and static-1 hazards. (b) Dynamic-0 and dynamic-1 hazards.

---

[4]. An exceptional case is that of the wire-tie. (wired-AND, wired-OR), where the node is treated as a logic gate itself and performs a logic function. This is illustrated in Fig.3.2, an open-collector TTL example.

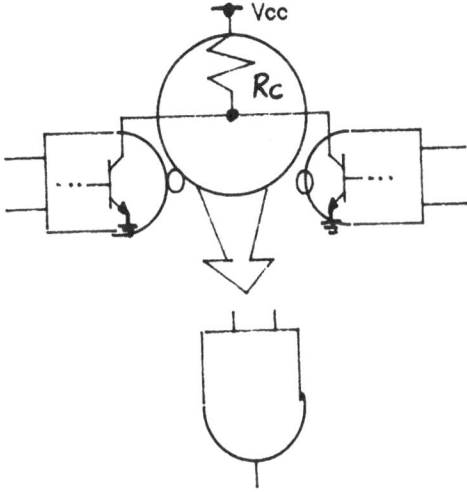

Fig.3.2 Open collector TTL structure and its equivalent logic model.

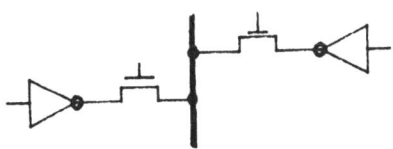

Fig.3.3 MOS transfer gates connected to a buss.

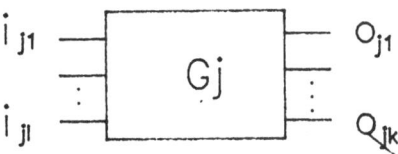

Fig.3.4 Port definitions for a logic gate.

In keeping with the assumptions above, it is assumed logic gates may have many outputs and many inputs, as shown in Fig.3.4. In a selective trace (or *event driven*) two-state logic simulator, the presence of a circuit hazard may be detected as follows. Since only one gate may drive a node, for single-output gates the marking, or *scheduling*, of a node to be processed is identical to scheduling the gate. For multiple-output gates, since it is a change at the input to the gate which may cause a signal change at the output of the gate, all outputs of the gate must be checked when the gate is scheduled.

```
for (each input pattern⁵, pᵢ ) {
    for (each scheduled gate, Cⱼ) {
        obtain the input states, iⱼ;
        obtain the present outputs, oⱼ-;
        for (each output, oⱼₖ) {
            compute the new output, oⱼₖ+;
            if (Cⱼ already processed for pᵢ) {
                for(each input changed at pᵢ) {
                    if( the input could cause an output transition
                        in the opposite direction to oⱼₖ+ ) {
                        print a hazard warning noting input transitions
                        and the direction of the output transition;
                    }
                }
            if ( oⱼₖ+  ≠  oⱼₖ- ) {
                update oⱼₖ;
                schedule fanouts of oⱼₖ to be processed at pᵢ;
                }
            }
        }
    }
}
```

This process may be enhanced to provide the type of hazard (static or dynamic, 1 or 0) and the number of hazards occurring at each node per pattern by maintaining a counter at each output which is incremented whenever a hazard occurs. The zero-delay hazard analysis outlined above is very conservative.

If any of the inputs p_i are outputs that were generated at p_{i-1}, then the simulator is operating as a synchronous logic simulator where each input pattern represents the next clock cycle. Two-state simulation has a number of limitations. Gates must be initialized prior to the analysis to either 0 or 1. If a sequential circuit is under analysis, storage nodes such as the output of flip-flops may not be known at initialization time. In the simulation of real hardware, there is generally a time when a signal is changing state during which its actual logic level is *indeterminate* or *unknown*. To model initialization and the transition condition, a third state may be added: the unknown state, X [31].

---

[5]. An input *pattern* is the set of external signals applied to the circuit inputs for which the circuit outputs are to be computed. The application of a new pattern results in the scheduling of all input gates whose inputs change.

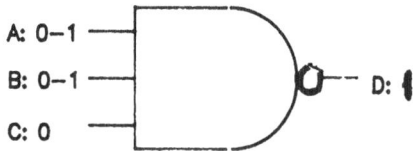

(a) While c=0, d=1 independent of a and b.

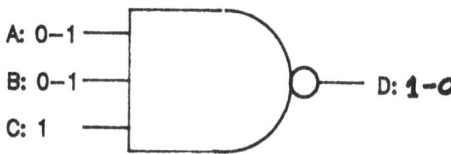

(b) D will execute the same transition independent of the order of A and B transitions.

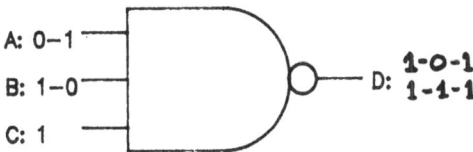

(c) If A changes first, D will be in the 0 state until B changes. If B changes first, D will remain 1 regardless of A. A hazard warning should be printed.

Fig.3.5 3-input NAND gate and possible input sequences.

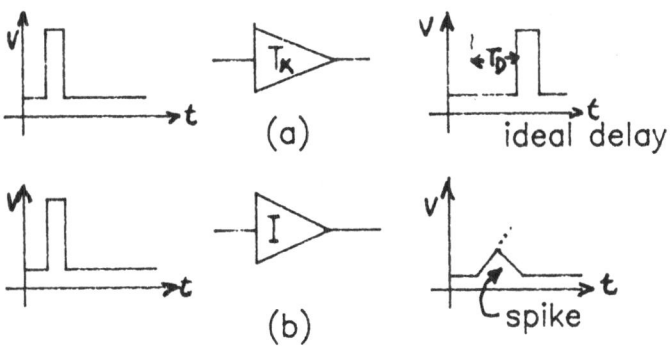

Fig.3.6 Logic delay models and their effect on a short pulse. (a) Transmission-line delay and (b) Inertial delay.

Since the X state may be used to represent a transition, it is necessary to introduce the concept of delay time. Early simulators used *unit delay* models to represent timing. In a unit delay simulator all gates have the same (unit) delay between signal transitions. For logic constructed from a single gate type which has similar rise and fall delays, the unit delay simulator can provide a useful analysis and lends itself to efficient implementation. Where more than one gate type is used, *assignable delays* provide a more accurate model. In the assignable delay simulator, the delay of the logic gates may be assigned an integer value, T_D. This delay is a multiple of the fundamental analysis timestep, or *minimum resolvable time* (MRT) as introduced in the previous section for timing analysis. Here, the MRT is the minimum non-zero delay of a logic gate and its value depends on the technology being simulated. MRT may be 1ns for NMOS, while a value of 100ps may be appropriate for ECL circuits. For logic circuits where rise and fall delays vary widely (such as single channel MOS) it is necessary to provide both assignable rise (T_{plh}) and fall (T_{phl}) delays for each gate.

More precise timing models may be characterized by delays with a min-max range in which an element may respond.[32] The minimum value could specify the shortest possible propagation delay, the maximum the longest possible delay. This effect can be implemented with the use of two additional transition states: *rising* (R) and *falling* (F).

There are two ways in which gate delays may be interpreted as illustrated in Fig.3.6. A *transmission line*, or group-delay model would propagate the input patterns directly to the output, delayed by T_{plh} or T_{phl} for rising and falling edges respectively. Even very short input pulses are propagated unaltered, as shown in Fig.3.6(a). The second approach is to use an *inertial delay* model, in which the "inertia" or response time of the gate is modeled. If an input event occurs in less time than the time required for the gate to respond, it will be lost as shown in Fig.3.6(b). Note that in this case output *spikes* may be generated. A spike is defined as a signal of shorter duration than necessary to change the state of an element. Spikes may be generated by input hazards or by very narrow input pulses to a gate. A spike may be propagated into the fanout gates as either a new state (S, or E for "error condition" [32]) or they may be deleted from the output and a warning message printed. The latter technique generally provides more information from the analysis since the spike is generally an error and will be removed by the designer. By not propagating the spike, more information may be obtained about the correct operation of the following circuitry.

With three static states (0, 1 and X) and assignable inertial delays, it is possible to detect most significant hazard and race conditions. Cyclic races will often appear as oscillations in the output, as illustrated in Fig.3.7. Hazards are generally detected as spikes in gate output signals. The analysis of a typical logic circuit may proceed as follows:

```
for (each unit of MRT at which gates are scheduled) {
    for (each gate scheduled, G_j) {
        obtain input states, i_j;
        obtain output states, o_j-;
        for (each output, o_jk) {
            compute the new output, o_jk+;
            if ( o_jk- = o_jk+ ) next ;
            compute new time to schedule fanouts, F_jk, t_s^n;
            if ( F_jk are not currently scheduled ) {
                schedule F_jk at t_s^n;
                update o_jk;
            }
            else {
                obtain current scheduled time t_s^o;
                if ( t_s^o ≥ t_s^n ) {
                    unschedule F_jk;
                    schedule F_jk at t_s^n;
                }
                update o_jk;
                issue spike report;
            }
        }
    }
}
```

Three-state simulation is not sufficient for the analysis of MOS circuits which contain transfer gates. For these circuits many gate outputs may be connected to a single node as shown in Fig.3.8, and it is necessary to determine which output is controlling the state of the node, or *buss*. If more than one gate is forcing the node, a *buss contention* warning must be generated by the simulator. It is possible to represent the condition where the output of $M1$ is not controlling the buss (G_1 is logic 0) by modeling the output of $M1$ as X in that case. If this technique is used there is no longer any distinction between the true unknown state and the *off* condition of the gate. With the addition of a fourth static state, *high impedance* (H) or *nonforcing*, the distinction is maintained. The four static states are now illustrated in Tbl.3.1. Here, the x-axis may be considered a voltage (or current) axis. High voltage is represented by logic 1, low voltage logic 0, and in between is unknown, X.

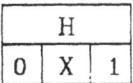

Tbl.3.1 State Table for 4-State Logic Simulation

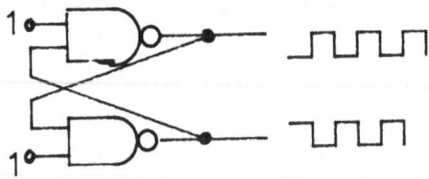

Fig.3.7 Simple cyclic race circuit illustrating oscillatory
output waveform.

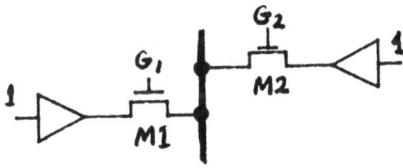

Fig.3.8 MOS transfer gates connected to a buss.

With the addition of the H state, buss contention can be predicted without
confusion. But what if all the gates driving a node are off? What is the input
to a fanout gate then? In MOS circuits, the last gate output is generally
"stored" on the parasitic capacitance at the node and held for some time.
This may be modeled by saving two states at each node, the present state
and the immediate past state. If the present state is H then the previous
state can be used to determine the input to fanout gates. (The previous
state is required to model accurately storage elements in any case).

Another approach that can be used is to add two more static states, as
shown in Tbl.3.2. The low impedance states are called *forcing* states: 1_f, X_f,
0_f, and there are now three high impedance states: 1_h, X_h, and 0_h, which
carry the information about the previous signal level as well.

0_h	X_h	1_h
0_f	X_f	1_f

Tbl.3.2 State Table for 6-State Logic Simulation

Consider the circuit of Fig.3.9. If $M1$ and $M2$ are both conducting, it is clear
that the state at node (2) cannot be determined from our simple model.
But what of nodes (1) and (3)? Since the transfer gates are in fact bidirec-

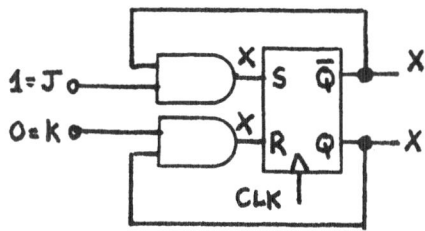

Fig.3.10 Unknown states at output will not allow this circuit to power up correctly.

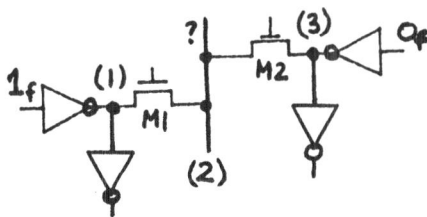

Fig.3.9 Potential error condition even for fanouts at nodes (1) and (3) unless extra states are used.

tional, the signal at node (2) may force nodes (1) and (3). In reality, the output impedance of the inverter is probably considerably lower than the output impedance of the transfer gate and hence the inverter output would dominate the node. To model this effect another three states may be added, called *soft* states: 1_s, X_s, and 0_s, which correspond to the output of the transfer gate when its gate node is on and its input is a forcing or soft state. These states are shown in Tbl.3.3. The LOGIS logic simulator [35] uses these nine static states.

Z

0_h	X_h	1_h
0_s	X_s	1_s
0_f	X_f	1_f

V

Tbl.3.3 State Table for 9-State Logic Simulation with voltage and impedance axes shown.

The y-axis of this state table may be considered an impedance axis. In fact, the output of a logic gate may be mapped into this state table by measuring its output voltage (or current) and output impedance. This technique may also be used to *coerce*, or translate, electrical outputs from timing analysis into logic inputs for mixed-mode analysis as described in Section 4. If 16 bits of data were used to represent the voltage and impedance at a node in timing analysis, this could be considered to represent $2^{16} - 1$ discrete states along each axis of Tbl.3.3. Additional delay times are now required to model the transition times between different states.

The nine-state simulator does not adequately model the interaction between transfer gates of different geometry, or the effect of parasitic capacitance variations on the charge sharing across a transfer gate. These effects could be represented by a finer resolution (more states) in Tbl.3.3. To model this behavior accurately timing simulation can be used. The bidirectional nature of transfer gates can be approximated by noting all the forcing states at the nodes of an arbitrary tree of transfer gates and tracing their fanouts through the tree, setting all affected nodes to the appropriate soft state unless a conflict is detected.

One more state is required to account for initialization in sequential circuits. Consider the J-K flip-flop circuit of Fig.3.10. If the outputs are assumed unknown at initialization, they will never be set by data at the inputs of the flip-flop due to the feedback of the X states. To overcome this problem a distinction is made between *initial* unknowns X_i and *generated* unknowns X_g. When an initial unknown is encountered during the simulation, it is ignored in the processing of the gate it is driving. The difference between initial and generated unknown states can also prove useful in determining those parts of a circuit not exercised during the simulation (still at X_i).

3.3. Models for Logic Analysis

For a logic simulator with many logic states, gate models can become quite complex. Whereas for simple gates and a few states a simple table look-up scheme can be used, when many states are involved a combination of program and table look-up techniques is generally used. A model for an MOS NAND gate in a nine-state simulation may consist of:

```
NAND ( G );
    o + = 0_f;
    for (each input i_j) {
        if (i_j = 0_f or i_j = 0_s or i_j = 0_h ) o + = 1_f;
    }
    t_s^n = T_d(o +, o - );
}
    end;
```

where G is a pointer to the data structure for the gate in question. G contains pointers to the gate inputs, i, the gate output o, and their associated logic states. It also contains a pointer to the gate model data which in turn would contain the table of state transition times T_d. This structure is illustrated in Fig.3.11.

Some simulators allow the definition of gate models as simulator input. Others require the addition of subroutines to the simulator at compile and load time for additional models. Most logic simulators permit nesting of models (*macros*) although nested models are expanded to the intrinsic gate level prior to analysis in almost all cases.

3.4. Scheduling Techniques

The heart of any efficient logic simulator is an event-control mechanism. If the inactivity of the digital network is to be exploited to the maximum degree possible, the overhead imposed by the event-control processor must be small.

A number of techniques have been employed for the control and sequencing of events in a logic simulator. The most straightforward approach is to sort the gates to be processed in order of precedence such that all signals on which a gate depends have been computed before it is processed. This technique is sometimes referred to as *static levelling*. Feedback loops are extracted and treated separately. Although this approach works well for synchronous models it is not as effective for asynchronous circuits where a variety of delays are present and almost every gate has a delay associated with it. In this case many "levels" of logic are required to simulate the network. An approach of this type is used in the SALOGS logic simulator [36].

An effective scheduling technique used for asynchronous logic simulation is the Time Queue (TQ) approach [37], illustrated in Fig.3.11. Each entry in the queue represents a discrete point in simulated time. Time moves ahead in fixed increments of MRT, determined by the user, which correspond to consecutive entries in the time queue. Each entry in the queue contains a pointer to a list of *events* which are to occur at that instant in time. An event is defined as the change of logical state of an output node of a gate or input signal source. The new state may or may not be the same as the state already held by the output. If the new state is different from the old one, all elements whose input lines are connected to this output node, the fanouts, must be processed to see if the change affects their outputs.

When an output is evaluated and the new value is the same as the value

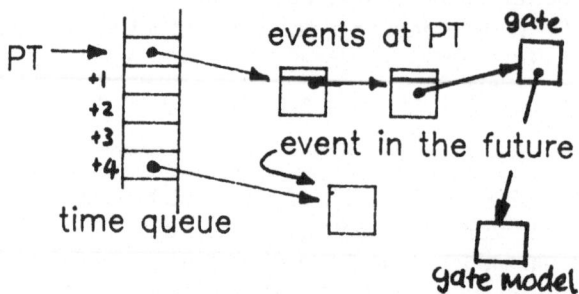

Fig.3.11 Schematic representation of time queue.

already held by the node, the event is cancelled and the fanouts are not added to the time queue. If the new value is different, the event is executed by adding the list of fanouts to the time queue. The time queue corresponds to a "bucket sort" of events in time and is very efficient for inserting and accessing events. It is clear that such a list would be too long for an entire simulation and hence the time queue is generally implemented as a circular list. If the length of the list is chosen to be an integral power, n, of 2 (a typical value for n might be 8) then the offset of a new event from the time queue origin, given it is to be inserted T_D units from present time (PT) is $(PT + T_D)mod2^n$. Since an integral power of 2 was chosen, this computation can be performed with a simple integer add an mask operation. Note that once the events at a particular time have been processed, the list is cleared at that time. The time queue technique is an example of a *dynamic levelling* approach, since the order of event processing is determined by the circuit conditions at that time. For events outside the range 0-2^n from PT, a separate mechanism must be used. A time-ordered list is one technique which may be used for these events as illustrated in Fig.3.12. Each list entry contains the time the event is to occur and a pointer to the event itself. This list has a relatively long list insertion time ($O(nlogn)$ for a list of n entries) but most events occur within a few MRT units of PT and hence it is generally only input source changes that are stored in this list.

There may be many events scheduled at one timepoint. They are linked as shown in Fig.3.11 and 3.13 and processed in order. Should a new event be scheduled at PT, it must be added to the end of the list. This process is expedited by using an extra pointer to indicate the end-of-list entry as shown in Fig.3.13.

The overhead incurred by an efficient implementation of the event scheduler is typically less than 10% of the simulation time. Scheduling techniques have been proposed which combine the speed advantage of synchronous scheduling with the time queue approach [38]. It is not clear this provides a significant advantage for asynchronous simulation at the gate level but may be useful for hybrid gate level and register transfer level simulation.

Other techniques which may be used to improve both the speed and memory savings of logic analysis include *backtracing* and hierarchical analysis.

In the backtracing approach, back pointers are used to trace back from output nodes to circuit inputs (fanin tables may be used here, just as fanout

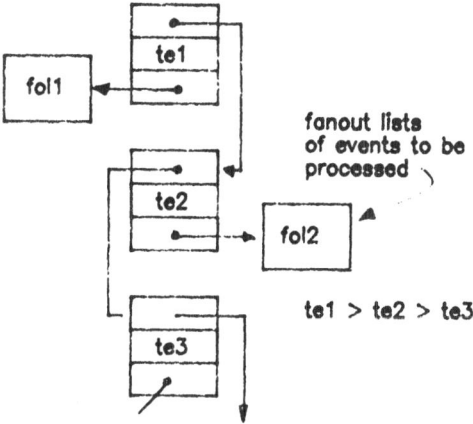

Fig.3.12 Time-ordered linked list for events outside the range of the time queue.

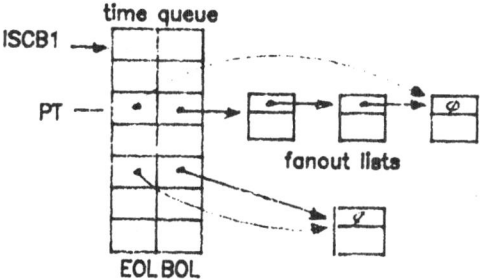

Fig.3.13 The time queue as implemented in the SPLICE program.

tables were used for forward tracing). If only outputs at selected nodes are required, the logic nett may be traced prior to analysis and only those gates which are on a topological path between the inputs and the particular outputs of interest need be sensitized for analysis. Using this approach, as the simulation proceeds, fanout elements which cannot influence the outputs of interest can be ignored and hence the simulation time may be reduced even further [47]. This technique does require the searching of a rather large circuit graph which can become expensive and may only be cost-effective for long simulation runs.

With this approach, first-order estimates of worst case or statistical delays along such paths can also be obtained without the need for detailed simulation. [48-50].

Rather than fully expanding nested macro definitions, it is possible to maintain the hierarchy of the description and keep only a single representation for the internal connection of a macro-cell, or subcircuit. Whenever the cell is *instantiated* or used, a *state vector* is established which contains the states of all internal nodes of that particular instance of the cell. In this way, duplicate copies of the cell connections are not generated and substantial memory savings can be achieved if a subcircuit is used many times.

4. MIXED-MODE SIMULATION

4.1. Introduction

For the analysis of most large MOS circuits neither electrical nor logic analysis is alone sufficient. The detailed waveform information of an electrical analysis is required for some parts of the circuit but an electrical analysis of the entire network would require an excessive amount of computer resources. A logic-level analysis is often sufficient for parts of the circuit and can be performed much more efficiently than an electrical analysis. Although timing analysis is generally much more efficient than circuit analysis there are circuits which cannot be simulated accurately using timing analysis, such as circuit blocks containing strong bilateral coupling between nodes. For these blocks, a detailed circuit analysis may be required.

A number of *mixed mode* simulators have been developed which combine analyses at more than one conceptual level. The SABLE system [39] addresses the high-end of the design and allows behavioral, RTL, and gate-level descriptions to be combined. Both the Diana program [40] and the SPLICE program [41] allow concurrent circuit, timing and logic analyses of different parts of the circuit. The MOTIS program [9] has recently been extended to combine timing, logic, and RTL level analyses [42].

In this section, the architecture of the SPLICE program will be described as an example of a table-driven, mixed-mode simulator. Recent extensions to the program and a detailed example of program operation will also be presented.

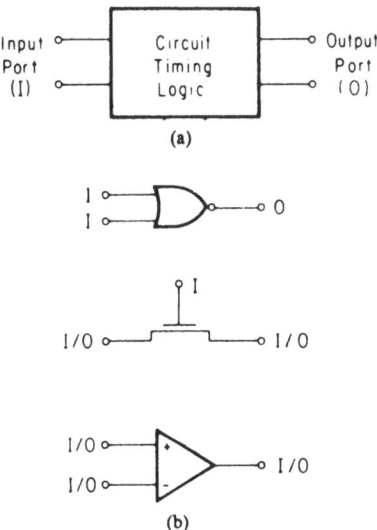

4.1 (a) Port definition and (b) examples for different elements in mixed-mode analysis.

4.2. Architecture of a Mixed-Mode Simulator

One approach to the design of a mixed-mode simulator is to combine existing circuit, logic, and timing simulators via data *pipes* such that the three forms of analysis can be performed concurrently and may pass values of circuit variables, with the appropriate transformations, between one another via the data pipes which connect them. Such an approach would result in a very inefficient simulation.

In a table-driven simulator, it is the structure of the data tables that makes for efficient simulation. Hence if a common data format can be determined for all types of circuit elements and circuit nodes, a single event scheduler can be used to process them all.

Techniques for using selective trace with logic analysis were described in Section 2. These techniques may be extended to mixed-mode analysis by defining elements of the network as having one or a combination of two types of ports:

a) *Input Ports* which sample the voltage or logic level at a node but play no part in determining its value.

b) *Output Ports* which are responsible for determining the voltage or logic level at a node but do not sample the signal at the node.

Elements may be of type logic (gates, flip-flops), timing (single transistors, macromodels), or circuit blocks described by a nodal matrix (operational amplifiers, sense amplifiers). Examples of these elements are illustrated in Fig.4.1. Note that ports may be both input and output. These ports are indicated as *I/O* ports in Fig.4.1.

Nodes of the network have four properties:

a) The *type* of the node (logic, timing or circuit), and hence the type of signal stored at the node (electrical, discrete),

b) The value of the voltage or logic level at the node itself.

c) *Fanin* elements, which play some part in determining the voltage or logic level at the node.

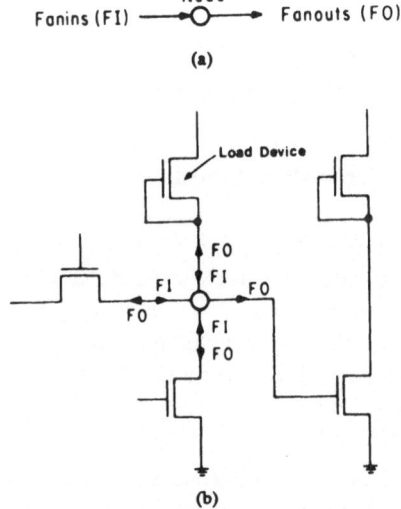

(a)

(b)

4.2 (a) Fanin/out definition and (b) example for timing analysis.

d) *Fanout* elements, which sample the signal level at the node but play no part in determining its value.

These properties are illustrated in Fig.4.2. For the timing node of Fig.4.2(b), the fanin and fanout elements are shown for a group of MOS transistors connected to the node. Note that for an electrical node an element can be both a fanin and a fanout element. For example the load device of Fig.4.2(b) is a fanin element because it may supply current to the node and thus play a part in determining the voltage at the node. It is also a fanout element because a change in voltage at the node may change the amount of current supplied by the load device.

It is necessary to modify the selective trace algorithm, described earlier, for combined electrical and logic analysis. With the above concepts in mind, at time t, the modified selective trace algorithm proceeds as follows:

```
for (each element, E, scheduled to be processed at t) {
    for (each output or I/O port, N, of E) {
        if (N has been processed at t) {
            if (N is of type logic) check for spike conditions;
            else if (node is scheduled at t+MRT) skip it;
                else schedule it at t+MRT;
            else {
                evaluate the contributions of fanins at N;
                evaluate the new signal level at N;
                if (the new level is significantly different from
                    the old level) {
                    D = the delay through the element in MRT⁶;
                    schedule the fanouts of N to be processed at time t+D;
                }
            }
        }
    }
}
```

If the next node to be processed is an external node of a circuit element, circuit analysis techniques are used to evaluate the new signal level at the node. Once the modified selective trace algorithm has determined which circuit block is to be processed for one unit of MRT, the timing element models are evaluated using the SOR-Newton algorithm, described in Section 2, for all devices connected to the external circuit nodes. This process is illustrated in Fig.4.3 and results in a Norton equivalent network model for the interface nodes to the remainder of the network. Since the coupling is bidirectional, for circuit blocks coupled directly by a single timing node per port, this process is almost identical to Node Tearing [8]. The resulting circuit block, which may have a number of internal nodes, is then analyzed using algorithms similar to those used in a conventional circuit analysis

---

6. $D = 0$ for timing and circuit elements but may be greater than zero for logic elements.

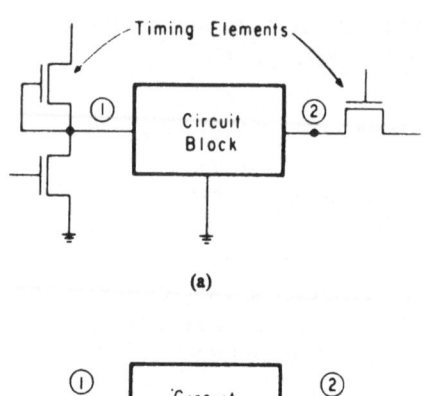

(a)

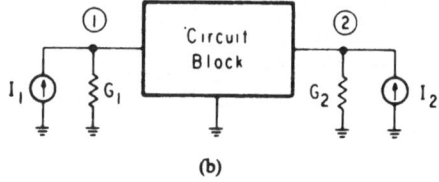

(b)

4.3 Processing of circuit block. (a) The original circuit
block and its timing interface (b) The Norton equivalent at
a timepoint.

program such as SPICE2[1]. In this case, however, a number of Newton-Raphson iterations are used at each timepoint to ensure local convergence to within 0.1% change in nonlinear branch currents between iterations. This provides a more accurate analysis for the internal nodes of the circuit block and a faster local asymptotic rate of convergence than would be obtained from the one-step SOR-Newton algorithm.

4.3. The Discrete-to-Electrical Interface

Timing analysis and circuit analysis may be coupled directly since they both use voltage and impedance to model the conditions at a node and hence an additional interface is not required. Discrete logic simulation does require an interface to and from the electrical analysis. This interface may be achieved by implicit signal *coercions*. That is, whenever an electrical element is connected to a logic node, and vice versa, an automatic signal transformation is implied. Alternately, special circuit elements may be used to perform the transformation. In either case, *thresholding* may be used to convert voltage and impedance to logic levels while *logic-electrical conversion* may be used for the reverse transformation.

The thresholding function may be performed directly using the impedance-voltage concept described in Section 3, where the effective impedance to ground and the open circuit voltage at the node are used to define a logic state. This can be performed via a direct mapping into one of the state tables of Section 3. The impedance and voltage boundaries separating the states may be defined globally for implicit coercions or as models for special thresholding circuit elements if a different mapping is required at specific nodes.

The logic-to electrical conversion can be performed using Boolean-Controlled Switches [40] as described earlier by De Man, or via programmable converters which convert a logic transition at their input to a voltage (or current) change, with a parameterized time delay, at their output. An example of such an element for 4-state simulation is shown in Fig.4.4. Note that when a high impedance (H) state is at the input of the converter, the output voltage remains constant as though the parasitic capacitance at the node was holding the signal value. Programmable changes in the node-to-ground impedance are also possible in a general conversion and the Boolean Controlled Switch can be viewed as a special case of this approach, where on a 0-1 transition the impedance switches rapidly from a low value to a very high value while the equivalent current source at the node may not change.

The implementation of such a rapid impedance change does require some care in a nodal analysis program, as described in [40]. A constant rate of change of voltage or impedance at the output of the converter is sufficient for most applications. More complex output waveforms, such as exponentials, may be generated by adding simple circuit elements (resistors, capacitors, MOS output stage) to the output of the converter.

4.4. The SPLICE Program

Program SPLICE implements the algorithms described above. It is a hybrid simulation program which can perform simultaneously circuit, timing, and logic analyses of MOS circuits. The modified selective trace algorithm is used to control all three types of analysis and hence only those parts of the

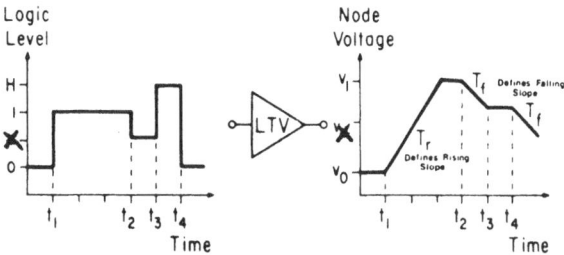

4.4 Example operation of Logic-Voltage conversion.

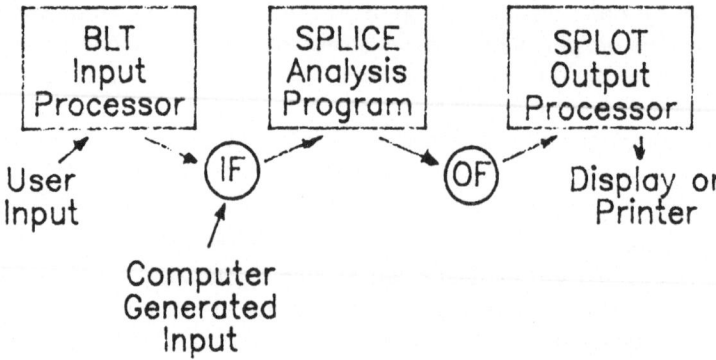

4.5 The block structure of program SPLICE

circuit with changing signal levels are processed.

At present the program can analyze circuits containing any of the common dynamic and static MOS logic gates and flip-flops, as well as some higher-level logic circuits. Both N-channel and P-channel transistors may be used in the electrical analysis.

Four-state logic simulation has been implemented: 1, 0, X, H. All data is scaled and stored in integer form except for circuit-block matrix entries. Voltages may have any one of ±32767 levels which span twice the range of supply voltages. This range is sufficient to accommodate most circuits which incorporate bootstrapping effects. At present SPLICE does not use a separate algorithm for dc steady-state analysis. Node voltages or logic levels may be preset by the user and the nonlinear transient analysis is used to obtain a steady state. For most digital circuits where a long transient analysis is to be performed the cost of the steady state analysis is only a small percentage of the overall cost of the analysis, as illustrated in Section 4.6.

The SPLICE package consists of three independent programs which communicate with one another via data files, as shown in Fig.4.5. The input description of the circuit to be simulated is read by the BLT program [43]. BLT performs subcircuit expansion for nested macro circuit definitions and processes forward model references. It also simplifies the description by translating node names into a contiguous set of node numbers and generates a relatively simple intermediate file (IF) to be read by the analysis program. As the simulation proceeds, user-requested outputs are written onto another file (OF) for post-processing by the SPLOT program [44]. SPLOT produces output plots of electrical or logic node signals, as well as performing a variety of other tasks such as producing truth-table output, searching for logic patterns, etc. The SPLOT post-processor can threshold electrical outputs to convert them to equivalent logic signals. With the program partitioned into these three parts it is relatively easy for a SPLICE user to provide different input and output processors, common to other programs at a particular site. The intermediate input file is also convenient as an input medium for machine-generated data, such as circuit topology extracted from integrated circuit mask data [45] or symbolic layout data of the form described by Hsueh in a later Chapter [46].

4.5. A Simple SPLICE Example

Since the SPLICE analysis is table-driven, the initial phase of the analysis involves setting up the data structures used during the analysis. This phase is referred to as SETUP. Following SETUP, the SOLVE phase of the simulation is entered and the analysis proceeds under the control of the event scheduler.

The simple MOS shift register cell of Fig.4.6 will be used to illustrate the various stages of SPLICE simulation and their associated data structures. The input data for this example is shown in Fig.4.7. Once the BLT program has translated the data, it appears in the IF file as shown in Fig.4.8. Note that the models are represented by numbers and the signal path names (numbers in this case) have been mapped to a contiguous set of numbers. The *node map* is output at the end of the data to allow the post processor to label plots with the original node names. Output requests are processed as elements, scheduled if the node voltage or state changes. When processed, they simply write the node voltage or state and the current time to the OF file. Hence output data is only generated when a node changes its voltage or state. Models are output first to eliminate the need for the analysis program to process forward referenced models. The analysis commands are output with a key number first and a list of parameters following, as are the circuit elements.

4.6 MOS shift-register cell example.

The SETUP portion of the SPLICE program reads this file and sets up the data structures shown in Fig.4.9. Fanout and fanin tables are generated, the nodes are labelled with a node *type* (1=logic, 2=timing, 3=external circuit node, 4=internal circuit node). Provision is made for storing the signal at the node and the node capacitance, for timing and external circuit nodes. The analysis begins by scheduling all the fanout tables, as shown in Fig.4.11(a). Note that the tables themselves are not moved. They are linked by adding a pointer as the first entry in the table. Fig.4.11(b) shows the time queue at some later time. Note the sources have scheduled themselves to be processed at their next transition times and the logic gate outputs are scheduled some time from PT.

Fig.4.11 shows the output waveforms obtained for this analysis. The same circuit was analyzed using an all-transistor representation (a total of 6 transistors and 3 sources). The resulting waveforms are shown in Fig. 4.11(b) for comparison. The analysis time for this circuit was somewhat longer than the mixed-mode case however the savings are not really evident in such a small example.

```
splice
mos latch example - using logic inverters
$
$...model definitions
model sourcel tsrc  : v0=0.0 v1=5.0 d=0 p=30ns    0ns 14ns 15ns 29ns
model phasel  tsrc  : v0=0.0 v1=5.0 d=0 p=15ns    0ns  7ns  8ns 14ns
model phase2  tsrc  : v0=5.0 v1=0.0 d=0 p=15ns    0ns  7ns  8ns 14ns
model mos  ntxg  : vt= 1.12 wol=1.0  gam=0.266 kp=79u phi=0.6 lam=0.02
model inv  inv   : tr=4ns tf=3ns
$
$...circuit description
tranl   store    input    phasel   mos
invl    output   outbar            inv
inv2    outbar   store             inv
tran2   output   store    phase2   mos
capl store gcapr : .10pf
$
sourcel input sourcel      ; input to latch
source2 clockl phasel      ; phase 1 of clock
source3 clock2 phase2      ; phase 2 of clock
$
$...analysis requests
$
opts 1 8.0                 ; analysis options
topts 0.1 0.05 1000 0.1    ; timing analysis options
llevels 2.0 3.0            ; voltage-to-logic conversion thresholds
vlevels 0.2 4.5 2.5        ; logic-to-voltage conversion values
time 1ns 200ns
$
$... output request
table clockl clock2 input store outbar output
$
go
$
end
```

4.7 SPLICE input data for this example

COMMENTS	DATA FILE	MEANING
MODELS		
mos	33	model type
	6	no. of parameters
	1.0	parameter
	7.9e−5	values
	⋮	⋮
inv	21	model type
	2	no of parameters
	4e−9	parameter
	3e−9	values
ELEMENTS	−1	end of models
tran1	1	model number
	3	no of nodes
	1	drain
	2	source
	3	gate
inv1	2	model number
	2	no of nodes
	4	output
	5	input
inv2	2	model number
	2	no of nodes
	5	output
	1	input
tran2	1	model number
	3	no of nodes
	4	drain
	1	source
	6	gate
ANALYSIS	−1	end of elements
	6	request type
	2	no of parameters
	1	values
	8.0	
NODE MAP	--1	end of requests
	store	node names
	input	in order
	phase1	
	output	
	outbar	
	phase2	
	−1	
	−1	

4.8 Data structure output from BLT.

ADDRESS	CONTENTS	MEANING
mod1:	33	table pointer
	par1–>	param pointer
par1:		table (400 entries)
mos1:	mod1–>	model pointer
	3	no of nodes
	2	no of outputs
	nod1–>	node
	nod2–>	pointers
	nod3–>	
log1:	mod2–>	model pointer
	1	no of nodes
	1	no of outputs
	nod5–>	node
	nod1–>	pointers
nod1:	fol1–>	fanout pointer
	fli1–>	finin pointer
	2	type: timing
	ts	scheduled time
	Vn	voltage
	Vn–1	old voltage
	Yn	conductance
	cap–>	capacitance pointer
fol1:		scheduler link
	mos1–>	element
	mos2–>	pointers
	log1–>	
fli1:	mos1–>	element
	mos2–>	pointers

ADDRESS	CONTENTS	MEANING
mod2:	21	table pointer
	par2–>	param pointer
par2:		gate delays
mos2:	mod1–>	model pointer
	3	no of nodes
	2	no of outputs
	nod4–>	node
	nod1–>	pointers
	nod6–>	
log2:	mod2–>	model pointer
	1	no of nodes
	1	no of outputs
	nod4–>	node
	nod5–>	pointers
nod5:	fol5–>	fanout pointer
	fli5–>	finin pointer
	1	type: logic
	ts	scheduled time
	Ln:n–1	logic values
fol5:		scheduler link
	log1–>	element
		pointers
fli5:	log2–>	element
		pointers

4.9 Node and element tables for the example

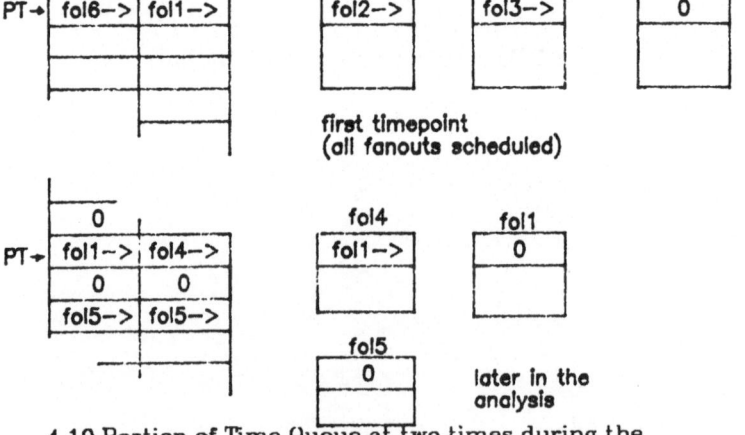

4.10 Portion of Time Queue at two times during the simulation.

227

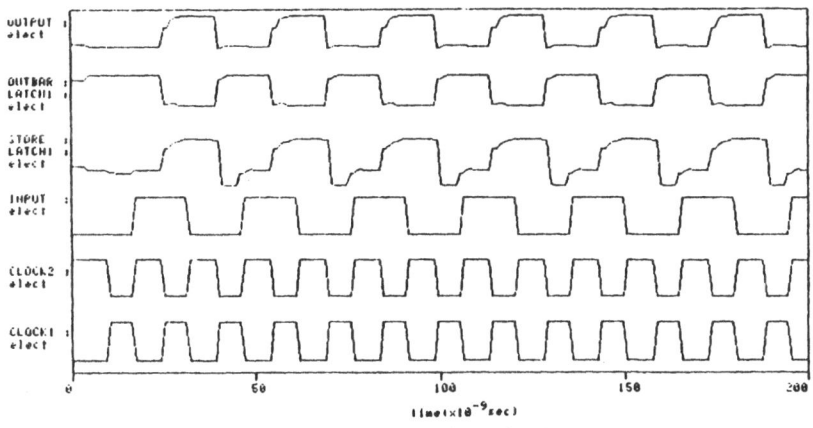

SPLICE 0.0A   aos latch - all timing elements

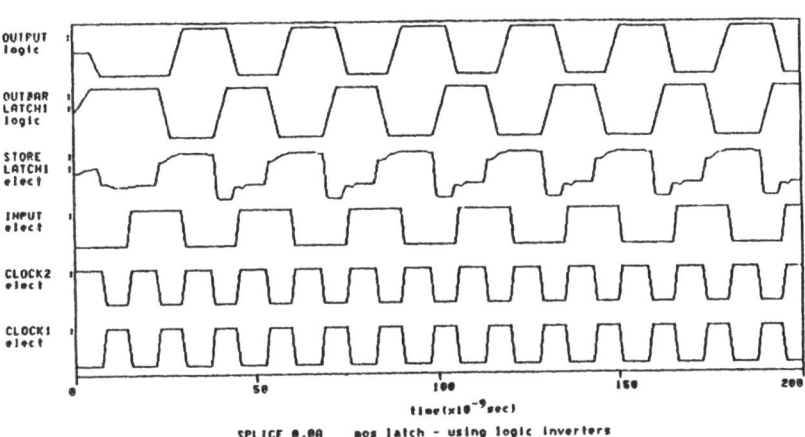

SPLICE 0.0A   aos latch - using logic inverters

4.11 Output waveforms from SPLICE for this example.
(a) All timing transistors (twice the cpu time c.f. the
mixed analysis) and (b) inverters at the logic level
Note that logic waveforms have been plotted for comparison.

4.6. Simulation Examples

This section presents the simulation results for a number of circuits using SPLICE. The results show that by using hybrid analysis for a large circuit, between one and three orders of magnitude speed improvement and between one and two orders of magnitude reduction in memory requirements can be obtained compared to conventional circuit analysis. The first example, a 256-by-1 bit dynamic RAM circuit, demonstrates the use of concurrent circuit, timing, and logic analyses. The second example, a 700 MOS transistor digital filter circuit, illustrates the speed improvement possible using an event-driven analysis.

A block diagram of the RAM circuit is shown in Fig.4.12. A schematic of the row and column decoders used in the analysis of this circuit is shown in Fig.4.13(a) and includes both logic gates and timing elements. The input/output circuits and storage transistors of the RAM are analyzed using a timing analysis and each sense amplifier is analyzed as a separate circuit block. The schematic for the sense amplifier and associated "dummy" cell is shown in Fig.4.13(b).

Tbl.4.1 contains a summary of the statistics provided by SPLICE for the RAM analysis. It also includes estimates for the memory and time requirements of SPICE2 for the same analysis. Fig.4.14 shows the waveforms produced in a "write-1-read-1" mode. The voltage at the storage node is indicated as A_{bit}.

The simulation time for this example was estimated to be approximately twenty times faster than for SPICE2. This estimate is based on the MOS model evaluation time and other overhead required in the circuit analysis [2]. The simulation could not be performed with SPICE2 as the memory requirements for the SPICE2 analysis exceeded 300,000 words on a CDC 6400 computer and other resources were not available. The reason that the analysis of this circuit does not show a better speed improvement is that a circuit-level analysis must be performed for each sense amplifier which is relatively time-consuming. Another factor is the parallel nature of the circuit. For the short period when the column line is switching, the fanouts of all the storage transistors in the column must be processed.

Since each sense amplifier is connected to one of these transistors via the row, for that period of time all sixteen sense amplifiers are analyzed by the program. The separate analysis of these circuits is more time-consuming than a single circuit analysis in this case due to overhead associated with the access to- and loading of- the separate sparse circuit matricies.

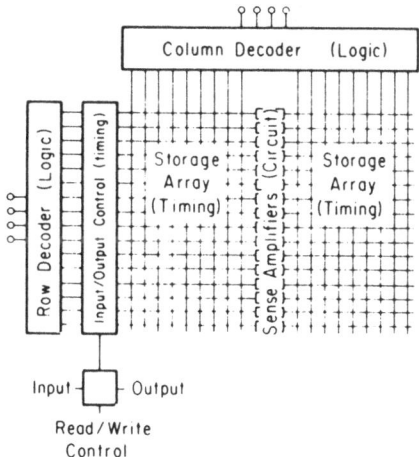

4.12 Block diagram of 256x1 bit RAM.

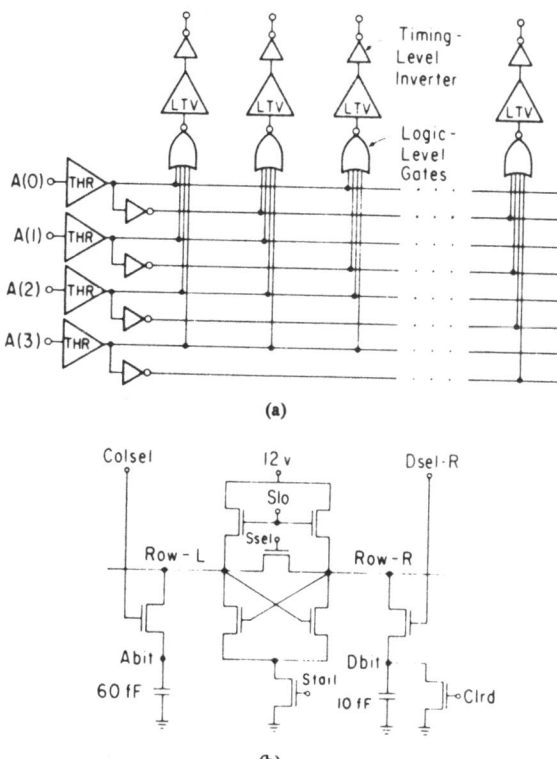

4.13 (a) Schematic of row and column decoders for RAM circuit.
Note the mixture of logic and timing elements.
(b) Schematic of RAM sense amplifier. Each sense
amplifier is processed as a separate circuit block

230

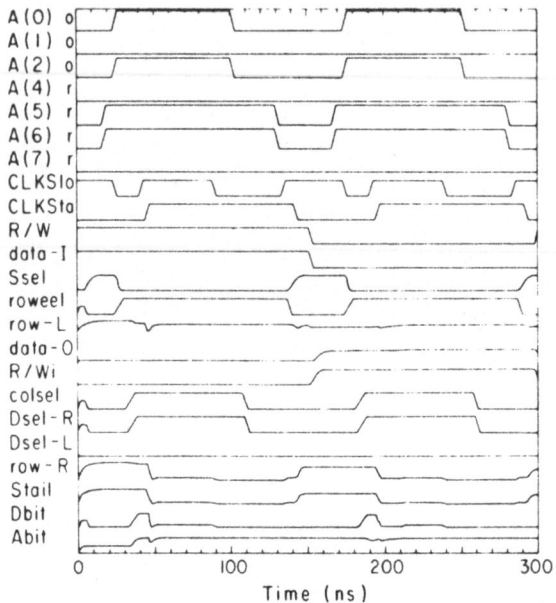

4.14 Waveforms prodoced by SPLICE for a write-1-read-1
operation in the RAM.

SPLICE STATISTICS FOR THE 256-BY-1 BIT DYNAMIC *RAM* AND
DIGITAL FILTER CIRCUITS

Dynamic RAM Circuit

459 nodes   96 circuit, 305 timing, 58 logic

565 elements 96 circuit, 411 timing, 58 logic

	SPLICE	SPICE 2
		(estimate)
Memory required for data (words)	8,805	220,000
Central processor time (seconds)	1,237	21,000

Digital Filter Circuit

393 nodes   393 timing

705 elements 705 timing

| Memory required for data (words) | 10,102 | 210,000 |
| Central processor time (seconds) | 475 | 50,000 |

Tbl.4.1 Statistics for RAM and Digital Filter analyses.
Note the times for SPICE are estimates.

Fig.4.15 shows the block diagram of an integrated circuit which performs a digital filtering function. This simulation involved the analysis of 700 MOS transistors which realize the blocks shown as solid lines in Fig.4.15 and was performed using a timing analysis. For this example, the delay functions (Z^{-1}) were implemented using RAM and were simulated with sources. The input data for SPLICE was derived directly from the integrated circuit layout file.

The waveforms produced by SPLICE are shown in Fig.4.16. The circuit is a serial, pipelined filter and the first 10 clock cycles are used to clear the circuit by clocking zeros into all the shift registers, adders, and multiplexers. A reset pulse is then issued and a data pulse is entered into the filter. Note that two power supplies (7 and 12 volts) are used in the circuit, hence the different levels seen at the output of the second adder $(Sum.B\,2)$.

Tbl.4.1 summarizes the statistics produced by SPLICE for this example. The simulation was performed for 4000ns with a 1ns MRT and the simulation time is estimated to be over 100 times less than would be required using SPICE2.

The savings obtained from the event scheduling scheme are illustrated in Fig.4.17. This plot shows the average number of events per node per unit of MRT and corresponds to the number of nodes processed by the scheduler at each timepoint. The time average of these events indicates that the circuit is less than 20% active and only during clock transitions does the circuit become highly active. This result supports the claim made earlier regarding the relatively low activity of large digital circuits even at the electrical level.

4.7. Extensions to SPLICE

A number of extensions to the SPLICE program are presently under development and some of them are outlined below.

As MOS technology advances and smaller circuit components can be made, the switching delay of an MOS transistor is decreasing. However, the average propagation delay along interconnect (both metal and polysilicon) is not decreasing at the same rate and in fact may increase. The interconnect can therefore play a large part in determining the circuit characteristics and must be modeled accurately in the simulator. With the table-driven scheme described above for SPLICE, the interconnect (or node) is already provided with its own data structure. What is necessary then is to extend these data structures to take into account the additional properties of

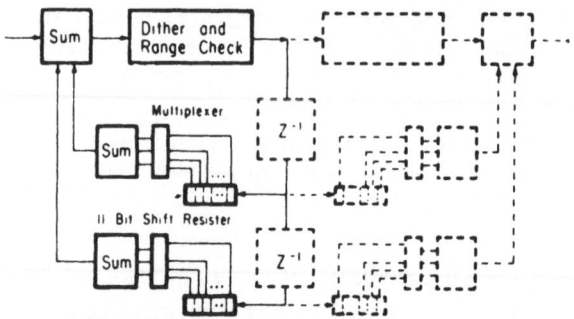

4.15 Block diagram of the Digital Filter circuit.

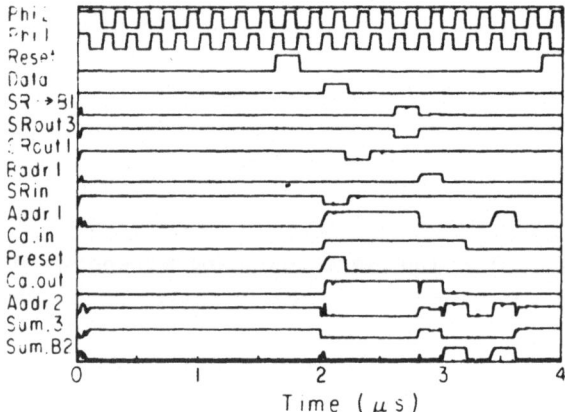

4.16 Output waveforms from SPLICE for the Digital Filter circuit

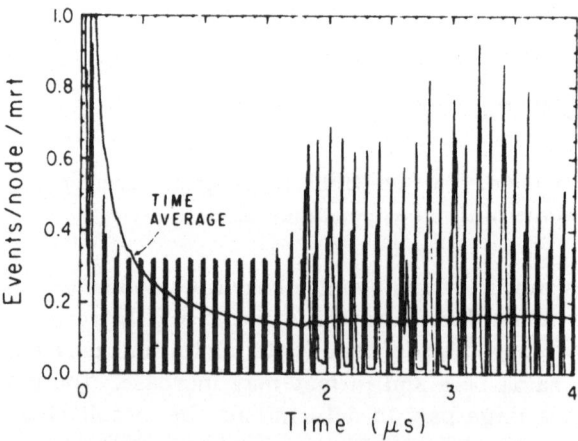

4.17 Circuit activity for the Digital Filter as observed by the SPLICE analysis.

nodes, such as independent delays between ports and decay times due to leakage for high impedance nodes. With the addition of a model table for each different type of node, the node data structure may be made identical to that of a timing or logic element. A simplified representation of the data structures for the circuit fragment of Fig.4.18(a) is shown in Fig.4.18(b). Note that although twice as many *elements* must now be scheduled for each set of events (nodes are now scheduled also), scheduler operation has been simplified substantially since only one fanout is connected to each node. The processing of an element e_{jk} would now proceed as follows:

```
obtain all input port signals, i_j ;
obtain all output port signals, o_j-^7 ;
evaluate the new outputs, o_j+ ;
for (each output, o_jk ) {
   if ( o_jk+ ≠ o_jk- ) {
      update o_jk ;
      check for hazards;
      schedule fanout node, n_jk , of o_jk ;
   }
}
```

And the processing of a node follows as:

```
obtain all fanin contributions, i_j^8 ;
obtain all fanout element inputs, o_j- ;
evaluate the new outputs, o_j+ ;
for (each fanout signal, o_jk ) {
   if ( o_jk+ ≠ o_jk- ) {
      update o_jk ;
      schedule its fanout element, e_jk ;
   }
}
```

An additional advantage of this modification is that signal coercions can now be handled directly by the node model. Since only one fanout element is present per node, the fanout port can have a signal type appropriate for the element it drives.

The SPLICE program is also being extended to nine-state logic simulation thus allowing more effective processing of MOS transfer gates at the logic level, as described in Section 3.

---

7. o_j is a set of two-element vectors now. Both the signal level and the impedance at the node represent the node output. Both values are used in all comparisons and operations.

8. i_j includes both voltage and impedance for electrical nodes, as for the element case above.

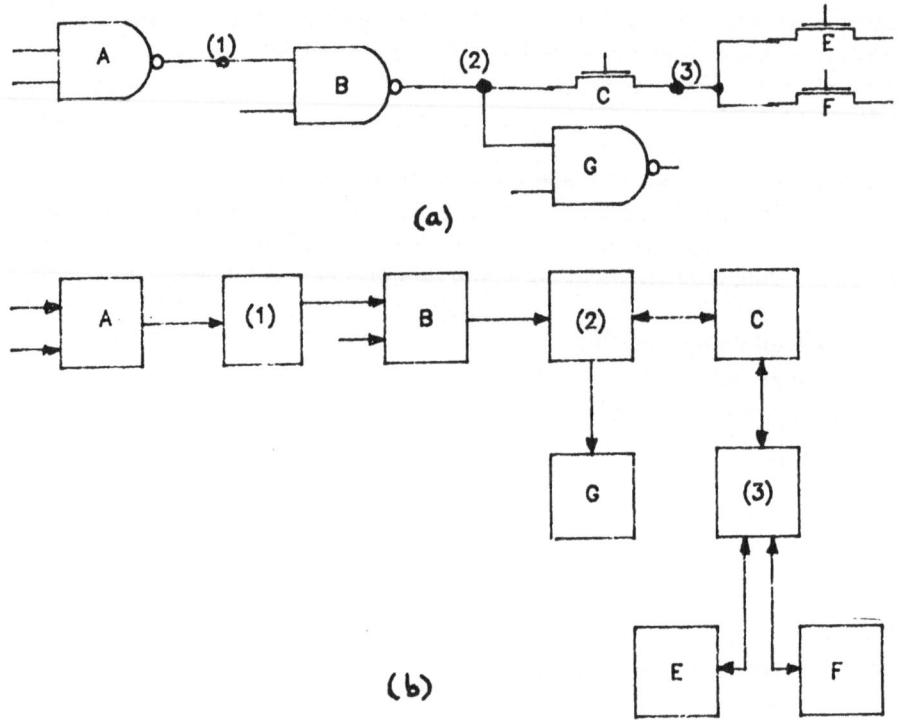

4.18 (a) Circuit fragment and (b) data structure schematic using node models.

5. ACKNOWLEDGEMENT

The author wishes to acknowledge the invaluable assistance in programming and preparing examples for this chapter provided by J. Straus and J Kleckner. Many thanks also to D. O. Pederson for his many useful suggestions and continuous encouragement.

6. REFERENCES

[1] L. W. Nagel, "SPICE2: A Computer Program to Simulate Semiconductor Circuits," ERL Memo No. ERL-M520, University of California, Berkeley, May 1975.

[2] A. R. Newton and D. O. Pederson, "Analysis Time, Accuracy and Memory Requirement Tradeoffs in SPICE2," 11th Asilomar Conference on Circuits, Systems and Computers, pp.6-9, Asilomar. Ca. Nov. 1977.

[3] N. B. Rabbat and H. Y. Hsieh, "A Latent Macromodular Approach to Large-Scale Sparse Networks," IEEE Trans. Circuits Syst., vol.CAS-23, pp.745-752, Dec.1976.

[4] N. B. Rabbat, A. L. Sangiovanni-Vincentelli and Hsieh, H. Y., "A Multilevel Newton Algorithm with Macromodeling and Latency for Analysis of Large-Scale Nonlinear Networks in the Time Domain," IEEE Trans. Circuits and Systems, Vol. CAS-26, No.9, Sept.1979.

[5] P. Yang, T. N. Trick and I. N. Hajj, Private Communication.

[6] A. L. Sangiovanni-Vincentelli, L-K. Chen, L. O. Chua, "An Efficient Heuristic Cluster Algorithm for Tearing Large-Scale Networks," IEEE Trans. Circuits and Systems, Vol.CAS-24, Dec. 1977.

[7] L. O. Chua and L. K. Chen, "Diakoptic and Generalized Hybrid Analysis," IEEE Trans. Circuits and Systems, Vol.CAS-23, No.12, pp.694-705, Dec.1976.

[8] A. L. Sangiovanni-Vincentelli, L-K Chen and L. O. Chua, "A New Tearing Approach - Node-Tearing Nodal Analysis," Proc. IEEE Int. Symp. Circ. and Syst., pp.143-147, Phoenix, Ariz. 1977.

[9] B. R. Chawla, H. K. Gummel and P. Kozak, "MOTIS - An MOS Timing Simulator," Trans. IEEE, Vol.CAS-22, No.13, pp.901-909, Dec. 1975.

[10] S. P. Fan, M. Y. Hsueh, A. R. Newton and D. O. Pederson, "MOTIS-C: A New Circuit Simulator for MOS LSI Circuits," Proc. IEEE Int. Symp. Circuits Syst., pp. 700-703, April 1977.

[11] N. Tanabe, H. Nakamura, and K. Kawakita, "MOSTAP: An MOS Circuit Simulator for LSI," Proc. IEEE Int. Symp. on Circ. and Syst., pp.1035-1039, Houston, Texas, April 1980.

[12] H. Shichman and D. A. Hodges, "Modeling and Simulation of Insulated Gate Field-Effect Transistor Switching Circuits," IEEE JSSC Vol.SC-3, pp.285-289, Sept. 1968.

[13] G.R.Boyle, "Simulation of Integrated Injection Logic," ERL Memo No. ERL-M78/13, University of California, Berkeley, March 1978.

236

[14] J. M. Ortega and W. C. Rheinboldt, "Iterative Solution of Nonlinear Equations in Several Variables," New York, Academic Press, 1970.

[15] A. R. Newton, "The Analysis of Floating Capacitors for Timing Simulation," Proc. Thirteenth Asilomar Conference on Circuits, Systems and Computers, Asilomar, Ca. Nov. 1979.

[16] G. De Micheli, A. R. Newton, and A. L. Sangiovanni-Vincentelli, "New Algorithms for Timing analysis of Large Circuits," Proc. IEEE Int. Symp. on Circ. and Syst, pp. 439-444, Houston, Texas, April 1980.

[17] M. Y. Hsueh, A. R. Newton and D. O. Pederson, "New Approaches to Modeling and Electrical Simulation of LSI Logic Circuits," Journees d'Electronique, pp. 403-413, Lausanne, Switzerland, 1977.

M. Y. Hsueh, A. R. Newton and D. O. Pederson, "The Development of Macromodels for MOS Timing Simulators," Proc. IEEE Symp. on Circuits and Systems, pp.345-349, New York, May 1978.

[18] A. E. Ruehli, A. Sangiovanni-Vincentelli, Private Communication.

[19] R. A. Rohrer, "Passive Algorithms and Stable Computation," to be published.

[20] C. A. Desoer and E. S. Kuh, "Basic Circuit Theory," McGraw-Hill, 1975.

[21] O. Nevanlinna and W. Liniger, "Contractive Methods for Stiff Differential Equations," IBM Research Report No. RC-7122.

[22] L. O. Chua and P. M. Lin, "Computer Aided Analysis of Electronic Circuits: Algorithms and Computational Techniques," Prentice Hall, NJ, 1975.

[23] "Standard Mathematical Tables," S. M. Selby, Ed., Chemical Rubber Co., Ohio, 1965.

[25] N. B. Rabbat, "Macromodeling and Transient Simulation of Large Integrated Digital Systems," Ph.D. Thesis, The Queen's University of Belfast, UK, 1971.

E. B. Kosemchak, "Computer Aided Analysis of Digital Integrated Circuits by Macromodeling," Ph.D. Thesis, Columbia University, USA, 1971.

[26] M. Y. Hsueh and D. O. Pederson, "An Improved Circuit Approach for Macromodeling Digital Circuits," Proc. IEEE Int. Symp. on Circ. and Syst., Phoenix, Arizona, pp. 696-699, April 1977.

[27] R. C. Baldwin, "An Approach to the Simulation of Computer Logic," 1959 AIEE Conference paper.

[28] S. Seshu and D. N. Freeman, "The Diagnosis of Asynchronous Sequential Switching Systems," IRE Trans. on Electronic Computers, pp.459-465, Aug. 1962.

J. H. Katz, "Optimizing Bit-Time Computer Simulation," Comm. ACM, Vol.6., No.11, pp.679-685, Nov. 1973.

[29] J. R. Duley and D. L. Dietmeyer, "A Digital Design System Language (DDL)," IEEE Trans. Computers, Vol.C-17, No.9, pp.850-860, Sept. 1968.

C. Y. Chu, "An ALGOL-like Computer Design Language," Comm. ACM, vol.8, no.10, pp.607-615, Oct. 1965.

M. Barbacci, "The ISPL Language," Carnegie Mellon University, Department of Computer Science, 1977.

Other references may be found in:

IEEE Computer, Vol.7, No.12, Dec. 1974

[30] L. Shalla, "Automatic Analysis of Electronic Circuits Using List Processing," Comm. ACM, Vol.9, pp.372-380, May 1966.

[31] J. S. Jephson, R. P. McQuarrie, and R. E. Vogelsberg, "A Three-Value Computer Design Verification System," IBM System Journal, No.3, pp.178-188. 1969.

[32] S.A.Szygenda and E.W.Thompson, "Digital Logic Simulation in a Time-Based, Table-Driven Environment. Part 1. Design Verification," IEEE Computer, March 1975, pp.24-36.

[33] M. A. Breuer, "General Survey of Design Automation of Digital Computers," Proc. IEEE, Vol.54, No.12, Dec.1966.

M. A. Breuer, "Recent Developments in the Design and Analysis of Digital Systems," Proc. IEEE, Vol. 60, No.1, pp12-27, Jan. 1972.

[34] R. C. Baldwin, "An Approach to the Simulation of Computer Logic," 1959 AIEE Conf. paper.

[35] LOGIS is a commercially available logic simulator written by F. Jenkins and available from the ISD timesharing company.

[36] G. R. Case, "SALOGS -- A CDC 6600 Program to Simulate Digital Logic Networks, Vol.1 - User's Manual," Sandia Lab Report SAND 74-0441, 1975.

[37] S. A. Szygenda and E. W. Thompson, "Modeling and Digital Simulation for Design Verification and Diagnosis," IEEE Trans. Computers, vol. C-25, pp. 1242-1253, Dec.1976.

[38] W. Sherwood, "A Hybrid Scheduling Technique for Hierarchical Logic Simulators or 'Close Encounters of the Simulated Kind'," Proc. 16th Design Automation Conference, pp249-254, San Diego, Ca. June, 1979.

[39] D. D. Hill and W. M. Van Cleemput, "SABLE: Multi-Level Simulation for Hierarchical Design," Proc. IEEE Int. Symp. on Circ. and Syst., pp.431-434, Houston, Texas, April, 1980.

[40] G. Arnout, H. De Man, "The Use of Threshold Functions and Boolean-Controlled Network Elements for Macromodelling of LSI Circuits," IEEE J. of Solid-State Circuits, Vol.SC-13, pp.326-332, June 1978.

[41] A. R. Newton, "The Simulation of Large-Scale Integrated Circuits," ERL Memo No. ERL-M78/52, University of California, Berkeley, July 1978.

A. R. Newton, "Techniques for the Simulation of Large-Scale Integrated Circuits," Trans. IEEE, vol.CAS-26, No.9, pp.741-749, Sept. 1979.

[42] N. N. Nham, et. al., "The Mixed Mode Simulator," Proc. 17th Design Automation Conference, Minneapolis, Minn., June 1980.

[43] J. D. Crawford, "A Unified Hardware Description Language for CAD Programs," ERL Memo No.UCB/ERL M79/64, University of California, Berkeley, May 1978.

[44] The SPLOT program was developed by J. Straus, Electronics Research Laboratory, University of California, Berkeley.

[45] B. T. Preas, B. W. Lindsay and C. W. Gwynn, "Automated Circuit Analysis Based on Mask Information," 13th Design Automation Conference, pp.309-317, 1976.

[46] M. Y. Hsueh and D. O. Pederson, "Computer-Aided Layout of LSI Circuit Building Blocks," Proc. IEEE Int. Symp. on Circ. and Syst., pp.474-477, Tokyo, Japan, 1979.

[47] "TESTAID-III Programming and Operating Manual," Manual Part No. 91075-93008, Hewlett Packard, Loveland Co. 80537.

[48] J. McClure, "Frequency Domain Simulation of Very Large Digital Systems," Proc. IEEE Int. Symp. on Circ. and Syst., pp.424-430, Houston, Texas, April, 1980.

[49] T. M. McWillaims, "Verification of Timing Constraints in Large Digital Systems," Proc. IEEE Int. Symp. on Circ. and Syst., pp.415-423, Houston, Texas, April, 1980.

[50] T. I. Kirkpatrick and N. R. Clark,"PERT as an Aid to Logic Design," IBM journal of Research and Development, pp.135-141, March 1966.

[51] Behavioural simulation may be performed using programming languages which have a notion of concurrency. Simula, Modula, Algol and Pascal have all been used in this mode.

[53] G. R. Case, "SALOGS -- A CDC 6600 Program to Simulate Digital Logic Networks, Vol.1 - User's Manual," Sandia Lab Report SAND 74-0441, 1975.

P. Wilcox and A. Rombeck, "F/LOGIC - An Interactive Fault and Logic Simulator for Digital Circuits," Proc. 13th Design Automation Conference, pp. 68-73, 1976.

Several large-scale logic and fault analysis simulators are commercially available, such as CC-TEGAS3, D-LASAR, and LOGCAP.

[54] W. T. Weeks, A. J. Jiminez, G. W. Mahoney, D. Mehta, H. Qassemzadeh, and T. R. Scott, "Algorithms for ASTAP - A Network Analysis Program," Trans. IEEE, vol.CT-20, pp.628-634, Nov. 1973.

SIMULATION OF ANALOG SAMPLED-DATA MOSLSI CIRCUITS

H. De Man, J. Vandewalle, J. Rabaey

Catholic University of Leuven, Belgium.

1. INTRODUCTION AND HISTORICAL OVERVIEW

Although the principle of using time variant networks containing switches, amplifiers and capacitors for filtering purposes has been proposed by Fettweiss [1] more than a decade ago, only recently it has been put into practice by exploiting MOSLSI technology [2 3].
Also, based on the use of digitally controlled MOS switches and capacitors, charge redistribution A/D and D/A convertors [4] are now used very intensively.
Based on the same principles also waveform generators [5 6] and analog computing systems [7] have been proposed.
Today analog MOS circuits are in extensive use in a wide variety of signal processing applications ranging from telecommunications to consumer products. It is not exagerated to speak of a breakthrough of analog LSI which is fully compatible with the digital circuitry on the same chip.
A nice example is the Intel 2920 analog microprocessor [7] which combines the best of analog and digital circuitry on the same chip.

The design of switched capacitor (S.C.) networks imposes a new world of thinking since such networks are time variant networks in which sampled and continuous operation is often mixed as is illustrated in Fig. 1.
The path ① through C_3 is continuous from input source $v_i(t)$ to output $v_o(t)$ whereas path ② is a sampled path.
Moreover the sampling frequency f_s, defined as the frequency of the clock $\emptyset(t)$ driving the switches S_1 and S_2, is usually orders of magnitude larger than the signal frequency f of $v_i(t)$.
All these characteristics would make traditional time domain simu-

lation of such networks very costly while at the same time sinusoïdal analysis in the complex plane is not as readily applicable.

Nevertheless, time, frequency, sensitivity and noise analysis
of such networks is of outmost importance since calculation of
such effects as stray capacitance, clock-phasing, finite op-amp
gain and bandwidth, aliasing effects, component sensitivity, noise
performance, non-linearities etc... is almost impossible to do by
hand calculations. At best only educated guesses are possible
which need extensive verification by CAD.

In this contribution a general unified theory for the computer aided analysis of S.C. networks as well as its practical implementation in the mixed-mode simulator DIANA will be discussed.

As is often the case when a new field such as S.C. networks
opens up, a number of parallel and often independent approaches
to the mathematical formulation of these networks have taken place.
A first formulation of modified nodal equations for S.C. networks
is due to Brglez [8 9]. Use is made of a Laplace transform formulation and frequency domain analysis requires the set up and mu-
tiplication of a number of complex switching matrices.
Liou and Kuo provide a state variable formulation of two-phase
S.C. networks and consider the possibility of time-domain solution
using one MNA matrix per switch state with branch relations for
all capacitors and voltage sources [10] . This leads to very large
(numerically ill conditioned) sparse matrices.
Almost simultaneously Y. Tsividis [11] proposed a general MNA-like
time-domain formulation with reduced number of variables together
with methods for general frequency domain solution based on the use
of switching matrices and their products. No computational aspects
for computer implementation of this method were considered but
recently Fang and Tsividis [12] have presented a method for generating compact forms of modified nodal time domain equations
without need for pivoting. On the other hand a practical, user
oriented CAD program for time and frequency analysis of switched
capacitors was first proposed by the authors [13 14 17]. The
program is based on a simple MNA formulation of switch branches in
terms of Boolean Controlled network elements [15]. Matrix set-up
is straightforward and compatible with all commonly used circuit
CAD techniques. Frequency domain simulation is based on Fast
Fourier Transform techniques (FFT) and encompasses all possible
cases in resistorless S.C. networks.
The MNA matrices can be L U decomposed without modification of the
original techniques mentioned in [16]. The technique has been implemented in DIANA [17] . Since the technique is compatible with
the normal time domain analysis mode of this program, also detailed
analysis of S.C. using resistive elements, nonlinear capacitances,
opamps etc... is feasible in so called mixed-mode operation [17].
Efficient frequency and sensitivity analysis are possible based on
the concept of adjoint S.C. network introduced by the authors [18].

At present also noise analysis is being implemented.
Brglez [19] has presented another CAD program for time and frequency domain called SCOP based on his theoretical work discussed in
[8]and[9]. It can perform optimization by a costly perturbation
method without use of sensitivities.

All the above techniques make use of Fourier transform methods
in one form or another but parallel to this way of thinking other
authors propose an 'equivalent' circuit approach. They derive an
equivalent circuit of impedances in z-transform and then solve
the resulting augmented circuit using AC analysis techniques [20]
[21 22]. This method has been generalized in [23] by introducing
a generalized circulator element. Since the theory in[23] covers
or generalizes most of the above results it will be taken as the
basis for the rest of this contribution.

In section 2 the modified nodal analysis time domain formulation of ideal and nonideal S.C. networks is discussed and illustrated with DIANA examples. A technique for mixed-mode S.C. simulation will be considered.

In section 3 frequency analysis techniques are discussed on
the basis of generalized z-transform methods. The concept of adjoint S.C. network is introduced in section 4 and efficient FFT
based frequency analysis is proposed and illustrated using the
DIANA program. Techniques for direct frequency analysis are considered. Also sensitivity and noise analysis based on the adjoint
network concept are considered. Sensitivity analysis using DIANA
will be demonstrated.

2. TIME DOMAIN MNA FORMULATION OF S.C. NETWORKS.

2.1. Ideal linear and nonlinear S.C. networks.

In this subsection we propose a formulation of ideal S.C. network
equations which are naturally suited for direct implementation in
MNA based circuit simulators. It is based on charge conservation
and Boolean Controlled switches [15]. No matrix multiplications [9]
nor topological partitioning [12] is necessary. Only one matrix
with Boolean switching elements is used. Matrix compaction follows
from using so called <u>composite switch branches</u>.

Let us start the discussion by first considering linear networks containing ideal switches, capacitors, independent voltage
and <u>charge</u> sources as well as the four dependent sources (VCVS,
QCVS, VCQS, QCQS). The nonlinear case will be discussed later.
The switches S (Fig. 2) are controlled by Boolean clock variables
[15] $\emptyset_S$ (t)$\in \{0,1\}$ whereby $\emptyset_S=0$ corresponds to an open and
$\emptyset_S=1$ to a closed switch. The time is partitioned into time slots
$\Delta_k = (t_k, t_{k+1}]$ such that the clock signals do not vary in Δ_k

244

i.e. $\emptyset_S(t) = \emptyset_{S,k}$ for $t \in \Delta_k$.

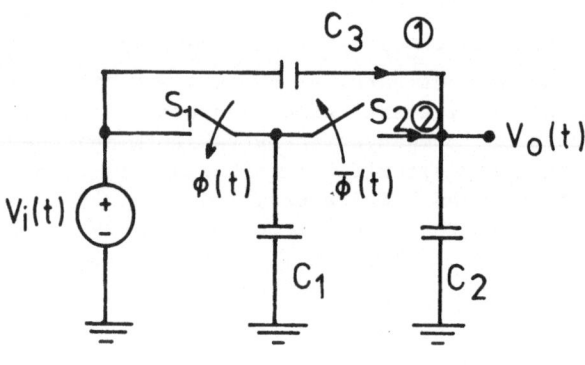

Fig. 1

Referring to Fig. 2., let $v_{i,k-1} = v_i(t_k^-)$ be the voltage at node i at the end of Δ_{k-1} and let $v_{i,k}(t)$ represents the voltage at node i for some $t \in \Delta_k$. Similarly let $q_{j,k}(t)$ be the charge transferred in branch j from $t=t_k^-$ to $t \in \Delta_k$ or :

$$q_{j,k}(t) = \int_{t_k^-}^{t \in \Delta_k} i_j(t) \, dt \qquad (1)$$

Now similar to the original MNA formulation [16] one can write Kirchoffs "charge" equations in all nodes and introduce branch relations for voltage sources and the Boolean controlled switches [15]. Referring to Fig. 2. one obtains e.g. for node i :

$$Cv_{i,k}(t) - Cv_{j,k}(t) + q_{A,k}(t) + q_{E,k}(t) + \emptyset_{s,k}q_{s,k}(t)$$

$$= Cv_{i,k-1} - Cv_{j,k-1} \qquad (2)$$

This gives rise to the first row in the matrix in Fig. 2b. Furthermore all ideal switches, voltage and controlled sources except the VCQS require a branch relation.

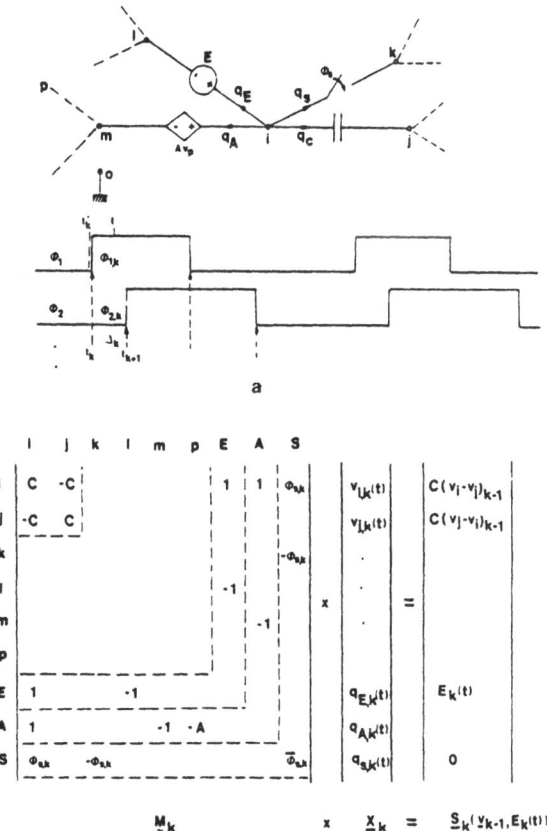

Fig. 2

The branch relations for E and A are trivial from Fig. 2b. Important is the implementation of switch branch S as follows :

$$\phi_{S,k} v_{i,k}(t) - \phi_{S,k} v_{j,k}(t) + \overline{\phi}_{S,k} q_{S,k}(t) = 0 \qquad (3)$$

ϕ_S denotes the Boolean complement of ϕ_S.
From eqs. (2) and (3) it is clear how S is implemented into the matrix M_k. As shown by the dotted lines in Fig. 2b. every element of the S.C. network has a simple entry (stamp) into the matrix M_k which is nothing but the widely used MNA [16] matrix now also including controlled switches and operating on a vector X_k of voltages and charges.
Fig. 3. shows a number of voltage controlled switch branches, their implementation in the M_k matrix and their description in the DIANA

246

program.

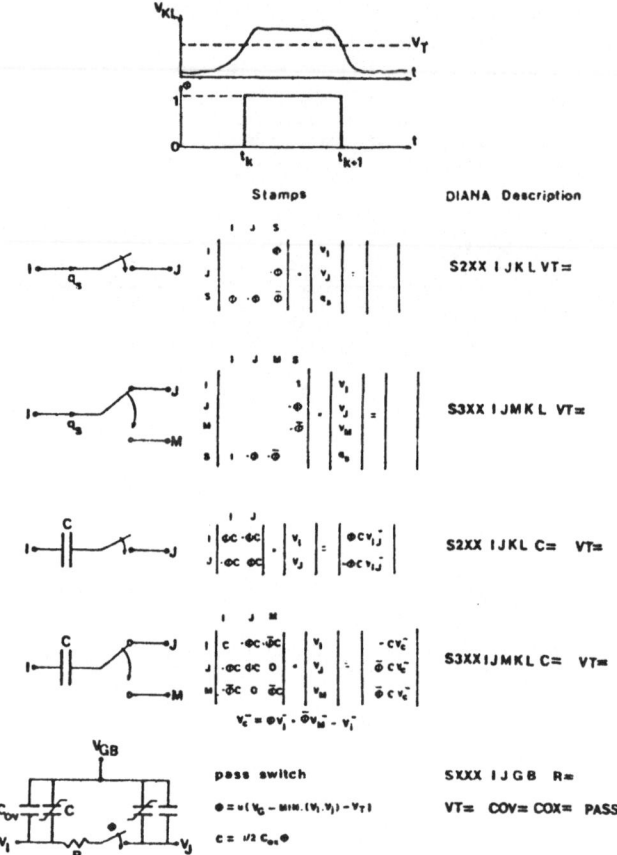

Fig. 3

The upper four branches are for SC networks. The pass switch is a
first order MOS triode switch model for studying clock feedthrough
and switching transient effects in classical transient analysis
(see further). In contrast to previously published methods these
switch models give a <u>direct</u> link between circuit design and <u>single</u>
MNA matrix. Clearly no restriction is imposed on clock sequences
which can be derived from e.g. digital circuit parts as was discussed in the paper on mixed mode simulation using DIANA [24]. The
introduction of all other elements discussed above is trivial.
Notice further that switches in series with capacitors give rise to
the elimination of a nodal <u>and</u> a branch equation and therefore lead

to very compact Boolean controlled matrices. We call such branches "composite switch branches". They are present in nearly all SC networks and should be preferentially used as they lead to much faster simulation. Often the matrices resulting from such a technique have a dimension which is smaller than the ones proposed in other methods [9 10 12]. Perhaps a combination of the method proposed in [12] with the composite branch, can still lead to more compact matrices but further research is necessary to prove this. In closed form the matrix equation can be written as :

$$\underset{\sim}{M}_k \underset{\sim}{X}_k = \underset{\sim}{S}(\underset{\sim}{v}_{k-1}, \underset{\sim}{s}(t)) \tag{4}$$

Eq. 4 is valid for a continuous independent source term : $\underset{\sim}{s}^T(t) = \left\{\underset{\sim}{w}^T(t), \underset{\sim}{u}^T(t)\right\}$. $\underset{\sim}{w}(t)$ is a charge excitation vector and $\underset{\sim}{u}(t)$ a voltage source excitation vector. A particular case of interest are so called piecewise constant sources which are constant within each Δ_k. It is clear that in such case the response is also piecewise constant and given by :

$$\underset{\sim}{M}_k \underset{\sim}{X}_k = \underset{\sim}{S}(\underset{\sim}{v}_{k-1}, \underset{\sim}{s}_k) \tag{5}$$

Returning now to Fig. 2b. for piecewise constant inputs it is clear that the capacitor stamp is equivalent to the widely used backward Euler integration companion model [25] with a unit time-step h = 1. This is also equivalent to the formulation of Liou and Kuo [10]. They use a "resistor" 1/C in series with a voltage source v_c^- instead of the usual "conductance" C in parallel with charge source $C v_c^-$ (this greatly increases their MNA matrix dimension). The above then leads to a very simple simulation procedure for any ideal S.C.-network with piecewise constant input :

Procedure 1 : any S.C. network with piecewise constant input can be simulated in the time domain in an existing MNA based circuit simulator as follows :

1. Set up MNA matrix as usual, introduce switch branches as discussed. Reorder for $\underset{\sim}{L} \underset{\sim}{U}$ decomposition as proposed in [16].

2. Use backward Euler with h = 1.

3. Start from initial values $v_c(o)$ for all C's. Default is zero.

4. Bypass all time step control. Real time coincides with clock transitions $t_1, t_2 \ldots t_k \ldots$

5. For all k set up and solve (5) by the usual sparse matrix techniques.

Procedure 1 is built into the DIANA program as an option.
A careful look at Fig. 2b. also reveals that the method is fully compatible with the normal transient mode of simulation if backward Euler integration with $h \neq 1$ is used and all q are taken as currents i.
Therefore in DIANA all modes of simulation from pure detailed transistor transient up to the above SC simulation are possible.

Since DIANA also allows for mixed analog, timing and logic simulation a full mixed analog-digital MOSLSI circuit can be simulated as will be demonstrated in the examples below.
Notice the following :
1. The above method can easily be extended towards networks containing non-linear capacitors and dependent sources. For example let the capacitor in Fig. 2a. be nonlinear and described by $q(v_c)$, then eq. (2) becomes :

$$q(v_{c,k}(t)) + q_{A,k}(t) + q_{E,k}(t) + \emptyset_{S,k} q_{S,k}(t) = q(v_{c,k-1}) \quad (6)$$

This is a non-linear equation in $v_{c,k}(t)$ which can be solved by the Newton-Raphson technique as used in all CAD programs. This leads to essentially the same method as for non-linear backward-Euler with unit time step for capacitors. Also non-linear gain of opamps can be introduced. A program which e.g. handles backward-Euler non-linear C's allows for direct simulation of non-linear SC networks using the procedure discussed above.

2. All switched capacitor networks for filtering are driven by T-periodic clocks i.e.

$$\emptyset_i(t+T) = \emptyset_i(t) \; \forall \, t, i \quad (7)$$

or all time slots $\Delta_{k+\ell N}$ for $\ell = 0,1,2,\ldots$ with the same relative position k in a period (called phase k) have the same duration and the clock values are the same i.e. $\emptyset_i(t) = \emptyset_{ik}$ for $t \in \Delta_{k+\ell N}$ for some ℓ. There are N phases per clock period T.
In such case it is clear that, for a linear S.C. network the matrices M_k are also T-periodic and in principle if we store the N $L \, U$ decompositions then only one forward and backward substitution per clock-phase is necessary for time domain simulation which would greatly enhance computational speed. This technique has not yet been implemented in DIANA since also nonlinear circuit solving is provided.

Applications and examples of time-domain simulation.

Time domain simulation of ideal S.C. networks has, amongst others, the following applications :

1. Computation of impulse response of filters;
2. Verification of A/D and D/A response;
3. Study of offset and leakage current drift;
4. Study of non-linear distortion in S.C.-circuits;
5. Study of clock-feedthrough effects (drift- and amplitude) etc...

Points 1, 3 and 4 are illustrated in the examples below. Points 2 and 3 are discussed in section 2.2.

A) Impulse response calculations-offset drift.
Fig. 4. shows a well known [26] 5th order elliptical PCM filter. All switches are 2 phase clocked and use is made of LDI integrators.

Fig. 4b. shows the input to DIANA. Notice the extensive use of
composite switch branches.   Op-amps are modeled by VCVS with a
gain A = 1000.
The matrix dimension is 22 x 22.   Matrix sparseness is 17 %. A dynamic memory reservation of 5.72 K bytes is needed. The impulse response is computed for a unit voltage excitation during the first
input phase with input switch connected to node ①.
Notice that the impulse response for the other clock phase is zero.
The number of time points requested NOUT = 2000.

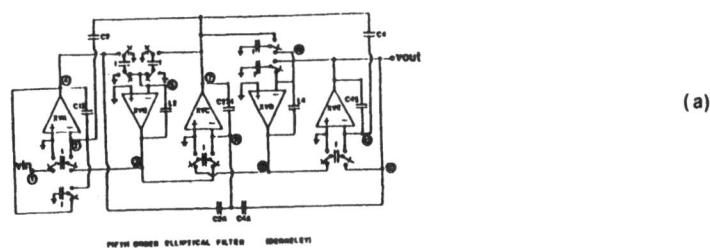

(a)

```
ESAT ELLIPTICAL FILTER
. TRAN NOUT=2000 DELTA=1
. OPTIONS -OUT
. SC IN=1 OUT=10 CL=CL  FREQ
, CLOCKING
INPUT CL V-T=0 0U 2 0.78125U 2 3.906250 0 4 6875U 0 7 8125U CYC
, CIRCUIT
C12 4 2 8.4701
C234 7 11 14.707
CL2 5 3 5.0436
CL4 8 9 6.848
C45 10 12 5.3381
C4 12 7 1.40804
C2 7 2 4.1475
C2A 11 4 4.1475
C4A 11 10 1.40804
XVA 4 0 0 2 1K
XVB 3 0 0 5 1K
XVC 7 0 0 11 1K
XVD 9 0 0 8 1K
XVE 10 0 0 12 1K
S31 0 4 2 CL 0 C=1 VT=1
S321 CA 1 0 CL 0 C=2 VT=1
S322 CA 3 2 CL 0 C=2 VT=1
S331 CC 0 4 CL 0 C=2 VT=1
S332 CC 5 0 CL 0 C=2 VT=1
S341 CD 0 7 CL 0 C=2 VT=1
S342 CD 5 0 CL 0 C=2 VT=1
S351 CH 9 0 CL 0 C=2 VT=1
S352 CH 3 11 CL 0 C=2 VT=1
S36 0 8 7 CL 0 C=1 VT=1
S37 0 8 10 CL 0 C=1 VT=1
S391 CG 9 0 CL 0 C=2 VT=1
S392 CG 10 12 CL 0 C=2 VT=1
END
```

(b)

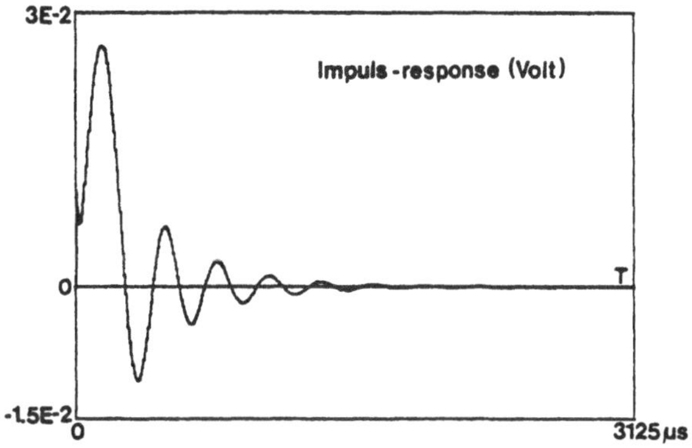

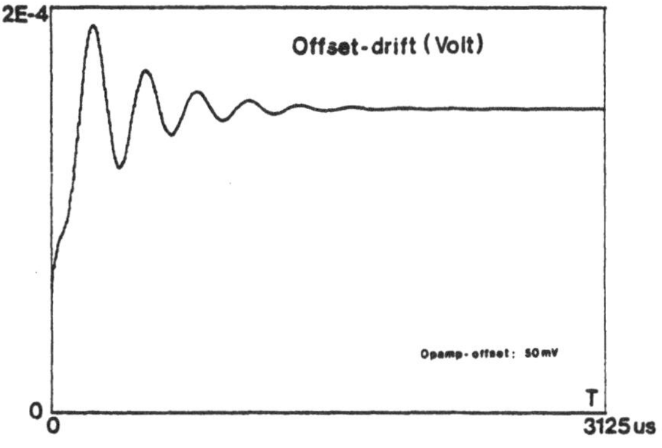

Impuls-response (Volt)

3E-2

0   T

-1.5E-2
0   3125 µs

(c)

Offset-drift (Volt)

2E-4

Opamp-offset: 50mV

0   T
0   3125 us

Fig. 4

Fig. 4c. shows the computed impulse response by the DIANA program.
The computation of the impulse response plays an important role
in the frequency response calculations to be discussed in next
section.
CPU time for this example is 2 min. 30 sec. on VAX 11/780.

B) Offset drift effects.

MOS op-amps usually have a large offset voltage (ca. 10...100mV).
This offset can cause drift of the operating point of op-amps in
a S.C. filter system. These effects are difficult to predict but
can easily be simulated in the S.C. time domain mode.
Consider for example the circuit in Fig. 5a. in which a potential
method to avoid   large capacitor ratios is tested.

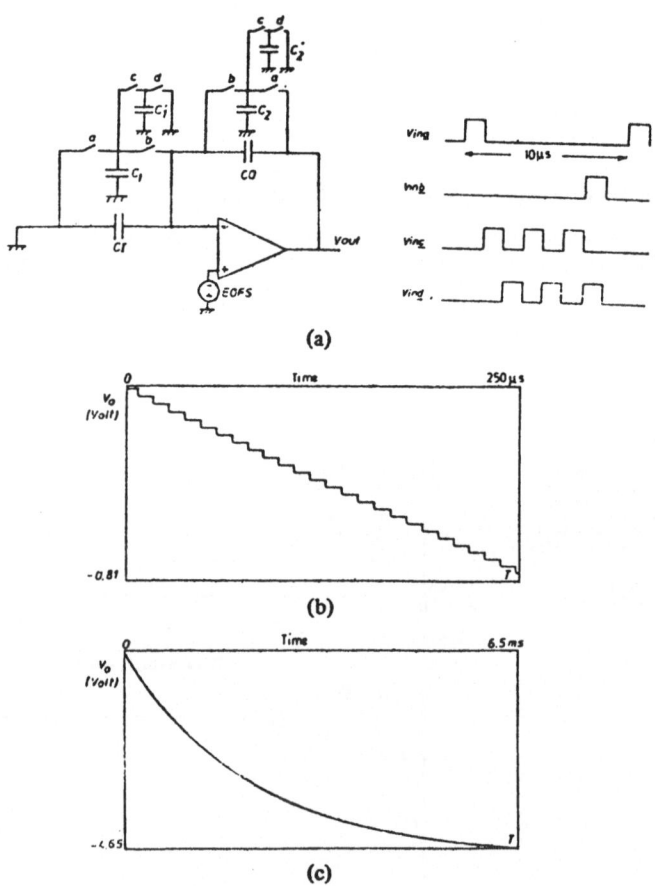

(a)

(b)

(c)

Fig. 5

Within the basic clock cycle ($V_{in\ a}$, $V_{in\ b}$) a charge redistribution using clocks ($V_{in\ c}$, $V_{in\ d}$) takes place on (C_1, C'_2) comparable in magnitude to C_I and C_0. We want to study the influence of op-amp offset EOFS = 50 mV.

Fig. 5b. shows the DIANA output plot of a <u>classical transient</u> analysis over 25 clock periods using PASS switch models (see Fig. 3.). 4o points/basic clockcycle have been used (1000 time steps).

Fig. 3c. shows the result of the new MNA method over 650 clock periods, calculating one solution per time slot. The simulation reveals −4.65V drift due to 50 mV offset of the op-amp. This result was discovered by CAD and is caused by inadequate feedback of output offset errors by redistribution on (C_2, C'_2). Notice that the new method is ca. 26 times faster than classical transient analysis as both run times are the same (20 sec. CPU on IBM 370-158). If a nonlinear MOS model had been used, the difference would be ca. 100...150. Notice also that no problem occurs due to the complicated multiphase clock sequence.

C) Simulation of non-linear S.C. networks.
As discussed above, the MNA method allows also for time domain simulation of nonlinear S.C. networks. This method can be of interest to study <u>drift effects due to non-linearities</u> or to compute the distortion from a discrete <u>Fourier analysis</u> of a steady state sinusoïdal response, as can be also done for linear circuits in SPICE.

A simple example of such a simulation using DIANA is shown in Fig. 6.

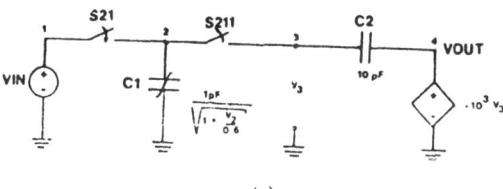

(a)

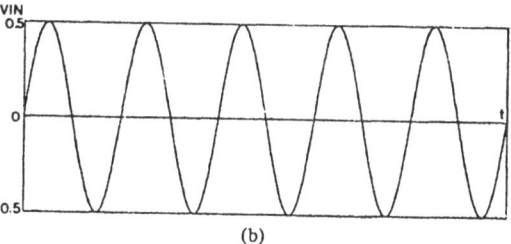

Fig.6

(b)

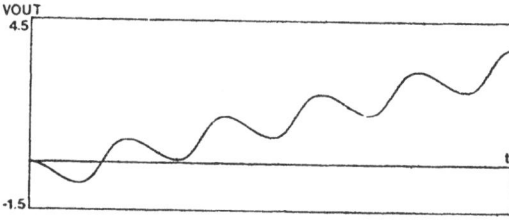

(c)

Shown is an integrator with a non-linear n-MOS junction
sampling capacitor taken from the built-in MOS model in DIANA.
The switches S_{21} and S_{211} are operating in opposite phase.
From the output plot we see that a 0.5 Volt amplitude sinusoïdal
input causes a drift and distortion in the output of the integrator. During the negative excursions (forward biased junction capacitor) more charge is exchanged than during the positive excursions. This example clearly demonstrates the possibilities and
ease of simulation of ideal S.C. networks in the time domain using
the controlled switch branch concept in the MNA matrix.

2.2. Time domain simulation of second order effects in non-ideal
      S.C. networks. Top down design.

The above discussed method is valid for S.C. networks not containing time constants introduced by resistive elements. In the
method it is basically assumed that immediately after a clock transition the circuit is in an equilibrium state to be sampled in the
next clock phase.
Notice that there exists a striking similarity between this situation and a synchronous digital system. Both are time partitioned
and therefore allow for a top-down design procedure whereby first
the principles of the system, built of "delay-less" (resistorless)
elements, is verified. For a digital system this can be done using
e.g. RTL simulation. For analog sampled circuits the above discussed simulation (one point/clock phase) can be used. We call
this top level design. Once this has been checked, the design can
be refined to include delay elements. For analog S.C. networks
this means resistors, transistor-models for switches, one-pole
op-amp models and perhaps, if necessary, the detailed op-amp
simulation in one or several filter sections.

In analogy to the concept of mixed-mode simulation from RTL
level to circuit level for digital circuits it would therefore
be advantageous to have mixed-mode simulation for analog sampled
MOSLSI circuits.

This can easily be done since the above MNA formulation for
ideal S.C. networks is fully compatible with traditional time
domain simulation.

This has been done by implementing the ideal S.C. simulation
mode within the mixed-mode simulation program DIANA [17] which
then has four operation modes shown in Fig. 7.
Mode I discussed above allows for conceptual verification. The
other modes have been discussed before [24]. Timing and logic mode
allow for simulation of S.C. networks together with digital control
circuitry. An example of this has been presented in [24].

Resistive effects such as finite bandwidth effects in op-amps and clock-feedthrough effects can be studied by using simulation mode II whereby of course now regular stepsize integration has to be used, which increases computer time.

DIANA PROGRAM

MODE	PRIMITIVES	VARIABLES	ANALYSIS
I) SAMPLED-DATA	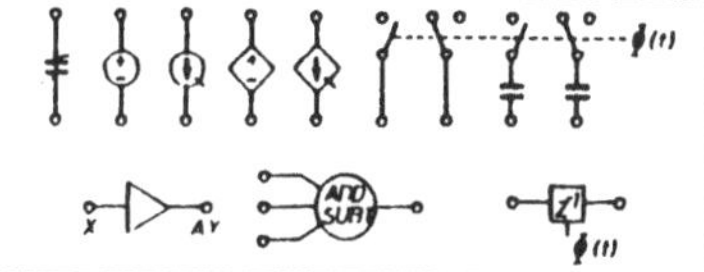	v, q, t, w	TIME (LIN. & NONLIN.) FREQUENCY SENSITIVITY NOISE (ROUNDOFF)
II) (SUB) CIRCUIT	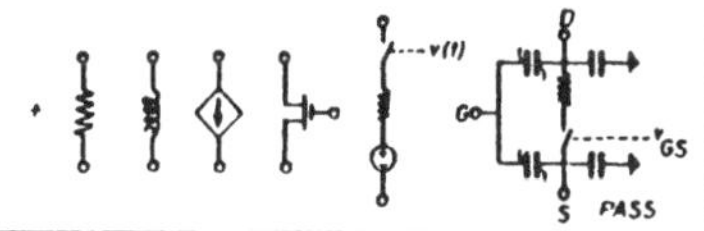	V, I, t	TIME, DC
III) TIMING	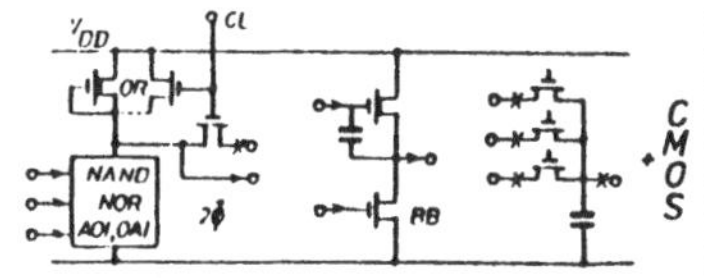	V, I, t	TIME
IV) LOGIC	(N)AND; (N)OR; (N)EXOR; AOI; OAI; J-K FF & MACRO'S	1, 0, X, U, D, t	TIME

Fig. 7

As an example Fig. 8. shows a simulation of the same Elliptical filter as shown in Fig. 4a. The input voltage is zero but all switches have been substituted by MOS pass-switches.available in circuit simulation mode. The op-amps are single pole amplifiers with gain A = 1000 and bandwidth f_n = 4.5 MHz. A clock of 128 kHz and a swing of (+5, -5V) is shown in the upper plot. It drives the pass switches and causes clock-feedthrough through the C_{gs} and C_{gd} of the MOS transistors.
Shown on Fig. 8. are simulated results of this feedthrough effects

254

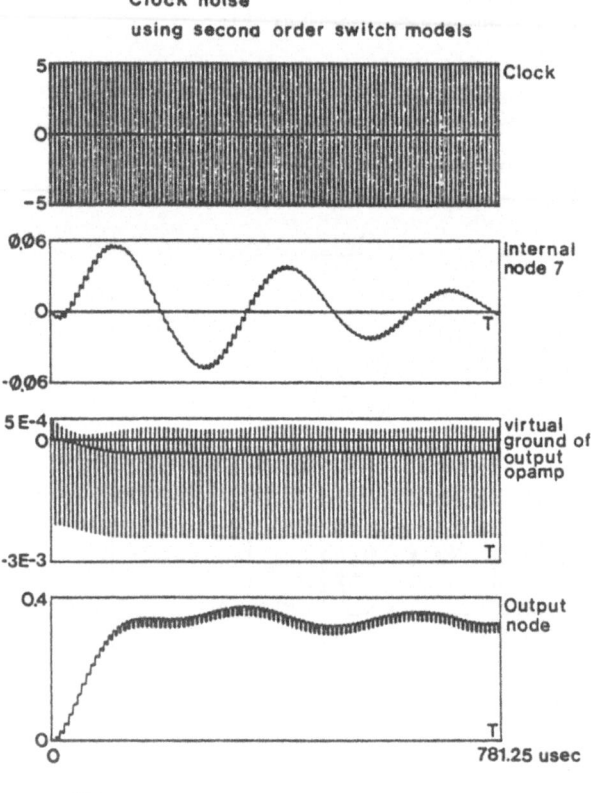

Fig.8

on internal node ⑦ (Fig. 4a.), the virtual'ground of op-amp
XVE (Fig. 4a.) and the output node ⑩.
One can clearly see that the clockfeedthrough causes an output DC
offset of ca. 0.35 Volts. The spikes at the input of the op-amp
result from instantaneous resistive voltage division when the MOS
transistors are switched on as well as from the finite response
time of the op-amp's.

Simulation over 1000 time points requires only 3 min. 30 sec.
CPU time on a VAX 11/780 under VMS operating system.

An other example of a <u>mixed-mode</u> simulation, whereby also
logic simulation mode IV is used, is shown in Fig. 9.

Fig. 9

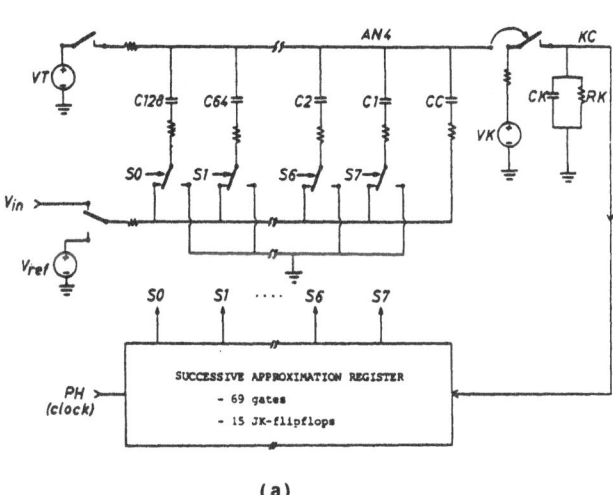

(a)

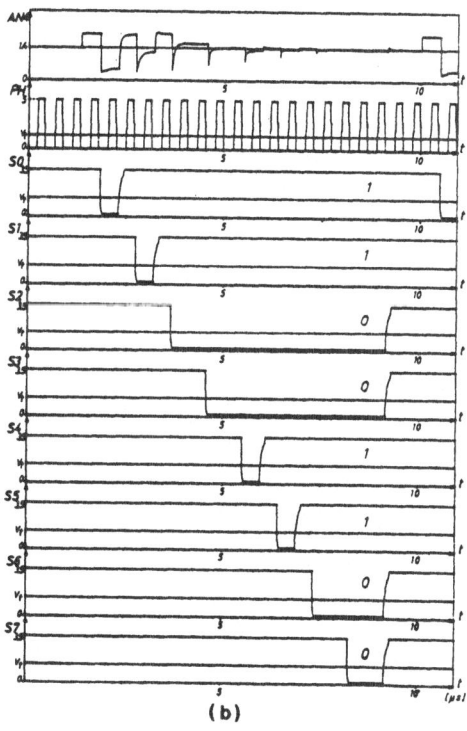

(b)

Fig. 9a. is an 8 bit charge redistribution A/D converter. In this example the switches are non-ideal resistive switches and on a conceptual level, the comparator is modeled by a controlled switch. The successive approximation switches $S_0...S_7$ are controlled by a successive approximation register consisting of 69 gates and 15 JK flip-flops available in mode IV in DIANA.
Fig. 9b. shows the simulation results for the comparator input AN4, the clock input and the logic output voltages $S_0...S_7$ driving the switches. One full conversion cycle of 12μs has been simulated over 600 time points in 8.3 sec. CPU on an IBM 3033.

From this example one can clearly see how the resistive effects of switches on the conversion cycle can be studied.

Sofar we have discussed time domain simulation of ideal and non-ideal S.C. networks, including mixed-mode analog-digital simulation. In the next paragraph frequency domain CAD analysis for T-periodic ideal and non-ideal linear S.C. networks is discussed.

3. THEORY OF FREQUENCY DOMAIN ANALYSIS TECHNIQUES FOR CAD.

In this section first the theoretical framework of the frequency analysis of ideal linear S.C. networks will be given. The theory is based on [23] which is a generalization of a number of other theories proposed earlier [8...11] , [20...22].
Starting from the z-domain equations for S.C. networks direct frequency domain as well as FFT based methods for CAD will be derived. Examples will be given in section 4 after the introduction of the concept of adjoint S.C. network.

3.1. The z-transform equations of S.C. networks.

Assuming the definitions and symbols defined for an ideal T-periodic linear S.C. network in section 2.1. and referring to MNA equations (4) and (5) as well as Fig. 2b. it is possible to reformulate equations (5) and (4) as follows :
Theorem 1 : if an ideal linear S.C. network is excited by a piece-wise-constant excitation ($w(t) = w_m$, $u(t) = u_m$, $t \in \Delta_m$), then its response in the time domain is also piecewise-constant $v(t) = v_m$, $q(t) = q_m$ in Δ_m and given by

$$\begin{bmatrix} A_k & B_k \\ C_k & D_k \end{bmatrix} \begin{bmatrix} v_{k+\ell N} \\ q_{k+\ell N} \end{bmatrix} = \begin{bmatrix} E_k \, v_{k+\ell N-1} \\ 0 \end{bmatrix} + \begin{bmatrix} w_{k+\ell N} \\ u_{k+\ell N} \end{bmatrix} \qquad (8)$$

where $v(t)$ (resp., $u(t)$) is the vector of the voltage responses at the nodes (resp., voltage sources in some selected branches) and where $q(t)$ (resp., $w(t)$) is the vector of the charges transferred in some selected branches (resp., injected by charge sources in the nodes) between $t_{k+\ell N}$ and t for $t \in \Delta_{k+\ell N}$, and where the contributions to A_k, B_k, C_k, D_k and E_k are given by the usual MNA stamps some of which

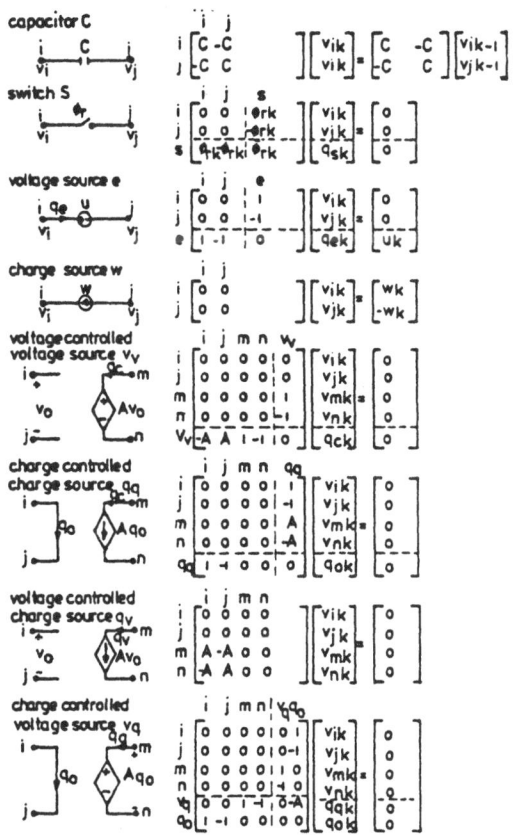

Fig. 10

are illustrated in Fig. 10. See also example Fig. 2 .

If the S.C. network is excited by an arbitrary excitation $\underset{\sim}{w}(t)$, $\underset{\sim}{u}(t)$, then the piecewise-constant part of the input and output $\underset{\sim}{w}_m = \underset{\sim}{w}(t^-_{m+1})$, $\underset{\sim}{u}_m = \underset{\sim}{u}(t^-_{m+1})$, $\underset{\sim}{v}_m = \underset{\sim}{v}(t^-_{m+1})$ and $\underset{\sim}{q}_m = \underset{\sim}{q}(t^-_{m+1})$ are determined by (8) and the remainder wave forms $\underset{\sim}{w}^*(t) = \underset{\sim}{w}(t)-\underset{\sim}{w}_m$, $\underset{\sim}{u}^*(t) = \underset{\sim}{u}(t)-\underset{\sim}{u}_m$, $\underset{\sim}{v}^*(t) = \underset{\sim}{v}(t)-\underset{\sim}{v}_m$, and $\underset{\sim}{q}^*(t) = \underset{\sim}{q}(t)-\underset{\sim}{q}_m$ for $t \in \Delta_m$ are related by

$$\left[\begin{array}{c|c} A_k & B_k \\ \hline C_k & D_k \end{array}\right] \cdot \left[\begin{array}{c} v^*(t) \\ q^*(t) \end{array}\right] = \left[\begin{array}{c} w^*(t) \\ u^*(t) \end{array}\right] \text{ for } t \in \Delta_{k+\ell N} \qquad (9)$$

Proof : As discussed in section 2.1. the first set of equations of (8,9) express the charge conservation and the second set are the constitutive relations of some selected branches. Equation (9) can be obtained by starting from eq. (4), using the decomposition of

258

a continuous signal in its piecewise constant part and its remainder wave form. Simple linear decomposition then gives (9).
Notice that eq. (8) is an algebraic -difference equation from one clock phase to the next whereas (9) is a pure algebraic equation for $t \in \Delta_{k+\ell N}$.

The physical significance of the above theorem can be expressed as follows :

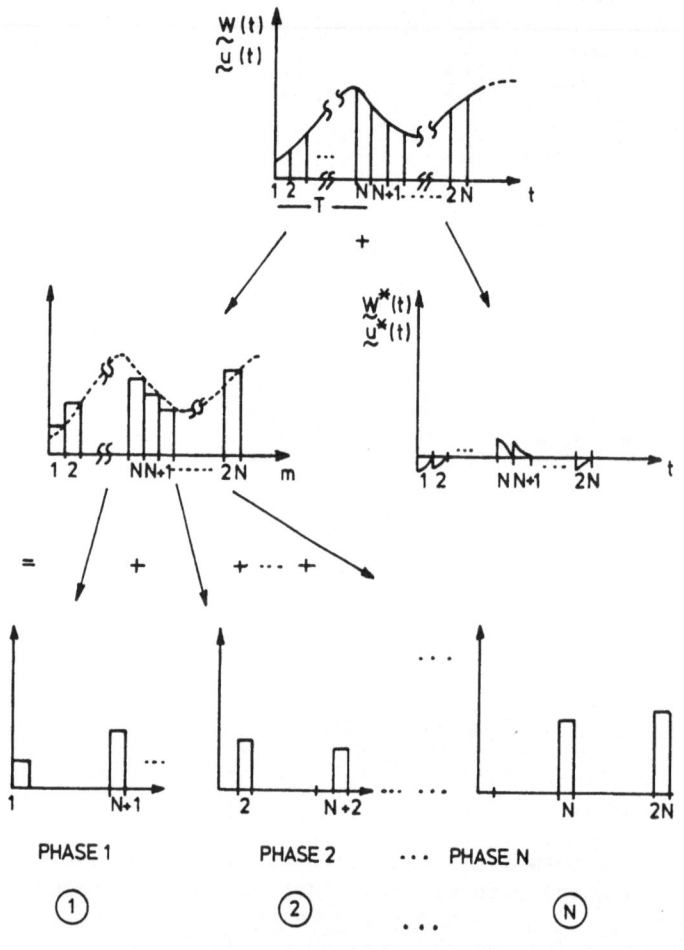

INPUT WAVEFORM DECOMPOSITION

FIG.11

Decompose the signals $w(t)$ and $u(t)$ into a piecewise-constant waveform such that the amplitude in each time slot is equal to the value at the end of this time slot and a remainder waveform (Fig. 11). The output of the S.C. network is then the sum of a piecewise-con-

constant waveform obtained by solving the N-periodic implicit
linear algebraic and difference equation (8) and a remainder output obtained by solving the linear algebraic equations (9).
Notice now :

1) the piecewise constant input solution of (8) is nothing
else than the time domain analysis technique discussed in 2.1.,
procedure 1;

2) equation (9) is an algebraic one and as a result :
the response at the end of an arbitrary time slot Δ_m where all
$\underset{\sim}{w}^*(t) = \underset{\sim}{w}(t^-_{m+1}) - \underset{\sim}{w}_m = \underset{\sim}{0}$ or $\underset{\sim}{u}^*(t) = \underset{\sim}{u}(t^-_{m+1}) - \underset{\sim}{u}_m = \underset{\sim}{0}$
is also zero i.e. the S.C. network behaves with respect to the re-
mainder waveform as a "resistive" (memoryless) network with "resistors" 1/C.
This is because no time constants are present in the network. This
property will be important to compute the frequency response of
S.C. networks for continuous inputs and direct capacitive I/O
coupling.
Based on the above theorem it will now be possible to formulate a
z-domain form of equations (8) and (9). This form is the basis
for direct and FFT based CAD methods.
We will first restrict ourselves to the case of piecewise constant
inputs but this restriction will be removed later.
We make again use of the linearity of the capacitors and the controlled sources. Since (8) is time-varying the techniques of the
z-transform are not readily applicable. Fortunately (8) is periodic
and hence we can adapt a method of Jury [27], p. 57 . We partition
the sequence of values at the end of each time slot $\underset{\sim}{v}_1$, $\underset{\sim}{v}_2$, $\underset{\sim}{v}_3$,
$\underset{\sim}{v}_{N-1}$, $\underset{\sim}{v}_N$, $\underset{\sim}{v}_{N+1}$ into N different sequences each having the same
phase : $\underset{\sim}{v}_1$, $\underset{\sim}{v}_{N+1}$, $\underset{\sim}{v}_{2N+1}$ ... and $\underset{\sim}{v}_2$, $\underset{\sim}{v}_{2+N}$, $\underset{\sim}{v}_{2+2N}$, ... etc. until
$\underset{\sim}{v}_N$, $\underset{\sim}{v}_{2N}$, $\underset{\sim}{v}_{3N}$ ... This is also shown in Fig. 11 where the piece-
wise constant signal is decomposed into waveforms ①, ② ... Ⓝ.
We take the standard z-transform of these N sequences.

$$\underset{\sim}{V}_k(z) \overset{\Delta}{=} \mathcal{Z}\{\underset{\sim}{v}_{m+\ell N}\} = \sum_{\ell=0}^{\infty} \underset{\sim}{v}_{m+\ell N} z^{-\ell} \qquad (10)$$

where m=1, 2, ... N. Analogously we obtain $\underset{\sim}{Q}_k(z)$, $\underset{\sim}{W}_k(z)$ and
$\underset{\sim}{U}_k(z)$. Remember also [27] that if $\underset{\sim}{v}_i = \underset{\sim}{0}$, $i \lesssim 0$

$$\mathcal{Z}\{\underset{\sim}{v}_{k+(\ell-1)N}\} = z^{-1} \underset{\sim}{V}_k(z) \qquad (11)$$

Based on these definitions it then possible to formulate the following theorem :
Theorem 2 : the above described linear T-periodic SC network is
described in the z-domain by

$$
\left[
\begin{array}{cccc|cccc}
\underset{\sim}{A}_1 & & & -\underset{\sim}{A}_1 z^{-1} & \underset{\sim}{B}_1 & & & \\
-\underset{\sim}{E}_2 & \underset{\sim}{A}_2 & & & & \underset{\sim}{B}_2 & & \\
& -\underset{\sim}{E}_3 & \underset{\sim}{A}_3 & & & & \underset{\sim}{B}_3 & \\
& & \ddots & -\underset{\sim}{E}_N \quad \underset{\sim}{A}_N & & & & \ddots \quad \underset{\sim}{B}_N \\
\hline
\underset{\sim}{C}_1 & & & & \underset{\sim}{D}_1 & & & \\
& \underset{\sim}{C}_2 & & & & \underset{\sim}{D}_2 & & \\
& & \underset{\sim}{C}_3 & & & & \underset{\sim}{D}_3 & \\
& & & \ddots \quad \underset{\sim}{C}_N & & & & \ddots \quad \underset{\sim}{D}_N
\end{array}
\right]
\left[
\begin{array}{c}
\underset{\sim}{V}_1 \\ \underset{\sim}{V}_2 \\ \underset{\sim}{V}_3 \\ \vdots \\ \underset{\sim}{V}_N \\ \hline \underset{\sim}{Q}_1 \\ \underset{\sim}{Q}_2 \\ \underset{\sim}{Q}_3 \\ \vdots \\ \underset{\sim}{Q}_N
\end{array}
\right]
=
\left[
\begin{array}{c}
\underset{\sim}{W}_1 \\ \underset{\sim}{W}_2 \\ \underset{\sim}{W}_3 \\ \vdots \\ \underset{\sim}{W}_N \\ \hline \underset{\sim}{U}_1 \\ \underset{\sim}{U}_2 \\ \underset{\sim}{U}_3 \\ \vdots \\ \underset{\sim}{U}_N
\end{array}
\right]
\tag{12}
$$

where the missing entries are zero.

Proof : By writing (8) in full for $k = 1, 2, \ldots N$ we obtain an augmented set of time-invariant algebraic and difference equations on which (10, 11) can be applied in order to obtain (12).
It will be shown below that eq. (12), which is very sparse, can be used as a basis for direct analysis in the frequency domain. Before discussing this, it is of interest to formulate the solution of (12) in terms of the inverse of the matrix in (12) which will be defined as the z-domain transfer matrix $\underset{\sim}{M}$.

$$
\left[
\begin{array}{c}
\underset{\sim}{V}_1 \\ \underset{\sim}{V}_2 \\ \cdots \\ \underset{\sim}{V}_N \\ \hline \underset{\sim}{Q}_1 \\ \underset{\sim}{Q}_2 \\ \cdots \\ \underset{\sim}{Q}_N
\end{array}
\right]
=
\left[
\begin{array}{cccc|cccc}
\underset{\sim}{G}_{11} & \underset{\sim}{G}_{12} & \cdots & \underset{\sim}{G}_{1N} & \underset{\sim}{H}_{11} & \underset{\sim}{H}_{12} & \cdots & \underset{\sim}{H}_{1N} \\
\underset{\sim}{G}_{21} & \underset{\sim}{G}_{22} & \cdots & \underset{\sim}{G}_{2N} & \underset{\sim}{H}_{21} & \underset{\sim}{H}_{22} & \cdots & \underset{\sim}{H}_{2N} \\
\cdots & \cdots & \cdots & & \cdots & \cdots & \cdots & \\
\underset{\sim}{G}_{N1} & \underset{\sim}{G}_{N2} & \cdots & \underset{\sim}{G}_{NN} & \underset{\sim}{H}_{N1} & \underset{\sim}{H}_{N2} & \cdots & \underset{\sim}{H}_{NN} \\
\hline
\underset{\sim}{K}_{11} & \underset{\sim}{K}_{12} & \cdots & \underset{\sim}{K}_{1N} & \underset{\sim}{L}_{11} & \underset{\sim}{L}_{12} & \cdots & \underset{\sim}{L}_{1N} \\
\underset{\sim}{K}_{21} & \underset{\sim}{K}_{22} & \cdots & \underset{\sim}{K}_{2N} & \underset{\sim}{L}_{21} & \underset{\sim}{L}_{22} & \cdots & \underset{\sim}{L}_{2N} \\
\cdots & \cdots & \cdots & & \cdots & \cdots & \cdots & \\
\underset{\sim}{K}_{N1} & \underset{\sim}{K}_{N2} & \cdots & \underset{\sim}{K}_{NN} & \underset{\sim}{L}_{N1} & \underset{\sim}{L}_{N2} & \cdots & \underset{\sim}{L}_{NN}
\end{array}
\right]
\left[
\begin{array}{c}
\underset{\sim}{W}_1 \\ \underset{\sim}{W}_2 \\ \cdots \\ \underset{\sim}{W}_N \\ \hline \underset{\sim}{U}_1 \\ \underset{\sim}{U}_2 \\ \cdots \\ \underset{\sim}{U}_N
\end{array}
\right]
\tag{13}
$$

or in short :

$$
\underset{\sim}{Y} = \underset{\sim}{M} \, \underset{\sim}{X}
\tag{14}
$$

The submatrices $\underset{\sim}{G}_{k\ell}$, $\underset{\sim}{H}_{k\ell}$, $\underset{\sim}{K}_{k\ell}$ and $\underset{\sim}{L}_{k\ell}$ allow a very simple but important interpretation, which is illustrated in Fig. 12.
Up to this point we consider the SC network as a discrete device which transforms the input sequences of samples at t_i^- , $i = 1, 2, \ldots$ of the voltage sources and charges sources into the output sequences of samples of voltages or charges at t_i^-, $i = 1, 2, \ldots$
If only nonzero voltages sources are applied during time slots ℓ, $\ell+N$, $\ell+2N$, $\ldots$ and if the output node voltages are only observed

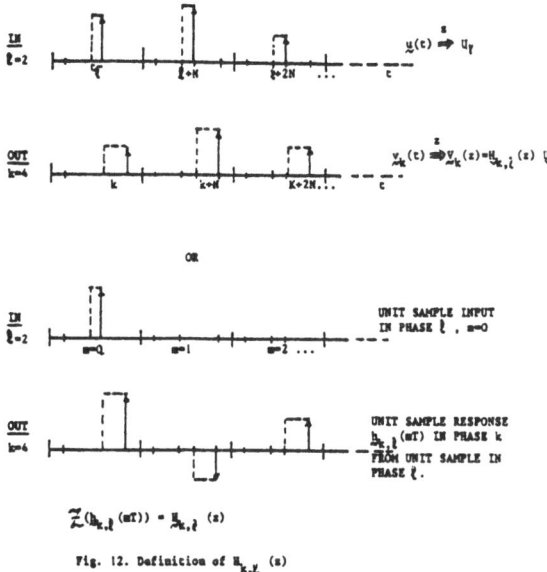

Fig. 12. Definition of $H_{k,\ell}$ (z)

during time slots k, k+N, k+2N, ... then $H_{k\ell}(z)$ relates the z-transform of this input sequence to this output sequence i.e. $V_k(z) = H_{k\ell}(z) \, U_\ell$ or it relates inputs at phase ℓ to outputs at phase k. In other words $H_{k\ell}(z)$ is the z-transform of $h_{k\ell}(mT)$, m=0,1, ... i.e., the node voltage responses observed during time slots k, k+N, k+2N ... to unit input voltages applied during time slot ℓ. Clearly $H_{i1}(z)$ is the z-transform of the node voltage response during time slots i, i+N, i+2N, ... to the same input, and can thus be derived from the same network analyses as $H_{k\ell}$ (z).

Thus, if we perform a time domain solution as in section 2.1. for a unit-sample impuls in phase ℓ and we observe the unit sample voltage responses $h._\ell$ (mT) to it in all phases k=1...N on all nodes then we can obtain one column $H._\ell$ (z) from (13) as follows :

$$H._\ell \, (z) = \sum_{m=0}^{\infty} h._\ell \, (mT) \, z^{-m} \tag{15}$$

Equation (15) shows a relationship between the terms in (13) and unit sample time domain analysis using the method discussed in 2.1. This property will be needed to derive the analysis method presently used in DIANA and to be discussed below.

Further, notice that the inverse of the matrix of (9) can easily be obtained from the z-domain transfer matrix. This implies

$$
\begin{bmatrix} \underset{\sim}{v}^{\boldsymbol{*}}(t) \\ \hline \underset{\sim}{q}^{\boldsymbol{*}}(t) \end{bmatrix} = \begin{bmatrix} \underset{\sim}{G}_{kk}(\infty) & | & \underset{\sim}{H}_{kk}(\infty) \\ \hline \underset{\sim}{K}_{kk}(\infty) & | & \underset{\sim}{L}_{kk}(\infty) \end{bmatrix} \cdot \begin{bmatrix} \underset{\sim}{w}^{\boldsymbol{*}}(t) \\ \hline \underset{\sim}{u}^{\boldsymbol{*}}(t) \end{bmatrix} \qquad \text{for } t \in \Delta_{k+\ell N} \qquad (16)
$$

The interpretation of (16) is quite important for S.C. networks with continuous inputs and direct capacitive I/O coupling. In fact eqs. (9) and (16) express direct capacitive feedthrough of the remainder waveforms $\underset{\sim}{u}^{\boldsymbol{*}}(t)$ and $\underset{\sim}{w}^{\boldsymbol{*}}(t)$ and therefore only direct responses ($z=\infty$) from input phase k to observation phase k can result. Notice also that e.g.

$$
\underset{\sim}{H}_{kk}(\infty) = \underset{\sim}{h}_{k,k}(0)
$$

This means that the terms of (16) can be found simply from the first terms of the unit sample responses (m=o) observed at the same phase k of the unit-sample input. Based on the above theory it is now possible to formulate the basis for frequency domain analysis.

3.2. Analysis in the frequency domain.

We describe in this subsection how the practically useful frequency properties can be derived from the time domain or z-domain analysis of the previous sections. In many applications such as in filtering one is mainly interested in the frequency components of the input and output voltage waveforms. So we will only derive the expressions for voltages. Since a SC network is a time-varying network, in general for a sinusoïdal excitation many frequencies appear at the output.
Consider a sinusoïdal excitation $\underset{\sim}{u}(t) = \underset{\sim}{U}e^{\,j\Omega t}$.
In steady state the output of an ideal S.C. network observed at all phases will be a waveform $\underset{\sim}{v}(t)$ such as the one shown in Fig. 13a.
We call $\underset{\sim}{V}(\omega)$ the spectrum of $\underset{\sim}{v}(t)$. One can also consider this waveform $\underset{\sim}{v}(t)$ as a sum over all phases k (k=1,2...N) of voltages $\underset{\sim}{\hat{v}}_k(t)$ which are the voltage $\underset{\sim}{v}(t)$ observed only during phase k and zero outside (dotted line in Fig. 13a.).
We call $\underset{\sim}{\hat{V}}_k(\omega)$ the spectrum of $\underset{\sim}{\hat{v}}_k(t)$.
Clearly :

$$
\underset{\sim}{v}(t) = \sum_{k=1}^{N} \underset{\sim}{\hat{v}}_k(t)
$$

and

$$
\underset{\sim}{V}(\omega) = \sum_{k=1}^{N} \underset{\sim}{\hat{V}}_k(\omega) \qquad (17)
$$

According to (17) it is sufficient to calculate $\underset{\sim}{\hat{V}}_k(\omega)$ which is the spectrum of the output waveform $\underset{\sim}{v}(t)$ only observed within phase k.

It is shown in an appendix that the following theorem holds :

Theorem 3. Given a voltage excitation $\underset{\sim}{u}(t)$ with Fourier transform $\underset{\sim}{U}(\Omega)$, the spectrum $\underset{\sim}{\hat{V}}_k(\omega)$ with $\omega = \Omega + n\omega_s$ of $\underset{\sim}{\hat{v}}_k(t)$, observed during phase k only, can be written as :

$$\underset{\sim}{\hat{V}}_k(\omega) = \sum_{n=-\infty}^{+\infty} \underset{\sim}{X}_k(\omega,\Omega) \, \underset{\sim}{v}(\Omega) \tag{18}$$

whereby :

$$\omega = \Omega + n\omega_s \tag{19}$$

and

$$\omega_s = \frac{2\pi}{T} \tag{20}$$

Furthermore $\underset{\sim}{X}_k(\omega,\Omega)$, which we will call the transmission function for phase k is given by :

$$\underset{\sim}{X}_k(\omega,\Omega) = \alpha_k(\omega) \sum_{\ell=1}^{N} \underset{\sim}{H}_{k\ell}(e^{j\Omega T}) \, \exp\left[j\,\Omega t_{\ell+1} - \omega t_{k+1}\right]$$

$$+ \left[\alpha_k(\omega-\Omega) - \alpha_k(\omega)\right] \, \exp\left[j(\Omega-\omega)t_{k+1}\right] \, \underset{\sim}{H}_{kk}(\infty) \tag{21}$$

with $\underset{\sim}{H}_{k\ell}$ defined by (8) and

$$\alpha_k(\rho) = 2\left\{\sin\left[\rho(t_{k+1}-t_k)/2\right] \exp\left[j\rho(t_{k+1}-t_k)/2\right]\right\} / T\rho \tag{22}$$

The procedure to prove this theorem is illustrated in Fig. 13b, c,d. The response $\underset{\sim}{v}(t)$ is, according to theorem 1 (eqs. (8) and (9)) first decomposed into the piecewise constant part and the remainder waveform $\underset{\sim}{v}_k^*(t)$ per phase k.

The piecewise constant part of the spectrum can be found from the terms $\underset{\sim}{H}_{k\ell}$ (z), $z=e^{j\Omega t}$ from the z transfer matrix and convolving it with a block $r_k(t)$ of width Δ_k centered on $(t_{k+1}-t_k)/2$ which causes the $\alpha_k(\omega)$ sinx/x type term (Fig. 13c).

Finally the spectrum of $\underset{\sim}{v}_k^*(t)$ is obtained from the spectrum of the product of :

$$\underset{\sim}{v}_k^*(t) = \underset{\sim}{H}_{kk}(\infty) \cdot (e^{j\Omega t} - e^{j\Omega t_{k+\ell N+1}}) \tag{23}$$

with a T-periodic block function of width Δ_k (Fig. 13d.).

Clearly then the first term in (21) takes into account the piecewise-constant signals and the second term the continuous coupling. From (17,18) it follows that a sinusoidal input $\underset{\sim}{u}(t)$ generates an output which has a line spectrum whose lines are ω_s a part. Mostly one is only interested in the contribution at the same frequency as the input. This is among others the case if the bandwidth of $\underset{\sim}{u}(t)$ is smaller than half the sampling frequency. For such cases we define the following practical frequency domain transfer functions $\underset{\sim}{H}(\omega)$ (resp., $\underset{\sim}{H}_k(\omega)$) [II] . Apply a sinusoidal excitation $\underset{\sim}{u}(t) = \underset{\sim}{U} e^{j\omega t}$ and compute or measure the frequency component at the same pulsation ω in the output $\underset{\sim}{v}(t)$ (resp., $\underset{\sim}{v}_k(t)$, then $\underset{\sim}{H}(\omega)$ (resp., $\underset{\sim}{H}_k(\omega)$) relates the phasor $\underset{\sim}{U}$ to that of the out-

264

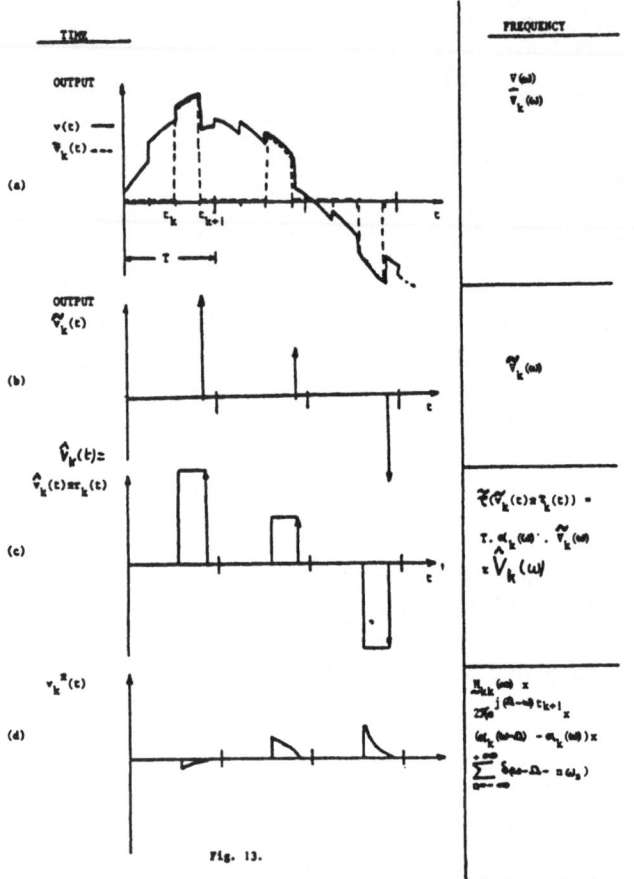

Fig. 13.

put $\underset{\sim}{v}(t)$ (resp., $\underset{\sim}{v}_k(t)$) at pulsation ω .
From theorem 3 it follows indeed :

$$\underset{\sim}{H}_k(\omega) = \alpha_k(\omega) \sum_{\ell=1}^{N} \underset{\sim}{H}_{k\ell}(e^{j\omega T}) \exp\left[j\omega(t_{\ell+1} - t_{k+1}) \right]$$

$$+ \left[\frac{t_{k+1} - t_k}{T} - \alpha_k(\omega) \right] \underset{\sim}{H}_{kk}(\infty), \tag{24}$$

$$\underset{\sim}{H}(\omega) = \sum_{k=1}^{N} \underset{\sim}{H}_k(\omega) \tag{25}$$

where α_k is given by (22).

3.3. FFT based frequency domain analysis.

It is interesting to note that eq. (24) can be interpreted as follows :
The transfer function $\underset{\sim}{H}_k(\omega)$ in time slot k is a superposition of :
a) The superposition of the N responses $\underset{\sim}{H}_{k\ell}(e^{j\omega T})$, $\ell = 1, 2 \ldots$ N
in phase k due to the sinusoidal excitation in the phases ℓ ,
$\ell = 1, 2 \ldots$ N. While considering the output there is a "time
delay" $t_{k+1} - t_{\ell+1}$ between output and input sample taken, which
causes the "phase" factor exp. $[-j\omega(t_{k+1} - t_{\ell+1})]$. Since we consider the output to be sampled and held within Δ_k one has to multiply with the $\alpha_k(\omega)$ function.

Notice that in order to compute $\underset{\sim}{H}_{k\ell}(e^{j\omega T})$, $\ell = 1, 2 \ldots$ N we need to
compute a row of the z-domain transfer matrix and not a column,
which, as discussed in 3.1., eq. (15), can be found from the FFT of
one unit sample response in the time domain.
So apparently, in order to find this first term for   m phases
it is necessary to perform   N unit sample responses and m x N
FFT's.   It will be shown in section 4 that this can be reduced
to only m analysis using the adjoint S.C. network concept (with m ≤ N)

b) The second part in (24) is due to the direct I/O coupling of
a continuous input. In fact one can consider $\underset{\sim}{H}_{kk}(\infty)$ = $\underset{\sim}{h}_{kk}(o)$ as
the direct feedthrough (gain) in phase k with respect to the in-
put. If the input is sinusoidal with unity amplitude then the
remainder output is the difference between a block modulated sine
wave of amplitude $\underset{\sim}{H}_{kk}(\infty)$ and duty cycle $(t_{k+1} - t_k)/T$ and its back-
wards folded sample of width $(t_{k+1} - t_k)$ which is represented by
$\alpha_k(\omega) . \underset{\sim}{H}_{kk}(\infty)$.
Clearly this part can easily be calculated from the first terms
$\underset{\sim}{h}_{kk}(o)$ of the unit sample impulse responses for input in phase k
with output in phase k.
The above discussion demonstrates a possible procedure for frequency domain calculations based on FFT techniques. We will come
back to it when discussing the adjoint network.

3.4. Direct frequency domain analysis.

Direct frequency domain analysis is possible in two ways :
a) Through the use of (12). Combining (12) with theorem 3 (eq.(21)
and (22)), follow the following procedure :
procedure 2 : To find $\underset{\sim}{X}_k(\omega, \Omega)$ with $\omega = \Omega + n\omega_s$ and given Ω
proceed as follows :

2a) In (12) put $\underset{\sim}{W}_i = \underset{\sim}{\varrho}$ (i=1,2...N), $\underset{\sim}{U}_\ell = \underset{\sim}{U} \alpha_k(\omega)$ exp j $(\Omega t_{\ell+1} -$
$\omega t_{k+1})$ and $z^{-1} = e^{-j\Omega T}$.
2b) Solve sparse matrix equation (12) for $\underset{\sim}{V}_k(\omega, \Omega)$.
2c) If continuous coupling is present, put $z^{-1} = o$, $\underset{\sim}{W}_i = o$, i=1,2...N
and $\underset{\sim}{U}_k = \underset{\sim}{U}$.

2d) Solve for $\underline{V}_k = \underline{H}_{kk}(\infty)$ and multiply with
$\left[\alpha_k(\omega-\Omega) - \alpha_k(\omega)\right] \exp\left[j(\Omega-\omega) t_{k+1}\right]$.

2e) Add the solutions under 2b) and 2d).
This procedure has to our knowledge not yet been implemented. It requires very efficient sparse matrix techniques presently under study. The method of Brglez [8 9] and Tsividis [11] is in fact based on a recursive solution of (12) which however involves a lot of matrix multiplications.

b) An other technique used in [20 21 22] is based on the construction of a time invariant network of impedances which has the same network matrix as [12]. The key idea in obtaining such an equivalent network is to convert the N phases of any branch into N different branches. This converts the different instances of <u>time</u> into different locations in <u>space</u>. We need the following <u>intrinsic N-port called a generalized circulator with constant G</u> which is defined by (Fig. 14).

$$
\begin{bmatrix} Q_1 \\ Q_2 \\ Q_3 \\ \cdot\cdot \\ Q_N \end{bmatrix} = \begin{bmatrix} 0 & & & -Gz^{-1} \\ -G & 0 & & \\ & -G & 0 & \\ & & \ddots & \ddots \\ & & & -G & 0 \end{bmatrix} \begin{bmatrix} V_1 \\ V_2 \\ \cdot\cdot \\ V_N \end{bmatrix}
\tag{26}
$$

Construction of the equivalent circuit $\mathcal{N}_e$ of a SC network $\mathcal{N}$:

1. For each of the N phases (i.e. N time slots in one period) a network is drawn with the switches in the correct position for this time slot.

2. The N networks are interconnected by generalized circulators as follows. For each capacitor C_i in the original circuit we need a circulator with constant C_i. Port 1 of this circulator is connected to the corresponding capacitor of the first circuit, port 2 to that of the second circuit and so on. Repeat for each capacitor.

It is easy to check that the resulting network is described by (12). This equivalent circuit reduces to those of [20 21 22], in the case that a clock period only contains two time slots. We have applied this algorithm to the circuit of Fig. 15a. with clock signals of Fig. 15b. and obtain the equivalent circuit of Fig. 15c.

4. THE ADJOINT S.C. NETWORK AND ITS APPLICATIONS [18 29].

In this section we derive the adjoint network of an ideal linear S.C. network and show how it can be used to reduce the computational complexity of the frequency, noise and sensitivity analysis.

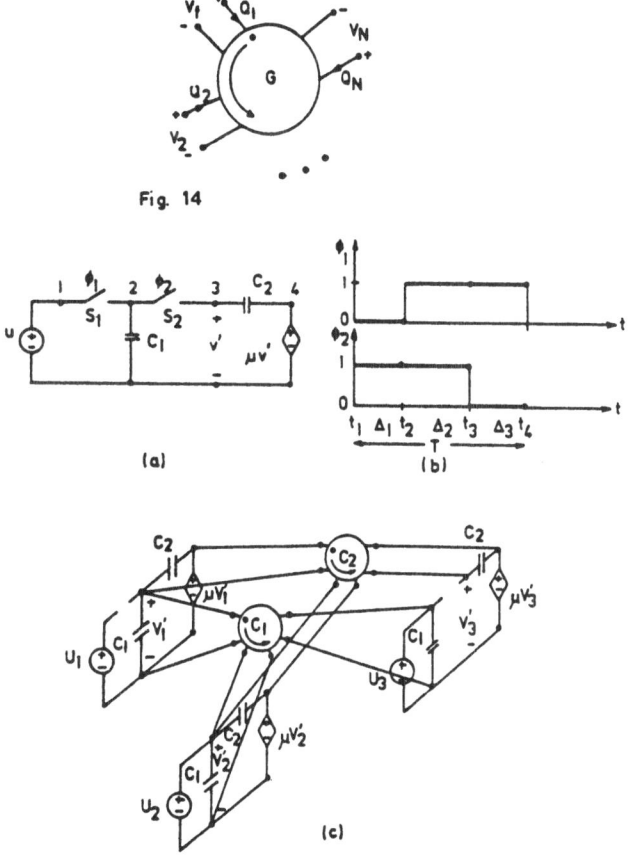

Fig. 14

(a)

(b)

(c)

Fig.15

The construction of the adjoint switched capacitor turns out to
be quite simple, even for switched capacitor networks with more
than two phases or with a continuous input-output path. Moreover
this construction can be performed equivalently in the time domain or on the equivalent circuit.
The concept has been used in DIANA for efficient frequency and sensitivity analysis and examples will be shown.

4.1. The concept of adjoint S.C. network.

The concept of adjoint network is well known for continuous linear
and nonlinear networks [28] and has lead to very efficient sensitivity and noise calculations. We will now define the adjoint
network for S.C. networks and show that even greater computational

savings are possible for such networks.

In analogy with continuous networks we propose the following construction of the adjoint S.C. network and then prove its useful properties :

Construction of the adjoint SC network $\widetilde{N}$ of a SC network N :

1. Construct a duplicate of N and denote the sources by $\widetilde{u}$ and $\widetilde{w}$.
2. Modify the nonreciprocal elements (controlled sources) as follows. Interchange the controlling and controlled ports. Replace a VCVS by a QCQS and vice versa. Multiply the controlling coefficient by -1 in the case of a QCQS or VCVS.
3. Reverse the time dependency of all time-varying elements in one period i.e. the new clock signal $\widetilde{\phi}_{rk}$ and the switching times $\widetilde{t}_i$ satisfy

$$\widetilde{\phi}_{rk} = \phi_{r\ N-k+1} \tag{27a}$$

$$\widetilde{t}_{\ell N+k} = t_{\ell N+1} + t_{(\ell+1)N+1} - t_{(\ell+1)N+2-k} \tag{27b}$$

for $\ell = 0, 1, 2 \ldots$ and $k = 1, 2, \ldots N$.

For example, by applying this algorithm on the SC filter section of Fig.16a. with clock signals (b) one obtains the adjoint S.C. network of Fig. 16c. with clock signals (d).

Observe that step 3 of the construction need to be executed if all clock signals are symmetric in one period. This is among others the case for clock signals with only two time slots of equal duration. Moreover, compared to the usual construction of the adjoint for linear time-invariant networks only step 3 is new and thus it takes into account that a SC network is a T-periodic network.

The useful property of the adjoint network of a linear time-invariant network is that its impedance matrix is the transpose of that of the original matrix. In the next theorem we prove that for a SC network one has to perform on the z-domain transfer matrix (13) additionally an operation which corresponds with the reversion of time in a period. Let the operation of block row and column reversion be defined and denoted as :

$$\tag{28}$$

$$
\begin{bmatrix}
G_{11} & G_{12} & \cdots & G_{1N} & H_{11} & H_{12} \cdots H_{1N} \\
G_{21} & G_{22} & \cdots & G_{2N} & H_{21} & H_{22} \cdots H_{2N} \\
\cdots & \cdots & & \cdots & \cdots & \cdots \cdots \\
G_{N1} & G_{N2} & \cdots & G_{NN} & H_{N1} & H_{N2} \cdots H_{NN} \\
K_{11} & K_{12} & \cdots & K_{1N} & L_{11} & L_{12} \cdots L_{1N} \\
K_{21} & K_{22} & \cdots & K_{2N} & L_{21} & L_{22} \cdots L_{2N} \\
\cdots & \cdots & & \cdots & \cdots & \cdots \cdots \\
K_{N1} & K_{N2} & \cdots & K_{NN} & L_{N1} & L_{N2} \cdots L_{NN}
\end{bmatrix}
\begin{bmatrix}
G_{NN} \cdots G_{N2} & G_{N1} & H_{NN} \cdots H_{N2} & H_{N1} \\
\cdots & \cdots & \cdots & \cdots \\
G_{2N} \cdots G_{22} & G_{21} & H_{2N} \cdots H_{22} & H_{21} \\
G_{1N} \cdots G_{12} & G_{11} & H_{1N} \cdots H_{12} & H_{11} \\
K_{NN} \cdots K_{N2} & K_{N1} & L_{NN} \cdots L_{N2} & L_{N1} \\
\cdots & \cdots & \cdots & \cdots \\
K_{2N} \cdots K_{22} & K_{21} & L_{2N} \cdots L_{22} & L_{21} \\
K_{1N} \cdots K_{12} & K_{11} & L_{1N} \cdots L_{12} & L_{11}
\end{bmatrix}
$$

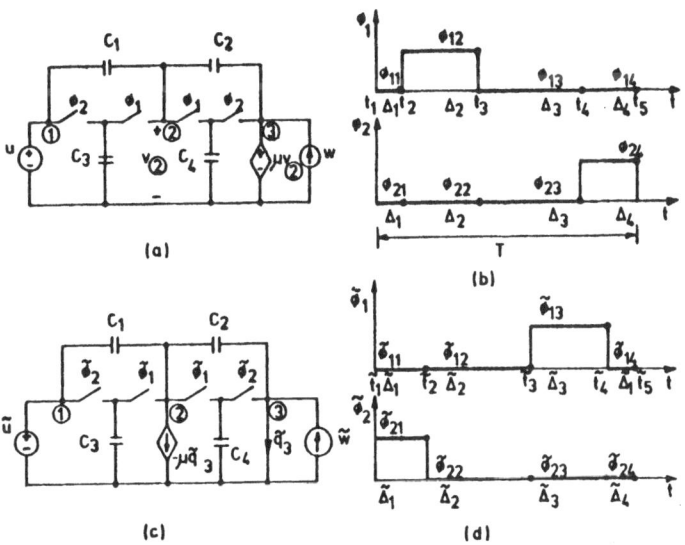

Fig. 16

<u>Theorem 4.</u> Given a switched capacitor network N with z-domain transfer matrix $\underset{\sim}{M}$ and its adjoint network $\tilde{N}$ with z-domain transfer matrix $\underset{\sim}{\tilde{M}}$, then

$$\underset{\sim}{\tilde{M}} = \underset{\sim}{M}^{T} \tag{29}$$

<u>Proof</u> : Without loss of generality assume that no composite bran-ches are used in the MNA equations (8) of N , i.e.

$$\underset{\sim}{A} \overset{\Delta}{=} \underset{\sim}{A}_1 = \underset{\sim}{A}_2 \cdots \underset{\sim}{A}_N \tag{30}$$

It is clear from Fig. 10 that all time-invariant stamps are trans-posed in the corresponding MNA equations of $\tilde{N}$. Moreover the time dependency satisfies (27). Hence

$$
\begin{bmatrix}
\underset{\sim}{A}^T & \vline & \underset{\sim}{C}^T_{N-k+1} \\
\hline
\underset{\sim}{B}^T_{N-k+1} & \vline & \underset{\sim}{D}^T_{N-k+1}
\end{bmatrix}
\begin{bmatrix}
\tilde{\underset{\sim}{v}}_{k+\ell N} \\
\tilde{\underset{\sim}{q}}_{k+\ell N}
\end{bmatrix}
=
\begin{bmatrix}
\underset{\sim}{E}^T \, \tilde{\underset{\sim}{v}}_{k-1+\ell N} \\
\underset{\sim}{0}
\end{bmatrix}
+
\begin{bmatrix}
\tilde{\underset{\sim}{w}}_{k+\ell N} \\
\tilde{\underset{\sim}{u}}_{k+\ell N}
\end{bmatrix}
\tag{31}
$$

for $k = 1, 2, \ldots N$ and $\ell = 0, 1, 2, \ldots$

Hence we obtain in the z-transform :

$$
\begin{bmatrix}
\underset{\sim}{A}^T & & & & & \vline & -\underset{\sim}{A}^T z^{-1} & \underset{\sim}{C}^T_N \\
-\underset{\sim}{E}^T & \underset{\sim}{A}^T & & & & \vline & & \underset{\sim}{C}^T_{N-1} \\
& -\underset{\sim}{E}^T & \underset{\sim}{A}^T & & & \vline & & & \underset{\sim}{C}^T_{N-2} \\
& & \ddots & \ddots & & \vline & & & & \ddots & \underset{\sim}{C}^T_1 \\
& & & -\underset{\sim}{E}^T & \underset{\sim}{A}^T & \vline & & & & & \\
\hline
\underset{\sim}{B}^T_N & & & & & \vline & \underset{\sim}{D}^T_N & & & & \\
& \underset{\sim}{B}^T_{N-1} & & & & \vline & & \underset{\sim}{D}^T_{N-1} & & & \\
& & \underset{\sim}{B}^T_{N-2} & & & \vline & & & \underset{\sim}{D}^T_{N-2} & & \\
& & & \ddots & \underset{\sim}{B}^T_1 & \vline & & & & \ddots & \underset{\sim}{D}^T_1
\end{bmatrix}
\begin{bmatrix}
\tilde{\underset{\sim}{V}}_1 \\ \tilde{\underset{\sim}{V}}_2 \\ \tilde{\underset{\sim}{V}}_3 \\ \vdots \\ \tilde{\underset{\sim}{V}}_N \\ \hline \tilde{\underset{\sim}{Q}}_1 \\ \tilde{\underset{\sim}{Q}}_2 \\ \tilde{\underset{\sim}{Q}}_3 \\ \vdots \\ \tilde{\underset{\sim}{Q}}_N
\end{bmatrix}
=
\begin{bmatrix}
\tilde{\underset{\sim}{W}}_1 \\ \tilde{\underset{\sim}{W}}_2 \\ \tilde{\underset{\sim}{W}}_3 \\ \vdots \\ \tilde{\underset{\sim}{W}}_N \\ \hline \tilde{\underset{\sim}{U}}_1 \\ \tilde{\underset{\sim}{U}}_2 \\ \tilde{\underset{\sim}{U}}_3 \\ \vdots \\ \tilde{\underset{\sim}{U}}_N
\end{bmatrix}
\tag{32}
$$

where the missing entries are zero. The matrix in (32) is that of
(12) after a transposition and block row and column reversion.
Hence the same is true for their inverses.

An interesting interpretation now directly follows from Theorem 4.

Corollary 3. Given a SC network N and its adjoint SC network $\tilde{N}$.
Then the voltage response at node i measured during time slots
k, k+N, k+2N ... on an excitation of N with a unit voltage impulse
in branch j during time slot ℓ and all other inputs zero, has
the same value as the charge response in branch j measured during
time slots N-ℓ+1, 2N-ℓ+1, ... on an excitation of N with a unit
charge impulse in node i during time slot k (Fig. 17) or :

$$
\tilde{\underset{\sim}{K}}_{N-\ell+1 \; N-k+1} (z) = \underset{\sim}{H}^T_{k\ell} (z)
\tag{33}
$$

As far as computing a solution of the adjoint is concerned, it is
important to observe that the LU factors of the adjoint can be immediately obtained from those of the nominal network. In fact in
the time-domain analysis the LU factors of N in Δ_k are the transpose of the LU factors of $\tilde{N}$ in Δ_{N-k+1}. Analogously in the z-domain analysis the LU factors of $\tilde{N}$ are LU factors of N after a
transposition and a block reversion of the rows of the first and
the columns of the second.

4.2. Application 1 : frequency domain analysis.

The use of SC networks relies on their abilities to generate
suitable frequency domain transfer function $H^{(ij)}_k(\omega)$ between the

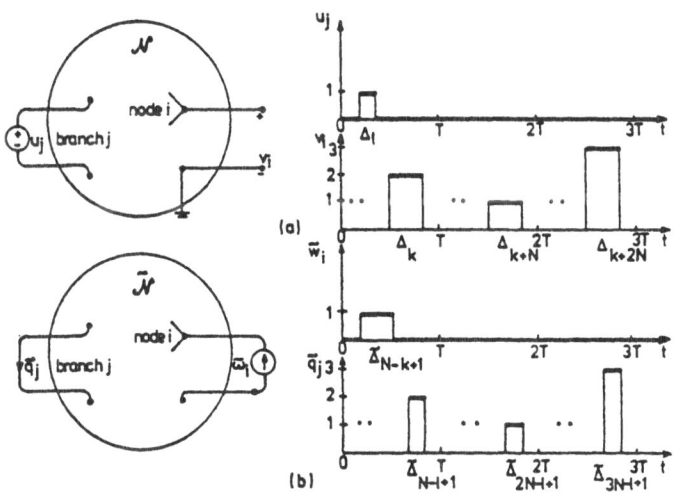

fig. 17

voltage source in branch j and the voltage $\hat{v}_k^{(i)}(t)$ at node i. If we use (24) in order to compute $H_k^{(ij)}(\omega)$, the entries $H_{k\ell}^{(ij)}(e^{j\omega T})$ for $\ell = 1, \ldots N$ are needed, which are obtained by performing N network analyses of the original circuit. Using (33) of Theorem 4 we can convert (24) into :

$$H_k^{(ij)}(\omega) = \alpha_k(\omega) \sum_{\ell=1}^{N} \tilde{K}_{\tilde{\ell}\tilde{k}}^{(ji)}(e^{j\omega T}) \exp\left[j\omega(t_{\ell+1}-t_{k+1})\right] +$$

$$\left[(t_{k+1}-t_k)/T - \alpha_k(\omega)\right] \tilde{K}_{\tilde{\ell}\tilde{k}}^{(ji)}(\infty) \tag{34}$$

with $\tilde{\ell} = N-\ell+1$ and $\tilde{k} = N-k+1$. Now $\tilde{K}_{\tilde{\ell}\tilde{k}}^{(ji)}(e^{j\omega T})$ for $\tilde{\ell} = 1, \ldots N$ can be obtained from one network analysis of the adjoint network as observed in the previous subsection. The same substitution can be performed in the transmission function $\underset{\sim}{X}_k(\omega,\Omega)$ (21) so that all effects of high frequency inputs on the baseband of the output (aliasing) can be studied using one network analysis of the adjoint network. Of course if one is interested in $H^{(ij)}(\omega)$ all N^2 entries $H_{k\ell}^{(ij)}(e^{j\omega T})$ for $k, \ell = 1, \ldots N$ are needed and N network analyses are needed if the original or the adjoint network is used.

Let us illustrate the computations of the transfer function and the aliasing effects with an example obtained using the DIANA program [13 14 17.]
In this program, by using the option .SC FREQ. the MNA matrix is transposed to get the one of the adjoint S.C. network and one can

272

specify a number of inputs and <u>one</u> output port. Automatically
the clock phases are inverted and an arbitrary $H_k^{(ij)}(\omega)$ can be com-
puted by computing all N appropriate unit sample responses in
<u>one</u> adjoint S.C. network analysis.
These responses are input files for a postprocessor FPR where their
FFT's are computed and their results combined according to (34).
The following are examples of such analysis :

<u>Example 1</u>. Elliptic filter.
We refer back to the elliptic filter of Fig. 4a. for which Fig. 4b.
is the input description. Notice the option .SC FREQ which causes
the adjoint analysis to take place.
The 2000 point unit-sample response (there is only one here) is
shown in Fig. 4c. It takes 2 min. 30 sec. CPU on a VAX 11/780.
The impulse response is then processed by FPR in 54 sec. CPU and
produces the amplitude Bode plot shown in Fig. 18a.

VAX CPU-TIMES: DIANA 2 min 30 sec
                FPR   54 sec

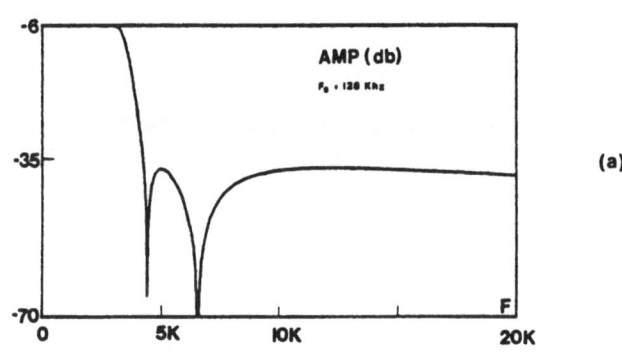

(a)

PASSBAND

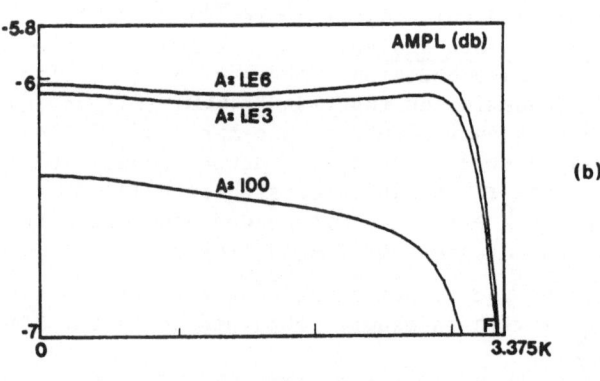

(b)

fig. 18

Fig. 18b. shows a detail of the pass-band of the filter which has been computed for three values of op-amp gain $A=10^6, 10^3, 10^2$.

This example clearly shows the potential of such simulations for the designer who can easily check the influence of all parameters, including stray capacitances, clock timing, sample and hold effects, observation time effect etc... on his design. Such influences are very difficult to get from breadboards, where often, due to scaling necessary for parasitics, no meaningful results can be obtained.
Fig. 18b. clearly shows that for a PCM quality filter, the op-amp gain should be in excess of 10^3 (60 dB) since otherwise the pass-band attentuation can not be tolerated.

Furthermore notice that once, after the DIANA run, the impulse response files are created, the user can then use all the options of FPR on the impulse files of the same circuit. For details we refer to the DIANA manual but we mention the following possibilities :
- frequency response observed on one   or several phases
- input sampled on one or several phases
- input continuous or piecewise constant
- influence of continuous I/O coupling
- aliasing effects.

An example of these facilities is illustrated in Fig. 19.
Fig. 19a. is a simple pole-zero S.C. filter and its input coding.
Fig. 19b. is one of the unit sample responses.
Fig. 19c. shows amplitude (dB) vs. frequency on a linear scale.
The input is sampled both during Cℓ1 and Cℓ2 and the clock duty cycles are both 50%. The following results are obtained using the options in FPR :

Curve A : piecewise constant input, no S/H effects.
Curve B : same but with S/H effects $(\alpha_k(\omega))$.
The zero of sin x/x occurs at $2 f_s$ due to double output phase sampling.
Curve C : continuous input, S/H effects included (most realistic response).
Curves A', B' and C' are simulations whereby input pulsations and their responses at $\omega = \omega_s - \Omega$ (or $f=f-f_s$) are considered (folding effect). Such simulations can be important to design e.g. anti-aliasing pre-filters.

4.3. Application 2.: sensitivity calculations.

In sensitivity analysis one is basically interested in the variation of the input-output relation due to a variation in one or several of the parameters of the network. This is of particular importance for filter design for the following practical reasons :
- computation of the effect of component tolerances on amplitude and phase response;

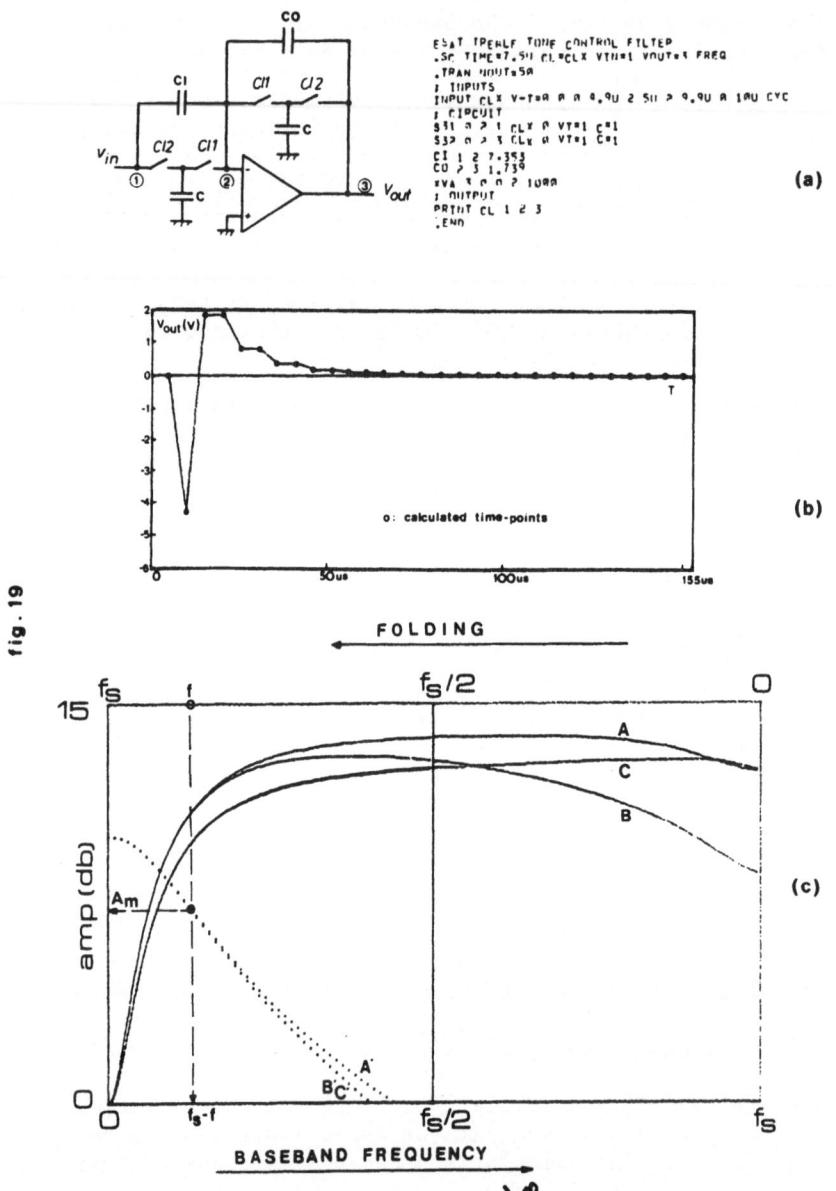

fig. 19

(a)

(b)

o: calculated time-points

FOLDING

(c)

BASEBAND FREQUENCY

- computation of group delay $t_d = -\dfrac{\partial \varphi}{\partial \omega}$;
- optimisation techniques which, to be efficient, always require all sensitivities. Perturbation methods tend to be very time consuming and inaccurate;
- worst case design techniques;
- design centering etc...

We will show now that the adjoint network can be used to reduce considerably the computational effort for sensitivity calculations.

First we obtain the derivative of an arbitrary entry $H_k^{(ij)}(z)$ of the z-domain transfer function (13) with respect to a parameter and combine these equations then in the computation of the sensitivities of the frequency domain transfer function $H_k^{(ij)}(\omega)$. Let the MNA matrix in (8,9) have a size sxs, then the transfer matrix of (13), which we denote by $\underset{\sim}{M}$, has a size NsxNs. From (12) and the circuit realization the derivative of the matrix $\underset{\sim}{M}^{-1}$ of (12) with respect to any parameter λ (e.g. a capacitance, or gain or temperature or a technological or geometric parameter) can be determined. By differentiating $\underset{\sim}{M}\,\underset{\sim}{M}^{-1} = \underset{\sim}{1}$ with respect to λ we obtain

$$\frac{\partial \underset{\sim}{M}}{\partial \lambda}\underset{\sim}{M}^{-1} + \underset{\sim}{M}\frac{\partial \underset{\sim}{M}^{-1}}{\partial \lambda} = \underset{\sim}{0}, \tag{35a}$$

or

$$\frac{\partial \underset{\sim}{M}}{\partial \lambda} = - \underset{\sim}{M}\frac{\partial \underset{\sim}{M}^{-1}}{\partial \lambda}\underset{\sim}{M} \tag{35b}$$

Clearly the right handside is easier to calculate since $\frac{\partial \underset{\sim}{M}^{-1}}{\partial \lambda}$ is known from the design and $\underset{\sim}{M}$ is known by inverting the matrix in (12) which implies Ns network analyses. If only one (or a few) entries of $\frac{\partial \underset{\sim}{M}}{\partial \lambda}$ are needed, the number of network analyses can be greatly reduced by using the adjoint network. If one is interested in the derivative of $H_{k\ell}^{(ij)}(z)$ (i.e. the z-transform transfer function from the voltage in branch j during phase ℓ to the voltage at node i and during phase k) with respect to the parameter λ we will show that <u>one original and one adjoint network analysis are sufficient.</u> Remember that one network analysis of $\mathcal{N}$ (resp., $\tilde{\mathcal{N}}$) produces one column of $\underset{\sim}{M}$ (resp., one column of $\underset{\sim}{\tilde{M}} = \underset{\sim}{M}^{ST}$). Substituting (29) in (35) generates

$$\frac{\partial \underset{\sim}{M}}{\partial \lambda} = - \underset{\sim}{\tilde{M}}\,\underset{\sim}{S}^T\frac{\partial \underset{\sim}{M}^{-1}}{\partial \lambda}\underset{\sim}{M} \tag{36}$$

Using this and the contributions of the components to the matrix of (12) we obtain the following derivatives :

a) For a capacitor C connecting node m to node n we have :

$$\frac{\partial H_{k\ell}^{(ij)}(z)}{\partial C} = -\sum_{r=1}^{N}(\underset{\sim}{\tilde{G}}_{\tilde{r}\tilde{k}}^{(mi)} - \underset{\sim}{\tilde{G}}_{\tilde{r}\tilde{k}}^{(ni)})\,(H_{r\ell}^{(mj)} - H_{r\ell}^{(nj)})$$

$$+ \sum_{r=1}^{N-1}(\underset{\sim}{\tilde{G}}_{\tilde{r}-1\,\tilde{k}}^{(mi)} - \underset{\sim}{\tilde{G}}_{\tilde{r}-1\,\tilde{k}}^{(ni)})\,(H_{r\ell}^{(mj)} - H_{r\ell}^{(nj)})$$

$$+ z^{-1}(\underset{\sim}{\tilde{G}}_{1\tilde{k}}^{(mi)} - G_{1\tilde{k}}^{(ni)})\,(H_{N\ell}^{(mj)} - H_{N\ell}^{(nj)}) \tag{37}$$

with $\tilde{r} = N-r+1$ and $\tilde{k} = N-k+1$

b) For a voltage controlled voltage source where the voltage in branch q is A times the voltage of node m with respect to node n we have :

$$\frac{\partial H_{k\ell}^{(ij)}(z)}{\partial A} = \sum_{r=1}^{N} \tilde{K}_{rk}^{(qi)} \left(H_{r\ell}^{(mj)} - H_{r\ell}^{(nj)} \right) \tag{38}$$

More sensitivities can be defined by the same technique. For a more detailed list see [29].
If composite branches are used in the MNA equations, only those terms in the sums (37, 38) have to be taken which physically appear in that phase. The derivatives with respect to the gain of QCVS, VCQS and QCQS can be determined analogously as those for a VCVS but are of less practical value and are hence omitted.

In practice one is interested in the sensitivities of the frequency domain transfer functions (24). We present here the sensitivities of $H_k^{(ij)}$ (i.e. at phase k) in terms of the derivatives (37, 38), which can be easily verified from (24) :

a) For a capacitor C connecting node m to node n we have :

$$S_C^{H_k^{(ij)}}(\omega) \triangleq \frac{C}{H_k^{(ij)}(\omega)} \frac{\partial H_k^{(ij)}(\omega)}{\partial C}$$

$$= \frac{C}{H_k^{(ij)}(\omega)} \left[\alpha_k(\omega) \sum_{\ell=1}^{N} \frac{\partial H_{k\ell}^{(ij)}(e^{j\omega T})}{\partial C} \exp\left[j\omega(t_{\ell+1} - t_{k+1}) \right] \right.$$

$$\left. + \left[(t_{k+1} - t_k)/T - \alpha_k(\omega) \right] \frac{\partial H_{kk}^{(ij)}(\infty)}{\partial C} \right] \tag{39}$$

b) For a voltage controlled voltage source where the voltage in branch q is A times the voltage of node m with respect to node n we have :

$$S_A^{H_k^{(ij)}}(\omega) = \frac{A}{H_k^{(ij)}(\omega)} \frac{\partial H_k^{(ij)}(\omega)}{\partial A}$$

$$= \frac{A}{H_k^{(ij)}(\omega)} \left[\alpha_k(\omega) \sum_{\ell=1}^{N} \frac{\partial H_{k\ell}^{(ij)}(e^{j\omega T})}{\partial A} \exp\left[j\omega(t_{\ell+1} - t_{k+1}) \right] \right.$$

$$\left. + \left[(t_{k+1} - t_k)/T - \alpha_k(\omega) \right] \frac{\partial H_{kk}^{(ij)}(\infty)}{\partial A} \right] \tag{40}$$

Formulas for sensitivity to clock switching time, as well as group delay can be found in [29].

It is easy to see that the computation of the practical sensitivities (39),(40) with respect to capacitors and op-amp gain require only N nominal and 1 adjoint network analysis.

If one is interested in the sensitivity of $H^{(ij)}(\omega)$ analogous equations can be derived. They have not been included for sake of brevity.

For these computations N nominal and N adjoint network analyses are needed. This figure should be compared to the Nxs network analysis if no adjoint was used. Also notice that sensitivity over the full frequency spectrum is obtained.

Examples.

The above technique has been implemented in DIANA as an option .SC SENS.

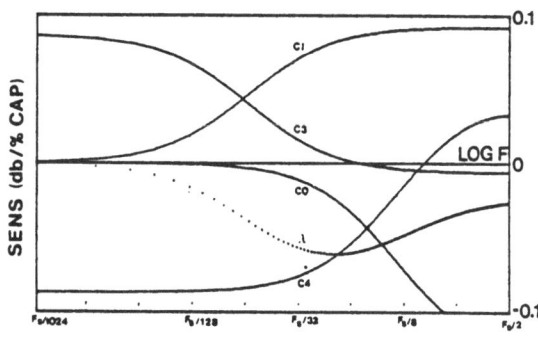

Fig. 20

As an example Fig. 20. shows the sensitivities of the filter in Fig. 19a.

The sensitivities shown are for the amplitude response expressed in dB with respect to percentage change of all capacitors. C3 is the input sampling capacitor and C4 the output sampling capacitor.

λ is a common increment factor for all capacitors :$C'=C+\lambda$. The calculated sensitivity is then $(\partial H^{(ij)}/\partial\lambda)/H^{(ij)}$. The sensitivity to a common tracking factor λ for all capacitors $C'=C.\lambda$ can also be calculated. However, for a S.C. circuit this sensitivity equals $\emptyset$.

In [29] an expression for this sensitivity is derived and it can be calculated using DIANA.

The expressions of sensitivities in dB/% can be obtained from (39, 40) by taking the real part and multiplying by 0.086858.

278

Since 4 phases are present all these sensitivities only require
4 nominal and 1 adjoint network analyses   followed by post-
processing with   FPR.
Finally Fig. 21. shows

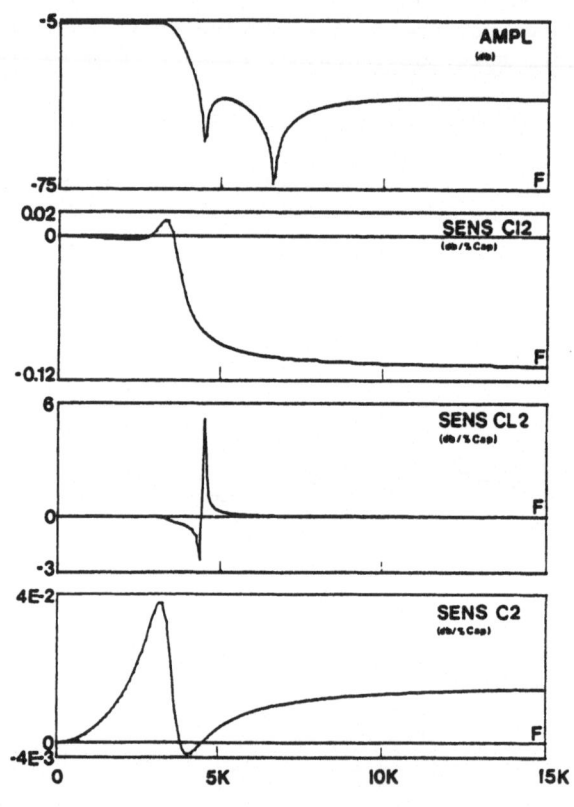

Fig. 21

the frequency response of the elliptic filter of Fig. 4a. together
with amplitude sensitivities in dB/% change of a few capacitor
values indicated in Fig. 4a. Outside the pass-band however, the
position of the zeros of the elliptical filter is much more sensitive to capacitor variations.

4.4. Application 3 : Noise calculations.

One of the most difficult to predict performance of an S.C. network is the noise behavior. Although not yet implemented in DIANA the theoretical basis to perform efficient noise analysis does exist and is given below.

Usually one is interested in the output noise voltage at one node i due to the noise in many (say p) components. We assume that the noise of each device can be modelled by a voltage source whose voltage is a stationary stochastic process. The vector $\underset{\sim}{u}(t)$ of the input noise voltage sources is then characterized by a spectral density matrix $\underset{\sim}{S}_u(\omega)$ which depends on the technology and which we assume to be known. Since the circuit is time-varying the output voltage $v^{(i)}(t)$ (resp., $v_k^{(i)}(t)$) is not a stationary stochastic process. By same computations as those of $[10]$ the average spectral density function is given by

$$S_{v_k^{(i)}}(\omega) = \sum_{n=-\infty}^{+\infty} \underset{\sim}{X}_k^{(i.)}(\omega,\omega-n\omega_s)\, \underset{\sim}{S}_u(\omega-n\omega_s)\, \underset{\sim}{X}_k^{(i.)*}(\omega,\omega-n\omega_s)$$

(41)

and

$$S_v(i)(\omega) = \sum_{n=-\infty}^{+\infty} \sum_{k=1}^{M} \underset{\sim}{X}_k^{(i.)}(\omega,\omega-n\frac{2\pi}{T})S_u(\omega-n\frac{2\pi}{T}) \sum_{j=1}^{N} \underset{\sim}{X}_j^{(i.)*}(\omega,\omega-n\frac{2\pi}{T})$$

(42)

where $*$ denotes complex conjugate and transpose, and

$X_k^{(i.)}(\omega,\Omega)$ is the i-th row of $X_k(\omega,\Omega)$ of (21)

If p is the number of noise voltage sources, then from (21) and (41, 42), $S_{v_k}(i)$ needs Np different $H_\ell(j)$ for $\ell = 1, 2...N$ and where j assumes p different values and hence it needs Np <u>different net</u>work analyses. The computation of $S_v^{(i)}$ requires $N^2 p$ different $\overline{H_k(j)}$ for $k, \ell = 1, 2, ... N$ and where j assumes p different values and hence it needs the same Np different network analyses as $S_{v_k}(i)$. Using Theorem 4 the i-th row of the transmission function is given by

$$\underset{\sim}{X}_k^{(i.)}(\omega,\Omega) = \alpha_k(\omega)\sum_{\ell=1}^{N} \underset{\sim}{\tilde{K}}_{\tilde{\ell}\tilde{k}}^{(.i)T}(e^{j\Omega T})\, \exp\left[j(\Omega t_{\ell+1}-\omega t_{k+1})\right]$$

$$+\left[\alpha_k(\omega-\Omega) - \alpha_k(\omega)\right]\, \exp\left[j\Omega-\omega t_{k+1}\right]\, \underset{\sim}{\tilde{K}}_{\tilde{k}}^{(.i)}(\infty)^T$$

(43)

where $\tilde{\ell} = N-\tilde{\ell}+1$ and $\tilde{k} = N-k+1$. This computation requires only one adjoint network analysis, since all entries of a column of the z-transform transfer matrix can be obtained from one network analysis. If we use (43) in the computation of $S_{v_k}(i)$ only <u>one adjoint network analysis</u> is needed and in the case of $S_v(i)$ only <u>N adjoint network analyses</u> are needed. If $p > 1$ this implies a considerable saving in computations.

5. CONCLUSIONS

In this contribution we have discussed and illustrated a number of
techniques to simulate S.C. filters and networks very efficiently
in the time and frequency domain.
It has been shown that, using the concept of the adjoint network
considerable savings are obtained in frequency domain analysis
as well as in sensitivity and noise calculations. The implementation of the FFT based technique in DIANA has been illustrated
by several examples.
The MNA approach used in DIANA is fully compatible with normal
continuous time domain simulation. Therefore, exploiting the
character of DIANA as a mixed mode simulator, a switched capacitor
circuit can be verified from concept all the way to the detailed
time domain behavior of the transistor circuit even including
steering logic circuitry. Since even models for delay blocks and
weighted multiplier models can be defined in the MDL lanuage
(see DIANA manual) also digital filter structures can be simulated
and processed for time, frequency, sensitivity and noise analysis
in the FPR postprocessor.
In this way a powerful program is obtained both for digital and
analog MOSLSI circuitry.

Appendix : Proof of Theorem 3. (see also Fig. 13.).

We calculate first the spectrum of the response for a sinusoïdal
excitation $\underset{\sim}{u}(t) = \underset{\sim}{U} e^{j\Omega T}$. Let $\underset{\sim}{\tilde{v}}_k(t)$ be the response sampled at
the end of the time slots $\Delta_k, \Delta_{k+N}, \ldots$ Remember that $\underset{\sim}{H}_{k\ell}(z)$ is the
z-transform of $\underset{\sim}{h}_{k\ell}(mT)$, $m = 0,1,\ldots$, then we have in general from
(13) for this sinusoïdal excitation

$$\underset{\sim}{\tilde{v}}_k(t) = \sum_{n=-\infty}^{+\infty} \delta(t-nT-t_{k+1}) \left[\sum_{\ell=1}^{N} \left(\sum_{m=0}^{\infty} \underset{\sim}{h}_{k\ell}(mT) \underset{\sim}{U} e^{j\Omega\left[(n-m)T+t_{\ell+1}\right]} \right) \right]$$

$$= \left(\sum_{n=-\infty}^{+\infty} \delta(t-nT-t_{k+1}) e^{j\Omega nT} \right) \left(\sum_{k=1}^{\ell} e^{j\Omega t_{\ell+1}} \underset{\sim}{H}_{k\ell}(e^{j\Omega T}) \underset{\sim}{U} \right) \quad (A1)$$

The Fourier transform of this sinusoïdally modulated pulse train is

$$\underset{\sim}{\tilde{v}}_k(\omega) = \sum_{n=-\infty}^{+\infty} \omega_s \, \delta(\omega-\Omega-n\omega_s) \sum_{\ell=1}^{N} e^{j\Omega t_{\ell+1}-\omega t_{k+1})} \underset{\sim}{H}_{k\ell}(e^{j\Omega T}) \underset{\sim}{U}$$
$$\quad (A2)$$

which has only contributions at the pulsations $\omega = \Omega + n\omega_s$.
In order to derive $\underset{\sim}{V}_k(\omega)$ we define a block waveform $s_k(t)$ which is
1 in Δ_k and zero outside. Using Theorem 2 and (9) we have

$$\hat{\underset{\sim}{v}}_k(t) = \sum_{\ell=-\infty}^{+\infty} s_{k+\ell N}(t) \, v(t)$$

$$= \sum_{\ell=-\infty}^{+\infty} s_{k+\ell N}(t) \, \underset{\sim}{v}_{k+\ell N} + \sum_{\ell=-\infty}^{+\infty} s_{k+\ell N}(t) \, \underset{\sim}{v}^*(t)$$

$$= \underset{\sim}{\tilde{v}}_k(t) * r_k(t) + \underset{\sim}{H}_{kk}(\infty) \, \underset{\sim}{U} \sum_{\ell=-\infty}^{+\infty} s_{k+\ell N}(t)(e^{j\Omega t} - e^{j\Omega t_{k+\ell N+1}})$$

$$\tag{A3}$$

where $r_k(t) = 1$ for $t_k - t_{k+1} < t \leqslant 0$. From (A2) we have for the first term of the right handside of (A3) :

$$\mathcal{F}\left\{\underset{\sim}{\tilde{v}}_k(t) * r_k(t)\right\} = 2\pi \underset{\sim}{\alpha}_k(\omega) \sum_{\ell=1}^{N} \underset{\sim}{H}_{k\ell}(e^{j\Omega T}) e^{j(\Omega t_{\ell+1} - \omega t_{k+1})} \sum_{n=-\infty}^{+\infty}(\omega - \Omega - n\omega_s)$$

$$\tag{A4}$$

The Fourier transform of the second term in (A3) can be obtained by standard techniques as

$$\mathcal{F}\left[\sum_{\ell=-\infty}^{+\infty} e^{j\Omega t} s_{k+\ell N}(t)\right] = 2\pi \underset{\sim}{\alpha}_k(\omega-\Omega) e^{j(\Omega-\omega)t_{k+1}} \sum_{n=-\infty}^{+\infty} \delta(\omega-\Omega-n\omega_s)$$

$$\tag{A5}$$

and

$$\mathcal{F}\left[\sum_{\ell=-\infty}^{+\infty} s_{k+\ell N}(t) e^{j\Omega t_{k+\ell N+1}}\right] = 2\pi \underset{\sim}{\alpha}_k(\omega) \, e^{j(\Omega-\omega)t_{k+1}} \sum_{n=-\infty}^{+\infty} \delta(\omega-\Omega-n\omega_s)$$

$$\tag{A6}$$

This implies that the spectrum $\hat{\underset{\sim}{V}}_k(\omega)$ for the sinusoïdal excitation $\underset{\sim}{u}(t) = \underset{\sim}{U} e^{j\Omega T}$ is

$$\hat{\underset{\sim}{V}}_k(\omega) = \underset{\sim}{X}_k(\omega,\Omega) \, \underset{\sim}{U} \, 2\pi \sum_{n=-\infty}^{+\infty} \delta(\omega-\Omega-n\omega_s) \tag{A7}$$

The linearity of the SC network then implies that the response to an arbitrary input $u(t) = \frac{1}{2\pi} \int_{-\infty}^{\infty} U(\Omega) \, e^{j\Omega T} \, d\Omega$ has a spectrum

$$\hat{\underset{\sim}{V}}_k(\omega) = \int_{-\infty}^{\infty} \underset{\sim}{X}_k(\omega,\Omega) \, \underset{\sim}{U}(\Omega) \sum_{n=-\infty}^{+\infty} \delta(\omega-\Omega-n\omega_s) \, d\Omega$$

This implies (18).

6. REFERENCES

[1] A. Fettweis : "Theory of Resonant-Transfer Circuits", in Network and Switching Theory (G. Birci, ed.), New York, Academic Press, pp. 382-446, 1968.

[2] B.J. Hosticka, R.W. Broderson, P.R. Gray : "MOS Sampled Data Recursive Filters using Switched Capacitor Integrators", IEEE Journal Sol. State Circuits, Vol.SC-12, pp. 600-608, Dec. 1977.

282

[3] J.T. Caves, M.A. Copeland, C.F. Rahim, S.D. Rosenblaum : "Sampled analog filtering using switched capacitors as resistor equivalents", IEEE Journ. of Solid State Circuits, Vol. SC-12, No. 6, pp. 592-600, Dec. 1977.

[4] J.L. McCreary, P.R. Gray : "All MOS charge redistribution Analog-to-Digital Conversion Techniques - Part I", IEEE Journal of Solid State Circuits, Vol. SC-10, pp.371-379, Dec. 1975.

[5] C. Hewer, M. de Wit : "NMOS Tone Generator", Proc. IEEE Int. Solid State Circuits Conference, Philadelphia, pp.30-31, Febr.77.

[6] E.K. Cheng, C.A. Mead : "Single chip character generator", Proc. IEEE Int. Solid State Circuits Conference, Philadelphia, pp. 32-33, Febr. 1975.

[7] M.E. Hoff et al. :"Single chip n-MOS microcomputer processes signals in real time", Electronics, March 1, 1979, pp. 105-110.

[8] F. Brglez : "Exact nodal analysis of switched capacitor networks with arbitrary switching sequences and general inputs - Part I", IEEE Proc. 12th Asilomar Conf. Circuits & Systems, Pacific Grove, Calif. Nov. 1978, pp. 679-683.

[9] F. Brglez : "Exact nodal anaysis of switched capacitor networks with arbitrary switching sequences and general inputs - Part II", IEEE Proc. Int. Symp. Circuits & Systems, Tokyo, July 1979, pp. 748-751.

[10] M.L. Liou, Y.L. Kuo : "Exact Analysis of Switched Capacitor Circuits with Arbitrary Inputs", IEEE Trans. Circuits & Systems, CAS-26, No. 4, pp. 213-223, April 1979.

[11] Y.P. Tsividis : "Analysis of switched-capacitive networks", IEEE Proc. Int. Symp. Circuits and Systems, July 1979, Tokyo, pp. 752-755.
Also : IEEE Trans. Circuits & Systems, Vol. CAS-26, no. 11, pp. 935-947, Nov. 1979.

[12] G.C. Fang, Y.P. Tsividis : "Modified Nodal Analysis with improved numerical methods for switched capacitive networks", IEEE Proc. Int. Symp. Circuits and Systems, Houston, April 1980, pp. 977-980.

[13] H. De Man, J. Rabaey, G. Arnout : "On the simulation of switched capacitor filters and convertors using the DIANA program", Proc. European Solid State Circuits Conference, Southampton, pp. 136-138, Sept. 1979.

[14] H. De Man, J. Rabaey, G. Arnout, J. Vandewalle : "Practical implementation of a general computer aided design technique for switched capacitor circuits", IEEE Journ. of Solid State Circuits, Vol. SC-15, No. 2, pp. 190-200, April 1980.

[15] G. Arnout, H. De Man : "The use of threshold functions and Boolean Controlled Network Elements for Macromodeling of LSI

circuits", IEEE Journal of Solid State Circuits, Vol. SC-13, No. 3, pp. 326-332, June 1978.

[16] C.W. Ho, A.E. Ruehli, P.A. Brennan : "The modified nodal approach to network analysis", IEEE Trans. Circuits Syst., Vol. CAS-22, pp. 504-509, June 1975.

[17] H. De Man, J. Rabaey, G. Arnout, J. Vandewalle : "DIANA as a mixed-mode simulator for MOSLSI sampled-data circuits", Proc. IEEE Symp. Circuits and Systems, Houston, April 1980, pp.435-438.

[18] J. Vandewalle, H. De Man, J. Rabaey : "The adjoint switched capacitor network and its applications", Proc. IEEE Symp. Circuits and Systems, Houston, April 1980, pp. 1031-1034.

[19] F. Brglez : "SCOP - A switched-capacitor optimization program", Proc. IEEE Symp. Circuits and Systems, Houston, April 1980, pp. 985-988.

[20] C.F. Kurth, G.S. Moschytz : "Nodal analysis of switched capacitor networks", IEEE Trans. Circuits & Systems, Vol. 26, Febr. 1979, pp. 93-104.

[21] C.F. Kurth, G.S. Moschytz : "Two port analysis of switched capacitor networks using four-port equivalent circuits in the z-domain", IEEE Trans. Circuits & Systems, Vol. CAS-26, no. 3, March 1979, pp. 166-180.

[22] Y.L. Kuo, M.L. Liou, J.W. Kasinskas : "Equivalent circuit approach to the computer aided analysis of switched capacitor circuits", IEEE Trans. Circuits & Systems, Vol. CAS-26, No. 9, Sept. 1979, pp. 708-714.

[23] J. Vandewalle, H. De Man, J. Rabaey : "A unified theory for the computer aided analysis of general switched capacitor networks", Accepted for publication and presentation at the ECCTD-80 Conference, Warshaw, Poland, Sept. 1980.

[24] H. De Man : "Mixed-mode simulation techniques and their implementation in DIANA", NATO ASI on Computer aids for VLSI circuits, Urbino, July 1980.

[25] L.O. Chua, P.M. Lin : "Computer aided analysis of electronic circuits; algorithms and computational techniques", Prentice-Hall, N.J. 1975.

[26] G.M. Jacobs, D.J. Allstot, R.W. Brodersen, P.R. Gray : "Design Techniques for MOS Switched Capacitor Ladder Filters", IEEE Trans. Circuits & Systems, Vol. CAS-25, No. 12, pp.1014-1021, Dec. 1978.

[27] E.I. Jury : "Theory and application of the z-transform method", John Wiley and Sons, New York 1964.

[28] S.W. Director, R.A. Rohrer, "The generalized adjoint network and network sensitivities", IEEE Trans. on Circuit Theory,

284

Aug. 1969, pp. 318-322.

[29] J. Vandewalle, H. De Man, J. Rabaey : "The adjoint switched capacitor network and its applications to frequency, noise and sensitivity", Report ESAT Labs., K.U.Leuven, Belgium 1980. Will also be published.

DESCRIPTION AND SIMULATION OF COMPLEX DIGITAL SYSTEMS BY MEANS OF THE REGISTER TRANSFER LANGUAGE RTS IA

H.-J. Knobloch

Inst. f. Datentechnik, Techn. Hochschule Darmstadt, Darmstadt, Germany

ABSTRACT. Register transfer languages serve as a medium to describe complex digital systems at early stages of logical design. The structure of the rt-language RTS Ia is presented; reasons for the incorporation of special language constructs are discussed. The application of RTS Ia as denotational system at several stages of digital design is demonstrated. Finally, the mode of operation of the RTS Ia simulation system is explained and an example of a complete simulation run is given.

1. INTRODUCTION

1.1 Motivation

The design of complex digital systems as e.g. digital computers requires modelling of the desired system at different levels of abstraction.

Although no generally accepted theory on digital system design exists, many auyhors (e.g. /1/) distinguish nine different digital system design interest levels.

Level	Objective
1.) System	Maximize total throughput by defining a network of suitable subsystems
2.) Programming	Prepare programs which may reside as firmware or system software

3.) Instruction Set	Develop an optimum instruction set
4.) Organisation	Specify the major facilities and data paths to best implement an instruction set
5.) Algorithm	Develop control algorithms whereby an organization can realize an intruction set
6.) Register	Detail organization and control algorithms without emphasizing the constrains of a particular technology
7.) Gate	Design logic networks with real gates, flip-flops etc.
8.) Circuit	Prepare electronic design of interconnected gates, flip-flops, etc.
9.) Fabrication	Specify physical placement of components, wiring lists, fabrication techniquies, etc.

Computer Aided Design tools are used to a wide extent in industry at the three lowest design levels, where a lot of algorithms have been developed to relieve the designer of cumbersome work. At the system- and programming-level general purpose simulation languages (GPSS, SIMULA,...) and high level programming languages (PASCAL, ALGOL, FORTRAN,...) are used and compilers exist as tools in the design process.

In contrast to this is the situation at level 3 to 6, where most of the design work is done without the assistance of existing computers. The generated documents are often incomplete system descriptions consisting of large reports, written in a natural language enriched with some informal flow charts or block diagrams.

A prerequisite for the developement of CAD-programs applicable at those levels is the existance of a formal denotational system, which allows the designer to express his ideas in an unambiguous computer-understandable form. Although many creative activities of the designer can not be automated, CAD-programs, based on such a denotational system, can be developed for

- automatic documentation
- verification by simulation and
- automatic transformation (logic synthesis).

Register transfer languages can be used as notational tools to express ideas in the early stages of logic design. Since 1964 more than 50 different language definitions have been published /2/. Their developement is based on Reed's recognition in 1952 /3/, that digital computers accept binary coded words as input and store this information in sets of registers. The contents of registers can only be changed at descrete points of time and new (next) information is only derived by transforming the present contents of register files. This behaviour can easily be described by simple Boolean transfer equations.

Normal programming languages (especially ALGOL and APL) have had great influence on rt-language developements and many syntactical forms have been adopted.

CDL from Chu /4/ can be viewed as the best known rt-language, because of its very simple and easily understandable structure.

DDL from Duley /5/ introduces for the first time the block concept of ALGOL in a consistent way.

CASSANDRE from Mermet /6/ gives new possibilities to express modular designs by introducing the unit-concept.

ISP /7/ was especially developed for specification of instruction sets and offers powerful features to express the transformations on information performed by computer instructions. However, it is difficult to describe computer structures in ISP.

AHPL /8/, APL*DS /9/ and MDL /10/ are derived from APL and use the powerful set of operations on binary vectors and matrices known from APL.

These and the many unnamed rt-languages did not find general acceptance by hardware designers in contrast to the acceptance of high level programming languages by programmers for the following reasons:

a) simulators - if developed for some rt-languages - were not generally available

b) rt-languages were mostly developed for educational purposes at universities; the set of operations and standard functions incorporated were often too small to obtain compact descriptions of real complex hardware designs

c) illconsidered adoption of too many language constructs from normal programming languages emphasizes system behaviour too much and neglects system structure

d) rt-languages are often difficult to learn because of many unnecessary exception rules

e) timing models are difficult to understand if rt-languages offer

constructs to describe asynchronous circuit behaviour; in
several cases only an intensive study of applied simulation
algorithms provides some insight into the intended semantics.

In view of this situation the following guidelines lead to the
developement of the rt-language RTS Ia /11/ which will be presented
in these lecture notes:

a) the language should be easy for hardware designers to learn and
easy for hardware designers to teach

b) in spite of its simple structure the language should be
applicable to all 4 design levels mentioned above; it should
support the stepwise design independently of the design method
applied (Outside in - Bottom up)

c) the design of the rt-language should contain a complete set of
syntactic and semantic rules which ensure, that any text
conforming to these rules describes a realizable piece of
hardware

d) it should support the compact and clear description of hardware
systems by offering special language constructs to specify
frequently used substructures appearing in hardware designs

e) there must be a complete simulation algorithm, which can be
applied to all valid (see c) RTS Ia descriptions without any
restrictions.

1.2 Basic features of RTS Ia

RTS Ia is a notational system to describe (to specify - not to
program) Mealy automatons with the accuracy used in mathematical
notational systems. All alphabets (input-, output-, state-)
consist of ordered sets of Boolean values (1 - 0, true - false,
high - low etc.). The simple timing model (before/after relation)
avoids the inconsistancies of other rt-languages that allow the
mixture of synchronous and asynchronous system behaviour in their
description.

RTS Ia offers a number of equivalence rules to describe one
and the same automaton in several different ways, thus clearly
showing special features, which the user is interested in during
specific design steps. It offers many constructs to describe very
complex automatons in a very compact and easily understandable way,
suppressing repetitive pieces of text.

Valid RTS Ia descriptions are always unambiguous automaton
descriptions, which also can theoretically be represented by

(translated into) large tables defining the total output- and
statetransition-function.

It is up to the user to decide at which level of detail he
wants to model his hardware system by a Mealy automaton.

a) leading edge triggered   b) computer
      D-flipflop

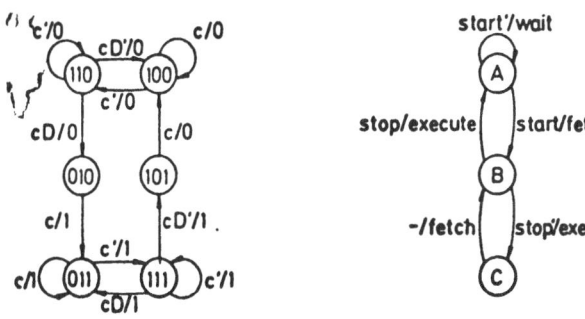

(inputvar. => <u>inputterminal</u> ; outputvar. => <u>outputterminal</u>
        statevariables => <u>register</u>)

Fig. 1.1 State diagrams at different levels of detail

In most cases complex automatons are realized by synchronous
digital systems. Although no equivalence rules between hardware
elements and language constructs are defined in RTS Ia, some simple
interpretations can be applied which yield a mapping of RTS Ia
descriptions into real hardware structures, especially if
synchronous hardware system designs are desired.

This general hardware-description-language approach has the
advantage of being very flexible, free of inconsistancies and
applicable at many levels of abstraction in the design process.
Furthermore, it is not restricted to a special technology or a
special design method.

It has the disadvantage that the user has to consider the
technical restrictions imposed by realizing an abstract automaton
description using special hardware elements (different
flipflop-types and gates etc.) and clocking schemes.

In spite of the theoretical approach in defining RTS Ia, the
language constructs incorporated allow an interpretation in such a
manner that descriptions can easily be mapped into isomorphic

hardware structures, consisting of flipflops, gates and MSI-IC's, or into flowcharts showing the way a hardware system is performing any desired function.

All statement constructs incorporated in RTS Ia are only extentions of notations used in switching theory and can all be transformed into a set of simple Boolean expressions on scalar input- and state-variables.

1.3 Methods used by RTS Ia to achieve compact descriptions of complex hardware systems

Multidimensional dataobjects and operations on them are introduced, so that many identical subnetworks can be described by one expression.

If already verified subdesigns are used as building blocks of larger systems (e.g. complex standard IC's), the specification of these subdesigns can be reduced to pure I/O behaviour description. A very powerful set of operations and standard functions lead to compact I/O behaviour descriptions.

RTS Ia supports the partitioning of complex automatons into small subautomatons, especially the seperation into control automaton and data processing automaton, thus reducing the complexity.

By nested conditional transfer and connection statements complex state transition or output functions can be defined stepwise by a series of simple functions.

1.4 Restrictions imposed by the RTS Ia Mealy automaton model

If a hardware system is modeled as a synchronous automaton in RTS Ia, the network may not contain direct feedback of gates (creating asynchronous behaviour). In practice, I/O pins of IC's driven by bidirectional bus drivers must be modeled by separate input- and outputlines. Synchronously driven flipflops, which are also driven via asynchronous set- and reset-lines, can only be modeled with restrictions.

The two-valued logic used in RTS Ia does not support direct modeling of tristate drivers.

The specification of real time behaviour of technical logic elements in cumbersome.

2. THE RT-LANGUAGE RTS IA

2.1 General remarks about RTS Ia

RTS Ia does not impose any format on the statements. Every line consists of at most 72 characters; blanks and comments (arbitrary strings inclosed in the symbols "....") can be inserted freely between any RTS Ia symbols. This feature should be used to create easy readable descriptions. All keywords of RTS Ia must be inclosed in -....-, they do not restrict the user in naming arbitrary identifiers. Any string of letters and digits starting with a letter can be used as a name; only the first 8 characters are significant.

2.2 Multidimensional data objects in RTS Ia

RTS Ia distinguishes two kinds of <u>data objects</u>: values and data carriers.

<u>Values</u> are ordered sets of simple Boolean values 0 and 1. A simple Boolean value is called a scalar; a linear order of scalars is a vector and a linear order of vectors is a field. The concept of RTS Ia is based on data objects of arbitrary dimension but the implementation of the simulation system only supports data objects with a dimension up to 2.

Two basic types of <u>data carriers</u> are available. They must be declared at the beginning of a RTS Ia description and can be named by arbitrary identifiers.

Carriers of <u>register</u>-type represent state-variables; carriers of <u>input</u>- and <u>outputterminal</u>-type represent I/O-variables of an automaton. <u>Terminal</u>-type carriers can be used to name signals which are outputs of combinatorial subnetworks. In the context of automaton descriptions, connections to <u>terminal</u> carriers can be viewed as identity declarations; thus they can all be eliminated by substitution of corresponding expressions, although in practical design situations no one would do so.

A basic feature of each <u>data object</u> is its dimension and the number of components in each <u>dimension</u>.

<u>Operations</u> on values, used in <u>expressions</u>, create new values. That is the reason why values in expressions can be represented by binary constants, by data carrier identifiers and by expressions themselves.

Transfer- and connection statements create a correspondence between ordered sets of data carriers and ordered sets of simple Boolean values.

A basic rule applied to Boolean operations on values and to transfers and connections is as follows:

Operations are performed between corresponding components, requiring that the data objects involved must be compatible in number of dimensions and number of components in each dimension.

2.3 Binary constants

Only scalar and 1-dimensional bitstrings can be represented directly in RTS Ia. Each binary constant starts with the symbol # and consists of three parts, the value (interpreting the bitstring as a binary number), the base of the number system used (B or R ≡ 2, O ≡ 8, D ≡ 10 (2-complement), H ≡ 16) and the length of the desired bitstring.

examples:	binary constant	bitstring created
	# 110 B 4	0110
	# -5 D 7	111 1011
	# F83 H 15	000 1111 1000 0011

(a shorter notation for scalar constants is #0 and #1).

2.4 Declaration of data carriers

Scalar data carriers are denoted by an arbitrary identifier; 1 and 2 dimensional data carriers by identifies followed by an indexlist inclosed in [.....]. The number of components in each dimension and their numbering scheme is expressed by a pair of decimal numbers separated by a colon, index pairs of different dimensions are separated by a semicolon.

examples of declared numbering schemes:

x ≡ component

```
    [7:5]              ┌───────┐
                       │ x x x │
                       └───────┘
                         7 6 5

    [0:4]              ┌───────────┐
                       │ x x x x x │
                       └───────────┘
                         0 1 2 3 4

  [7:6;0:2]           7 ┌───────┐
                        │ x x x │
                      6 │ x x x │
                        └───────┘
                         0 1 2
```

Data carrier declaration statements start with a keyword characterizing the special carrier type (-REGISTER-, -TERMINAL-, -INPUTTERMINAL-,-OUTPUTTERMINAL-), followed by a list of identifiers and terminate with a semicolon.

examples: -REGISTER- OVR, AK[0:3], MEM[0:1023; 15:0];
           -TERMINAL- CARRY, SUM[7:4], COMP[4:1;2:3];

2.5 Expressions

Expressions are used to describe the behaviour and/or structure of combinatorial circuits.

Although the normal operators known from Boolean algebra - extended to multidimensional Boolean values - allow the expression of any mapping, RTS Ia makes a set of standard functions available for compact description of frequently used mappings.

The operations may be classified into those applicable to values of any dimension (a) and those which interpret a scalar or bitstring as a binary number (b) and thus cannot be applied to 2 dimensional values.

Another possible criteria for classifying the operations could be to distinguish between operations which can be interpreted as a network of gates (c) and others, which only restructure multidimensional values (d).

A technical realization of the second kind of operations does not need any special hardware elements, only the interchange of a few wires is required. Many operations can be nested in one expression. The sequence of evaluation (the order of connecting gates) is determined by explicit bracketing or implicit operator priority.

Table 1 gives a complete overview of all operations defined in RTS Ia reflecting the above classification. All standard functions can be transformed into expressions using only the operations NOT, AND, OR, Indexing and Concatenation. However, their incorporation in RTS Ia aids the specification of complex functions in a short and easily understandable way. The functions dealing with binary arithmetic are especially easy to understand and are often used by hardware designers, while logic networks performing these functions can only be defined by extensive iterative equations, if basic Boolean operators are used exclusively.

examples of equivalence rules:

```
    F( C, 4)  ≡  (C , C , C , C)
 MINT( A, 3)  ≡  A[0]'*A[1]'*A[2]*A[3]
```

294

operation / symbol		number of Bool opnds	din of Bool opnds	dimension of result	example	result of example
OR	+	any	any (all eq)	eq opnds	A+B	0111
EXclusiv OR	++	any	any (all eq)	eq opnds	A++B	0110
Equivalence	**	any	any (all eq)	eq opnds	A**B	1001
AND	*	any	any (all eq)	eq opnds	A*B	0001
NOT	'	1	any	eq opnd	A'	1100
NAND	NAND	any	any (all eq)	eq opnds	NAND(A,B)	1110
NOR	NOR	any	any (all eq)	eq opnds	NOR(A,B)	1000
EQual	EQ	2	any (all eq)	0	EQ(A,B)	0
Not Equal	NE	2	any (all eq)	0	NE(A,B)	1
Address	[ ]	2	any[0,1]	1.opnd	B[A]	1
INCrement	INC	1		1	INC(A)	0100
DeCRement	DCR	1	1	1	DCR(A)	0010
ADD modulo	ADD	2	1	1	ADD(A,B)	1000
SUBtract mod	SUB	2	1	1	SUB(A,B)	1110
ADd w. Carry	ADC	3	1,1,0	1	ADC(A,B,C)	01001
SuB w. Borrow	SBB	3	1,1,0	1	SBB(A,B,C)	11101
Less Than	LT	2	1	0	LT(A,B)	1
Less or Equal	LE	2	1	0	LE(A,B)	1
Greater Than	GT	2	1	0	GT(A,B)	0
Great.or Equal	GE	2	1	0	GE(A,B)	0
MIN Term	MINT	1Bool,1dec	1	0	MINT(A,3)	1
MAX Term	MAXT	1Bool,1dec	1	0	MAXT(A,3)	0
COde	COD	1Bool,1dec	1	1	COD(D,3)	011
DECode	DEC	1Bool,1dec	1	1	DEC(A,6)	000100

(a) Bool. opnds interpreted as ordered sets of B. values

(b) Bool. opnds interpreted as binary numbers

values of operands used in the examples:

A[0:3]= #0011 B 4, B[0:3] = #0101 B 4,   C = #1,   D = #00010 B 5,   E = #0

Table 1 (part 1): RTS Ia operations which can be realized by gate networks (c)

operation / symbol		number of Bool opnds	dim of Bool opnds	dimension of result	example	result of example
Left SHift	LSH	2Bool,(1dec)	1,0	1	LSH(A,C,2)	1111
Right SHift	RSH	2Bool,(1dec)	0,1	1	RSH(E,A)	0001
ROtate Left	ROL	1Bool,(1dec)	1	1	ROL(A)	0110
ROtate Right	ROR	1Bool,(1dec)	1	1	ROR(A,3)	0110
TRanspose	TR	1	2	2	TR([101 / 001])	10 / 00 / 11
index	[ ]	1	any>0	any	A[3:0,1,3]	110001
concatenate in bit dim	(,,)	any	any	any>0	(A,B,C)	001101011
in worddim	(;;)	any	any	any>0	(A;B)	0011 / 0101
Fan out	F	1Bool,1dec	any	any>0	F(C,4) / F(A;3)	1111 / 0011 / 0011 / 0011

values of operands used in the examples:

A[0:3] = #0011 B 4, B[0:3] = #0101 B 4, C = #1, D = #00010 B 5, E = #0

Table 1 (part 2): RTS Ia operations which re-structure Boolean values (d)

$$A' \equiv (A[0]' , A[1]' , A[2]' , A[3]')$$
$$ROL(A) \equiv A[1:3,0]$$
$$NAND(A, B) \equiv (A*B)' \equiv ((A[0]*B[0])' , (A[1]*B[1])' ,$$
$$(A[2]*B[2])' , (A[3]*B[3])')$$

The universal definitions of the operations indexing, concatenation and addressing are very important and useful. They can be applied to operands which are themselves expressions.

Indexing consists in the selection and re-ordering of arbitrary subsets of components of a data object in any dimension.

Concatenation supports the composition of a n-dimensional data object from numerous smaller objects in any dimension.

Addressing concerns the selection of a component of a data object in the highest dimension by a scalar or vector Boolean value, which always requires the presence of a decoding network in a technical realization.

Many other rt-languages are too restrictive in their definition of these operations.

The combination of addressing and concatenation supports a short specification of the behaviour of multiplexers; other rt-languages sometimes try to express such behaviour by conditional expressions. E.g. the RTS Ia expression (A ; B)[S] means if S then B else A .

Each expression can be translated into an isomorphic combinatorial network, but it is up to the user of RTS Ia to explain by a comment that he has such an interpretation in mind or that he is not dealing with network structures at the moment.

The following examples should demonstrate the flexibility of RTS Ia expressions when specifying some combinatorial TTL-circuits.

-TERMINAL- A[4:1], B[4:1], C0, S, G;

"7400" NAND(A, B)   quadruple 2 input nand gates

"7483" ADC(A, B, C0)   4 bit binary full adders
        (the output carry C4 is in the first bitposition
        of the 5 bit outputvector)

"74157" F(G',4)*(A ; B)[S]   quadruple 2 line to 1 line
                                  data selector/multiplexer
        (the value of S determines whether vector A or B is
        selected and the value of not G is gating this
        vector to the output)

-TERMINAL- A, B, C, D, G1, G2;
"74154" (DEC((D,C,B,A), 16) * F(G1'*G2', 16)))'
                          4 line to 16 line
                          decoders/demultiplexers
         (signals D,C,B and A are concatenated to a 4-bit
         binary vector, whose binary value determines which
         of the 16 output lines of the decoder is raised
         to one; all 16 outputlines are gated by the value of
         "not G1 and not G2" and the whole output vector is
         inverted)

The last example should demonstrate how different RTS Ia
expressions for the same total function can be used to describe the
structure or the behaviour of some interconnected MSI-circuits.
The network consists of two multiplexers 74157, one full adder 7483
and 4 EXOR gates 7486.

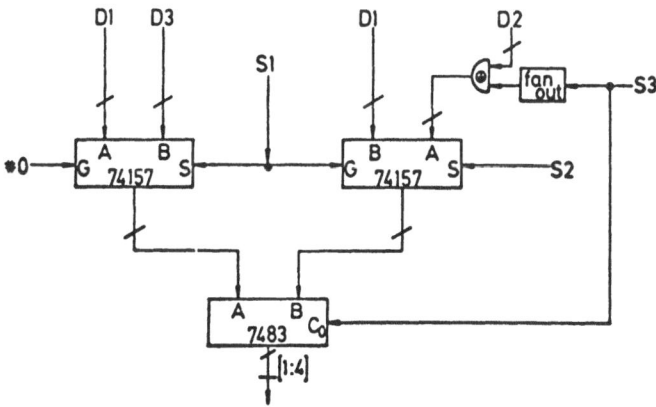

Fig. 2.1 Example of a MSI/TTL-network

structural description:
    -TERMINAL- D1[1:4], D2[1:4], D3[1:4], S1, S2, S3;
    ADC(F(#0',4)*(D1 ; D3)[S1] ,
         F(S1',4)*(D2++F(S3,4) ; D1)[S2] ,
                           S3   ) [1:4]

behavioural description:
"0"	(ADD(D1, D2) ;
"1"	SUB(D1, D2) ;
"2"	LSH(D1, #0) ;
"3"	LSH(D1, #1) ;
"4"	D3 ;
"5"	INC(D3) ;
"6"	D3 ;
"7"	INC(D3))[(S1 , S2 , S3)]

298

While the first expression directly shows how the input and output lines of the IC's are to be connected, the second expression would help a microprogrammer to understand the performed function when he determines the required code for the control signals S1, S2, S3 (the equivalence of both expressions can be proven by selective evaluation of the operands gated to the full adder when choosing all possible values of the control signals S1, S2, S3).

2.6 The register transfer statement

While expressions can be used to describe combinatorial network behaviour, the sequential behaviour of some circuits can only be specified in RTS Ia by using transfer statements. They determine how the next state value of a state variable depends on the present state and present input signal values.

An automaton specified by the following state diagram can be

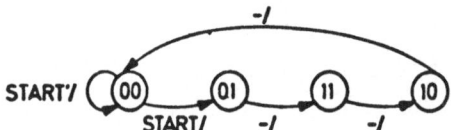

Fig. 2.2 State diagram of a Johnson counter

completely described in RTS Ia by the transfer statement

    Z <= (Z ; (Z[2],Z[1]')) [START+Z[1]+Z[2]]

if Z is declared as -REGISTER- Z[1:2]; and START as -INPUTTERMINAL- START;

It says that the next state of Z is the concatenation of (Z[2],Z[1]') if START+Z[1]+Z[2] is 1; otherwise, the new state value is equal to the old state value.

Keeping in mind that clocked flipflops store their old value as long as no new clock pulse arrives, the realization of the automaton by a synchronous machine would be more clearly described by a conditional transfer

    -IF- START+Z[1]+Z[2] -THEN- Z <= (Z[2],Z[1]') -FI-
  or shorter /START+Z[1]+Z[2]/ Z <= (Z[2],Z[1]') ;

It says that the current state value changes according to the source expression if and only if the conditional (control) expression evaluates to 1.

Thus if synchronous system design is intended, a simple
conditional transfer in RTS Ia is a basic statement which can be
directly implemented by clocked flipflops; the following figure
gives a set of possible hardware interpretations of this statement
type.

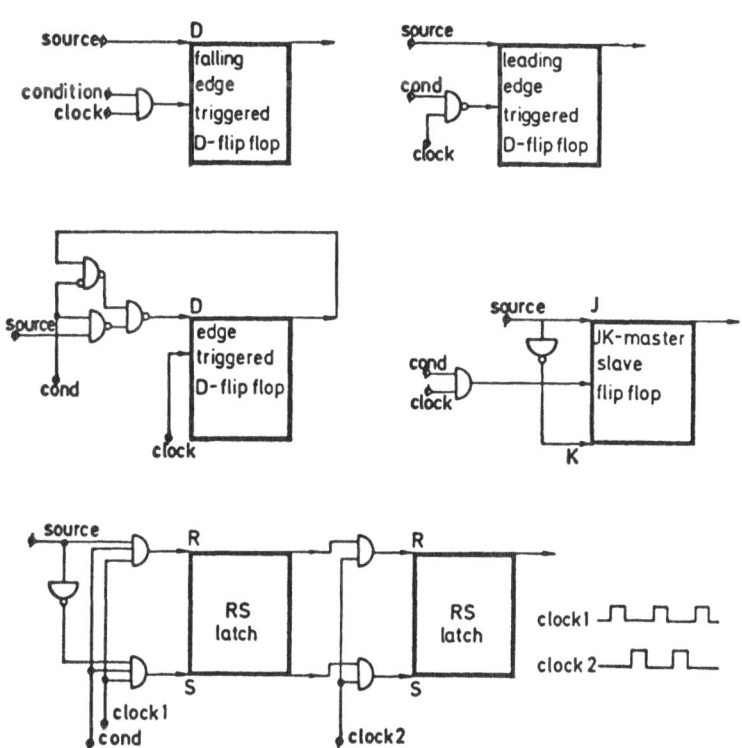

Fig. 2.3 Proposals for the realization of a simple
          conditional transfer

It should be emphasized that the determination of a special
flipflop type for a technical realization of a conditional transfer
statement is not part of the RTS Ia language. The designer has to
check whether his real timing conditions of the source-, condition-
and clocksignal yield the state transition behaviour desired.

Several different conditional transfers with the same target
register can be used in RTS Ia to describe complex state transition
functions step by step.

```
-REGISTER- Z;   -TERMINAL- COND1, COND2, EXP1, EXP2;
     /COND1/ Z <= EXP1;
     /COND2/ Z <= EXP2;
```

They clearly show the events under which the current state is changing, which is the preferable way of expressing the behaviour of a sequential machine. To get an unambiguous state transition function, it is necessary that COND1 and COND2 never evaluate to 1 (true) at the same time -- a condition which can sometimes only be proven dynamically during simulation if the interrelationship of inputsignals is not known in advance.

For a technical realization these two transfers can be transformed into one conditional transfer statement of the following form:

/COND1+COND2/ Z <= EXP1*COND1 + EXP2*COND2;

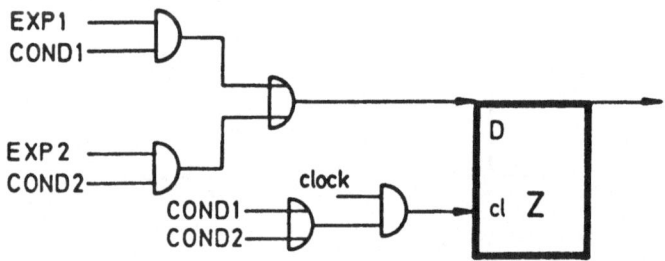

Fig. 2.4 Proposal for realization of above transfer

The concept of several conditional transfers serves to partition complex functions into smaller subfunctions and the assumption of not changing the state value if no condition is true reduces the volume of complex system descriptions considerably.

To give a feeling of the flexibility of RTS Ia, the following extentions to the conditional transfer concept should be noted:

a) The target register carrier of a transfer can be a
    single register carrier
    an indexed register carrier
    an addressed register carrier
    the concatenation of all three alternatives
b) one condition can control a whole set of transfer statements, in this case all governed statements are separated by commas.
c) conditional transfers can be nested to any depth and an additional "else"-part can be appended
    (-IF- ... -THEN- ... -ELSE- ... -FI-   or
      / ... / ... : ... ;   )
d) in addition to the IF-statement, a CASE-statement allows the enumeration of several different conditions, depending on

the value of a vectoriell Boolean expression interpreted
as a binary number
```
(-CASE- <b expr>
         :<dec number>:   <statement list>
         :<dec number>:   <statement list>
                   etc.
                               -ESAC-  )
```

All these advanced language constructs can be transformed back into
simple conditional or even unconditional transfers.

2.7 The terminal connection statement

Connection statements are used in RTS Ia to describe the
combinatiorial portion of a digital system. Their structure is
composed of

- a set of target terminal carriers   followed by
- the connection symbol =   followed by
- a source expression

Although their syntactical form is nearly identical to that of
the transfer statement, their meaning differs in several ways.
While register-type carriers when used as operands in expressions
make available only the old state value, terminal-type carriers
used as operands immediately pass new values to the next
expression. They do not store any information. In digital systems
they can be interpreted as wires being connected to the outputs of
gates or flipflops.

An unconditional connection to a terminal can be translated
directly into an isomorphic gate network if a structural
description is intended.

```
        -TERMINAL- A, B, X;
X = NAND( NAND( A, B'), NAND( A', B))
```

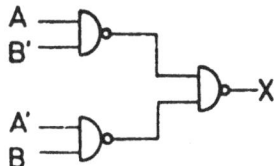

Fig. 2.5 Isomorphic gate network of above connection statement

A conditional connection like /A/ X = B´; should be read as
"if A is 1, X has the value of not B otherwise X has the value 0";
thus the statement is equivalent to the unconditional connection
X = A * B´ . If several conditional connections are defined with
the same target carrier, the same restriction mentioned about
conditional transfers will hold: only one condition may be active
at a time to guarantee unambiguous function specification.

Conditional connections serve as a medium for stepwise
definition of arbitrary Boolean functions, which can even be
written as a table of Boolean constants like

```
            -CASE- (A , B)
                 :0:  X = #0
                 :1:  X = #1
                 :2:  X = #1
                 :3:  X = #0    -ESAC-
```

describing the same EXOR-function as above. During the
developement of microprograms this specification method may be
useful. Connection and transfer statements can be freely
intermixed in the statement lists of IF- or CASE-statements if the
same condition governs both.

2.8 The structure of a complete RTS Ia description

In contrast to the extensive set of operations and rules for
building up expressions, the structure of a complete RTS Ia
description is very simple due to its recursive syntax.

All data carriers used must be declared initially (in an
arbitrary sequence). The digital system to be described is defined
by a statement list (single statements separated by commas) and the
description is completed by the keyword -FINIS-. The sequence of
statements is without meaning because parallel activities in a
network are being described. Four different statement types are
available and have already been mentioned:

- register transfer
- terminal connection
- IF-statement (in a short and long notation)
- CASE-statement

The last two statement types may contain one or several statement
lists again, thus offering the possibility to clearly separate a
hierarchically structured control flow from normal data processing
activities in the description of complex digital systems.

To enhance the readability it is sometimes advisable to use
the macro-facilities of RTS Ia:

If some part of a statement list is used often in a description, it can be declared as a "micro operation" immedeately after the declaration of data carriers.

(format: -MICRO- <name> -IS- <statem. list> ;)

This declared macro can then be "called" in any statement list by @<name>, which means that the whole declared statement list is substituted at every point of evocation. It does not extend the class of digital systems, which can be described within RTS Ia.

3. THE DEVELOPEMENT OF A MULTIPLIER-UNIT, USING RTS IA

3.1 Basic representation of a digital system

It has already been mentioned that – using RTS Ia for digital system description – the system is always modelled as Mealy automaton and that a variety of description forms can be generated for one and the same system, all defining the same automaton behaviour. In normal applications a designer must deal with different problem areas and he would assume an extended meaning to the language statements used, depending on his actual interests.

Two basic representation forms can be distinguished. One explicitly shows the sequential functioning (behaviour) of a system and the other explicitly shows the hardware structure used to build up the system.

3.2 The algorithmic description form

An algorithmic description form must explicitly express which information transports in a network are performed concurrently (often in one clock cycle) and which information transports are carried out in sequence. Normally flowcharts (extented by symbols to express parallelism) or state diagrams are used to denote such properties. In RTS Ia a state diagram can be represented directly by one large CASE-statement controlled by the contents of an explicit "state"-register. Under the different labels, created by decoding the state register contents, all operations performed simultaneously in parallel can be grouped together. State transitions can be expressed by transfering new (constant) values to the "state"-register. Many other rt-languages offer symbolic labels and GOTO-statements for this purpose; both methods give special prominence to exclusive operations in a network.

A state diagram of an automaton, performing the multiplication of two 4-bit binary numbers according to the "add and shift" algorithm is found in fig. 3.1. It assumes the existance of two

304

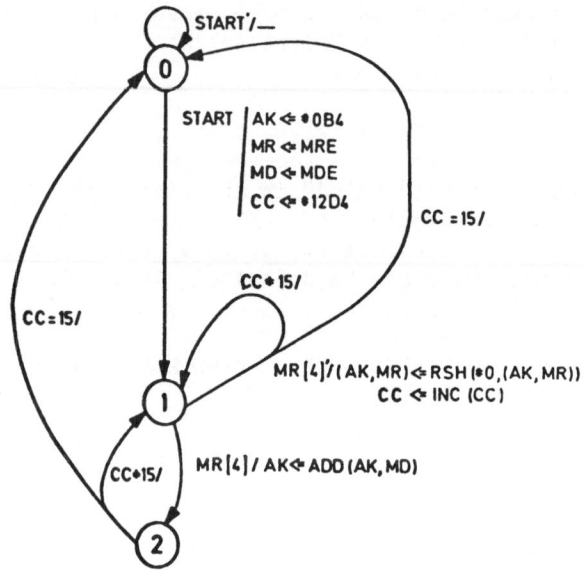

Fig. 3.1 State diagram of multiplication unit

"ALGIRITIM OF 4 BIT MULTIPLICATION
=================================="

```
-INPUTTERMINAL- START, MRE[1:4], MDE[1:4];
-REGISTER-      AK[1:4], MR[1:4], MD[1:4], CC[1:4], ST[1:2];

-CASE- ST
:0: -IF-START-THEN- AK <= #0B4,
                    MR <= MRE,
                    MD <= MDE,
                    CC <= #12D4,   ST <= #1B2    -FI-
:1: -IF-MR[4]-THEN- AK <= ADD( AK, MD),
                    ST <= #2D2
            -ELSE- (AK,MR) <= RSH( #0, (AK,MR)),  CC <= INC( CC),
                   -IF- EQ( CC, #15D4) -THEN- ST <= #0B2
                                       -ELSE- ST <= #1B2 -FI--FI-
:2: (AK,MR) <= RSH( #0, (AK,MR)),  CC <= INC( CC),
    -IF- EQ( CC, #15D4) -THEN- ST <= #0B2
                        -ELSE- ST <= #1B2  -FI-
-ESAC-
-FINIS-
```

Fig. 3.2 Algorithmic RTS Ia description of multiplication unit

data registers MR and MD for multiplier and multiplicand, one
accumulator AK, one count-register CC to control the number of
add-shift cycles and one "state"-register .ST to distinguish the
three different states. The multiplication unit is activated by a
pulse on inputterminal START; the data operands are taken over
from inputlines MRE and MDE.

The RTS Ia description in figure 3.2 contains the complete
information of the state diagram in a one-to-one correspondence.
It clearly separates the data operations of initial register
loading, register addition and register shift, which are activated
at different points of time.

The use of normal RTS Ia expressions to specify the required
data operations always guarantees unambiguity and the designer can
be certain of the existance of some hardware structure to implement
the hardware algorithm defined.

3.3 The structural description form

A structural RTS Ia description form cannot be defined in an
equivalently simple way like the algorithmic form. The concept of
structure depends on the level of detail a designer is interested
in. In particular, if he is designing networks of
LSI/MSI-circuits, he is only dealing with the interconnections of
whole circuits and not with interconnections inside a chip that are
already fixed by the IC-producer. Because the same language
constructs of RTS Ia are used for functional and structural
descriptions, a designer has to mark by comments the parts of his
description which should be interpreted in the first and which in
the second way.

Pure structural RTS Ia descriptions are composed of only two
types of statements: unconditional connections and simple
conditional transfers without nesting. In addition, every terminal
or register carrier may only be used once as target carrier. This
results in a unique outputname for a combinatorial network, if it
is defined by an unconditional connection, or in an unambiguous
wiring scheme for data and clock inputs of flipflops (see examples
in fig. 2.1), if simple conditional transfers are considered.

Inputlines to combinatorial networks (always described by
RTS Ia expressions) do not receive explicit names; every operand
represents an inputline, which should be connected with the
appropriate outputline, defined by other subexpressions or explicit
carrier names.

In real design situations a mixture of both description forms
is often used, especially if the structure of the data processing

```
        "STRUCTURE OF DATA PROCESSING PART OF
              4 BIT MULTIPLICATION
        ======================================="


-INPUTTERMINAL- START, MRE[1:4], MDE[1:4];
                                    -INPU- D;  "DON'T CARE"
-REGISTER-      AK[1:4], MR[1:4], MD[1:4], CC[1:4], ST[1:2];
-TERMINAL- SUM[1:5], MPX1[1:4], MPX2[1:4], CCEQ15, CCE[1:4];
           "CONTROL SIGNALS"
-TERMINAL- S1,S2,S3,S4,S5,S6,S7,S8,S9,S10, CLEAR,LOAD,ENT,ENP;
           "NEXT STATE VALUE"
-TERMINAL- NXST[1:2];


  SUM  = ADC( AK, MD, S1),                            "7483"
  MPX1 = F( S2',4)*( SUM[2:5] ; (#0,AK[1:3]))[S3],    "74157"
/ S4 / AK <= MPX1; ,                                  "74175"

  MPX2 = F( S5',4)*( MRE ; (AK[4],MR[1:3]))[S6],      "74157"
/ S7 / MR <= MPX2;                                    "74175"
/ S8 / MD <= MDE;                                     "74175"
/ S9 / -IF- CLEAR' -THEN- CC <= #0B4               "74163"
                -ELSE- -IF- LOAD' -THEN- CC <= CCE
                              -ELSE- -IF- ENT*ENP -THEN-
                                CC <= INC(CC) -FI--FI--FI-;,
        CCEQ15 = MINT(CC,15),
/ S10 /  ST <= NXST; ,                              "1/2*74175"


      "PERMANENT CONNECTIONS"
CCE = #12D4,   S5 = #0,  CLEAR = #1,  ENT = #1,  S9 = S4, S10 = S4,
 S1 = #0,
              "TABLE-LIKE DEFINITION OF CONTROL SIGNALS
              ========================================="


/ST[1]'*ST[2]'*START'/ (S4, S2, S3, S6, S7, S8, LOAD, ENP, NXST )
            = (#0, D, D, D, #0, #0,   D,   D, (D,D));
/ST[1]'*ST[2]'*START / (S4, S2, S3, S6, S7, S8, LOAD, ENP, NXST )
            = (#1, #1, D, #0, #1, #1,  #0,   D, #1D2);
/ST[1]'*ST[2] *MR[4]'*CCEQ15'/
                 (S4, S2, S3, S6, S7, S8, LOAD, ENP, NXST )
            = (#1, #0, #1, #1, #1, #0,  #1,  #1, #1D2);
/ST[1]'*ST[2] *MR[4]'*CCEQ15 /
                 (S4, S2, S3, S6, S7, S8, LOAD, ENP, NXST )
            = (#1, #0, #1, #1, #1, #0,  #1,  #1, #0D2);
/ST[1]'*ST[2] *MR[4]/ (S4, S2, S3, S6, S7, S8, LOAD, ENP, NXST )
            = (#1, #0, #0,  D, #0, #0,  #1,  #0, #2D2);
/ST[1]*ST[2]'*CCEQ15'/ (S4, S2, S3, S6, S7, S8, LOAD, ENP, NXST )
            = (#1, #0, #1, #1, #1, #0,  #1,  #1, #1D2);
/ST[1]*ST[2]'*CCEQ15 / (S4, S2, S3, S6, S7, S8, LOAD, ENP, NXST )
            = (#1, #0, #1, #1, #1, #0,  #1,  #1, #0D2);
-FINIS-
```

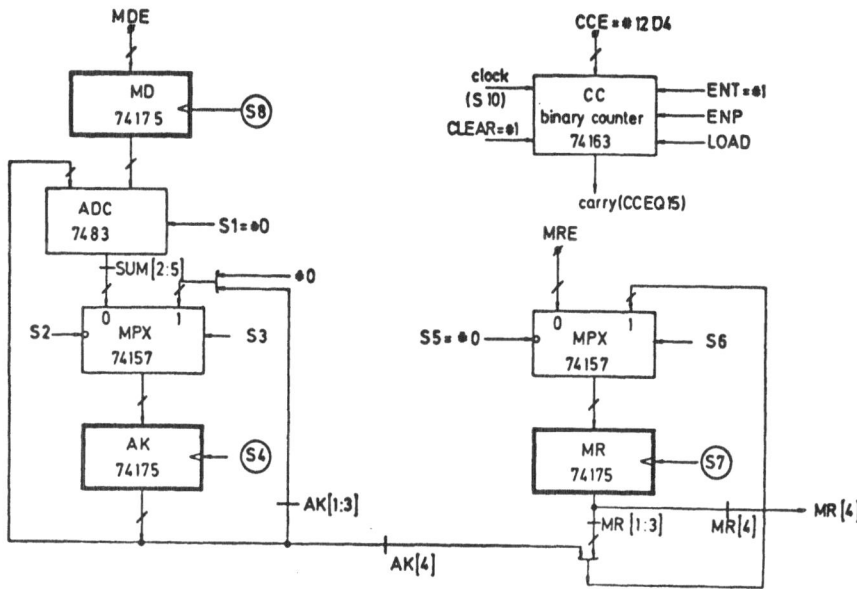

Fig. 3.3 Structure of the data processing part
a) RTS Ia description   b) MSI/TTL network

part of a system is developed first. Estimations of effectiveness
may sometimes result in modifications of the desired hardware
algorithm that can easily be documented, if the control part is
still described in an algorithmic form.

In the special example of the multiplier unit, the designer
has to find one combinatorial network for every data register
input, which can be controlled to perform all different information
transports, required by the algorithm. Referring to the AK
register, a simple structure of a multiplexer 74157 with a
fulladder 7483 at one data input can perform all three transfers
used in the algorithm, if control signals S2, S3 and S4 are chosen
according to the following table:

S2	S3	S4	
–	–	0	AK remains unchanged
1	–	1	AK <= #0B4
0	0	1	AK <= ADD(AK, MD)
0	1	1	AK <= RSH(#0, AK)

To implement the transfers MR <= MRE and MR <= RSH(AK[4], MR)
only one MPX 74157 must be used; register MD can be loaded
directly.

Operations concerning the step counter CC can all be performed by the IC 74163. Thus only a behavioural description of this IC is necessary, which shows the dependency between the operations and control signals involved.

During this design phase a set of new control signals is introduced and the required behaviour of the control unit can first be defined in a table-like form. The table could be interpreted as the contents of a microprogram storage, if a microprogrammed control unit is desired. A complete RTS Ia description of the results of this design step can be found in fig. 3.3.

If a hardwired control unit is developed, only the last part of the description must be replaced by the description of the gate network, which generates the control signals and the next state value (see fig. 3.4).

```
              "HARD WIRED CONTROL UNIT
              ======================="

LOAD = NAND(ST[1]',ST[2]'),  S6 = LOAD,     S2 = LOAD',
  H1 = NAND(S2,START),        S8 = H1',     S4 = NAND(S2,H1),
 ENP = NAND(MR[4],ST[2]),     S3 = ENP,     S7 = NAND(S4,ENP)',
   NXST[1] = ENP',       NXST[2] = NAND(CCEQ15',ENP)'
-FINIS-
```

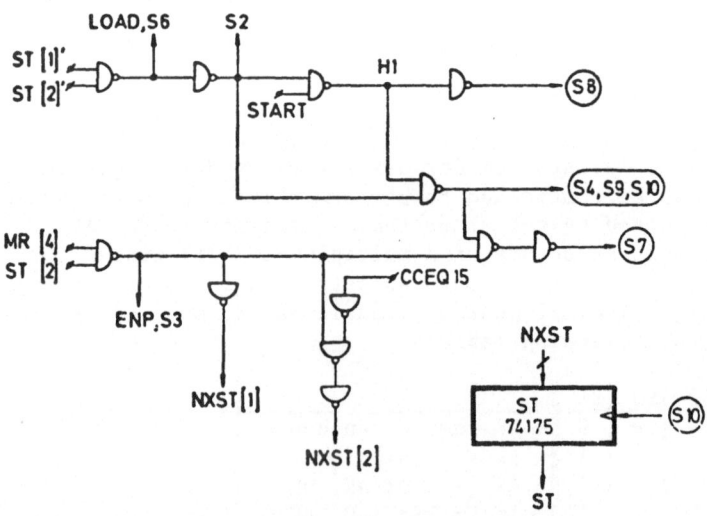

Fig. 3.4 Structure of hard wired control unit
a) RTS Ia description b) gate network

This small design example should have demonstrated how different forms of rt-desriptions of the same system can be used

for documentation of design ideas at several levels during the
design process. It should be emphasized that when a simulator is
used for verification, all different description forms should
generate identical simulation results, which can be compared to
detect logical design errors.

4. SIMULATION OF RTS IA DESCRIPTIONS

4.1 General remarks

Efficiency as a general goal of rt-simulation system developement
depends on the different requirements imposed upon such a system by
the main application areas.

Rt-simulation can be used at the instruction set level to
develop software for not yet existing computers. It can also be
used to evaluate the performance of total system designs by running
typical instruction sequences, but the most obvious application is
the early verification of not yet complete hardware subsystem
designs.

While in the first two cases, the underlying rt-descriptions
are mostly correct, the third application has to deal extensively
with incorrect rt-descriptions. That is why rt-simulators used for
verification should not only possess good run time efficiency but
also should offer a powerful set of tools for easy error detection
and error localization.

While formal errors can be detected automatically (e.g. by a
compiler) according to the rt-language definition rules,
"intention" errors (formally correct descriptions generating
unintended simulation results) can only be recognized by the
hardware designer himself, who should be assisted by a powerful
dynamic debugging system.

4.2 Essential tasks of the rt-simulation system

Besides evaluating cycle per cycle the state transition and output
functions defined by some rt-description, the simulation system
must be able to perform the following basic tasks at user's demand:

- raising the digital system described to any initial state, by
  storing arbitrary values in the register-type carriers
- imitating the behaviour of the environment of the digital system
  to be simulated by supplying inputterminal-type carriers with
  arbitrary sequences of values
- printing out values of arbitrary rt-carriers, which are

310

calculated during simulation
- terminating a simulation test experiment.

Simulation control commands serve to direct the simulator's
activity when it is occupied with these auxiliary tasks.

4.3 Structure and basic features of the RTS Ia simulation system

Simulation of RTS Ia is performed by a two stage method, which
means that a compiler reads a complete RTS Ia description, checks
it for syntactic and semantic errors and translates it into some
internal data structure (IDS). On this internal data structure the
interpreter calculates cycle by cycle the state transitions
described in RTS Ia. A user has the ability to interact with the
interpreter and control its activity by means of simulation control
commands.

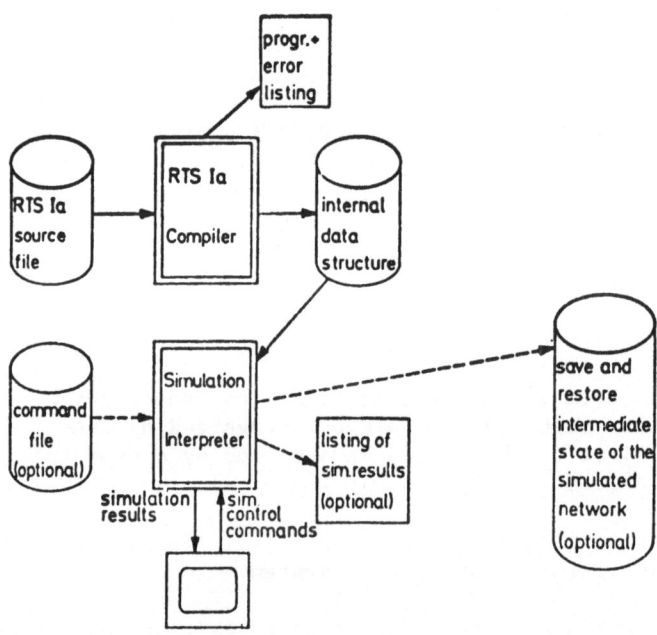

Fig. 4.1 Data flow between programs and I/O devices

Fig. 4.1 shows the necessary and optional data flow between the
two programs and the I/O devices of the host computer.

While the RTS Ia compiler is working like a batch program
(error checking can be done without human intervention), the

interpreter program is based on heavy human intervention, offering the designer tools to adapt his test strategy to intermediate simulation results.

A strict separation between hardware system description in RTS Ia and simulation control commands has been considered for the following reasons:

- there may by other programs working on rt-descriptions; simulation is just one of many possible applications
- it may be necessary to run different simulation test experiments, all with the same rt-description
- the hardware designer should not be burdened with simulation problems while he is describing his design ideas in RTS Ia.

Because rt-simulation should be applied particularly for such complex hardware systems where gate simulation would consume too much computer time, the following considerations were taken into account to gain a runtime efficient simulation system:

1) While the compiler analyses a rt-description only once, the interpreter has to calculate state transitions repetitively for many hundreds of cycles. Thus interpreter speed primarily determines total simulation system speed.

a) That is why in defining the internal data structure special attention has been paid to the fact that the compiler should perform all data independent calculations on the rt-description only once. The interpreter is relieved of many unnecessary repetitive and data independent search tasks if he has direct access to all required information in appropriate tables and data structures.

b) Detection of formal errors is also mostly delegated to the compiler. It checks every RTS Ia description extensively on syntactic and semantic rules, distinguishing more than 100 different error messages. It should be emphasized that nearly 50 checks deal with operand compatibility in all aspects where RTS Ia expressions are analyzed. Only four different error situations have to be checked by the interpreter during each simulation cycle:

1) register- or inputterminal-type carriers are not initiated before their value is used in a calculation
2) the value of an address expression points to a position outside the declared bounds of the addressed carrier
3) the value of the argument of the COD function is not a binary vector with one bitposition being 1 and all other positions being 0
4) during one simulation cycle two or more different transfer/

connection statements try to change the value of the same
target carrier bit position. This most serious error
situation often arises if rt-descriptions are composed of
subsystem descriptions, partly structural partly algorithmic.
It notifies ambiguous automaton function definitions.

2) During the simulation of one cycle the interpreter has to
evaluate rt-expressions most of the time (conditional and source
expressions). To reduce the amount of evaluation time per
cycle, three measures were taken.

a) Offering a list inside the IDS, which contains the sequence of
substitutions of terminal-type carriers, every expression need
be calculated at most once per cycle (no stability criterion
must be observed).

b) Using the information on hierarchical control structures in
RTS Ia (IF- and CASE-statements), results of control
expressions determine which subsets of dependent expressions
must be calculated. That is why in contrast to DDL no
transformations of rt-descriptions to a canonical form are
performed in advance of the simulation.

c) While evaluating one complete rt-expression, high-order
standard RTS Ia functions are not reduced to a series of
primitive Boolean operations but are calculated directly, using
the powerful instruction set of the host computer.

To obtain a simulation system, which is relatively independent
of the host computer, most of the programs are coded in FORTRAN IV
and some computer attributes are kept as parameters (e.g.
bitlength of one integer word, characters per line and lines per
page used on the local line printer). Only about 100 lines of
assembler code performing character and bit handling have to be
recoded when the system is to be transfered to another host
computer.

Each of the two programs consists of nearly 4000 statements.
With respect to to the interpreter, only 1300 statements are
necessary for pure rt-simulation, the rest is intended for
simulation control command interpretation. This relation evidently
shows that pushing a lot of tasks into the compiler and choosing
the appropriate internal data structure results in a simple
simulation algorithm.

4.4 The system - user interface of the RTS Ia compiler

As mentioned earlier, the compiler translates RTS Ia descriptions
without human intervention. When the user starts the program, he

only has four parameter options to control:

a) the amount of information contained in the compiler listing
   (RTS source text, internal data structure, compiler statistics)

b) the unit to receive the listing (line printer/data set)

c) the generation of IDS

d) the selecton of zero- or one-origin for the implicit index
   numbering scheme.

4.5 The system - user interface of the RTS Ia interpreter

The basic tasks of the interpreter have already been described in
4.2. As RTS Ia descriptions are event oriented, the mode of
operation of the interpreter is also event oriented in the
following sense:

   Modeling the behaviour of the program by a state diagram (fig.
4.1), the user has to distinguish between two different modes of
operation.

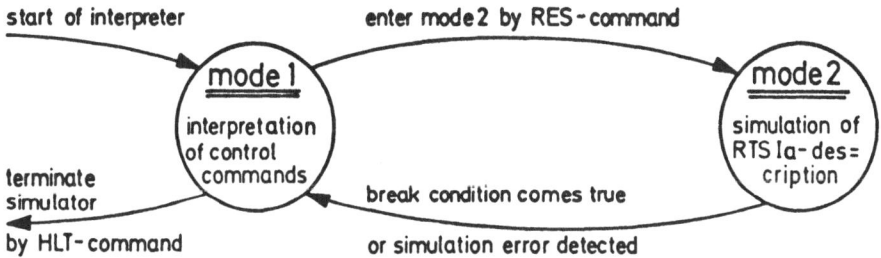

Fig. 4.2 Modes of the RTS Ia interpreter

   While in mode 1, the interpreter reads any sequence of control
commands offered by the user, the real rt-simulation is done in
mode 2 by calculating the state transition defined in the RTS Ia
description for any number of cycles. The user has the option of
inducing the interpreter to enter mode 2 for rt-simulation with the
RES(ume) command. Once the interpreter is in that mode, it
simulates an arbitrary number of cycles until one of several user
defined break conditions comes true, causing it to return to mode 1
to accept a new sequence of control commands.

   While erronous control commands are rejected immediately in
mode 1, the detection of a simulation error in mode 2 is sent to

the user at the end of the actual cycle, forcing the interpreter to enter mode 1. The user then has the opportunity to search for the cause of the error situation by exploring the actual state of his digital system decription.

4.6 Form and effect of some simulation control commands

Control commands have to satisfy a special format; they may be 4 to 72 characters long; the first three characters represent the command code; every command string is terminated by the symbol &. If a command has one or several "operands", they are separated from the command code and from each other by a semicolon.

4.6.1 Break conditions.

Break conditions may depend on simulation time or on values of arbitrary RTS Ia-carriers.The interpreter keeps track of the simulation time by consecutively numbering every cycle starting with 0. By means of the STP-command the user can define an

absolute time break at cycle n1 by #A=n1 and a

relative time break n2 cycles from now by #R=n2 and a

periodical time break starting at cycle n3 and every n4 cycles later by #P=n3, n4

(n1, n2, n3, n4 are positive integers).

Value dependent break conditions are defined by

- the name of a RTS Ia carrier
- the symbol = or =′ (to be read equal resp. not equal)
- a value represented in the binary (B,R), octal (O), decimal (D) or hexadecimal (H) number system.

Finally by #B=n5 the user can ask for a break if a transfer or connection in line number n5 (the position of the transfer or connection symbol <=, = in the RTS Ia source text) is evaluated, a condition which is only of interest in algorithmic RTS Ia descriptions.

Example:

STP; #R=5; READY=1; AK=′13D&

This command causes the interpreter to interrupt rt-simulation

- 5 cycles from now,
- always when the scalar carrier READY is 1 and

- always when the carrier AK does not have the decimal value 13.

While break conditions #A and #R are erased automatically if they become true, all others remain active until they are explicitly disabled by the user with the command DES. The identification of a special break condition in this command is performed with the same format of operands used in the STP-command.

4.6.2 Supplying rt-carriers with values.
To initiate <u>register</u> carriers or to supply <u>inputterminal</u> carriers with any signal waveform, the <u>SET-command</u> is used. Scalar and 1-dimensional carriers are denoted directly by their identifier, 2-dimensional carriers by identifier and index number of a special "word". The = sign separates name and value part; values of bitstrings can be represented in the four different numbering systems mentioned above. To reduce the length of the SET-command, the following abbreviations are accepted:

The value of a bitstring containing only 0's can be omitted, the carrier identifier is sufficient in this special case.

If several words of a 2-dimensional carrier get the same bitstring value, the total range of words involved may be specified.

<u>Inputterminal</u> carriers retain the value supplied by the SET-command until a new SET-command is interpreted. Thus to specify input waveforms, only changes of values must be given by SET-commands.

Example:

SET; START=1; AK=FF H; DATAIN; M(3:5)=1001 B &

4.6.3 Output of rt-carrier values.
The print out of simulation results is bitstring oriented in the same way as the input of values, which means that scalar and 1-dimensioal values may be displayed directly in one of four possible numbering systems while 2-dimensional values can only be printed word by word.

Two different outputstreams are at the user's disposal:

a) For immediate observation of results on the terminal, he may use the <u>display command DSP</u>. The operand list may contain an arbitrary sequence of rt-carrier identifiers, with the option of each identifier is being followed by the desired output number representation (binary is standard). If all scalar and 1-dimensional carriers are to be displayed (hexadecimal dump), the operand list of the DSP-command must be empty. The same command can also be used to display

- the actual simulation cycle number (#)
- all active value-dependent break conditions ($),
- all active time-dependent break conditions (#A #R #P).

Example:

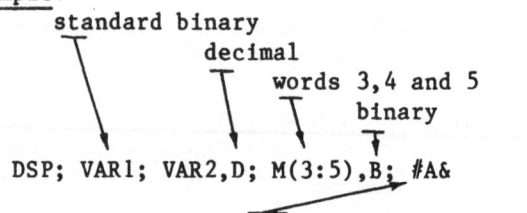

standard binary

decimal

words 3,4 and 5

binary

DSP; VAR1; VAR2,D; M(3:5),B; #A&

the active absolute time break condition

Every DSP-command causes a single output operation, when it is interpreted.

b) If continuous recording of results at the line printer during a whole simulation session is desired, the commands TBI or TBS should be used. They may be given only once and cause the interpreter to automatically generate a clearly arranged table with up to 30 columns (containing up to 30 different rt-carrier values).

TBI causes one line of values every simulation cycle to be printed, while TBS only causes a line to be printed if a break condition is true. Thus the second form should be used in extensive simulation sequences where values are only of interest at special points of time.

Example:

TBI; VAR1; VAR2, D; M(3:5), B &

It should be mentioned that in both cases the user is relieved from output format specification, the layout of the printed results is arranged by the interpreter.

4.6.4 Summary of remaining commands.
The resume-command RES momentarily stops the interpretation of a sequence of commands and restarts rt-simulation.

The halt-command HLT finishes a simulation session.

The NEW-command restarts a total simulation session without reloading the interpreter program.

The save-command SAV is useful if someone wants to keep the momentary state of a simulated rt-description on secondary memory.

This state can be restored later on (in the same or in another simulation session) by the unsave-comand UNS.

The ONT- and OFT-command are used to switch on and off the trace mode of the interpreter. Trace mode means that the interpreter prints a list of all transfer and connection statements (the soure text line numbers), which are evaluated in one cycle. Analyzing such a listing, the user can retrace which conditions of his (algorithmic) rt-description were true at which time.

Finally, sequences of simulation control commands can be predefined in a data set (e.g. initialization commands for a rt-description, which are used in many simulation tests).

The CCI-command (change command input) causes the interpreter to read the following commands from the data set instead of from the terminal until a second CCI-command is read there and the user gets back control at his terminal for further input.

The rewind-command REW supports the multiple use of a predefined command sequence data set in one simulation session.

For completeness it should be mentioned, that after starting the RTS Ia interpreter, the user can influence total interpreter behaviuour by changing some parameters.

E.g. - line printer paper length and width for table-like output can be specified
      - all rt-carrier values can implicitly be set to zero
      - a maximum number of intended simulation cycles can be fixed to prevent infinite loops and
      - immediate termination of simulation can be forced, when simulation errors are detected.

Simulation control command interpretation starts for the first time when these parameters are fixed.

4.7 A brief simulation example

Fig 4.3 shows a complete sequence of simulation control commands, which generate the table-like output of simulation results, when applied to one of the three descriptions of the multiplication unit presented in 3.

The first line declares the headline of the table. All values will be printed in binary form with leading zeros (,R).

```
TPI; AK,R; MR; MD; CC; ST; START&
SET; AK=FH; MR=FH; MD=FH; CC=7H; ST; START; MDE=5D; MRE=11D&
STP; #R=1; ST=0; START=1&
RES&
SET; START=1&
RES&
SET; START&
RES&
HLT&
```

CYNR I	AK I	MR I	MD I	CC I	ST I	START I
1 I	1111 I	1111 I	1111 I	0111 I	00 I	0 I
2 I	0000 I	1011 I	0101 I	1100 I	01 I	1 I
3 I	0101 I	I	I	I	10 I	0 I
4 I	0010 I	1101 I	I	1101 I	01 I	I
5 I	0111 I	I	I	I	10 I	I
6 I	0011 I	1110 I	I	1110 I	01 I	I
7 I	0001 I	1111 I	I	1111 I	I	I
8 I	0110 I	I	I	I	10 I	I
9 I	0011 I	0111 I	I	0000 I	00 I	I

Fig. 4.3 Simulation control commands and simulation results

The second line sets all registers and inputterminals to arbitrary values; only the state register ST and START receive 0.

The third line declares 3 break conditions, one time dependent and two value dependent. The first and second will be used to generate a start impulse, the third to detect the end of the multiplication operation.

The RES in line 4 start the rt-simulation. When the first break condition ($\#R=1$) is true after one simulation cycle, START is raised to one (line 5) and rt-simulation restarted (line 6). One cycle later break condition 2 (START=1) becomes true; the START signal is set back to zero (line 7) and rt-simulation is restarted again (line 8). 6 cycles later the state register ST receives the value zero. The third break condition becomes true (ST=0) and the simulation session can be concluded with the HLT-command in line 9.

The table of simulation results clearly shows the sequence of add- and shift operations in registers AK, MR, the operation of the step counter CC and the sequence of states ST when number 5 is multiplied with number 11 by the unit designed.

REFERENCES
1. Dietmeyer, D.L., Duley, J.R.: Register transfer languages and their translation, in: Breuer, M.A., Editor: Digital system design automation, Computer Science Press, 1975
2. Special issue on Computer Hardware Description Languages, IEEE Computer, Vol.7 Nr.12, 12/1974
3. Reed, I.S.: Symbolic Sythesis of Digital Computers, Proc. ACM, 9/1952, pp 90-94
4. Chu, Y.: An ALGOL-like computer design language, CACM Vol.8, 10/1965, pp 607-615
5. Duley, J.R.: DDL - A Digital System Design Language, PhD dissertation, University of Wisconsin, Madison, 1967
6. Bressy, Y.: The Language CASSANDRE, Description and Processing, (Manual), Grenoble, 1975
7. Bell, G., Newell, A.: Computer Structures: Readings and Examples, Mc Graw - Hill, 1971
8. Hill, F.J., Peterson, G.R.: Digital Systems: Hardware Organisation and Design, Wiley, New York, 1973
9. Franta, W.R., Giloi, W.K.: APL*DS: A Hardware Description Language for Design and Simulation, pp 45-52, 1975 International Symposium on Computer Hardware Description Languages and Their Application - Proceedings, Graduate Center, City University of New York, New York City, Sept.3-5, 1975 (IEEE Nr. 75CH1010 - 8C)
10. Lipovski, G.J.: On Conditional Expressions in Digital Hardware Description Languages, Proc. 2nd Workshop on Hardware Description Languages, Darmstadt, 1974, (ACM German Chapter Lectures W-1974)
11. Knobloch, H.-J.: RTS Ia - ein System zur formalen Beschreibung und Simulation komplexer Schaltwerke, D 17 Darmstaedter Dissertationen, 1978

FUNCTIONAL SIMULATION

J. MERMET
Maître de Recherche at C.N.R.S.
Director of the Research Laboratory of the National
School of Engineering in Applied Mathematics and
Informatics of Grenoble (E.N.S.I.M.A.G.)
- with collaboration of D. BORRIONE, EOUTTERIN, Y.BRESSY
C. LE FAOU,

A - RTL SIMULATION : CASSANDRE AND LASCAR

I - CHOICE OF A HARDWARE DESCRIPTION LANGUAGE

During the design process a number of choices will have to be
tried ; constant modifications will be brought, as the project
evolves and new problems are being faced, by hardware and software people ; several types of models will have to be built. For
these reasons, we decided not to use ad-hoc simulators which require too much time and debugging efforts for changes. We then
need a general simulation language, that possesses some indispen-
sable properties to be well suited to the characteristics of our
evaluation method.
1) The language should be a descriptive one ; the structure of the
   system, and all the connexions between the components must
   appear.
2) The language must be modular. We must be able to decide, test,
   and simulate the components of the system separately. This will
   permit us to modify a component, or replace it by a functionally identical one (but internally different) without altering
   the overall simulation model.
3) Description at different levels of detail must be possible :
   hardware, firmware, functions, algorithms.
4) Simultaneous events and parallelism must be easy to express, and
   automatically managed.
5) The language must give the possibility to simulate within the
   same experiment performances,conflicts, resource management
   algorithms, etc...
6) When the same component is described at different levels of
   detail, it should be easy to validate the most abstract model

from the most detailed one.

7) We need to process large benchmarks. The simulator has to be efficient.

8) It is preferable to have a language that can be run both interactively, for quick correction or modification of a model, andon batch processing, for long simulation runs.

The usual way of being satisfied when so exacting, is to define one's own language, when no existing one fulfills all the required conditions.

The development of the language CASSANDRE was a first step to fulfill these requirements because

- it possesses the properties 1, 2, 4, 8.

- it is a computer aided design tool for hardware description and verification. It is a high level, Algol-like, language, whose meaningful variables are 'REGISTERS', 'SIGNALS', 'CLOCKS' and 'STATES', with a wide variety of operators on both scalars and arrays. It has two nice properties, from the user's point of view :

- subsets of a description, called 'UNIT', can be written and checked for correctness separately ; modification and debugging are therefore easier.

- the state of the model can be saved and restored as often as necessary during simulation ; from a given state it is therefore possible to see the evolution of the model, according to different values of input signals, without having to start again from the begining of the simulation.

However, as we wish to use higher level models running on long benchmarks, which is not what the language was made for, we meet important shortcomings of CASSANDRE from this standpoint :

- when one does not want to describe everything in detail, the only way to do it is to tabulate the outputs of a unit from its inputs ; this can be quite costly in storage.

- big descriptions, or descriptions that own many signals, are very slow to simulate ; it would be unreasonable to simulate long streams of instructions.

The idea was therefore to extend CASSANDRE. Nothing that formely existed has been changed, so as not to disturb the users who are not interested in our modifications. But new features have been implemented, that will permit to describe functionally some units, when one is not interested in their hardware detailed description ; simulation time will then be greatly improved. The result of this extension was LASCAR.

II. CASSANDRE

The syntax of CASSANDRE has been chosen closely connected with a known form of expression whose formalism facilitates its understanding and use.

The semantics of a model written in CASSANDRE are the description

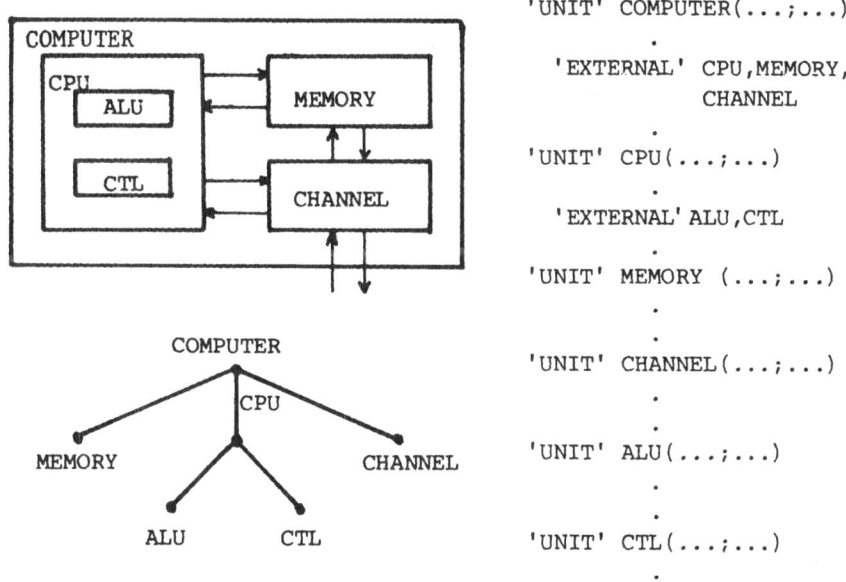

Fig. 1. Nested structure of UNITS and CASSANDRE Description.

of both system behavior and system structure :
- system behavior
during simulation of the model, the user can verify the sequence
of internal states and output signals as a response to any input
stream.
- System structure
The description itself can give all necessary information on the
expected physical implementation, depending upon the constructs
of the language which have been used to write the model.
The unit concept introduced in CASSANDRE enables system description not to be considered as a whole, but as a set of elements.
Their complexity is left to the designer's choice. This enables
the iterative structure of many logic systems to be outlined.
Figure 1 is an example of a CASSANDRE description (a nested
structure of UNITS).

1. The UNITS

Unlike many other languages, where a logic system description
appears as a whole, a CASSANDRE description can be segmented into
as many elements, as small or as complex as necessary. These parts
are called UNIT (see fig. 1), this segmentability results from
three essential constraints :
- The length of a text should be limited, so that it be possible
  for it to be written easily by the designer and then introduced
  into the machine and compiled.

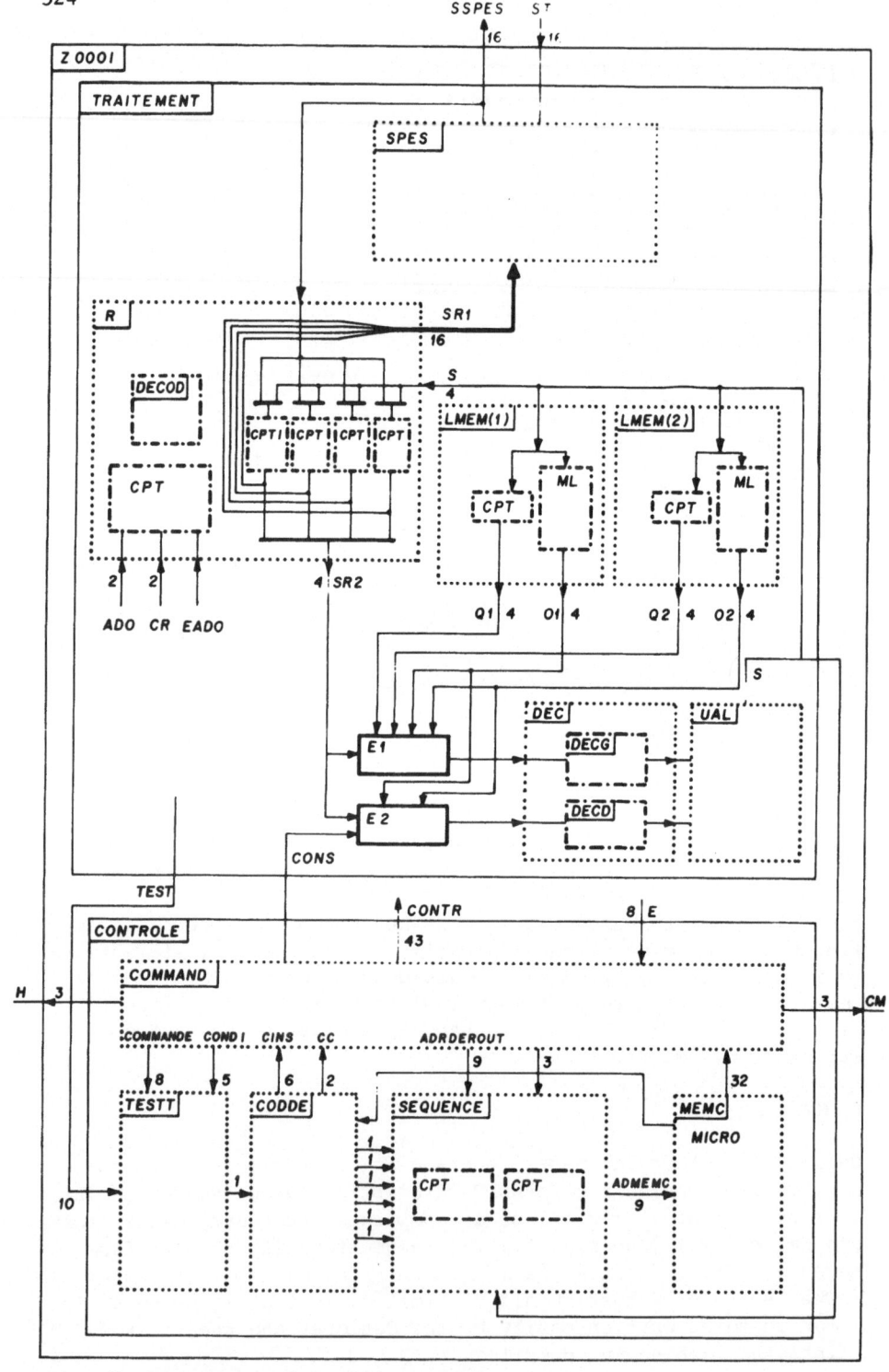

Fig. 2. Structure of a mini-computer.

- Logic system implementation methods require that this system be simple, if they are manual, in order to stay within the reach of the designer's understanding, if they are automatic, to give the results within a reasonable time.
- Logic system implementation technology, in its turn, imposes a partition of the language into subsets of a hierarchy varying according to the complexity of the material implementing them (example, figure 2). [4]

From the formal viewpoint, a unit communicates with the outside world only through its input-output interface.

Interface variables are the only objects of a unit which are accessible by its environment. The inputs and outputs are either signals or clocks.

Units used as components of units of a greater size are declared 'EXTERNAL' and described independently (see fig. 1).

Connections of instances of these units describe how they are used. The description of a single unit falls into three parts :

- Declarations and specifications
- Permanently valid instructions
- List of unit states and instructions controlled by each state.

The first part calls upon the basic elements of the language and subunits in order to list the hardware components of a given unit.

2. Basic elements of the language.

There are three distinct element families in this language :

- Logic variables (storage elements, boolean connection elements) REGISTER, SIGNAL
- Synchronization variables CLOCK, MASTERCLOCK
- Automata state variables STATE

    2.1. Logic variables : Registers and signals are to be distinguished.

    Any variable technically representing an element capable of information storage is declared as a REGISTER or MEMORY. Any wire representing the output of either a combinatorial circuit or a storage element is declared SIGNAL.

    Registers, memories, signals may be arrays of any dimension given in the declarations.

    Example : 'REGISTER' RA (0:7), RB (0:7, 0:63, 0:63) RC;

    This statement declares RC as a flip-flop, RA as an 8 bit Register and RB as a 3 dimensions memory having 64 x 64 eight bit words.

    It is to be noted that this declaration indicates no technical implementation of the storage elements, which may be designed from ferrite stores as well as RS or D Flip-Flop, themselves implemented with gates, themselves implemented with MOS transistors.

    Some expressions can be made up from registers, memories, signals and constants by using the following operators (Fig 3)

- boolean operators (component to component)
- sequential operators : delay, derivation
- non boolean operator : proper when their operands are arrays
  (figure 3).

Example 'REGISTER' R1 (1:8), R3 (1:6), R4;
       'SIGNAL'   S1 (1:8), S2 (0:10) ;

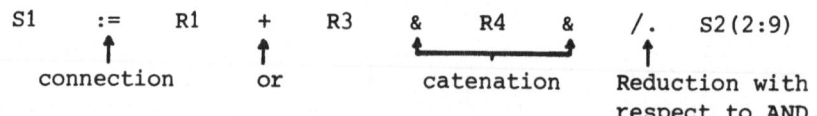

S1   :=   R1   +   R3   &   R4   &   /.   S2(2:9)

    connection   or   catenation   Reduction with
                                                  respect to AND.

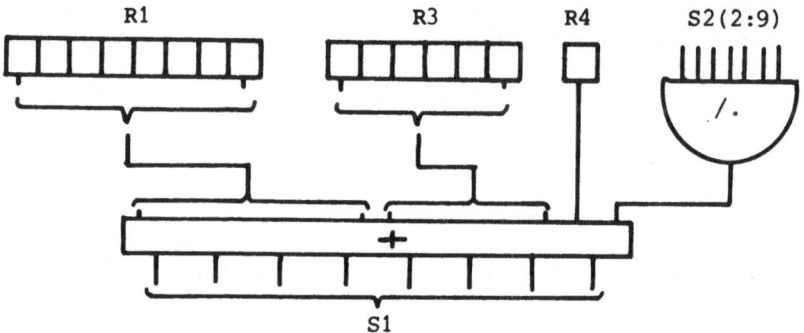

Fig. 3.

2.2. Synchronization variables :

2.2. Synchronization variables : any synchronization variable (formally considered as a pulse) must be specified or declared as a CLOCK.

It is reasonable to give a special treatment to these variables which in synthesis play the particular role of energy supplier and synchronizer element.

Some expressions can define new clocks, either by derivation of logic signals or, as a union or a delay of existing clocks. Clocks may also be arrays.

In order to make those definitions accurate the following should be noted :

a. Duality between variables of the REGISTER and SIGNAL type. The value of a SIGNAL is independent of the CLOCK type variable.

The value of a REGISTER changes only under the influence of a CLOCK.

b. Duality between variables of SIGNAL and CLOCK type.

The value of a SIGNAL variable is fixed during a certain time : it is a level.

The value of a pulse variable is supposed to be instantaneous and infinitely short (ideal hypothesis).

A SIGNAL is a boolean function of a certain number of REGISTER outputs.

2.3. Automata states : The STATE specification establishes
the existence of state registers without fixing their dimensions. A state variable is thus a boolean variable not yet
encoded. However it can be loaded under the influence of a
pulse like a Register variable in order to describe the sequencing of automata. When no STATE statement appears in the
declarative part of a UNIT, an automaton is however associated
with the UNIT. The description of the UNIT can mention transition between states of this implicit automaton using the "ENA-
BLE" intructions and the **state** labels (see bellow).

3. Elementary operators

3.1. Boolean operators : boolean operators must all have
compatible dimensions. The following operations are applied
bit by bit :
- . logical AND
- + logical OR
- - negation
- = conjunction
- $\neq$ disjunction
- > greater than
- $\geq$ greater or equal
- < less than
- $\leq$ less or equal

EXAMPLE :
    If OP1 (1:4) contains 0011
    and OP2 (1:4) contains 1010

the result of	- OP1	is	1100
"	OP1.OP2	"	0010
"	OP1+OP2	"	1011
"	OP1=OP2	"	0110
"	OP1$\neq$OP2	"	1001
"	OP1>OP2	"	0001
"	OP1$\geq$OP2	"	0111
"	OP1<OP2	"	1000
"	OP1$\leq$OP2	"	1110

3.2. Non boolean operators
a) Reduction /op.
The notation /op. is given when op. is one of the logical operators . + = $\neq$ > $\geq$ < $\leq$
The reduction applies to boolean variables of any dimension.
It works in the first direction. After this operator, the
first dimension is degenerated, that is, reduced to one element - the structure of the variable is thus changed.
Signification
/. result = 0 if at least one bit of the operand is 0,
            otherwise 1
/+ result = 1 if at least one bit of the operand is 1,
            otherwise 0   etc...

328

EXAMPLE :
. the branching:S   := /+ R(1:4) corresponds to the figure 4

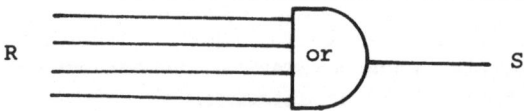

R   or   S

Fig. 4

. with the array A (1:4, 1:2), containing 0110
                                           1100
  /.A is reduced to the vector $\begin{pmatrix} 0 \\ 0 \end{pmatrix}$

REMARK
The result of a vector reduction is a scalar and that of an n-
dimensional array an n-1 dimensional array.
b) Catenation &
This operator allows two variables with the same number of dimensions to be coupled together according to the first dimension. All the dimensions other than the first must be compatible.
EXAMPLES :
. Let A (2:7, 0:7, 1:36)
  and B (1:10, 1:8, 11:46)
A & B is an array (16,8,36)
. Two vectors may be catenated :
     S(1:8) := R(1:5) & T(1:3)
corresponding to figure 5

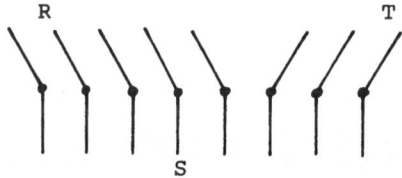

Fig. 5

c) *D Shift
This operator, denoted *D|n|, and applied to array variables,
acts on the first dimension.
It causes this dimension to be shifted by n positions.
- to the left if n is positive (no sign)
- to the right if n is negative (-sign)
The overrunning positions are lost. The result is shorter by
n positions.
EXAMPLE :
S(1:7) := *D|3| R (1:10)
is a shift to the left of
3 bits of the vector R
(Figure 6)

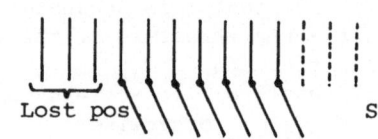

Fig. 6

IS a shift with zero fill is to be realized, this must be explicitly written using the catenate operator

EXAMPLE :

     R1 := ⊬D|3| R2 & OOO

d) *R Rotation

This operator is similar to the preceding one in every point, but the positions freed on the left or right are replaced by the positions previously lost (circular shift).

EXAMPLE :

S(1:5) := *R|-1| T(1:5) is a straight rotation of one position from vector T (figure 7).

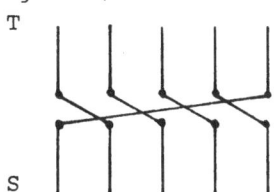

T

S

Fig. 7

e) *P Permutation (or transposition)

The permutation operation, denoted *P|n_1||n_2|, used on variable: with at least two dimensions, transposes dimensions n_1 and n_2 of the array.

EXAMPLE :

S(1:5, 1:2) := *P|1||3| R(1:2, 6, 1:5) ;

When the variable has only two dimensions, they need not be designated.

EXAMPLE :

S(1:5, 1:2) := *P R(0:1, 0:4) ;

REMARK

In the case of an array variable, the permutation allows the use of the reduction, catenation shift and rotation at any level, by bringing it back to the first position.

EXAMPLE :

  *P(*D|2| (*P S(1:4, 1:8))) is a shift to the left of two positions on the second dimension of S.

f) % Delay

This operator may only be used in an asynchronous description. %|n| introduces in CASSANDRE the notion of duration, by delaying by n time units the logical signals or expressions to which it is applied.

EXAMPLE :

     S := %|3| A

At the S assignment, the value of A three time units previously, must be considered.

The figure 8 shows the evolution of S in terms of an evolution of A.

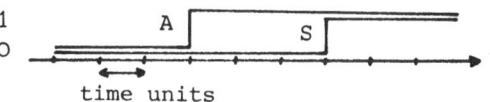

time units

Fig. 8

330

It is possible to accumulate these delays ; that is, several
transitions may be stored in the delay unit.
EXAMPLE :
The expressions $\%|n_1|\ (A\leftharpoonup\%|n_2|\ B)$
$\qquad\qquad\%|n_1|A + \%|n_2 \rightarrow n_1|B)$
$\qquad\qquad\qquad\qquad$ are equivalent.

g) Adressing operator

This operator, whose hardware equivalent is a decoder, is sym-
bolized $\cancel{S}$. Applied to a vectorial variable, it gives the numeric value of the binary code contained in this variable.
EXAMPLE : Let signal S be connected to A ($\cancel{S}$(B)). If B(1:3)
contains 101, S is connected to A(5)   (figure 9).

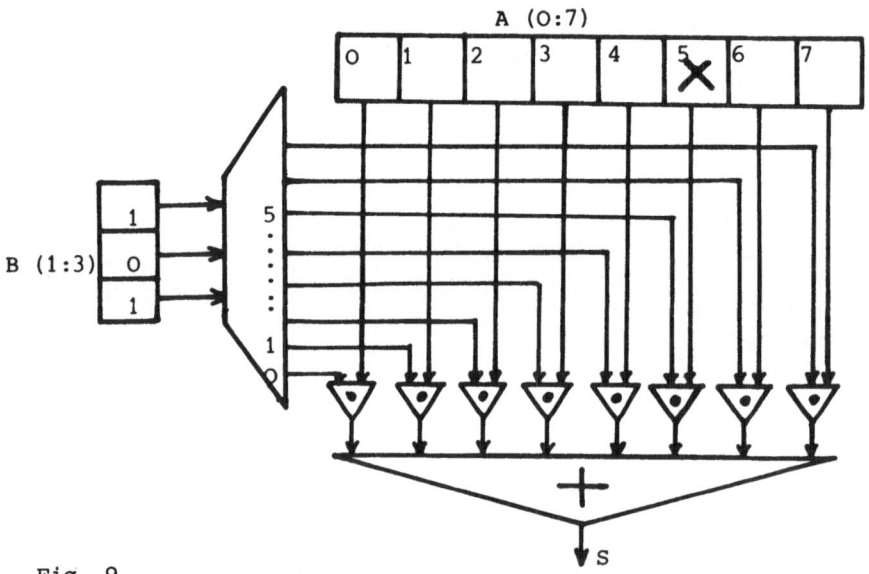

Fig. 9

h) Derivation operator

The derivation operator $\odot$ is specific of the asynchronous
version. It allows the passage from a signal variable to a
clock. More generally this operator creates a pulse on the
leading edge of a combinatorial network or a signal output.
EXAMPLE : S := $\odot$(T), S must be declared CLOCK

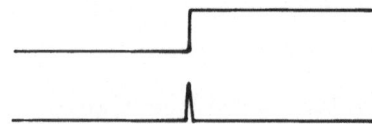

Fig. 10
REMARK
This operator allows the operation of certain flipflops to be
described, where loading is controlled by a signal and not by

a clock.
When falling edge control is wanted, the desired effect is
obtained by negating the control variable.
EXAMPLE :   S:= $\textcircled{a}$ (-T)

4. Instructions

   4.1. Connection statements indicate that signals are connected to the output of a logical expression and at the input or
output of a unit.
The connection of a wire or of a sheet of wires to the output
of a combinatorial circuit corresponds to the physical action
of soldering the output of a circuit-B on to a wire A. This
branching is presented as :

        A := B ;

In CASSANDRE, the expression on the right hand side connection symbol corresponds to a combinatorial circuit. An
expression results from the operands :
- signals, registers and constants
and from the operators :
- boolean operator, permutation, reduction, shift, rotation,
  transposition etc.
An expression may also be conditional.
The connection symbol := signifies that the network outputs,
corresponding to the expression in the right part, are connected bit wise to the signal in the left part. Compatibility
constraints must be respected in order that the branching
be correct : the dimensions of the two sides of the ":=" symbol must be the same.
Likewise, the elements of the right hand side expression must
all have compatible dimensions.
   4.2. Assignment statements load registers. A clock determines the instant at which some events are valid.
A clock expression (or a clock) placed between the symbol
<and > indicates the presence of the synchronisation clock.
This quantity must be reduced to a scalar.
EXAMPLE :
    <H> REG (1:8) <= A(1:2) & B(0:4) & /.C
This operation loads bit by bit in the storage element REG
(register) and under the impulse action, the corresponding
bit of the right hand side expression.
   4.3. Unit connection : the unit connection describes the
use made of the unit by linking its inputs and outputs. The
interface variables of the inner unit are connected to signals
or clocks, or expressions, or clock expressions of the enclo-
sing unit.
The unit connection is thus rather like a multiple connection
of all its inputs/outputs, and it follows the connection
rules.

The connection format recalls the heading and the external declaration of the unit in question. The unit name is followed by the list of expressions which feed the inputs and the list of signals connected to the outputs, according to their declared position (figure 11).
EXAMPLE :

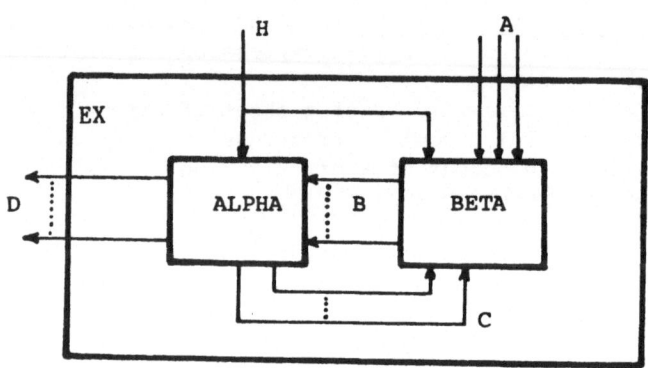

Fig. 11 Connection of units ALPHA and BETA

```
'UNIT' EX (A(1:3),H ; D(0:7)) ;
'MASTERCLOCK' H ;
'EXTERNAL' ALPHA ('CLOCK', (0:11);(0:5),(0:7)),
          BETA ('CLOCK', (0:3),(0:5) ; 0:11)) ;
'SIGNAL' B(0:11), C(1:6) ;
    :
    ALPHA (H,B ; C,D) ;
    BETA  (H,A,C ; B) ;
```

4.4. Conditional instructions
a) Simple Conditional instruction
Syntax
   simple-cond-inst → If C then statements else statements

- the condition C must be reduced to a scalar. Its value, 0 or 1, conditions the statement following the THEN or the ELSE with the usual conventions.
EXAMPLE :
        'IF' /+A(1:8) 'THEN' B:=C 'ELSE' B:=D
B is connected to D if all the bits of A are at 0. Otherwise it is connected to C.
b) Vectorial IF
A set of logical conditions may be regrouped in a vector (variable or expression) in such a way that each of its components conditions one or several statements.
- if the condition vector has n components, there must be exactly n groups of statements after the THEN and the ELSE clauses

EXAMPLE :
```
'IF' TEST 'THEN'  (A := B ; R := S)                    (a)
                  (ADD(*,OUT) := W ; H := H1)          (b)
                  (U := V)                             (c)
          'ELSE'
                  (Z := O)                             (d)
                  (H := H2)                            (e)
                  (U := 1)                             (f)
```
TEST is a 3 bit vector

If TEST contains for example 101, the statement groups (a), (e) and (c) are valid.
- the part 'ELSE' .... is optional.
- if a component of the condition commands no action, a pair of empty brackets must be joined on.
EXAMPLE :
```
'IF' COND14 'THEN'  (A := L)                           (a)
                    ( )
                    (R := 1)                           (b)
                    (D := O)                           (c)
          'ELSE'
                    (S := T)                           (d)
                    (U := OO1)                         (e)
                    ( )
                    ( )
```
If COND4 contains, for example 0110, only the groups (d) and (b) determine an action.

4.5. Automata and states : (statements under states) A control automaton is described in CASSANDRE by a sequence of states and by a set of statements allowing them to be activated. A state is a labelled group of statements. These groups are written after the "always valid" statements.
The range of one label extends to the next label encountered or to the end of the description.
The implicit state register (with the same name as the unit) always contains the current state of the automaton.
REMARK
At the beginning of every simulation, the automaton is found in the first state.

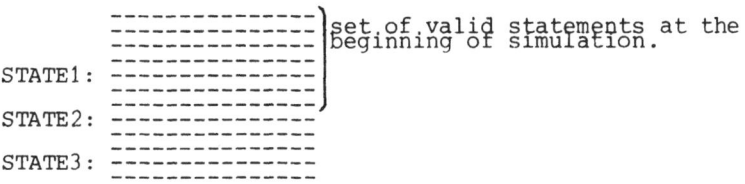

The statement ENABLE only may modify the contents of the implicit state register. This operation is an assignment, therefore it may only take place within the range of a clock.

334

EXAMPLE :
                <H> 'IF' C 'THEN' 'ENABLE' E1 'ELSE' 'ENABLE' E2 ;

4.6. DO statement : The statement executes the set of actions
controlled by the state indicated, without leaving the current
state. It is similar to a sub-program. The DO statement must
not appear within the range of an impulse since there is no
register assignment.
EXAMPLE :
                'DO' SET ;
                'DO' 'IF' BIT 'THEN' LAB1 'ELSE' REG(16) ;

4.7. Instruction Validation : There is no given sequential
chaining between the instructions in CASSANDRE. However instruction validity is determined by a set of conditions. Consequently several instructions may be valid at the same time
and description order is irrelevant. All units are also supposed to operate in parallel.
The way the signals are set up is irrelevant.
We always suppose them available at subsequent clock pulse.
Only clock controlled actions are supposed to be validated at
an exact instant.
In this way, the iterative network description presents no
difficulty. It is to be noted that in the case of a register
it is the content thereof before the clock pulse which will
enter the instruction computations.
An assignment in the register, on the contrary, will define
its value after the clock pulse.
4.8. Iterative structures description facilities : Several
copies of a unit may be used (figure 12). They are differen-
tiated by a number called duplication number placed immediately after the name. These numbers are CASSANDRE integers or
arithmetic expressions.
EXAMPLE

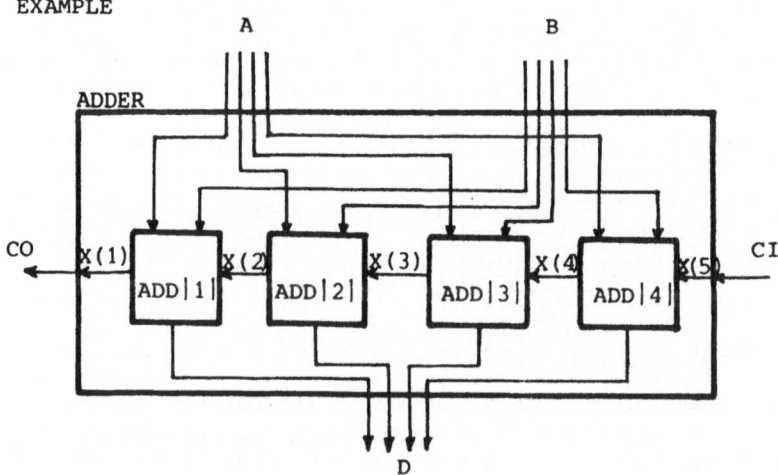

Fig. 11

Connexions :
```
        ADD|1|  (A(1),B(1),X(2) ; D(1),X(1));
        ADD|2|  (A(2),B(2),X(3) ; D(2),X(2));
        ADD|3|  (A(3),B(3),X(4) ; D(3),X(3));
        ADD|4|  (A(4),B(4),X(5) ; D(4),X(4));
```
Or, with the FORALL instruction :
```
        'FORALL' I 'FROM' 1 'TO' 4
        'BEGIN'
            ADD|I|  (A(I),B(I),X(I+1) ; D(I),X(I));
        'END'
```

III - THE LASCAR EXTENSION

1. The type 'INTEGER' :

We define two new types of variables : 'INTEGERS' and 'COUNTERS'.
Both will be implemented as a full machine-word (4 bytes on the
IBM 360); the first bit is a sign bit. Integers and counters can
take positive and negative values. They are considered as storage
elements. They must be declared like the other variables. An integer can be a scalar, a vector or an array. A counter is a special
case of scalar integer, we shall discuss its interest later on.
EXAMPLE :
```
        'INTEGER' X, Y;
        'ARRAY' 'INTEGER' A(0:7), ADTABLE(0:3,0:15);
        'COUNTER' K1, K2, COMPT;
```

2. Operators on integers :

Operators are defined on scalars as well as arrays ; in the second
alternative, operations are executed between the elements of identical rank, as for CASSANDRE boolean operators.
Arithmetic operators : + - . / 'REM' 'ABS'
Logical operators : $<$ $>$ $\leq$ $\geq$ $=$ $\neq$

3. Conversion operators :

We go from a boolean value to an integer value using the operator
$. Conversely we go from an integer value to a boolean value using
the operator !
EXAMPLE :
```
        'REGISTER' A(1:20), B(0:47);
        'INTEGER' C;
        A <= ! (C) A receives the binary value that codes the
                    value of C
        C $ = $ (B) C receives the integer value coded by the
                    boolean vector B
```
Rules : - an integer always has one less dimension than a register
        or a signal, because it implicitly has a first dimension

of 32 bits.

- the conversion result depends on the length of the first dimension of the boolean variable. High order bits may be truncated, or zeros may be added to their left.

4. Integer assignment :

We define a new assignment symbol for integers : $ = . An integer is considered as a storage element. An assignment is therefore under the range of a clock pulse. An integer may receive the value of any integer expression with compatible dimensions. An integer expression is made of constants, integers, arithmetic variables defined by a 'FORALL' loop, and boolean values after conversion.
EXAMPLE :
Let us take the declarations in the first example, to which we add :
```
        'REGISTER' T(0:47, 0:3) ;
        we can write :
        <H> 'FOR' I = 0 'TO' 5 'BEGIN'
        A(I) $ = (I + X) * 5 'END',
        ADTABLE (,0) $ = $ (T),
        X $ = Y 'REM' 3 ;
```

5. The 'INIF' conditional instruction :

It is quite identical to the 'IF' conditional instruction in CAS-SANDRE. One or several groups of instruction are validated by a vector of comparisons between integer expressions.
EXAMPLE :
```
        <H> 'INIF' A (0:1) < ADTABLE (1:2, 15) 'THEN'
        (X $ = X + 3', 'ENABLE' STATE 1)
        (Y $ = 0)
        'ELSE'
        (X $ = 0, 'ENABLE' STATE 2)
        (Y $ = 1, A (2) $ = A (1) + A (2) * 2) ;
```

6. Operators on counters :

It is very useful, when simulating a system, to place event counters which spy its operation. These counters are, to our description, like a hardware monitor in a real system. They should not be mistaken with actual elements of the model, and their introduction should have no noticeable effect on the simulation time. For these reasons, we decided to name them differently, and to implement highly performant operators on them, to increase and decrease them of one unit. Thus we defined the operators 'PLUSONE' and 'MINUSONE', which can only work on counter variables (these operators can be abbreviated 'P1' and 'M1'). Counters can also be tested, and be assigned with an integer expression. This is very useful to introduce delays into an entirely algorithmic unit.

EXAMPLE : STATE 1 :   H 'P1' (COMPT)
                      'INIF' COMPT = 20 'THEN' 'ENABLE' STATE 2 ;

3.7. Subprogram call :

We found it necessary, in order to accelerate simulation time, to
introduce the possibility of call a routine written in an algorithmic language. For convenience in use, we decided to take the
FORTRAN conventions for parameter transmission. Due to the specific
data structure of our language, the only acceptable parameters to
a FORTRAN routine are 'COUNTERS' and 'INTEGERS'. However, any user
who knows the LASCAR system well enough can write a routine in
assembly language, that will accept any LASCAR variable as a parameter. Such a routine must be declared before being used. The number of expected parameters must appear in the declaration, for subsequent verification when a CALL is issued (however, it is the
user's responsibility to declare the correct number of parameters)
EXAMPLE : Let us suppose we are making a trace driven simulation
of a processor. At a hardware level, we need to simulate the main
storage, and the fetch of instructions from it. If the number of
instructions is large, the size of the simulated memory will
exceed the available storage on our host computer. At the LASCAR
level, fetching an instruction from main memory can be simulated
with a routine call, that will read one instruction on a disk (or
tape) file. Moreover, these instructions may have been previously
preprocessed, so as to accelerate the simulation of decoding and
operands address calculation.
    'INTEGER' OPCODE, OPER2, ADDRESS ;
    'ROUTINE' NEXT (4) ;
          ·
          ·
    FETCH : H 'CALL' NEXT (OPCODE, OPER1, OPER2, ADDRESS) ;

3.8. A few comments on these extensions :

We wish to insist on some choices that we made. First we introduced
a whole set of new key words and symbols. We did not intend to
bother the user with a more complicated syntax, but to oblige
him to be always aware of the concepts he manipulates ; when he
writes a model using the LASCAR extensions, the level is not that
of a hardware description, but a more abstract one. Second, we did
not alter the meaning of inputs and outputs of a 'UNIT'. They still
are limited to clocks and signals (6). The reason is that, to prove
the functional equivalence of two units, it is sufficient to prove,
that, under identical inputs, their outputs are identical. Once
again, we want the user to demonstrate this fact before he replaces
a unit by a differently modeled one. Last we allow that different
'UNITS' in the same model be described at different levels. This
may be quite useful when one is interested in the detailed behavior
of only a subpart of the system. Also, in a top down system design,

the LASCAR level can serve as a specification language. Then, the designer can describe in more details each unit, one after the other, down to the pure CASSANDRE level. Therefore, at any stage of the project, some units may be known down to the gate level, and others merely functionally described. It is therefore quite necessary that several levels of description may coexist in the same model. We believe that LASCAR meets this 'HIERARCHY' requirement.

We are aware that the implementation of integers and counters is machine dependant, as soon as we allow conversions from strings of bits to integer values and backward. Again, the use of special operators is here to remind the user he has to be very careful of what he is doing. However, since our purpose was to accelerate simulation time, we thought the most efficient way to achieve this goal was to directly use the hardware of the host machine, which always involves restrictions. Finally, we stress the fact that the more hardware (or micro-programmed) operators are being replaced by simple integer operations, the greater improvments in simulation time are obtained. Compared to CASSANDRE descriptions, LASCAR descriptions are simulated 50 to 200 times faster.

IV.SOFTWARE ARCHITECTURE OF THE CASSANDRE AND LASCAR SYSTEMS

CASSANDRE system is a set of programs whose main purpose is to enable the use of a CASSANDRE language implementable version (figure 13).

The system is written in IBM 360 Macro-Assembler.

The program making up the processing of the various phases are fast, but rather bulky (of the order of 150 k bytes in the present version).

LASCAR, built on CASSANDRE, has the same structure.

REMARK : The second part of the processing, belonging to the simulation proper, is clearly separated of the system remainder by compiler generation of an intermediate language chain.

1. The Editor

This module reads a CASSANDRE and LASCAR source text and gives a chain which the blanks have been eliminated out of, but containing all coded basic symbols.

2. The Verifier

This module reads the chain thus edited and gives a completely coded chain without any declaration included in it. The declarations are presented as a list. This module additionally checks for :
- Syntax errors
- Correct use of declared symbols
- Dimension compatibilities.

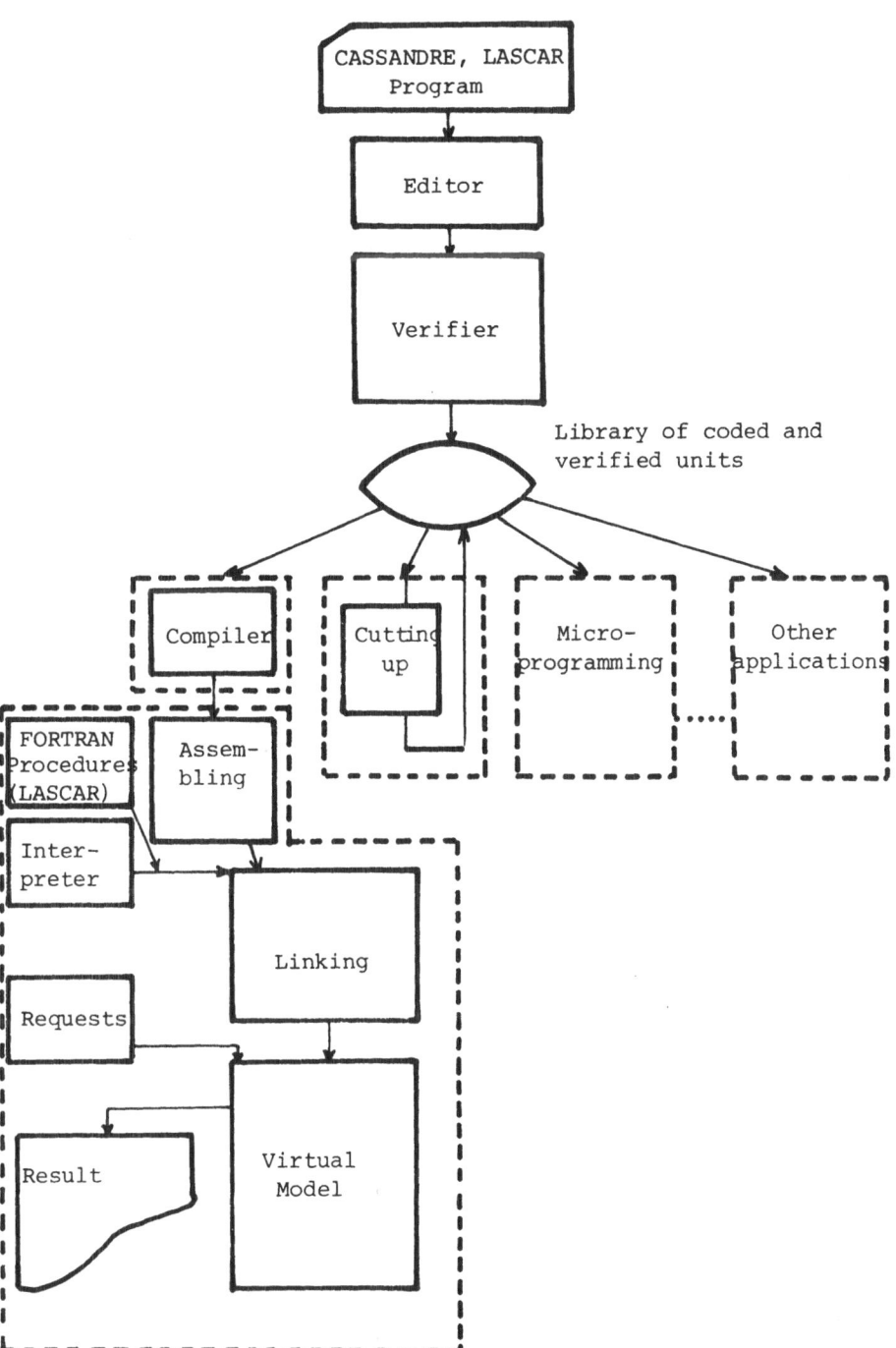

Fig. 13. Structure of the system

340

Thus a library of coded and verified units is obtained.

3. The Compiler

The compilation process includes :
- a description compilation from the unit library which results
  in an interpretable chain written in the intermediate language.
- a macro-assembling of this chain in order to obtain the reservation areas intended to contain the various beings of the logic
  description and an interpretable binary chain.

4. The Simulator

The simulation can be interactively directed (verification, cycle
by cycle, of the behaviour of the model), or batch. The simulator
interprets parts of the coded string of the model, according to
the orders issued by the user. The command language of the simulator allows one to :
- assign values to the registers, input signals, counters and
  integers of the model (at the beginning of simulation, all
  variables are set to zero)
- ask that the value of any variable be printed
- ask that one or more clock cycles be executed
- test the boolean, integer, counter and state variables of the
  model
- load and execute personal routines
- save or restore the state of the model
- define macro-commands from a group of the above elementary orders.
There exist two different versions of the simulation package with
different kinds of timing :
- a synchronous simulator to perform logical operations at the
  level of elementary boolean functions and algorithms.
- an asynchronous simulator to compare the technological realization to the logical network taking account of the elementary
  delays of the operators.

4.1. The synchronous simulation : In this mode all the
memory elements are loaded by external pulses. The described
system is not allowed to create new pulses. But it can gate
external pulses with logical conditions.
Each simulation cycle which the user asks enables the action
of the pulses for memory and state variables. Loading conflicts and instability of the network are detected at this
moment.

4.2. The asynchronous simulation : The timing is different
and so is the significance of "CYCLE" in this mode. Time is
cut off into intervals corresponding to dates 0, 1, 2, 3,...
etc... The action of the pulses is now enabled at different
dates of the simulated time, dynamically determined. The
asynchronous version possesses two more operators than the
synchronous one :

- the "Delay" operator which is the CASSANDRE equivalent of technological delays. These operators are associated with durations which are the basis of the computing of the successive dates of the general timing.
- the "Derivation" operator which allows the simulated system to create its own pulses from existing signals.
REMARK : LASCAR exists in synchronous version only.

5. System support

In its present version, the CASSANDRE and LASCAR system may be used with as system support
   . CASSANDRE
      Synchronous version : In batch processing without
                            OS/360,370
                            In conversational mode with
                            CP/CMS-360 or VM/CMS-370
      Asynchronous version : In conversational mode only
   . LASCAR : as the synchronous version of CASSANDRE.
The systems are distributed by the French MICADO Association.

B - PRINCIPLE OF AN INTEGRATED C.A.D. SYSTEM FOR LOGIC AND
    ELECTRONIC CIRCUITS
I - LASSO
1. Introduction

When a computer system (hardware and software) architecture
is being designed, its constituent modules are usually defined in several phases, by successive refinements. Its is
therefore very important for the designer to have access to a
coherent set of description and simulation languages, to write
models with various levels of details.
Asynchronous CASSANDRE, Synchronous CASSANDRE, and LASCAR had
already been developped at Laboratoire I.M.A.G., with the objective of a continous covering of all digital design phases.
However, these languages did not suffice, as the more abstract
of them, LASCAR, is still too close to hardware implementation. An intermediate tool, between LASCAR and general process simulation languages like SIMULA was lacking ; LASSO has
been defined to fill the gap.
The LASSO language is intended to be used as a specification
level to the other languages. Thus, in a LASSO model, the
structure of module interconnections and components functional behavior are described, but not their detailed implementation nor micro-synchronization. As a consequence a number
of fundamental principles of CASSANDRE and LASCAR have been
generalized.

2. Modularity in LASSO

2.1. The entity and unit concepts

A model written in LASSO is made of a fixed number of units, interconnected by physical links (control signals and data paths), or
logical ones (exchange of messages). A unit should be viewed as
a processor, capable of executing instructions or implementing
several functions, rather than a process. In particular, units
are not dynamically created or destroyed, their life time is equal
to that of the whole model.
A unit can itself be a network of other interconnected units, and
this decomposition can be repeated down to an arbitrary depth.
We make a distinction between the definition of the properties and
functional behavior of a unit, which is done in an entity declaration, and its use as a model component. An entity is an object
generator, whose definition may be parameterized. A unit is an
instance of an entity, for which all parameters have been fixed.
This concept of UNIT is an extension of the CASSANDRE UNIT.
EXAMPLE :
A memory generator can be defined by the following entity declaration :   ENTITY memory (INTEGER n,a,b);
              :
              :
         END %memory% ;

where n,a,b are integer parameters respectively for the size
of the memory in words, the reading time of a word and
the writting time of a word.
In a LASSO description subdivided into several units, every units
must be an instance of an already declared entity. The definition
of the generating entity can be :
- local to the model, in which case the text of the entity belongs
  to the declaration part of the model, and is unknown outside
- external to the model : the text of the entity is predefined, and
  can be found in a library. The entity can be shared among several
  models.
Each unit of a model must be declared, as being derived from a
specific entity, and be assigned a specific identifier.
At the point of unit declaration, all parameters (if any) must be
bound. Several identical units can be declared as an array : they
will then be referenced using the usual indexing mechanism.
EXAMPLE :
        ENTITY computer

              .
              .
              .

              .

          ENTITY cpu ;   . . . .   END %cpu%;
                            % cpu is local to computer %
          EXTERNAL ENTITY memory (INTEGER n,a,b);
                            % memory is external to computer %

          UNIT proc [1:4] : cpu ; % declaration of 4 instances of
                                  cpu %
          UNIT m1, m2 : memory (512, 6, 8) ; % declaration of two
                              identical instances of memory, with 512
                              words, reading time= 6 units of time,
          .   writing time= 8 units of time %
          .
          .
          .

          .

        END %computer% ;

2.2. Structure of an entity description

Every entity description is divided into three parts, the last
one being optional ; each part starts with a key-word.
        2.2.1. The interface : The interface part holds declarations
        of control input-output signals, and associated data, by
        which every unit derived from this entity will be connected
        to other components of a model.
        2.2.2. The entity BODY : In the entity body, internal objects
        are declared, and the entity functional behavior is described.
        One or more of the following points are found :

1- declaration of local or external entities
2- if point 1 is not empty, declaration of component units
3- declaration of internal variables, signals and procedures
4- if point 2 is not empty, connexion of internal units
5- a set of statements describing the input/output behavior of the entity.

2.2.3. The entity SPECIFICATIONS : This part, if present, indicates the properties which every unit derived from the entity must satisfy, so that the description is meaningful (relations on parameters, exclusion on input signals, etc..). In addition, the expected behavior can be indicated, which the interpretor will have to verify : input-output signals dependencies and time constraints.

2.3. Example of a total memory description

```
ENTITY memory (INTEGER n,a,b) ;
  INTERFACE
    INTEGER address,data ;
    INPUT askr (address) %request for read, transmits address%,
          askw (address,data) ; %request to write data at address%
    OUTPUT okread (data), %acknowledge reading, transmit data%
           okwrite ; %acknowledge end of write%

  BODY
    INTEGER table [1:n] ; %the memory holds n elements of type
                          integer%
    ? ready ? SELECT (askr,askw) VALIDATE (read,write) ;
            %initial state, switch according to input signals%
    ? read ? BEGIN    %action of reading, duration 'a' units of
                      time%
              data := table[address];
          END    DELAY (a)    VALIDATE (endread) ;
    ? write ? BEGIN    %action of writing, duration 'b' units of
                       time%
              table[address] := data ;
          END    DELAY (b) VALIDATE (endwrite) ;
    ? endread ?    VALIDATE (ready, okread) ;
    ? endwrite ?   VALIDATE (ready, okwrite) ;
  SPECIFICATION
    a < b ; %reading time less than writing time%
    ¬ (askr ∧ askw) ; %no simultaneous request to read and write%
END    %memory%
```

3. Communication and synchronization between UNITS

In a description, the component units are asynchronous. Two units may communicate if one is internal to the other, or if the two are internal to a same third one, at the same imbrication level·

Then a connection must be established between their interface signals.

One or several signals, and their associated data if any, are sent to the interface of a unit by execution of the VALIDATE statement in the unit body. According to the established links, transmission to other interfaces will be made either in zero time, or after the number of time units which have been indicated with the DELAY statement at the point of connection.

4. Internal synchronization and control of an entity :

In a model, instances of an entity interpret their input signals and associated data. This interpretation induces execution of one or several (sequential or parallel) actions.

4.1. Actions

An action is represented by a bloc. The bloc is prefixed by a condition, which can be either an internal or an input signal, written between question marks. The execution of the action is started when the condition signal is true. When execution is over, the condition signal is reset, and in general another signal is set, which may be internal or output. A DELAY statement specifies the duration of the action ; if absent, the action is supposed to be instantaneous.

The action itself is a piece of algorithm, which may invoke functions and procedures, written in an extention of PASCAL, to which array and conversion operators have been added.

```
Example :
INTEGER w,x[0:3], y[0:3];
BOOL   z[0:23];
EXTERNAL FUNCTION decode (INTEGER n; BOOL t[0:n]) RESULT INTEGER;
?a?   BEGIN   IF   cond   THEN   x := y+(1,1,0,0)   ENDIF;
              w := decode (23,z);
      END   DELAY (5)   VALIDATE (b);
```

4.2. Control primitives

The control part is used to validate output signals, and internal signals initiating actions. These validations depend upon the values of input signals, signals conditionning and ending actions, and tests on variables. Control can be represented by a transition graph, in which places represent signals and boxes represent action and control primitive transitions (see 7).

Signals hold logical values 0 and 1. Transitions portray the most
widely used decision procedures in logical circuits. The primitive
transitions are given now :

- AND transition
  ?a1 ∧ a2 ∧ ..∧ an? VALIDATE (b1, b2, ...,bm)
  If all ai are 1 and all bj are 0, the transition fires : all bj
  are set and all ai are reset ; otherwise wait (n positive,
  m non-negative).
  The "AND" transition is used to synchronize the completion of
  parallel asynchronous actions when it has more than one input
  ("JOIN") and to initialize several parallel asynchronous actions
  when it has more than one output ("FORK").

- TEST transition
  ?a? TEST p VALIDATE (b, c) ;

  p is a scalar boolean expression, on interface or internal data
  variables. If signals b and c are 0 and a is 1, p is evaluated.
  If result is 0, b is set to 1, otherwise c is set to 1 ;
  in both cases a is reset to 0.
  The "TEST" transition can be viewed as a switch to which a decision function is associated.

- INDEX transition
  ?a? INDEX i FROM m TO n VALIDATE (bm, bm+1, ..., bn)
  or
  ?a? INDEX i VALIDATE (b0, b1, ... bk)

  i is an integer scalar expression whose result must belong to
  the [m:n] interval. The number of output signals of the INDEX
  primitive must be equal to n−m+1.
  If all output signals are 0, and a is 1, i is evaluated, and
  the result is used as an index to select the output signal that
  is set. a is reset. If m and n are omitted, m takes the default
  value 0, and n the number of output signals minus 1.
  A typical application of the "INDEX" transition is the decoding
  of operations in a processor.

- SELECTP transition
  ?e? SELECTP (a1, a2, ... an)
       VALIDATE (b1, b2, ... bn) ;
  This transition tests two or more signals with decreasing priority from left to right. If all output signals are 0 and e is 1,
  the highest priority input signal is selected. The corresponding output signal is set to 1, and the selected input is reset
  together with e.

- SELECT transition
  ?e? SELECT (a1, a2, ...an)
  VALIDATE (b1, b2, ... bn) ;
  This transition is similar to the preceding one, except that all

signal are given equal priority, and only one ai input is allowed
to be 1 at a time, otherwise error.
If all output signals are 0 and e is 1, and one ai is 1, the corresponding bi is set to 1, and the selected ai is reset together
with e.
"SELECT" and "SELECTP" transitions can directly portray an EXCLU-
SIVE OR bus connection, and a priority selection scheme, in a multi-
plexor for instance.

- Delayed transitions
  If no DELAY statement is specified, a transition is supposed to
  be instantaneous. However, as for actions, it is possible to
  state that execution of a control primitive takes some (simulated) time. This is uniformly done using the DELAY statement with
  an integer constant or expression parameter, immediately preceding the VALIDATE statement.

Example :
INTEGER x, y ;
?a? BEGIN ..... END DELAY(5)VALIDATE(b);
?b   c? DELAY (x+y) VALIDATE(d, e, f);

348

II - IMAG3

INTRODUCTION :

The IMAG3 program performs DC,AC and transient analysis of nonlinear circuits consisting of resistors, capacitors, inductors and
any component described by a model. It allows also study of sensitivity and optimization of these circuits.
IMAG3 is a general CAD program. It performs DC,AC and transient
analysis and optimization of nonlinear circuits. Its main typical
features are :

1. Power and flexibility of the input language

- Any kind of nonlinearity can be practically described.
- It is possible to define a wide class of models for physical
  devices.
- These models once described can be stored into a library and
  used in the same way as resistors, capacitors, inductors. It
  is possible, for example, to use them in the building of new
  models.
- Easy and accurate description of optimization requirements and
  constraints.

2. Simulation possibilities

a) DC analysis using a modified Newton-Raphson iteration.
b) Transient analysis. The user has the choice of several integration methods, including implicit methods suitable for circuits
   with widely separated time constants.
c) AC analysis. This is a small signal circuit analysis. The DC
   bias is computed using the dc analysis of paragraph a).
d) For DC,AC and transient analysis, it is possible to obtain
   sensitivity with respect to any circuit parameter.
e) In cases a), b) and c), IMAG3 allows optimization of circuit
   parameters values, in order to satisfy any requirement in the
   presence of a given set of constraints (e.g. minimizing a gate
   delay time for given ranges of operating points and D.C. power
   consumption values).
f) Finally, it is possible for the user, to link the preceding
   functions as he wishes and to modify any circuit parameter
   without having to redefine the whole circuit.

3. Interactive use of the program

- At the level of the input language, the user can compile instructio
by instruction, with the possibility, if an error occurs, to correct immediatly only the erroneous instruction.

- At run time, the user can control the convergence of the algorithms (this is especially useful for optimization).

III - IMAG4

I. Introduction

Present integrated circuits may only be realized with the aid of
electric simulation tools. The designers of integrated circuits
all agree that the electronic simulation tools have an insufficient
Simulation capacity, bearing in mind the circuits which must be
realized. Likewise when Simulation is possible, it is too expensive.
Great progress has been made during these last few years in electronic Simulation. New ways explored. One of these, inspired by the
fact that the results of electronic Simulation are often too precise, is to work on a much simpler differential system (for example, instead of 50 differential variables in a model of operational amplifier, only 4 or 5 are kept, the "most significant").
This idea leads directly to macromodelisation.
Two approaches are possible :
1.1. Electric macromodelisation plus "Classic" simulation : This
uses the technique of the equivalent circuit : the designer elabo-
rates an electric model which is much simpler than the initial
circuit and which functions similarly in a certain domain. He may
then use this model in a classic Simulation program.
1.2. Mathematic macromodelisation plus functional simulation :
On a p pin circuit, access is supposed to be had to external potential and current variables only on each of these p pins. The
mathematic model consists of a set of p differential equations
which connect these variables, their derivatives and the time.
The characteristics of such circuits will be measured, supplied
by the constructor or calculated by a Simulation program
Based on these purely mathematical models (system of differential
equations), a program of functional Simulation which accepts as
data the proposed formalism (p pin box to which is linked a system of differential equations) enables the take-over of circuits
whose basic elements are interconnected components.
Our experience in the field of mathematical modelisation as well
as in that of realizing CAD programs(1,6) has lead us to chose
the second way and to build the Simulation program : IMAG4.

2. Functional simulation : IMAG4

2.1. Specifications of description language - Compatibility with
     IMAG3
We have in fact wanted the functional simulator to take over not
only interconnected macromodels but also electronic components

(which allow the maximum precision to be maintained at critical places).

All IMAG3 language description may therefore be taken over by IMAG4

The compatibility also stretches to the command language and to the simulation possibilities.

In order to represent circuits, the electricians use diagrams which are none other than geometric representations of networks

The description languages of IMAG3 enable these geometric representations to be described easily. In the network, each component (articulation) is entirely determined by the following criteria :

- its nature,
- its name,
- its position (i.e. the list of its connectors)
- a list of parameter values (physical or electric).

The description language of IMAG4 is an extension of that of IMAG3 with which it is totally compatible.

Introduced into the language has been the possibility of describing a class of supplementary components which will be called macrocomponents.

In the network each macrocomponent is presented as an articulation made up of a certain number of connectors which appear as output pins. The description of these macrocomponents is taken from the one used in IMAG3 to describe equivalent schemas.

A class of macrocomponent is thus defined by :

- MACRO : Symbol (Connections) list $

where :

. Symbol is an alphanumeric number chosen by the user to name the macrocomponent class.

. Connections is a list of formal parameters.

. List is a series of formal parameters.

eg. MACRO : AMP (1, 2, 3) RE, G, $

In the body of the definition primitives enable the specification of supplementary variables as well as the list of functioning equations.

Definition of supplementary variables VAR : list $
Definition of functioning equations   EQU : list $

where list is a series of separate equations separated by commas. The equations may be implicit or explicit with regard to any variable.

The electric variables linked to the list of external nodes of the macrocomponent are declared implicity in the program in the following way :

At each CI connection, four identifiers are generated

- VCI   = nodal tension of node CI
- ICI   = current entering at node CI
- DVCI   = d(VCI)/dt
- DICI   = d(ICI)/dt.

Similarly for each variable introduced, its derivative is defined implicitly:
EXAMPLE : Circuit RC introduced, in the form of macromodel :

$$\begin{array}{rl}
\text{MACRO} : & \text{RC (M, 1, 2) R, C } \not\!\!\!\!5 \\
\text{VAR} : & \text{VE, VS, } \not\!\!\!\!5 \\
\text{EQU} : & \text{VE = V1-VM, VS = V2-VM,} \\
& \text{VE-VS = R } \star \text{ 11,} \\
& \text{C } \star \text{ DVS = 11 + 12 } \not\!\!\!\!5 \; \not\!\!\!\!5
\end{array}$$

Macrocomponents (generated from their class) may be used in a circuit due to the primitive :
- Name (Connections) List $\not\!\!\!\!5$
where : Name is the name of its class followed by an alphanumeric
        label enabling its design.
        Connection is the list of macrocomponent connectors in the
        network.
        List is the series of effective parameters of the macro-
        component.
Eg. AMP 1 (2, 3 GND) 10, 100 K $\not\!\!\!\!5$
The circuit designers, familiar with the description languages
of IMAG3 will thus have no difficulty in using the description
language of the functional simulator.

2.2. Specification of the functional simulator

At the present moment, the numeric algorithms used are similar
to those existing in the programs of classical simulation. It
mainly concerns resolution algorithms of non-linear algebraic systems and of non-linear differential algebraic systems. The system
of equations which the numeric algorithms seek to solve may be
written :

$$(I) \quad 0 = F(X, Y, \frac{dY}{dt}, t)$$

Being iterative, these algorithms will very often have to evaluate
F ; the set of literal expressions enabling F to be calculated,
will thus be one of the invariants sought.
The system (I) has its eigenvalues of orders of very different
size and the resolution methods of such differential systems
will necessarily have to be implicit.
The Jacobian will therefore have to be evaluated a number of times.
Similary the system J.Z = B will have to be solved very often.
This had lead us to determine a new invariant which is the set
of expressions enabling J to be calculated.
The system J.Z = B is solved by using resolution techniques of
very sparse systems. This leads to two new invariants being
determined :
- operations of triangularisation of J
- operations of resolution of associated triangular systems.
In the search for efficiency these invariants are determined up to
the generation in machine code of the literal expressions.

IV - TOWARD A UNIFIED INTEGRATED CAD SYSTEM

We have said some words on tools developped at Grenoble to simulate
hardware at circuit, gate, RTL, or system level. Several other
software exist at other places which would be very interesting
to analyse. But we are supposed to present only our works.
We have thought that the advance of knowledge in the definition
of high level languages for example expendable languages, dynamic
type specification ... was a new opportunity for developping an
integrated approach of the different levels of modelling in which
we accumulated experience. Part of this reflexion was carried
over by the CONLAN group. An expendable basiclanguage mechanism
résulted of its work, which covers only the digital levels of
modelling.
An other part of the concepts emerged from our project on socio-
economic modelling called ODYSSEE. In this project, like in hybrid
simulation of circuits, the main difficulty is to make modules of
discrete models cooperate during the simulation process with
modules of continuous models. A whole language, not far from
real time description languages, was set up to describe the impact
of events on the behaviour of each part.of the models.
But the main impulse to the development of a unified integrate system was given by the necessity of rewriting for a new machine
(HB 68 - MULTICS) the LASSO, LASCAR, CASSANDRE and IMAG programs.
In front of such a large work we had to study in depth the common
concepts present at every level. We were lucky enough to discover
in these languages developed by different persons a significant
common kernel. This kernel has been included in two sublanguages
available at every simulation level :
- the prefixe language, to describe event conditions and the validation structure of any model [9] [10]
- the network description language to define the structure of
any model as a network of nested interconnected boxes and their
interfaces.
For the rest of the different languages we have made an extension
of "Base CONLAN" to include real variables, differential and
difference equations, non oriented connections which are of necessity for circuit simulation.
The syntax of existing languages can be defined entirely by using
these three components. We have of course taken this opportunity
to include new improvements at every levels. But, we accepted the
constraint of being completely compatible with the existing languages, because a lot of investment has been made in their use by
some firms. They must not have to change standards again.
In the following table we have mentionned 8 levels of modelling
This is an arbitrary choice which results from our own experience
but the main caracteristic of the new system will be to allow each
user to define this own set of primitives, according to the particular application he wants to develop, and the particular environ-

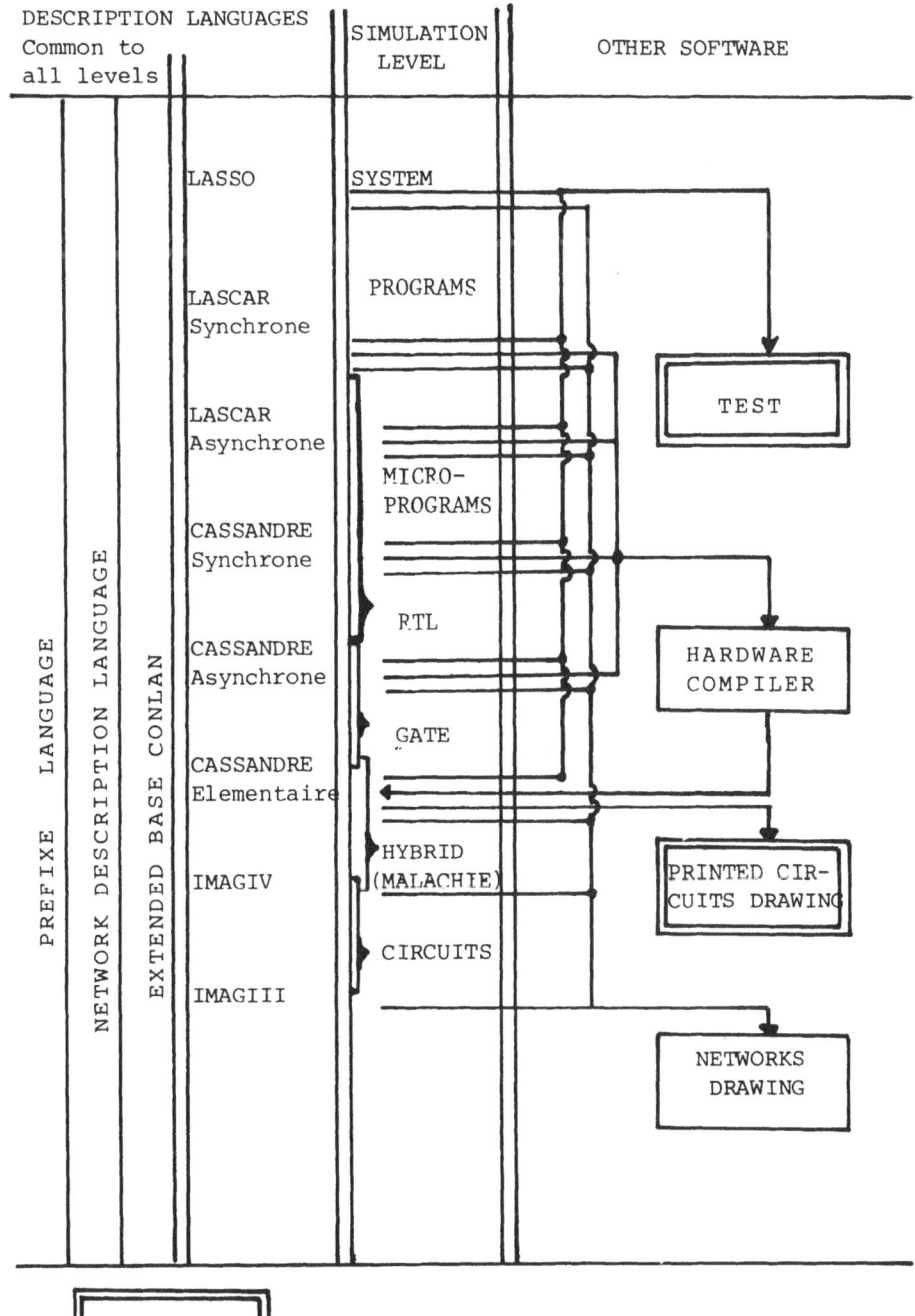

Not to be developped in our group.

ment in which he works.

Among these 8 levels, LASCAR asynchrone, CASSANDRE elementaire and
IMAGIV don't correspond to already used packages.

The first one is an accelerated version of RTL simulation, in
order to save money while simulating big logical descriptions
many times.(because of rough simplifications in the description
of a CASSANDRE model, LASCAR is 50 to 100 times faster).

The second one is a simplification of CASSANDRE in order to
provide a fast, gate level, asynchronous simulation. Boolean
variables will have eventually more than two values (for example
TRUE, FALSE, RISING, FALLING, UN KNOW$_N$)according to a truth table
given by the user. Delays will be taken into account. Unstable
logic will be detected.

IMAGIV is a macro-model level of circuit simulation, mainly based
on interconnection of boxes defined by differential equations
When used in a common simulation with CASSANDRE elementaire
(some modules being described in IMAG, others in CASSANDRE).

IMAGIV will provide hybrid simulation tool oriented toward LSI
which we have called MALACHIE (Models Analogous and Logical Applied to Highly Integrated Circuits Evaluation).

Implémentation of the integrated system :

The architecture of the software system implementing the integrated
CAD tool described above is rather complex, contains several compilers and is being developped by a team of about 10 persons.

The main modules of the compiling and simulation process are described in the following figure.

a) For every user's language a syntactic analyser is generated
by a general program.

This analyser invokes of lexicographical analyser. It performs
a syntactic and semantic verification of the correctness of the model
unit by unit. Error messages allow the user, in an interactive dia-
logue, to improve his description. A category of functional or
behavioural errors do not appear of course at this level, but
the possibility for a user to define its own types of basic elements and associated specifications on them provides a considerable consistency checking tool. During this phase the internal
structure of the model begins to be build. The network description
language for example is interpreted to generate the skeleton
of this structure.

b) The second phase of compiling is itself divided in several steps.
At first the "prefixe language" is treated. This compiling replaces
all the notions of this language by a list of boolean equations
defining the nested accumulated conditions and a condition-tree
which reflects the nesting of the original model (this tree is
used during the simulation process for optimization purposes,
for example, in the "Ordering Module").

Each leaf of this tree is an elementary condition which influences
an elementary action, or a block of them. After this work some

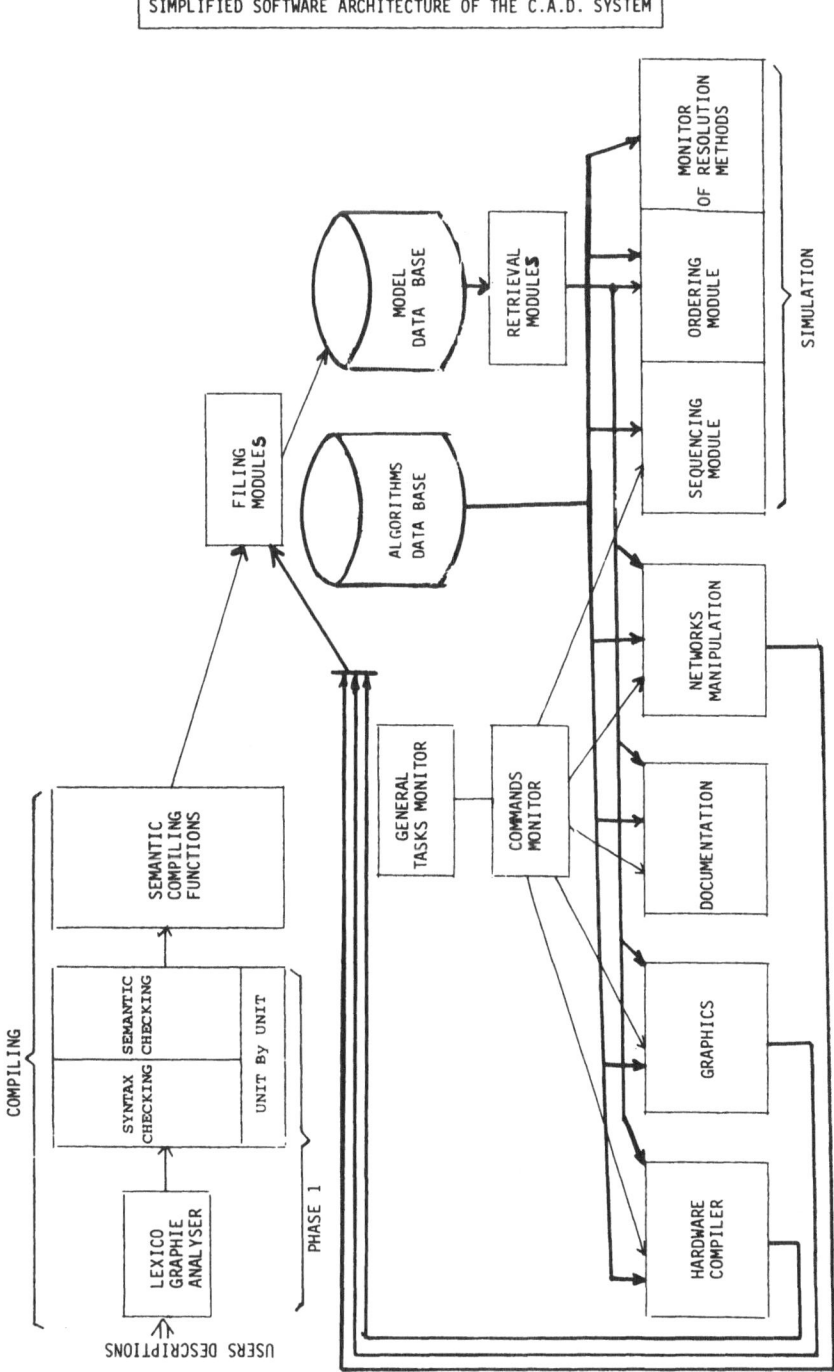

SIMPLIFIED SOFTWARE ARCHITECTURE OF THE C.A.D. SYSTEM

transformations are made on the analized description of units
used in the being processed model to divide equations when sub-
variables of a global variable (arrays for example) appear...
The module to be simulated is then compiled by semantic compiling
functions into a set of routines and data structures directly usable
by the compiler.
A "filing module" fills into the simulation data base the result
of the compiling.
c) The simulation process is in its turn performed by several
modules.
A retrieval module loads the simulation with the compiled material
necessary for a given run. It is composed essentially of a large
set of equations influenced by changing conditions. A complex
recursive module   tries first to find for the equations an execution order. It is recursive because it keeps track of the nested
structure of units of the description. This process is generally
unsuccessfull but the result of the program is the best possible
ordering due to the fact that non-ordered parts are isolated in
new units, as small as possible. The interest is clear : implicit
resolution methods (speaking of real equations) or cycles of
stabilisation (speaking of discrete equation) apply on parts as
small as possible, the rest of the system being ordered.
One must be aware that this ordering module has to be activated
as soon as the change of a condition may induce a change of the
previous execution order (in the worst case, at each step of the
simulation).
The simulation run is driven by a sequencing module which makes
the analysis of the events by interpreting their description in the
model and builds dynamically the next event from the present
state of the model (this next event can result from the resolution method or from the model description its date being then
determined by extrapolation).
The whole process is asynchronous, the mathematical resolution
methods and the discrete events caused by changes in conditions
alternatively producing the next step of the equations evaluation.
This is mainly why for example multi-level (hybrid) simulations
can be envisaged.

BIBLIOGRAPHY :

-M. JACOLIN, C. LE FAOU, B. PIMORT, M. VERAN :   [1]
"Simulation de Circuits Electroniques : programme IMAG2".
Colloque International sur la Microélectronique Avancée,
Paris, Avril 1970

-J. KUNTZMANN : "Théorie des réseaux", Dunod, 1972   [2]

-BOGO G., GUYOT A. ; LUX A., MERMET J., PAYAN C.,   [3]
"CASSANDRE and the computer aided logical Systems design",
Congrès de l'I.F.I.P., 23-28 Aout 1971

-J. MERMET   [4]
"Etude methodologique de la conception assistée par ordinateur
des systèmes logiques : CASSANDRE"
Thèse, Grenoble, 1973

-D. BORRIONE   [5]
"LASCAR : a LAnguage for Simulation of Computer ARchitecture"
International Symposium on Computer Hardware Description lan-
guages and their Applications
New York, Sept. 1975

-C. LE FAOU and J.C. REYNAUD :   [6]
"Functional Simulation of Complex electronic circuits"
ESSCIRC'77, September 20-22 1977

-D. BORRIONE, J.F. GRABOWIECKI   [7]
"Informal introduction to LASSO : a language for Asynchronous
Systems Specification and Simulation"
Proc. Euro I.F.I.P. 79, London, Sept. 1979

-R. PILOTY, M. BARBACCI, D. BORRIONE, D. DIET MEYER, F. HILL, [8]
P. SKELLY :
"CONLAN - A formal construction method for hardware description
languages : - basic principles,
             - language derivation,
             - language application".
3 papers in
National Computer Conference 1980,
Anaheim (CA), AFIPS
Conference proceedings, volume 49.

MERMET J.   [9]
Définition d'un Langage Préfixe pour exprimer les conditions
logiques et les évènements . Laboratoire I.M.A.G.
Rapport de Recherche, Décembre 1978

BRESSY Y., MERMET J., UVIETTA P.   [10]
Mécanismes de simulation discrète et continue, liés aux modèles
de systèmes dynamiques. Laboratoire I.M.A.G.
Rapport de Recherche, Juin 1980

DESIGN FOR TESTABILITY

T. W. Williams

IBM
Data Systems Division
Boulder, Colorado

Abstract

This section discusses the basic facts of design for testability.
A short review of testing is given along with some reasons why one
should test. The different techniques of design for testability
are discussed in detail. These include techniques which can be
applied to today's technologies and techniques which have been
recently introduced and will soon appear in new designs.

1. Introduction

Integrated Circuit Technology is now moving from Large Scale
Integration to Very Large Scale Integration. This increase in gate
count, which now can be as much as factors of three to five times,
has also brought a decrease in gate costs, along with improvements
in performance. All these attributes of Very Large Scale
Integration, VLSI, are welcomed by the industry. However, a
problem never adequately solved by LSI is still with us and is
getting much worse: the problem of determining, in a cost
effective way, whether a component, module or board has been
manufactured correctly.[1-3, 47-63]

The testing problem has two major facets:

   1. Test generation.[69-93]

   2. Fault simulation.[94-108]

With the vast increase in density, the ability to generate test
patterns automatically and conduct fault simulation with these

patterns has drastically waned. As a result, some manufacturers are foregoing these more rigorous approaches and are accepting the risks of shipping a defective product. One general approach to addressing this problem is embodied in a collection of techniques known as "Design for Testability".

Design for Testability initially attracted interest in connection with LSI designs. Today, in the context of VLSI, the phrase is gaining even more currency. The collection of techniques that comprise Design for Testability are, in some cases, general guidelines; in other cases, they are hard and fast design rules. Together, they can be regarded essentially as a menu of techniques, each with its associated cost of implementation and return on investment. The purpose of this work is to present the basic concepts in testing, beginning with the fault models and carrying through to the different techniques associated with Design for Testability which are known today in the public sector. The design for testability techniques are divided into two categories. The first category is that of the Ad Hoc technique for solving the testing problem. These techniques solve a problem for a given design and are not generally applicable to all designs. This is contrasted with the second category which is the structured approaches. These techniques are generally applicable and usually involve a set of design rules by which designs are implemented. The objective of the structured approach is to reduce the sequential complexity of a network to aid test generation and fault simulation.

The first category is that of the Ad Hoc techniques, the first of which is partitioning. Partitioning is the ability to disconnect one portion of a network from another portion of a network in order to make testing easier. The next approach which is used at the board level is that of a Bed of Nails and adding extra test points. The third Ad Hoc approach is that of Bus Architecture Systems. This is similar to the partitioning approach and allows one to divide and conquer--that is, to be able to reduce the network to smaller subnetworks which are much more manageable. These subnetworks are not necessarily designed with any design for testability in mind. The forth technique which bridges both the structured approach and the Ad Hoc approach is that of Signature Analysis. Signature Analysis requires some design rules at the board level, but is not directed at the same objective as the structure approaches are--that is, the ability to observe and control the state variables of a sequential machine.

For structured approaches, there are essentially four categories which will be discussed--the first of which is a multiplexer technique, Random Access Scan, that has been recently published and has been used, to some extent, by others before. The next techniques are those of the Level Sensitive Scan Design, LSSD approach and Scan Path approach which will be discussed in detail.

These techniques allow the test generation problem to be
completely reduced to one of generating tests for combinational
logic. Another approach which will be discussed is that of the
Scan/Set Logic. This is similar to the LSSD approach and the Scan
Path approach since shift registers are used to load and unload
data. However, these shift registers are not part of the system
data path and all system latches are not necessarily controllable
and observable via the shift register. The fourth approach which
will be discussed is that of Built-In Logic Block Observation,
BILBO, which has just recently been proposed. This technique has
the attributes of both the LSSD network and Scan Path network, the
ability to separate the network into combinational and sequential
parts, and has the attribute of Signature Analysis--that is,
employing linear feedback shift registers.

For each of the techniques described under the structured
approach, the constraints, as well as various ways in which they
can be exploited in design, manufacturing, testing and field
servicing, will be described. The basic storage devices and the
general logic structure resulting from the design constraints will
be described in detail. The important question of how much it
costs in logic gates and operating speed will be discussed
qualitatively. All the structured approaches essentially allow
the controllability and observability of the state variables in
the sequential machine. In essence, then, test generation and
fault simulation can be directed more at a combinational network,
rather than at a sequential network.

1.1 What Is A Test?

Some terms associated with testing may, at first glance, appear to
have similar meanings. Consider "fault" and "defect". As defined
in the IEEE Standard Dictionary,[9] a fault in a network will yield
incorrect results for a given input pattern. A defect, on the
other hand, may or may not cause a fault. Some means must exist to
tell whether a network contains any faults or not. A first-pass
method might be to use actual electrical network diagrams with
resistors, capacitors, inductors, etc., in order to model the
network with all possible defects which could cause faults.
Examples might be an open net, two-shorted nets, missing resistor,
insufficient β, etc. Clearly, there would be many such defects per
logic gate. As many as 30 defects could be associated with one AND
gate. Thus, a 1,000-gate chip could contain 30,000 defects which
could cause faults if they were assumed to occur one at a time.
Defects could appear two at a time, or even n at a time. This
multiplicity of defects tremendously complicates the modeling
problem, but even if one could model the problems, the next step
would be to generate a test for each of these faults and then fault
simulate them--a formidable task.

A model of faults which does not take into account all possible

362

defects,[1,2,3] but is a more global type of model, is the Stuck-At model. This is the most common model throughout the industry. The Stuck-At model assumes that a logic gate input or output is fixed to either a logic 0 or a logic 1. Figure 1(a) shows an AND gate which is fault-free. Figure 1(b) shows an AND gate with Input "A" Stuck-At-1 (S-A-1). The faulty AND gate perceives the "A" input as 1, irrespective of the logic value placed on that input. The pattern applied to both the fault-free and the faulty AND gate in Figure 1 has an output value of 0 in the fault-free machine (good machine), since the input is 0 on the "A" input and 1 on the "B" input, and the ANDing of those two leads to a 0 on the output. The pattern in Figure 1(b) shows an output of 1, since the "A" input is perceived as a 1, even though a 0 is applied to that input. The 1 on the "B" input is perceived as a 1, the results are AND'ed together to give a 1 output. Therefore, the pattern shown in Figures 1(a) and 1(b) is a test for the "A" input S-A-1, since there is a difference between the faulty gate (faulty machine) and the good gate (good machine).

In general, a necessary and sufficient condition to test for a faulty machine from a good machine is to apply a sequence of patterns to both the good and the faulty machine, so that a difference is detected between the good machine output and the faulty machine output. Furthermore, it makes no difference whether the faulty machine is one that consists of a representation of the good machine with one or more Stuck-At faults induced, or whether the faulty machine is a representation of the good machine with some defects which cause faults which may not necessarily be modeled by the Stuck-At fault. Hence, a test exists if, and only if, a difference can be found between the good and faulty machine

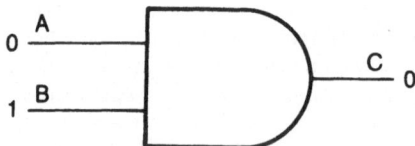

(a) Fault Free AND Gate (Good Machine)

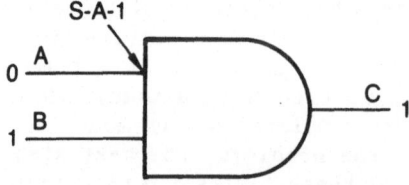

(b) Faulty AND Gate (Fault Machine)

Figure 1. Test for input stuck at fault

with a given input sequence of patterns. Figure 2 shows an input sequence to both good and faulty machines and an output sequence which distinguishes itself between the good and faulty machines. By virtue of the fact that the output sequences are different, the input sequence is a test for the faulty machine, whatever collection of abnormalities the faulty machine may possess.

Two or more faults may exhibit equivalent behavior.[32, 34, 37, 38, 42, 45] For example, the test pattern for an AND gate whose output is S-A-0 is a 1,1 pattern shown in Figure 3(a). However, the output for the same AND gate whose first input is S-A-0 also has the same test pattern of 1,1, with the same response see Figure 3(b). There is no pattern which will test one fault and not the other. Thus, the two faults depicted in Figure 3 cannot be distinguished from one another.

A diagnostic message is associated with a failing output pattern. This diagnostic lists all the single Stuck-At faults which could be detected on the output pattern. In the case of the two faults in Figure 3, they can only be detected with a 1,1 pattern, and there is no way to tell whether the S-A-0 is at the "A" input, at the "B" input or at the "C" output in the course of the test. At the gate level, this is not usually a significant problem, however, at the board level, the inability to diagnose to the failing component is a significant problem. As shown in Figure 4, for example, the good machine output should be a 0, however a 1 is being observed. There are three possible modules on the board which may give rise to the incorrect output--Module 1, Module 2 and Module 3. Now the ability to diagnose the fault to the proper module has significance and is reflected in the repair cost of the board.

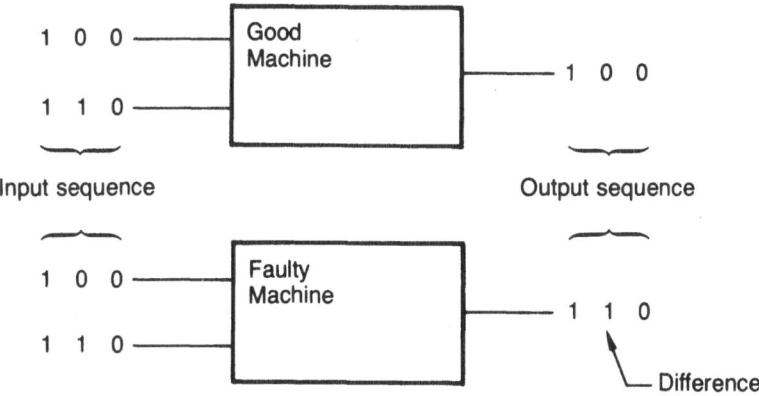

Figure 2. Results of input sequences of test pattern.

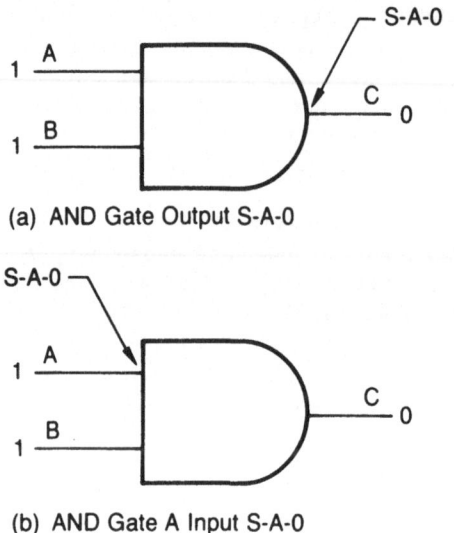

(a) AND Gate Output S-A-0

(b) AND Gate A Input S-A-0

Figure 3. Example of two faults with equivalent behavior at the output.

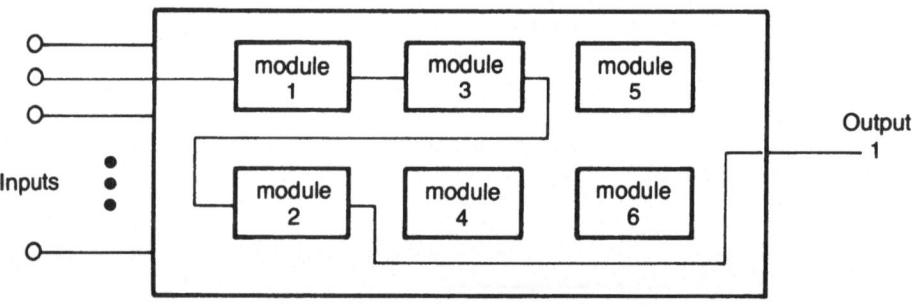

Figure 4. Diagnosis is the problem of determining which module of several is the cause of a faulty output.

The Stuck-At model does not accurately represent all possible defects which may cause faults. For example, the model does not include shorts (bridging faults).[39] Historically, a test with a high level of single Stuck-At fault coverage has resulted in low defect levels being shipped to the next level of assembly due to shorts.[39] Thus, the measure of high, single Stuck-At fault coverage has been sufficient as a measure to protect one against the shipment of shorts. In no way should it be construed that a high level of single Stuck-At faults guarantees that no shorts or bridging faults exist.

The phrase "single Stuck-At fault" needs to be explained.[1-3, 7] In general, when generating tests for a faulty machine, the faulty machine consists of the good machine with one, and only one, Stuck-At fault induced into it. This, at first inspection, does not seem to be an adequate representation of the real world. For example, a network may have more than one fault at a time--for example, 10 faults. Do the tests generated for a single fault in fact test for those networks which have 10 or more faults?

Consider a network containing M nets. Any net may be good, S-A-1, or S-A-0. Thus, all possible network state combinations would be 3^M. A network with 100 nets, then, would contain 5×10^{47} different combinations of faults, whereas, with the Stuck fault assumption, only 200 faults are assumed--that is, each net, one at a time, Stuck-At 1, and again, one at a time, Stuck-At 0. Again, at first inspection, this would appear to be a very inadequate test for all possible fault combinations that may exist, not even considering all of the possible defect combinations which may exist. However, history has proven out that the single Stuck-At fault assumption in prior technologies has been adequate. This assumption would have been discarded if the defect levels, passed on to the next level of packaging, were not acceptable. Again, history has shown that the single Stuck-At fault model does, in fact, result in acceptable defect levels shipped to the next level of packaging. The term "coverage" or "fault coverage" refers to the percentage of possible single Stuck-At faults that the test set will expose. A percentage is often associated with this fault coverage which would be the faults that are tested divided by the total number of assumed single Stuck-At faults.

1.2 The VLSI Testing Problem

The VLSI testing problem is the sum of a number of problems. All the problems, in the final analysis, relate to the cost of doing business (dealt with in the following section). There are two basic problem areas:

1. Test generation.

2. Fault simulation.

With respect to test generation, the problem is that as logic networks get larger, the ability to generate tests automatically is becoming more and more difficult. Figure 5 shows a graph of the ability of automatic test generators to generate tests as a function of the number of gates of a "general sequential network". The graph shows that as the size of a general sequential network gets in the range of 1,000 to 2,000 gates, the ability to generate tests decreases to unacceptable levels. Some test generation algorithms, which are commercially available, may claim a higher

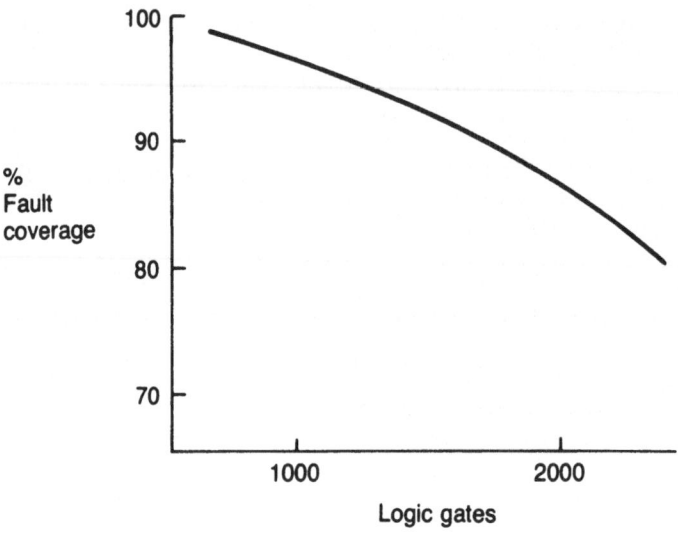

Figure 5. Trend of fault coverage obtained in practical cases versus network size.

upper limit on automatic test generation. This is a function of the sequential complexity of the network and the use of interactive "help". The more sequentially complex the network, the more difficult it is to generate tests.

Certainly, some algorithms are more superior than others. If the network is combinational (latch-free), on the other hand, the test generation algorithms, which are designed for combinational networks, are adequate for large networks (e.g., 5,000 gates and above).[70, 86, 87, 90] This latter fact will be exploited when the techniques for Design for Testability are discussed, in which the sequential networks are reduced, in essence, to combinational networks.

If the test patterns generated by an automatic test generator are inadequate, then the procedure employed most often is to supplement the automatically generated patterns with manually generated ones. The manually generated ones are directed at those faults which still remain untested by the automatically generated test pattern. This can be a time-consuming process which has the severe impact of lengthening the already too long design cycle and, as a result, adversely impacting market penetration.

The second facet of the VLSI testing problem is the difficulty in fault simulating the test patterns. Fault simulation is that process by which the fault coverage is determined for a specific set of input test patterns. In particular, at the conclusion of the fault simulation, every fault that is detected by the given

pattern set is listed. For a given logic network with 1,000 two
input logic gates, the maximum number of single Stuck-At faults
which can be assumed is 6,000. Some reduction in the number of
single Stuck-At faults can be achieved by fault equivalents.
However, the number of single Stuck-At faults needed to be assumed
is about 3,000. Fault simulation, then, is the process of applying
every given test pattern to a fault-free machine and to each of the
3,000 copies of the good machine containing one, and only one, of
the single Stuck-At faults. Thus, fault simulation, with respect
to run time, is similar to doing 3,001 good machine simulations.

Techniques are available to reduce the complexity of fault
simulation, however, it still is a very time-consuming, and hence,
expensive task.[94, 98, 99, 101, 104, 106-108]

It has been observed that the computer run time to do test
generation and fault simulation is approximately proportional to
the number of logic gates to the power of 3; hence, small increases
in gate count will yield ever-increasing run times. Equation 1

$$T = KN^3 \qquad\qquad\qquad (1)$$

shows this relationship, where T is computer run time, N is the
number of gates, and K is the proportionality constant. The
relationship does not take into account the fall-off in automatic
test generation capability due to sequential complexity of the
network. It has been observed that computer run time just for
fault simulation is proportional to N^2 without even considering
the test generation phase.

When one talks about testing, the topic of functional testing
always comes up as a feasible way to test a network.
Theoretically, to do a complete functional test ("exhaustive"
testing) seems to imply that all entries in a Karnaugh map (or
excitation table) must be tested for a 1 or a 0. This means that if
a network has N inputs and is purely combinational, then 2^N
patterns are required to do a complete functional test.
Furthermore, if a network has N inputs with M latches, at a minimum
it takes 2^{N+M} patterns to do a complete functional test. Rarely is
that minimum ever obtainable; and in fact, the number of tests
required to do a complete functional test is very much higher than
that. With LSI, this may be a network with N = 25 and M = 50, or
2^{75} patterns, which is approximately 3.8×10^{22}. Assuming one had
the patterns and applied them at an application rate of one
microsecond per pattern, the test time would be over a billion
years (10^9)!

If a complete functional test is not feasible, then how much
testing is enough? A set of patterns generally referred to as
'functional patterns" usually is inadequate as a good Stuck-At

fault test, since the testability is low (approximately 40% to 60%). If the board contains a complex function such as a micro-processor, the only way to test the board may be to run a stored program or manually-generated patterns. The options are severely limited in testing a micro-processor, since the gate level design is often not available for anyone (or even anyone). Again, history has been a good guide to determining how much testing is enough.

If we have a high level of single Stuck-At fault coverage at the board level, we feel confident of shipping a very low defect-level product to the next level of assembly. (High means between 85% and 95% testability.)

Once the test patterns have been obtained from, say, an automatic test pattern generator and have been fault simulated, there are two by-products: the first is the overall fault coverage known with a given set of test patterns. The second is the diagnostics which can be obtained every time a fault is detected on certain outputs for a certain pattern. This information is then used to help locate the failing components. The fewer the number of diagnostic messages per pattern, the higher the diagnostic resolution--i.e., the better a fault can be isolated to a failing module. The diagnostics which are part of the test data are called pre-calculated diagnostics. The diagnostics which are generated at the tester (e.g., guided probe data[52]) are learned or measured diagnostics. In any event, the ability to point to the failing module is directly related to the cost of testing and manufacturing, covered in the next section.

1.3 Cost of Testing

One might ask why so much attention is now being given to the level of testability at chip and board levels. The bottom line is the cost of doing business. A standard among people familiar with the testing process is shown in Table 1.

TABLE 1

Cost of detecting and diagnosing a fault at different levels.

Chip	Board	System	Field
30¢	$3.00	$30.00	$300.00

Some would argue that the level of difficulty would go up at a higher rate than a factor of 10 per level, due to denser integration at each level. Thus, if a fault can be detected at a chip or

board level, then some significantly larger costs per fault can
be avoided at subsequent levels of packaging.

There is a relationship between the defect level, a level of
fault coverage and the yield of a chip process. If one assumes
faults are uniformly distributed and independent, then it can be
shown that a test with fault coverage T ($0 \leq T \leq 1$) and a process
with a yield Y ($0 \leq Y \leq 1$) has a Defect Level, DL, ($0 \leq DL \leq 1$)
after testing with the relationship shown in Equation (2). The
yield is the percent good chips from a process.

$$DL = 1 - Y^{(1-T)} \tag{2}$$

Figure 6 shows a plot of the Defect Level, DL, as a function of
the fault coverage T for four different yields, Y = 0.15, 0.35,
0.55, and 0.75. With no testing at all--that is, fault coverage
T = 0, and a process yield of .15, the Defect Level would be 85%.
At a fault coverage of approximately 75% with the same yield,
this would give rise to a Defect Level of 40%. Clearly, if a
Defect Level of one or two percent is required, then a fairly
high level of fault coverage is required.

This same relationship can be used to predict board level Defect
Levels if the same uniform distributions and independent assumptions
are made. For example, assume a board contains 40 modules, each
with a 1.5% Defect Level. The probability of obtaining a good
board assuming perfect assembly is $Y_B = (1 - 0.015)^{40} = 0.546$.
Hence, 45.4% of the boards will have one or more faulty components.

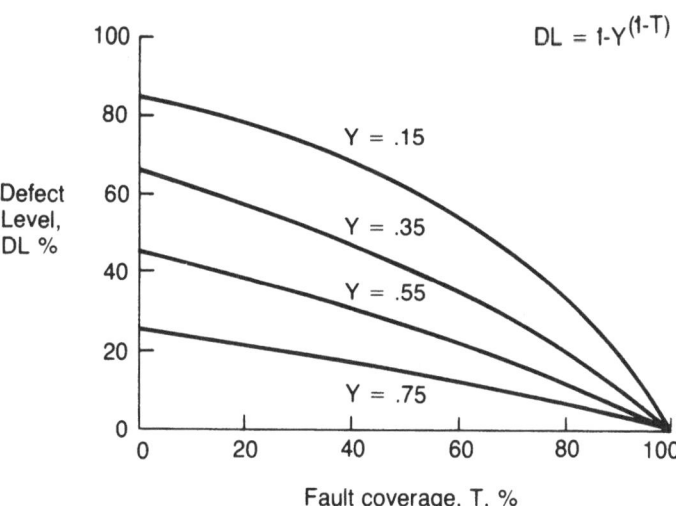

Figure 6. Graph of defect level as seen at the end of a manufacturing step including testing
as a function of fault coverage. (Plots for four component yields Y are shown.)

370

The Defect Level of the boards (DL_B) as a function of the fault coverage is

$$DL_B = 1 - (0.546)^{(1-T)} \tag{3}$$

Using Equation (3), to obtain boards with a Defect Level of 1.0% will require a level of fault coverage of 98%. This again assumes that each of the 40 modules placed on this board have a Defect Level of 1.5%.

With VLSI, the inadequacy of automatic test generation, and fault simulation, there is considerable difficulty in obtaining the level of testability required to achieve acceptable Defect Levels. If the Defect Level of boards are too high, the costs of field repairs are also too high. These costs and, in some cases, the inability to obtain a sufficient test have led to the need to Design for Testability.

2.0 Design for Testability

There are two key concepts in Design for Testability: controllability and observability. Control and observation of a network are central to implementing its test procedure. For example, consider the case of the simple AND block in Figure 1. In order to be able to test the "A" input Stuck-At 1, it was necessary to control the "A" input to 0 and the "B" input to 1 and be able to observe the "C" output to determine whether a 0 was observed or a 1 was observed. The 0 is the result of the good machine, and the 1 would be the result, if you had a faulty machine. If this AND block is embedded into a much larger sequential network, the requirement of being able to control the "A" and "B" inputs to 0 and 1 respectively and being able to observe the output "C", be it through some other logic blocks, still remains. Therein lies part of the problem of being able to generate tests for a network.

A third important concept derives from the controllability issue and from the fact that most networks of interest today are sequential networks. This is predictability: can we easily place the network under test into a known state from which all future states can be predicted*? Many networks are rated as untestable because they are difficult to initialize when working properly. They may become impossible to initialize in the presence of a fault.

---

* The concept of predictability contains the problem of initialization and non-deterministic behavior. The latter is becoming a problem in digital/ analog hybrid designs.

Examining the predictability issue illustrates the need for
controllability. Consider the simple network in Figure 7, a
Divide By Two counter. This network will require some special
attention to test, because it will start up from an unknown
state. This is a case of a network which does not have a
synchronizing sequence--i.e., an input sequence to leave a
network in a single state irrespective of its initial state.

There are two approaches to testing this network. One is to
apply a clock train of known frequency to the input and measure
the output frequency. The other is to perform a conditional test
such as the following:

```
If (Q = 0) Then            % Q = 0   initial case
Begin

Test (CLK = 1, Q = 1);

Test (CLK = 0, Q = 1);

Test (CLK = 1, Q = 0);

End Else

Begin                      % Q = 1   initial case

Test (CLK = 1, Q = 0);

Test (CLK = 0, Q = 0);

Test (CLCK = 1, Q = 1);

End
```

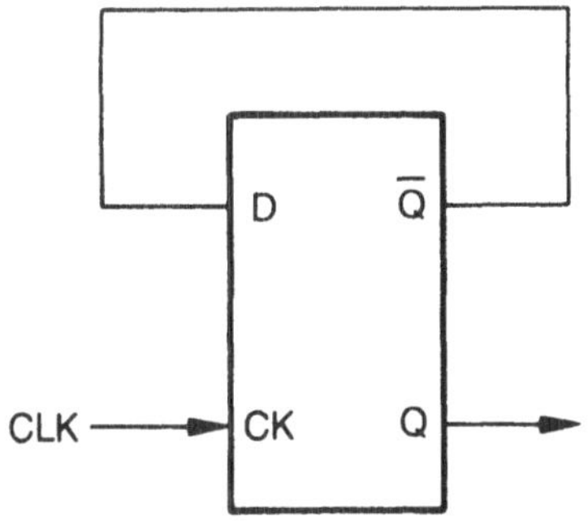

Figure 7. A flip-flop configuration that has an unknown initial state and no synchronizing
          sequence.

372

However, most networks of interest contain far too many possible initial states for us to create a test that runs from any one. Thus, the approach taken is to choose one from which the tests can proceed. Then, our problem is to move the network from some known state to that state selected as the start of the test. To do this, we will have to write a program that repeatedly examines the network's output and applies input, based upon those observations. Formally, this approach is known as an Adaptive Homing Sequence.[4] It is possible for such a homing program to fail to terminate, due to a fault in a network that defeats the homing strategy (for example, consider the circuit in Figure 7 with a Clock input open).

There is no problem with this approach to testing, if the only information desired from the test is to determine whether the network works--i.e., Go/No-Go information. However, if more precise diagnostic resolution is required, then the entire test must be run, which the failure of the Homing Sequence may prevent.

Years of practical experience and testing have identified the lack of a synchronizing sequence as the single most-destructive network characteristic with respect to its test. Yet given that one exists for a network, some additional subtleties exist which can trap the unwary. Consider the simple network in Figure 8. This network has the simplest possible synchronizing sequence, CLR = 0, which zeroes Q. Considering the following test:

CLR	0	1	1	1	1
CLK	0	0	1	0	1
Q	0	0	1	1	0

Time ⟶

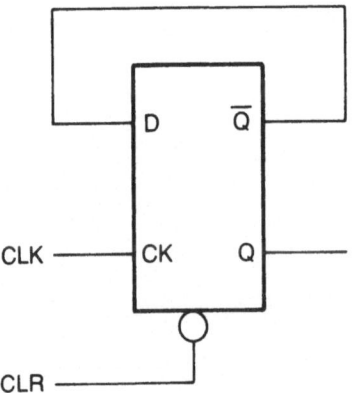

Figure 8. A synchronizable flip-flop circuit with an unsynchronizable failure mode.

By simple inspection, it appears that the test completely checks out the network, yet there are some faults that can slip through the test: the failure of the CLEAR line to clear the network (due, for example, to an open net on the CLEAR line). Consider this testing scenario:

1. Receive network at test station.

2. Put network on tester, apply power.

3. Run test. It passes.

4. Ship network to user (such as next level of production).

5. User returns network, complains about intermittent fault.

6. Go to 2.

The reason the test passes is twofold: (1) When power is applied to the network, it powers up in the $Q = 0$ state (with some probability); and (2) The test leaves $Q = 0$ upon completion; thus, rerunning or cycling the test still yields a passing result. The user, however, saw intermittent results, due to different Power Up state. A better test for this circuit would be the following:

CLR	0	1	1	1	1	1	1	0
CLK	0	0	1	0	1	0	1	1
Q	0	0	1	1	0	0	1	0

TIME ———→

This would be readily apparent to the user of a simulator who would see the following behavior for Q_{good} the correctly operating network (good machine) and Q_{faulty} (fault machine), the network with a non-operating CLEAR line.

Unknown Initial State

CLR	X	0	1	1	1	1	1	1	0
CLK	X	0	0	1	0	1	0	1	1
Q_{good}	X	0	0	1	1	0	0	1	0
Q_{faulty}	X	X	0	1	1	0	0	1	1

TIME ———→

After test 1 clears the network, the faulty network is still in an unknown and <u>possibly</u> different state represented by the X.[97,99]
At test time, there is a probability that the tester will see

Q = 1 and flag the fault, but also a probability that it will see
Q = 0 and continue the test. A simulator can take advantage of
this by continuing the simulation after this point with
$Q_{faulty} = Q_{good}$. Thus, the next occurrence of the CLEAR function
will produce an unconditional or solid detection, as indicated.
The fault simulator user is alerted about the conditional detection
of this fault and is told they must continue simulation until an
unconditional detection is achieved.

To summarize the issue of predictability, it is extremely
desirable for networks to be designed so that they are easy to
place into known states (and that they behave deterministically).
Formally, they should possess synchronizing sequences. Since a
large majority of components on the market today contain CLEAR or
PRESET functions, the synchronization criteria are, in general,
easily met. But as pointed out, it will be necessary to utilize
the synchronization function more than once in order to reliably
detect failures in its associated circuitry. This is a strike
against "uncontrollability" synchronization schemes, such as
automatic "power up", "clear" functions. As a rule, networks
should possess a short-length, clearly-evidenced synchronizing
sequences. Ideally, all networks should contain a master reset
signal which, in one pattern, uniquely determines the state of
every flip-flop in the network.

Controllability and observability, beyond this issue of
predictability, are pretty much self-evident criteria.
Observability deals with the ease with which internal states of a
network can be examined. Controllability deals with the ease with
which a network can be "steered" through its various functions.
Ease, in this case, usually means three things: how many patterns
will it take, how long will these take to apply, and how difficult
was it to derive them? Testing a 32-bit counter is an easily-
stated task, but at 10 MHz, it's going to take over seven minutes
if the test is not designed with some controllability and
observability.

Because of the need to determine if a network has the attributes of
controllability and observability that are desired--that is, it's
easy to generate tests for this network. A number of programs have
been written which essentially give analytic measures of
controllability and observability for different nets in a given
sequential network.[64-68]

After observing the results of one of these programs in a given
network, the logic designer can then determine whether some of the
techniques, which will be described later, can be applied to this
network to ease the testing problem. For example, test points may
be added at critical points which are not observable or which are
not controllable, or some of the techniques of Scan Path or LSSD

can be used to initialize certain latches in the machine to avoid the difficulties of controllability associated with sequential machines. The popularity of such tools is continuing to grow, and a number of companies are now embarking upon their own controllability/observability measures.

3.0 Ad Hoc Design for Testability

Testing has moved from the afterthought position that it used to possess to part of the design environment in LSI and VLSI. When testing was part of the afterthought--that is, given a sequential network--certain techniques had to be applied in order to make the network testable. It was a very expensive process. Because of the number of difficulties that were incurred in doing business in this mode and the fact that several products were discarded because there was no direct way to test them, testing became part of the design environment.

There are two basic approaches which are prevalent today in the industry to help solve the testing problem. The first approach categorized here is Ad Hoc, and the second approach is categorized as a Structured Approach. The Ad Hoc techniques are those techniques which can be applied to a given product, but are not directed at solving the general sequential problem. They usually do offer relief, and their cost is probably lower than the cost of the Structured Approaches. The Structured Approaches, on the other hand, basically are trying to solve the general problem with a design methodology, such that when the designer has completed his design from one of these particular approaches, the results will be test generation and fault simulation at acceptable costs. Again, the main difference between the two approaches is probably the cost of implementation and hence, the return on investment for this extra cost. In the Ad Hoc approaches, the job of doing test generation and fault simulation are usually not as simple or as straightforward as they would be with the Structured Approaches, as we shall see shortly.

A number of techniques have come forth in the category of the Ad Hoc approaches, and we will now discuss them.

3.1 Partitioning

Because the task of test pattern generation and fault simulation is proportional to the number of logic gates to the third power, a significant amount of effort has been directed at approaches called "Divide and Conquer".

There are a number of ways in which the partitioning approach to Design for Testability can be implemented. The first is to mechanical partition by dividing a network in half. In essence, this would reduce the test generation and fault simulation tasks by 8. Unfortunately, having two boards rather than one board can be a significant cost disadvantage to a program.

Another approach that helps the partitioning problem, as well as helping one to "Divide and Conquer" is to use jumper wires. These wires would go off the board and then back on the board, so that the tester and the test generator can control and observe these nets directly. However, this could mean a significant number of I/0 contacts at the board level which could also get very costly.

Degating is another technique for separating modules on a board. For example, in Figure 9, a degating line goes to two AND blocks that are driven from Module 1. The results of those two AND blocks go to two independent OR blocks--one controlled by Control Line 1, the other with Control Line 2. The output of the OR block from Controller Linr 1 goes into Module 2, and the output of Control Line 2 goes into Module 3. When the degate line is at the 0 value, the two Control Lines, 1 and 2, can be used to drive directly into Module 2 and Module 3. Therefore, complete controllability of the inputs to Modules 2 and 3 can be obtained by using these control lines. If those two nets happen to be very difficult nets to control, as pointed out, say, by a testability measure program, then this would be a very cost-effective way of controlling those two nets and hence, being able to derive the tests at a very reasonable cost.

A classical example of degating logic is that associated with an oscillator, as shown in Figure 10. In general, if an oscillator is free-running on a board, driving logic, it is very difficult, and sometimes impossible, to sync the tester with the activity of the logic board. As a result, degating logic can be used here to block the oscillator and have a pseudo-clock line which can be controlled by the tester, so that the DC testing of all the logic on that board can be synchronized. All of these techniques require a number of extra primary inputs and primary outputs and possibly extra modules to perform the degating.

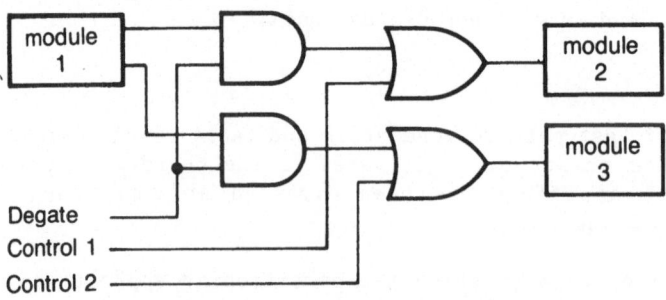

Figure 9. Use degating logic for logical partitioning

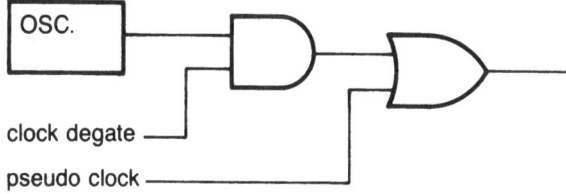

Figure 10. Degating lines for oscilator.

Thus, there are several ways to take the network and divide it, so that the test generation and fault simulation task are more manageable.

3.2 Test Points

Another approach to help the controllability and observability of a sequential network is to use test points.[21, 22] If a test point is used as a primary input to the network, then that can function to enhance controllability. If a test point is used as a primary output, then that is used to enhance the observability of a network. In some cases, a single pin can be used as both an input and an output.

For example, in Figure 11, Module 1 has a degate function, so that the output of those two pins on the module could go to non-controlling values. Thus, the external pins which are dotted into those nets could control those nets and drive Module 2. On the other hand, if the degate function is at the opposite value, then the output of Module 1 can be observed on these external pins. Thus, the enhancement of controllability and observability can be accommodated by adding pins which can act as both inputs and outputs under certain degating conditions.

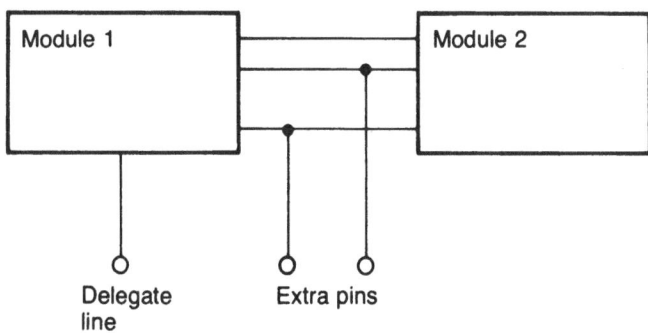

Figure 11. Test points used as both Inputs and Outputs

Another technique which can be used for controllability is to have a pin which, in one mode, implies system operation, and in another mode takes N inputs and gates them to a decoder. The output of the decoder is used to control certain nets to values which otherwise would be difficult to obtain. By so doing, the controllability of the network is enhanced

As mentioned before, predictability is an issue which is as important as controllability and observability. Again, test points can be used here. For example, a CLEAR or PRESET function for all memory elements can be used. Thus, the sequential machine can be put into a known state with very few patterns.

Another technique which falls into the category of test points and is very widely used is that of the "Bed of Nails"[27] tester. The Bed of Nails tester probes the underside of a board to give a larger number of points for observability and controllability. This is in addition to the normal tester contact to the board under test. The drawback of this technique is that the tester must have enough test points to be able to control and observe each one of these nails on the Bed of Nails tester. Also, there are extra loads which are placed on the nets and this can cause some drive and receive problems. Furthermore, the mechanical fixture which will hold the Bed of Nails has to be constructed, so that the normal forces on the probes are sufficient to guarantee reliable contacts. Another application for the Bed of Nails testing is to do "drive/sense nails"[27] testing, which, effectively, is the technique of testing each chip on the board independently of the other chips on the board. For each chip, the appropriate nails and/or primary inputs are driven so as to prevent one chip from being driven by the other chips on the board. Once this state has been established, the isolated chip on the board can now be tested. In this case, the resolution to the failing chip is much better than edge connector tests, however, there is some exposure to incomplete testing of interconnections. Also, the nails are an expensive way to test, as well as introducing a lack of reliability into the test. This technique has also been alluded to as "in-circuit testing." Usually, contact reliability is one of the largest drawbacks of this approach. Also, it clearly does not have any growing potential--that is, if, in fact, tomorrow's chips are today's boards, then we do not have enough I/O's at the chip level to accommodate this kind of testing approach. Therefore, it is suited to today's problems, and not tomorrow's.

3.3 Bus Architecture

An approach that has been used very successfully to attack the partitioning problem by the Micro-Computer designers is to use a bus structure architecture. This architecture allows access to critical buses which go to many different modules on the computer board. For example, in Figure 12, you can see that the data bus is involved with both the micro-processor module, the ROM module, the RAM module and the I/O Controller module. If there is external access to the data bus and three of the four modules can be turned off the data bus--that is, their outputs can be put into a high impedence state (three-state driver)--then the data bus could be used to drive the fourth module, as if it were a primary input (or primary output) to that particular module. Similarly, with the address bus, access again must be controlled externally to the board, and thus, the address bus can be very useful to controlling test patterns to the micro-computer board. These buses, in essence, partition the board in a unique way, so that testing of subunits can be accomplished. In particular, the RAM and the ROS can be tested, which, if they were previously embedded into the logic, would make them very difficult to test--that is, it would go through the address bus, as well as being able to read and write any kind of data values into the RAM. This technique takes a number of control lines and faults on the bus are difficult to isolate to.

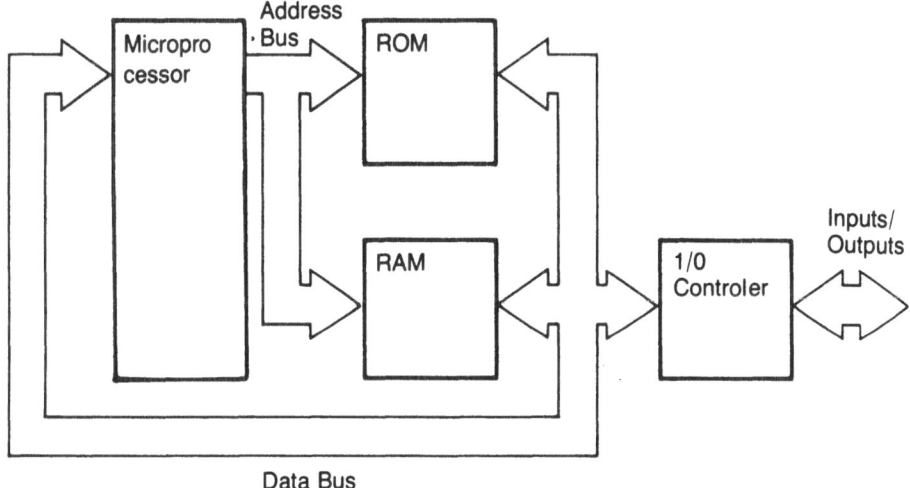

Figure 12. Bus structured microcomputer

3.4 Signature Analysis

This technique for testing, introduced in 1977,[25, 29, 50] is heavily reliant on planning done in the design stage. That is why this technique falls between the Ad Hoc approaches for Design for Testability and the Structured Approaches for Design for Testability, since some care must be taken at the board level in order to ensure proper operation of this Signature Analysis of the board.[10] Furthermore, Signature Analysis is best-suited to bus structure architectures, as mentioned above, and in particular,

those associated with micro-computers. This will become more
apparent shortly.

The integral part of the Signature Analysis approach is that of a
linear feedback shift register. Figure 13 shows an example of a
three-bit linear feedback shift register. This linear feedback
shift register is made up of three shift register latches. Each
one is represented by a combination of an L1 latch and an L2 latch.
These can be thought of as the master latch being the L1 latch and
the slave latch being the L2 latch. An "A" clock clocks all the L1
latches, and a "B" clock clocks all the L2 latches, so that turning
the "A" and "B" clocks on and off independently will shift the
shift register one bit position to the right. Furthermore, this
linear shift register has an Exclusive OR gate which takes the
output, Q2, the second bit in the shift register, and Exclusive
OR's it with the third bit in the shift register, Q3. The result
of that Exclusive OR is the input to the first shift register. A
single clock could be used for this shift register, which is
generally the case, however, this concept will be used shortly when
some of the structured design approaches are discussed which use
two non-overlapping clocks.

Figure 14 shows how this linear feedback shift register will count
for different initial values. Figure 14(a) shows the counting
sequence, if the initial value for the three shift registers is
000. If you take the Exclusive OR of Q2 with Q3, which is 00, the
output of the Exclusive OR will be 0. Thus, after shifting to the
right 1, Q1 will go to 0, Q2 and Q3 will just be the transposition
of Q1 and Q2 from 00's; and thus, the result will be a 000 in the
three bit positions of the shift register. Therefore, continuous
clocking of the "A" and "B" clocks will result in all zeros in the
shift register and no count. This is similar to an absorbing
state. If, however, on the other hand, the initial values are any
of the values shown in Figure 14(b)--for example, 001--then a
count will proceed of 7, counting down through the different
values. Notice that no value is repeated until you get to 011.
From 011, you go back to the initial value of 001 and then continue
the count. This count is of length 7, because this linear feedback
shift register happens to be of maximal length, which is 2^N -1,
where N is the number of shift registers in the shift register
sequence. There are only 3 shift registers, so we have 2^3 -1 which
is 7. The -1 represents the absorbing state, thus, we will count
from 1 to 7, and then back to 1 again.

For longer shift registers, the maximal length linear feedback
shift registers can be configured by consulting tables to
determine where to tap off the linear feedback shift register to
perform the Exclusive OR function. Of course, only Exclusive OR
blocks can be used, otherwise, the linearity would not be
preserved.

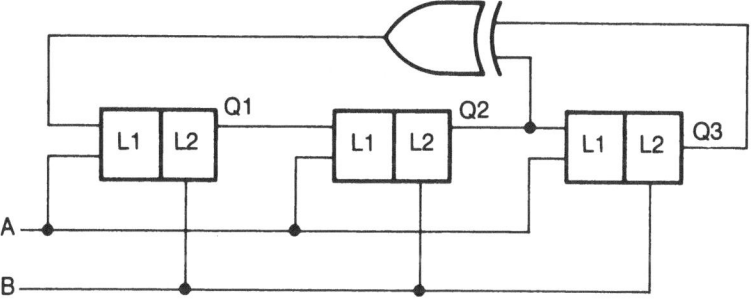

Figure 13. Linear feedback shift register.

	Q1	Q2	Q3
If initial value =	0	0	0

(a) Absorbing state

	Q1	Q2	Q3
If initial value =	0	0	1
	1	0	0
	0	1	0
	1	0	1
	1	1	0
	1	1	1
	0	1	1

$$2^n - 1 = 2^3 - 1 = 7$$

(b) Count of seven which is maximal for a three bit linear feedback shift register.

Figure 14. Counting capabilities of a linear feedback shift register.

The key to Signature Analysis is to design a network which can stimulate itself. A good example of such a network would be micro-processor-based boards, since they could stimulate themselves using the intelligence of the processor driven by the memory on the board.

The Signature Analysis procedure is one to have the shifting of the shift register in the Signature Analysis tool, which is external to the board and not part of the board in any way, synchronized with the clocking that occurs on the board, see Figure 15. A probe is used to probe a particular net on the board. The result of that probe is Exclusive OR'ed into the linear feedback shift register. Of course, it is important that the linear feedback shift register be initialized to the same starting place every time, and that the

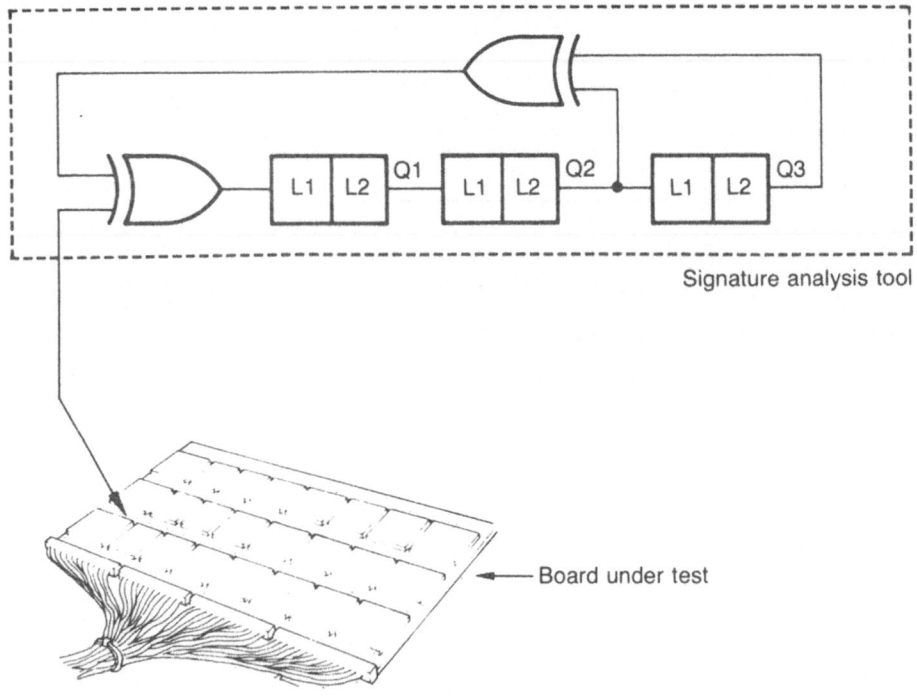

Figure 15. Use of Signature Analysis Tool

clocking sequence be a fixed number, so that the tests can be repeated. The board must also have some initialization, so that it can be repeated as well.

After a fixed number of clock periods--let's assume 50--a particular value will be stored in Q1, Q2 and Q3. It is not necessarily the value that would have occurred if the linear feedback shift register was just counted 50 times--Modulo 7. The value will be changed, because the values coming from the board via the probe will not necessarily be a continuous string of 1's; there will be 1's intermixed with 0's.

The place where the shift register stops on the Signature Analysis Tool--that is, the values for Q1, Q2 and Q_3 is the Signature for that particular node for the good machine. The question is-If there were no errors present at one or more points in the string of 50 observations of that particular net of the board, would the value stored in the shift register for Q1, Q2 and Q_3 be different than the one for the good machine? It has been shown that with a 16-bit linear feedback shift register, the probability of detecting one or more errors is extremely high.[50] In essence, the signature, or "residue", is the remainder of the data stream after

division by an irreduceable polynomial. There is considerable
data compression--that is, after the results of a number of
shifting operations, the test data is reduced to 16 bits, or, in
the case of Figure 15, three bits. Thus, the result of the
Signature Analysis tool is basically a Go/No-Go for the output for
that particular module.

If the bad output for that module were allowed to cycle around
through a number of other modules on the board and then feed back
into this particular module, it would not be clear which module was
defective--whether it was the module whose output was being
observed, or whether it was another module upstream in the path.
This gives rise to two requirements for Signature Analysis. First
of all, closed-loop paths must be broken at the board level.
Second, the best place to start probing with Signature Analysis is
with a "kernel" of logic. In other words, on a micro-processor-
based board, one would start with the outputs of the microprocessor itself and then build up from that particular point, once
it has been determined that the micro-processor is good.

This breaking of closed loops is a tenant of Design for Testability
and, in this case, a Design for Testability techniques for
Signature Analysis. There is a little overhead for implementing
Signature Analysis. Some ROM space would be required (to stimulate
the self-test), as well as extra jumpers, in order to break closed
loops on the board. Once this is done, however, the test can be
obtained for very little cost. The only question that remains is
about the quality of the tests--that is, how good are the tests
that are being generated, do they cover all the faults, etc.

Unfortunately, the logic models--for example, micro-processors--
are not readily available to the board user. Even if a microprocessor logic model were available, they would not be able to do
a complete fault simulation of the patterns because it would be too
large. Hence, Signature Analysis may be the best that could be
done for this particular board with the given inputs which the
designer has. Presently, large numbers of users are currently
using the Signature Analysis technique to test boards containing
LSI and VLSI components.

4.0 STRUCTURED DESIGN FOR TESTABILITY

Today, with the utilization of LSI and VLSI technology, it has
become apparent that even more care will have to be taken in the
design stage in order to ensure testability and produceability of
digital networks. This has led to rigorous and highly-structured
design practices. These efforts are being spearheaded, not by the
makers of LSI/VLSI devices, but by electronics firms which possess
captive IC facilities and the manufacturers of large main-frame
computers.

Most structured design practices[12-14, 16-19, 23, 27, 28, 30, 31] are built upon the concept that if the values in all the latches can be controlled to any specific value, and if they can be observed with a very straightforward operation for both controllability and observability, then the test generation, and possibly, the fault simulation task, can be reduced to that of doing test generation and fault simulation for a combinational logic network. A control signal can switch the memory elements from their normal mode of operation to a mode that makes them controllable and observable.

It appears from the literature that several companies, such as IBM,[12-14, 16-19, 27, 28, 31] Fujitsu Ltd., Sperry-Univac and Nippon Electric Co., Ltd. have been dedicating formidable amounts of resources toward Structured Design for Testability. One notes simply by scanning the literature on testing, that many of the practical concepts and tools for testing were developed by main-frame manufacturers who do not lack for processor power. It is significant, then, that these companies, with their resources, have recognized that unstructured designs lead to unacceptable testing problems. Presently, IBM has extensively documented its efforts in Structured Design for Testability, and these are reviewed first.

4.1 Level Sensitive Scan Design, LSSD

This section will be devoted primarily to describing the design constraints for LSSD and the various ways they can be exploited[13, 14, 16-18, 20, 31] during design, manufacturing, testing and field servicing. The basic storage device and the general logic structure resulting from the design constraints will also be described in detail. The important question of how much it costs for logic gates and operating speeds will also be discussed qualitatively.

The design rules essentially combine two concepts that are almost independent. The first is to design such that correct operation is not dependent on rise time, fall time or minimum delay of the individual circuits. The only dependence is that the total delay through a number of levels be less than some known value. This method is called Level Sensitive Design. The second concept is to design all internal storage elements (other than memory arrays), such that they can also operate as shift registers.

The concept of Level Sensitive Design is one method that will provide reliable operation without strong dependence on hard-to-control AC circuit parameters. This design method can be defined as follows:

Definition: A logic subsystem is level-sensitive if, and

only if, the steady state response to any allowed input state change is independent of the circuit and wire delays within the subsystem. Also, if an input state change involves the changing of more than one input signal, then the response must be independent of the order in which they change. Steady state response is the final value of all logic gate outputs after all change activity has terminated.

It is clear from this definition that level-sensitive operation is dependent on having only "allowed" input changes; thus, a Level Sensitive Design method will, in general, include some restriction on how these changes occur. In the detailed design rules, these restrictions and input changes are implied mostly to the clock signal. Other input signals have almost no restrictions on when they may change.

The principal objective in establishing design constraints is to obtain logic subsystems that are insensitive to AC

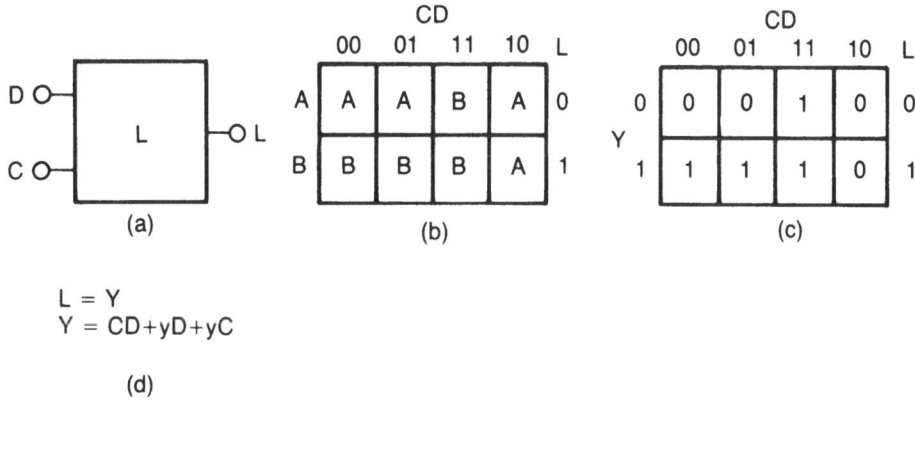

$$L = Y$$
$$Y = CD + yD + yC$$

(d)

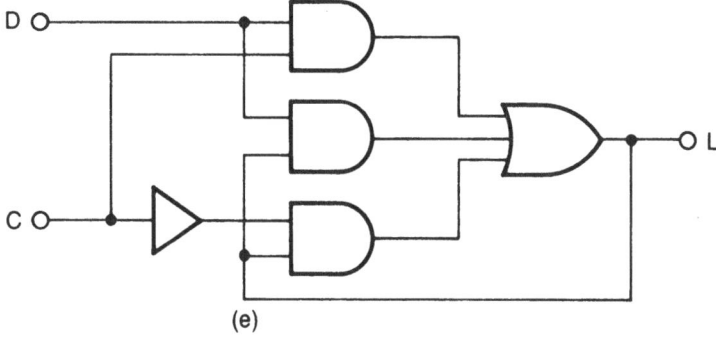

(e)

Figure 16. Hazard-free polarity-hold latch. (a) Symbolic representation. (b) Flow table. (c) Excitation table. (d) Excitation equations. (e) Logic implementation.

characteristics, such as rise time, fall time and minimum circuit delay. Consequently, the basic storage element should be a level-sensitive device that does not contain a hazard or race condition. The hazard-free polarity hold latch shown in Figure 16 satisfies these conditions--that is, no matter how slow the rise time of the clock pulse is or how slow the fall time of the clock pulse is, this particular latch will latch reliably, as long as the data does not change during the clock transition period.

The basic building block of this structured approach is that of a Shift Register Latch, SRL, and its method of interconnection will be described.

The polarity hold Shift Register Latch, SRL, is shown in Figure 17. It consists of two latches, L1 and L2. As long as the shift

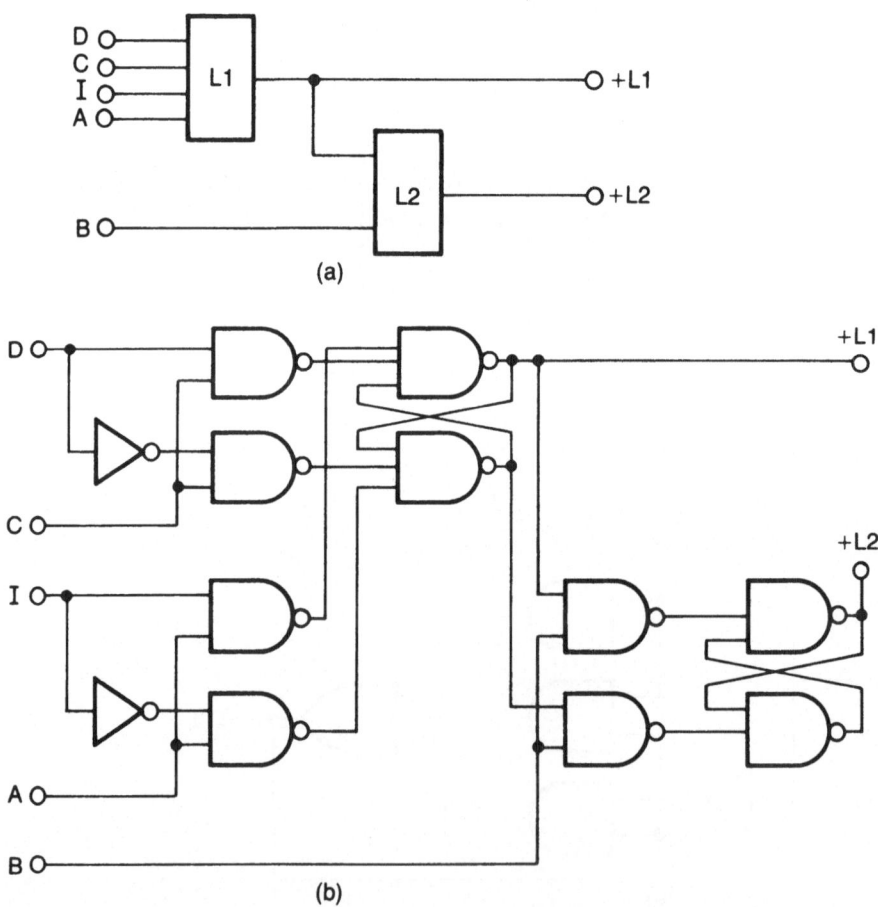

Figure 17. Polarity-hold SRL. (a) Symbolic representation. (b) Implementation in AND-INVERT gates.

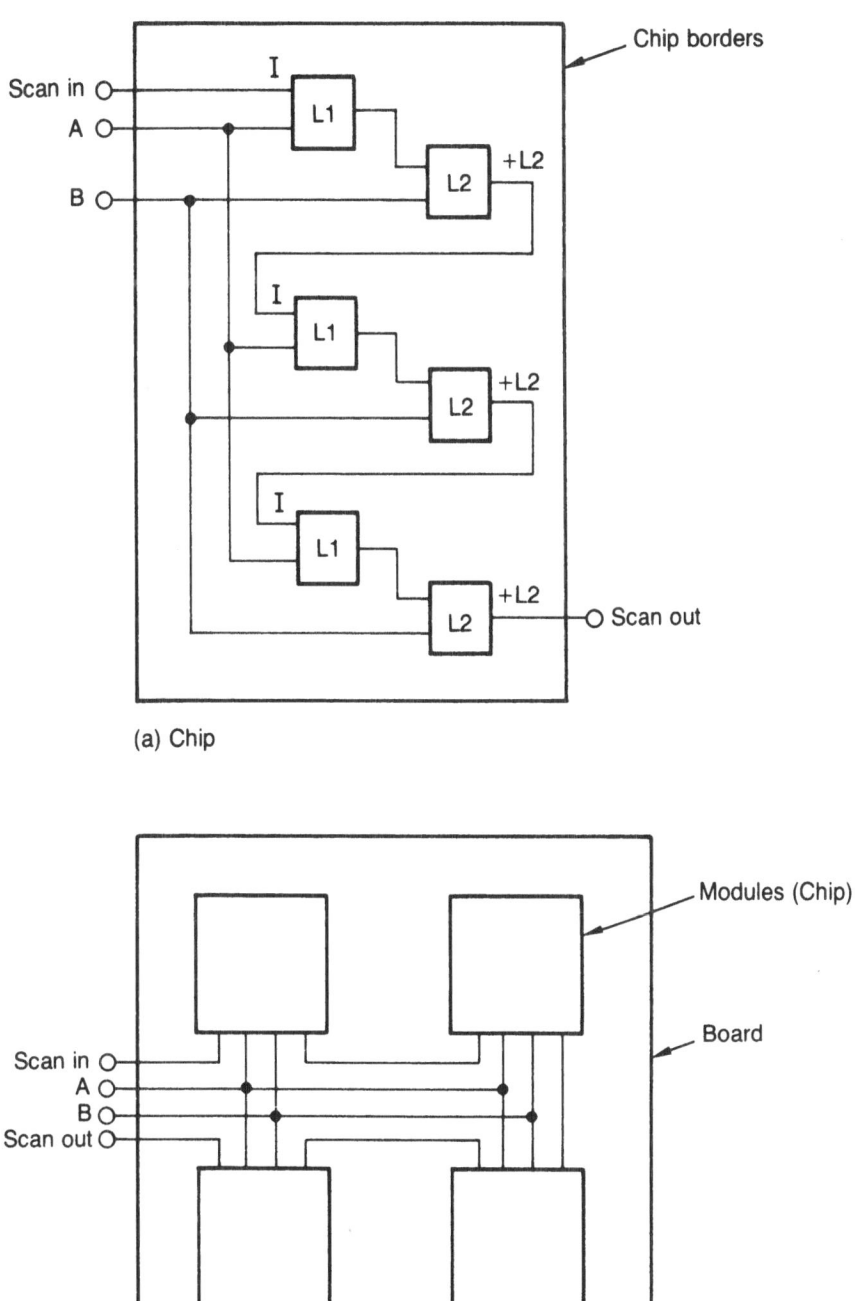

(a) Chip

(b) Board

Figure 18. Interconnection of SRLs on a chip and board

388

signals A and B are both 0, the L1 latch operates exactly like a polarity hold latch. Terminal I is the input for the shift register, and L2 is the output. When the latch is operating as a shift register, data from the proceding stage is gated into the polarity latch, L1, via I by a change of the A signal to 1 and then back to 0. After A has changed back to 0, the B signal gates the data in the latch L1 into the output latch connected to the output terminal +L2. Clearly, A and B can never both be 1 at the same time if the shift registers are to operate properly.

The modification of the polarity hold latch L1 to include shift capability essentially requires adding a clock input to the latch and a second latch L2 to act as intermediate storage during shifting. Thus, the basic cell is somewhere between two and three times as complicated as the polarity hold latch. Even so, the overall effect on chip design may be relatively small, if the circuit design is efficient. If the shift register gates are not used during normal operation, one can allow implementations that are efficient in both power and space. The interconnection of three SRL's into a shift register is shown in Figure 18(a), and the interconnection of four modules is shown in Figure 18(b). The shift signals are connected in series and the I (input) and +L2 (output) signals are strung together in a loop. At most, four additional I/O terminals are required at each level of packaging.

Figure 19 shows the classical model of the sequential network utilizing a shift register for its storage. The normal operation is to load data broadside into the SRL's, take the output of the SRL's and feed them back around into the combinational logic.

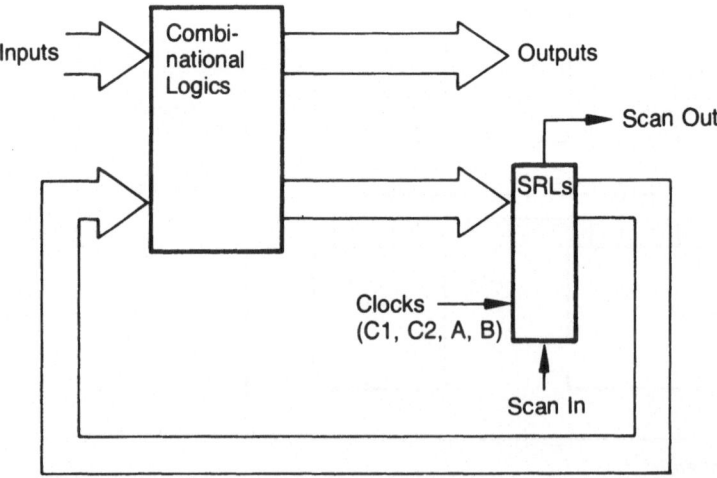

Figure 19. Classic model of a sequential network utilizing a shift register for storage.

Passing of data from the input side of the SRL's to the output side of the SRL's is controlled by the clocks--in particular, the system clock, C1, and the C2 clock which will be used to clock the L2 latches, along with the "B" shift clock. When the state of the sequential machine is to be observed, data is shifted out in a serial fashion where the "A" clock and the "B" clock are pulsed one after the other in order to cause the shift register action. This data can be loaded in via the Scan In port and unloaded via the Scan Out port.

Together, the level-sensitive concept plus the scan concept yields a design structure defined as Level Sensitive Scan Design, LSSD. The rules for this structure are given below.

1. All internal storage is implemented in hazard-free polarity hold latches.

2. The latches are controlled by two or more non-overlapping clocks such that:

   i) A Latch X may feed the data port of another Latch Y if, and only if, the clock that sets the data into Latch Y is not clocked by Latch X.

   ii) A Latch X may gate a Clock C_i to produce a gated Clock C_{ig} which drives another Latch Y if, and only if, Clock C_{ig} does not clock Latch X, where C_{ig} is any clock produced from C_i.

3. It must be possible to identify a set of clock primary inputs from which the clock inputs to SRL's are controlled, either through simple powering trees or through logic that is gated by SRL's, and/or non-clock primary inputs. In addition, the following rules must hold:

   i) All clock inputs to all SRL's must be at their "Off" state when all clock primary inputs are held to their "Off" state.

   ii) The clock signal that appears at any clock input of an SRL must be controlled from one or more clock PI's, so that it is possible to set the clock input of the SRL to an "On" state by turning any one of the corresponding PI's to its "On" state, and also setting the required gating condition from SRL's and/or non-clock PI's.

   iii) No clock can be AND'ed with either the true value or the complement value of another clock.

4.   Clock primary inputs may not feed the data inputs to latches, either directly or through combinational logic, but may only feed the clock input to the latches or the primary outputs.

A sequential logic network designed in accordance with Rules 1 through 4 will be level-sensitive. To simplify testing and minimize primary inputs and outputs, the network must also be able to shift data into and out of the latches in the system. Two more rules must, therefore, be followed:

5.   All system latches are implemented as part of an SRL. All SRL's must be interconnected to one or more shift registers, each of which has an input, an output and shift clocks available at the terminals of the package.

6.   There must exist some primary input sensitizing condition referred to as the "scan state", such that:

i)   Each SRL or scan-out PO is a function of only the single preceding SRL or scan-in PI in its shift register during the shifting operation.

ii) All clocks, except the shift clocks, are kept "Off" at the SRL inputs.

iii) Any shift clock to the SRL may be turned "On" or "Off" by changing the corresponding clock primary input for each clock.

If the design constraints are followed, a logic subsystem with two clock signals will have the structure shown in Figure 20. It is evident from the figure that the two clock signals are not to be active at the same time, so that no race will occur. This design technique is called a Double Latch Design, since all system inputs to network N are taken from the L2 latch. Hence, the system data flow is into the L1 latch, out of the L1 latch, to the L2 latch, out of the L2 latch and wrapped back around into the input of network N. Making use of the L2 latch greatly reduces the overhead associated with such a design method.

The operation of the subsystem is controlled by two clock signals. When C_1 is on, C_2 is off, and the inputs and outputs of network N are stable, assuming that the external inputs are also stable. The clock signal C_1 is then allowed to pass to the SRL system clock inputs. The system clock C_1 may be gated by signals from the network N, so that C_1 reaches the SRL if, and only if, the gate is active. Thus, some of the latches may change at C_1 time.

As long as C_1 is active long enough to latch the data values into

the L1 latches, proper operation will occur. The data on the
output of the L1 latches will now be ready to be latched into the
L2 latches. This will occur with the C_2 clock which comes in on
the "B" shift clock line. C_2 is turned on and then off. At the
time C_2 is turned off, the results on the L2 latches, assuming
again that C_2 is on long enough to latch up the value on its
inputs. The values on Y_1, Y_2. . .and Y_N are now available to be fed
back into the combinational network N. The network N is also ready
for new inputs on the primary inputs.

The shifting operation will occur when the C_1 clock is turned off,
the proper value is put on the scan input, and the "A" and "B"
clocks are turned on and off alternately--first "A", then "B", then
"A", then "B", etc. The result will be that any value desired can
be loaded into the SRL's. The value specified may come from the
test generator--that is, the value necessary for a particular test
of the combinational network N. Once the latches have been set up,
the system clock C1 can be turned on and off, thus, the results of
the tests for the combinational logic network N are now stored in
the L1 latches. A "B" clock is then turned on and off, thus, the
values that were in all the L1 latches are now duplicated in the L2
latches. The values in the SRL's are now ready to be shifted out.
This can be accomplished by applying the "A" clock, then the "B"
clock, the "A" clock, and then the "B" clock, etc. After each pair
of "A"/"B" clocks, a measurement is taken at the scan-out primary
output.

The shift register must also be tested, but this is relatively
simple and can be done by shifting a short sequence of 1's and 0's
through the SRL's.

If the combinational network, N in Figure 20, can be partitioned
into smaller independent subnetworks then test generation and
fault simulation can be carried out for each of the subnetworks.
Any partitioning of the general structure shown in Figure 20 will
result in a number of substructures that can be tested in the same
way--that is, all logic gates can be given combinational tests by
applying the appropriate test patterns at the primary inputs and at
the SRL outputs, by shifting in serially. The output pattern can
be obtained from the response outputs X_1, X_2,. . .and X_N, by
shifting out the bit pattern stored in the SRL's. Thus, the same
method can be used to test at any packaging level. Furthermore,
this technique is an integral part of the "Divide and Conquer"
technique which is similar to partitioning, except this partitions
combinational networks into smaller combinational networks. Thus,
using this design technique, the test generation and fault
simulation can be reduced to that of combinational test generation
and fault simulation on smaller networks.

One advantage of these design constraints is that the correct operation of the logic subsystem is almost independent of any transient or delay characteristics of the logic circuits. This fact can be seen by considering the operation of the structure in Figure 20. At the time C_2 occurs, some of the L2 latches of the SRLs may change state as a result of the signals stored in the L1 latches. These changes must propagate through the combinational network N and stabilize at X_1, X_2, ..., X_n before C_1 can occur. Thus the signals from the L2 latches must propagate fully through N during the time between the beginning of C_2 and the beginning of C_1.

The only delay requirement, then, is that the worst-case delay through N must be less than some known value. There is no longer any need to control or test rise time, fall time, or minimum network delays; only the maximum network delay need be controlled and measured. Moreover, individual gate dealys are not important; only the total delay over paths from the input to the output of network N need be measured.

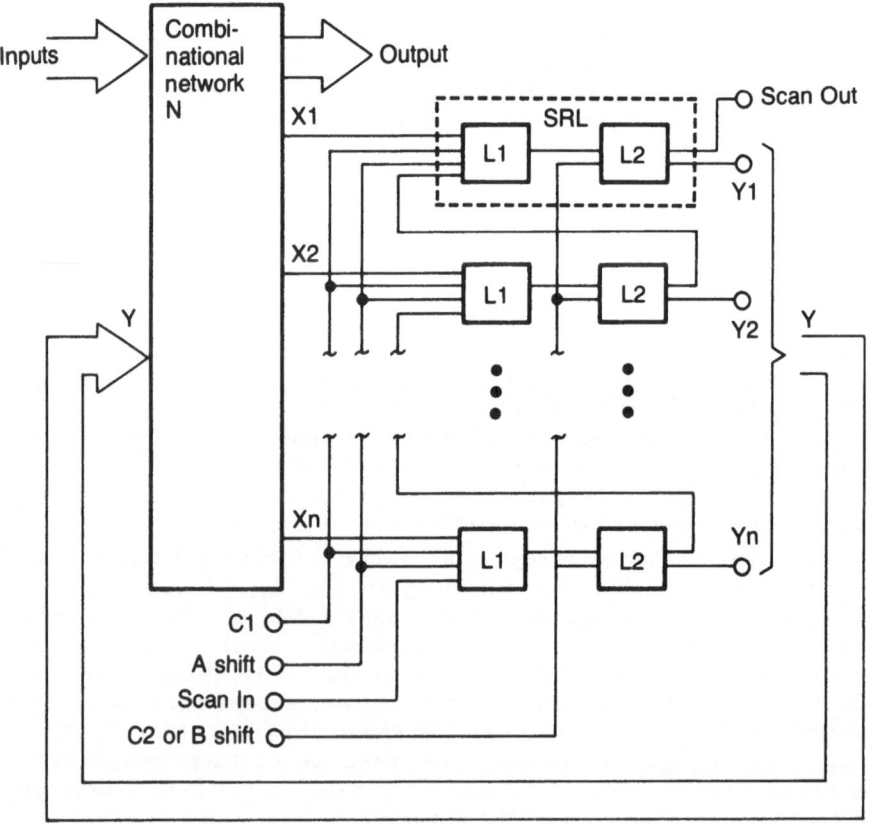

Figure 20. LSSD Double Latch Design

Probably the most important reason for using the SRL's shown in Figure 20 is that it provides the ability to make dynamic measurements of the nets buried within the chip. More specifically, it enables the technician to debug the machine or the Customer Engineer servicing it to monitor the state of every latch in the logic subsystem. This can be done on a single-cycle basis by shifting all the data in the latches out to a display device. This will not disturb the state of the subsystem, if the data is shifted back into the latches in the same order as it was shifted out. Thus, the state of all the latches can be examined after each clock signal or cycle.

In considering the cost performance impacts, there are a number of negative impacts associated with the LSSD design philosophy. First of all, the polarity hold latches in the shift register are, logically, two or three times as complex as simple latches. Up to four additional primary inputs/outputs are required at each package level for control of the shift registers. External asynchronous input signals must not change more than once every clock cycle. Fianlly, all timing within the subsystem is controlled by externally-generated clock signals.

In terms of additional complexity of the polarity hold latches, the overhead from experience has been in the range of 4% to 20%. The difference is due to the extent to which the system designer made use of the L2 latches for system function.

It has been reported in the IBM System 38 literature that 85% of the L2 latches were used for system function. This drastically reduces the overhead associated with this design technique.

With respect to the primary inputs/outputs that are required to operate the shift register, this can be reduced significantly by making functional use of some of the pins. For example, the scan-out pin could be a functional output of an SRL for that particular chip. Also, overall performance of the subsystem may be degraded by the clocking requirement, but the effect should be small.

The LSSD structured design approach for Design for Testability eliminates or alleviates some of the problems in designing, manufacturing and maintaining LSI systems at a reasonable cost.

4.2 Scan Path

In 1975, a survey paper of test generation systems in Japan was presented by members of Nippon Electric Co., Ltd.[19] In that survey paper, a technique they described as Scan Path was presented. The Scan Path technique has the same objectives as the LSSD approach which has just been described. The Scan Path technique similarities and differences to the LSSD approach will be presented.

The memory elements that are used in the Scan Path approach are shown in Figure 21. This memory element is called a raceless D-type flip-flop with Scan Path.

In system operation, Clock 2 is at a logic value of 1 for the entire period. This, in essence, blocks the test or scan input from affecting the values in the first latch. This D-type flip-flop really contains two latches. Also, by having Clock 2 at a logic value of 1, the values in Latch 2 are not disturbed.

Clock 1 is the sole clock in system operation for this D-type flip-flop. When Clock 1 is at a value of 0, the System Data Input can be loaded into Latch 1. As long as Clock 1 is 0 for sufficient time to latch up the data, it can then turn off. As it turns off, it then will make Latch 2 sensitive to the data output of Latch 1. As long as Clock 1 is equal to a 1 so that data can be latched up into Latch 2, reliable operation will occur. This assumes that as long as the output of Latch 2 does not come around and feed the system data input to Latch 1 and change it during the time that the inputs to both Latch 1 and Latch 2 are active. The period of time that this can occur is related to the delay of the inverter block for

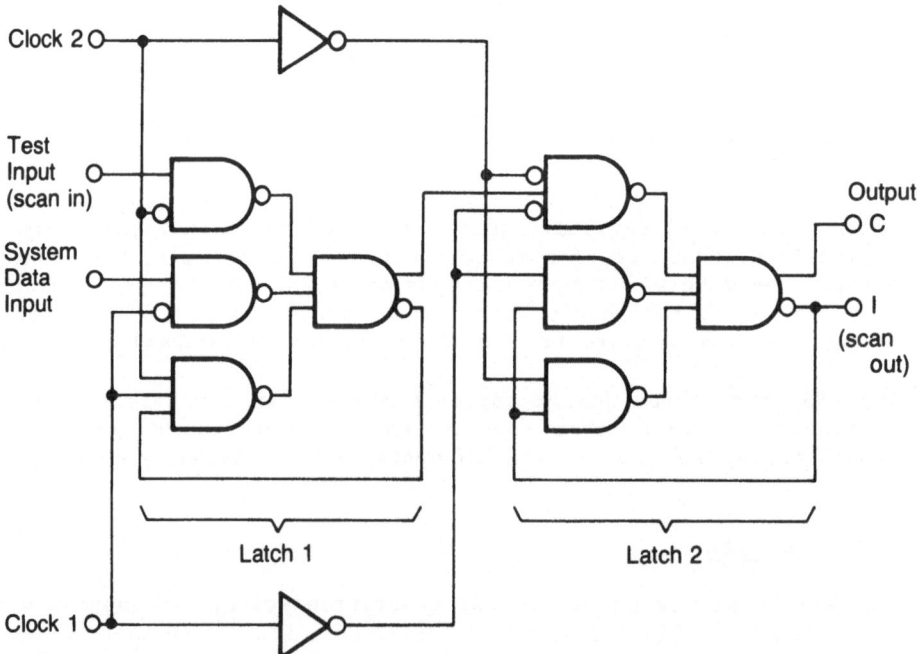

Figure 21. Raceless D-type flip-flop with Scan Path

Clock 1. A similar phenomena will occur with Clock 2 and its associated inverter block. This race condition is the exposure to the use of only one system dock.

This points out a significant difference between the Scan Path approach and the LSSD approach. One of the basic principles of the LSSD approach is level-sensitive operation--the ability to operate the clocks in such a fashion that no races will exist. In the LSSD approach, a separate clock is required for Latch 1 from the clock that operates Latch 2.

In terms of the scanning function, the D-type flip-flop with Scan Path has its own scan input called test input. This is clocked into the L1 latch by Clock 2 when Clock 2 is a 0, and the results of the L1 latch are clocked into Latch 2 when Clock 2 is a 1. Again, this applies to master/slave operation of Latch 1 and Latch 2 with its associated race with proper attention to delays this race will not be a problem.

Another feature of the Scan Path approach is the configuration used at the logic card level. Modules on the logic card are all connected up into a serial scan path, such that for each card, there is one scan path. In addition, there are gates for selecting a particular card in a subsystem. In Figure 22, when X and Y are both equal to 1--that is, the selection mechanism--Clock 2 will then be allowed to shift data through the scan path. Any other time, Clock 2 will be blocked, and its output will be blocked. The reason for blocking the output is that a number of card outputs can

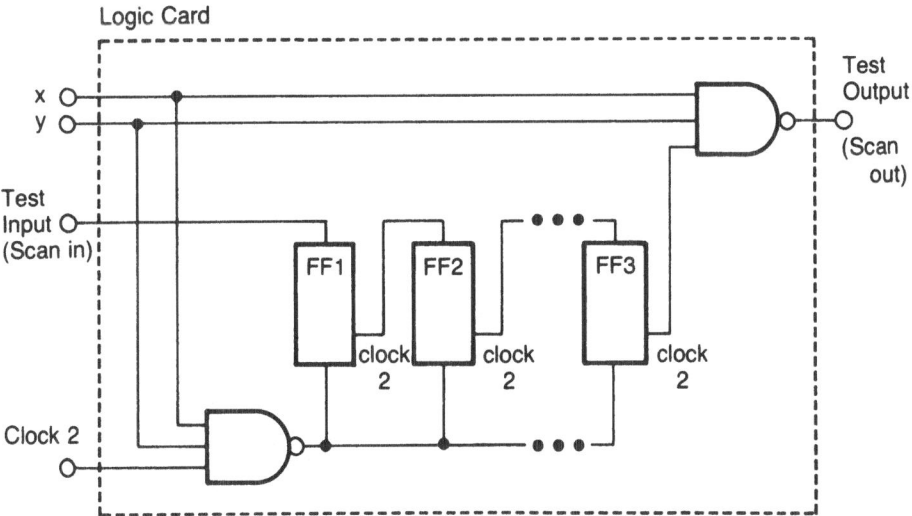

Figure 22. Configuration of Scan Path on Card

then be put together; thus, the blocking function will put their output to non-controlling values, so that a particular card can have unique control of the unique test output for that system.

It has been reported by the Nippon Electric Company that they have used the Scan Path approach, plus partitioning which will be described next, for systems with 100,000 blocks or more. This was for the FLT-700 System, which is a large processor system.

The partitioning technique is one which automatically separates the combinational network into smaller subnetworks, so that the test generator can do test generation for the small subnetworks, rather than the larger networks. A partition is automatically generated by backtracing from the D-type flip-flops, through the combinational logic, until it encounters a D-type flip-flop in the backtrace (or primary input). Some care must be taken so that the partitions do not get too large.

To that end, the Nippon Electric Company approach has used a controlled D-type flip-flop to block the backtracing of certain partitions when they become too high. This is another facet of Design for Testability--that is, the introduction of extra flip-flops totally independent of function, in order to control the partitioning algorithm.

Other than the lack of the level sensitive attribute to the Scan Path approach, the technique is very similar to the LSSD approach. The introduction of the Scan Path approach was the first practical implementation of shift registers for testing which was incorporated in a total system.

4.3   Scan/Set Logic

A technique similar to Scan Path and LSSD, but not exactly the same and certainly not as rigorous, is the Scan/Set technique put forth by Sperry-Univac.[27] The basic concept of this technique is to have shift registers, as in Scan Path or in LSSD, but these shift registers are not in the data path. That is, they are not in the system data path; they are independent of all the system latches. Figure 23 shows an example of the Scan/Set Logic, referred to as bit serial logic.

The basic concept is that the sequential network can be sampled at up to 64 points. These points can be loaded into the 64-bit shift register with a single clock. Once the 64 bits are loaded, a shifting process will occur, and the data will be scanned out through the scan-out pin. In the case of the set function, the 64 bits can be funneled into the system logic, and then the appropriate clocking structure required to load data into the system latches is required in this system logic. Furthermore, the

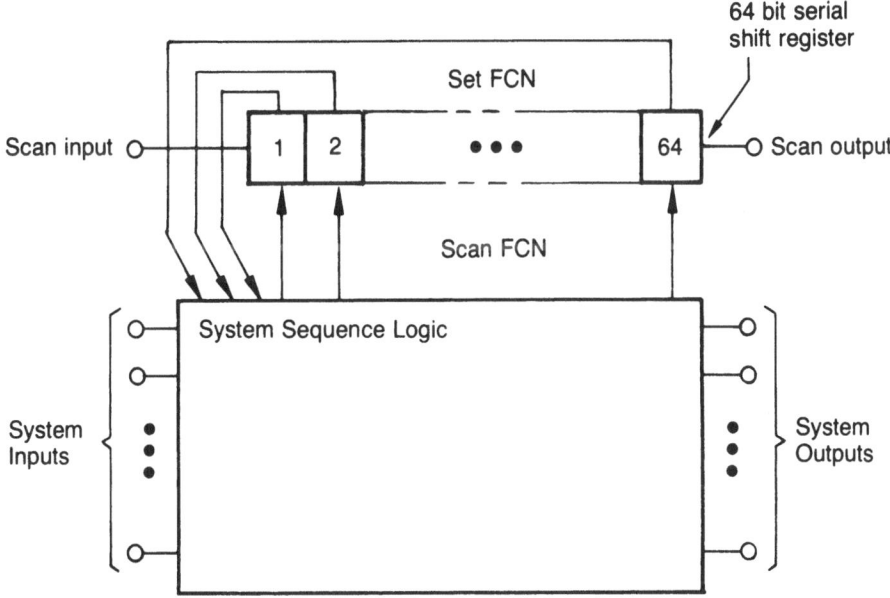

Figure 23. Scan/Set Logic (Bit-serial)

set function could also be used to control different paths to ease the testing function.

In general, this serial Scan/Set Logic would be integrated onto the same chip that contains sequential system logic. However, some applications have been put forth where the bit serial Scan/Set Logic was off-chip, and the bit-serial Scan/Set logic only sampled outputs or drove inputs to facilitate in-circuit testing.

Recently, Motorola has come forth with a chip which is T^2L and which has I^2L logic integrated on that same chip. This has the Scan/Set Logic bit serial shift registers built in I^2L. The T^2L portion of the chip is a gate array, and the I^2L is on the chip, whether the customer wants it or not. It's up to the customer to use the bit serial if he chooses.

At this point, it should be explained that if all the latches within the system sequential network are not both scanned and set, then the test generation function is not necessarily reduced to a total combinational test generation function and fault simulation function. However, this technique will greatly reduce the task of test generation and fault simulation.

Again, the Scan/Set technique has the same objectives as Scan Path and LSSD--that is, controllability and observability. However, in terms of its implementation, it is not required that the set function set all system latches, or that the scan function scan all system latches. This design flexibility would have a reflection in the software support required to implement such a technique.

Another advantage of this technique is that the scan function can occur during system operation--that is, the sampling pulse to the 64-bit serial shift register can occur while system clocks are being applied to the system sequential logic, so that a snapshot of the sequential machine can be obtained and off-loaded without any degradation in system performance.

4.4   Random Access Scan

Another technique similar to the Scan Path technique and LSSD is the Random Access Scan technique put forth by Fujitsu.[12] This technique has the same objective as Scan Path and LSSD--that is, to have complete controllability and observability of all internal latches. Thus, the test generation function can be reduced to that of combinational test generation and combinational fault simulation as well.

Random Access Scan differs from the other two techniques in that shift registers are not employed. What is employed is an addressing scheme which allows each latch to be uniquely selected, so that it can be either controlled or observed. The mechanism for addressing is very similar to that of a Random Access Memory, and hence, its name.

Figures 24 and 25 show the two basic latch configurations that are required for the Random Access Scan approach. Figure 24 is a

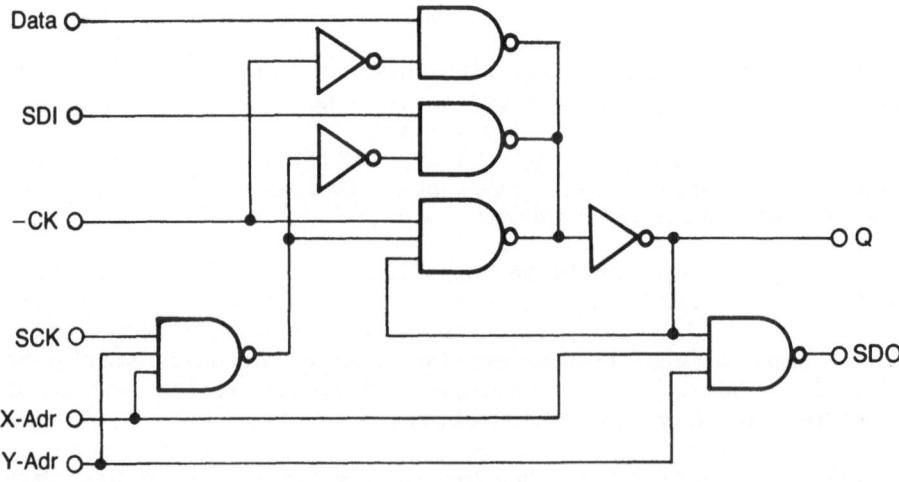

Figure 24. Polarity-hold type Addressable Latch.

single latch which has added to it an extra data port which is a
Scan Data In port (SDI). This data is clocked into the latch by
the SCK clock. The SCK clock can only affect this latch, if both
the X and Y address are one. Furthermore, when the X address and Y
address are one, then the Scan Data Out (SDO) point can be
observed.

System data labeled Data in Figure 24 and Figure 25 is loaded into
this latch by the system clock labeled CK.

The set/reset-type addressable latch in Figure 25 does not have a
scan clock to load data into the system latch. This latch is first
cleared by the CL line, and the CL line is connected to other
latches that are also set/reset-type addressable latches. This,
then, places the output value Q to a 0 value. For those latches
that are required to be set to a 1 for that particular test, a
preset is directed at those latches. This preset is directed by
addressing each one of those latches and applying the preset pulse
labeled PR. The output of the latch Q will then go to a 1.

The observability mechanism for Scan Data Out is exactly the same
as for the latch shown in Figure 24.

Figure 26 gives an overall view of the system configuration of the
Random Access Scan approach. Notice that, basically, there is a Y
address, an X address, a decoder, the addressable storage
elements, which are the memory elements or latches, and the
sequential machine, system clocks and CLEAR function. There is
also a Scan Data In (SDI) which is the input for a given latch,
Scan Data Out (SDO) which is the output data for that given latch,
and a scan clock. There is also one logic gate necessary to create
the preset function.

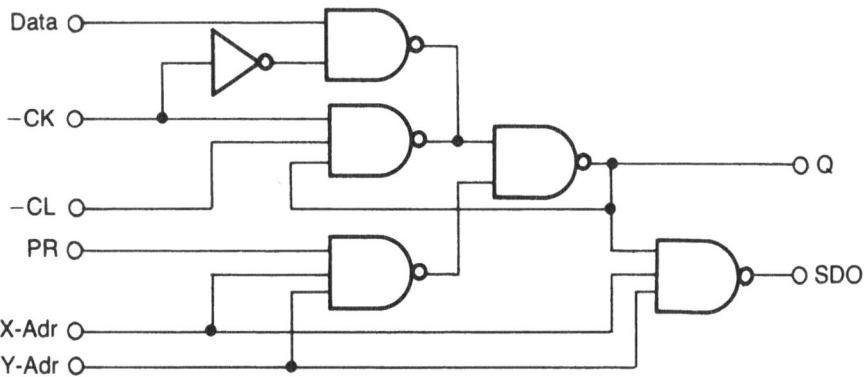

Figure 25. Set/Reset type Addressable Latch

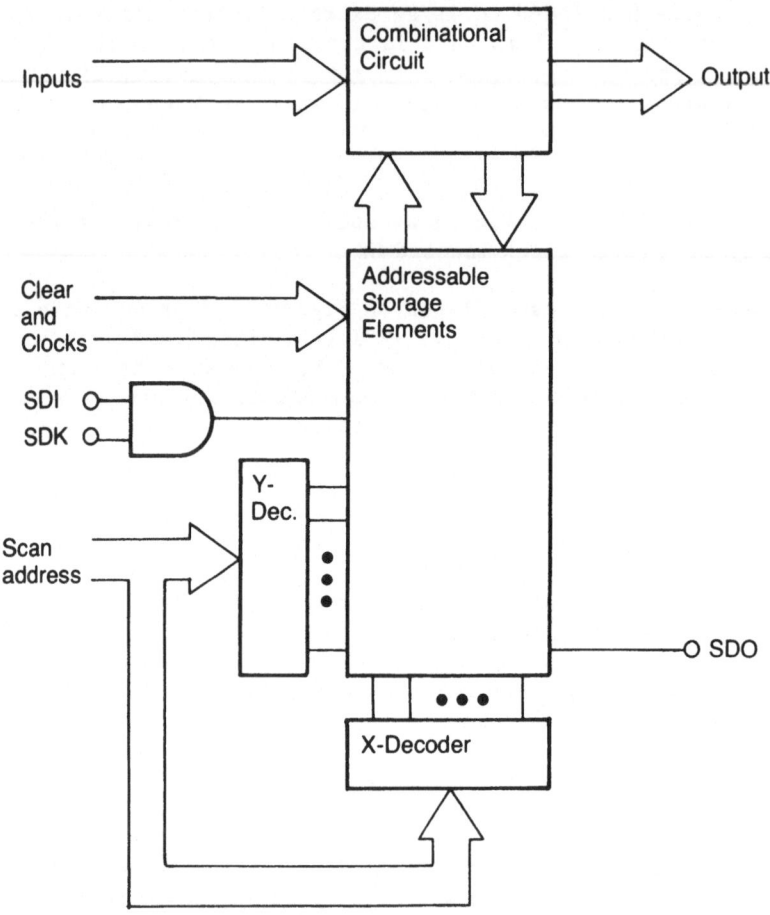

Figure 26. Random Access Scan network.

The Random Access Scan technique allows the observability and controllability of all system latches. In addition, any point in the combinational network can be observed with the addition of one gate per observation point, as well as one address in the address gate, per observation point.

While the Scan Path approach and the LSSD approach require two latches for every point which needs to be observed, the overhead for Random Access Scan is about three to four gates per storage element. In terms of primary inputs/outputs, the overhead is between 10 and 20. This pin overhead can be diminished by using the serial scan approach for the X and Y address counter, which would lead to 6 primary inputs/outputs.

4.5 Built-In Logic Block Observation, BILBO

A technique recently presented takes the Scan Path and LSSD concept and integrates it with the Signature Analysis concept. The end result is a technique for Built-In Logic Block Observation techniques, BILBO.[23]

Figure 27 gives the form of an eight-bit BILBO register. The block labeled L_i, (i = 1, 2, ..., 8) are the system latches. B_1 and B_2 are control values for controlling the different functions that the BILBO register can perform. S_{IN} is the scan-in input to the eight-bit register, and S_{OUT} is the scan-out for the eight-bit register. Q_i (i = 1, 2, ...,8) are the output values for the eight system latches. Z_i (i = 1, 2, ..., 8) are the inputs from the combinational logic. The structure that this network will be embedded into will be discussed shortly.

There are three primary modes of operation for this register, as well as one secondary mode of operation for this register. The first is shown in Figure 27(b)--that is, with B_1 and B_2 equal to 11. This is Basic System Operation mode, in which the Z_i values are loaded into the L_i, and the outputs are available on Q_i for system operation. This would be your normal register function.

When $B_1 B_2$ equals 00, the BILBO register takes on the form of a linear shift register, as shown in Figure 25(c). Scan-in input to the left, through some inverters, and basically lining up the eight registers into a single scan path, until the Scan-out is reached. This is similar to San Path and LSSD.

The third mode is when $B_1 B_2$ equals 10. In this mode, the BILBO register takes on the attributes of a linear feedback shift register of maximal length with multiple linear inputs. This is very similar to a Signature Analysis register, except that there is more than one input. In this situation, there are eight unique inputs. Thus, after a certain number of shift clocks, say, 100, there would be a unique signature left in the BILBO register for the good machine. This good machine signature could be off-loaded from the register by changing from Mode $B_1 B_2$=10 to Mode $B_1 B_2$=00, in which case a shift register operation would exist, and the signature then could be observed from the scan-out primary output.

The fourth function that the BILBO register can perform is $B_1 B_2$ equal to 01, which would force a reset on the register. (This is not depicted in Figure 27.)

402

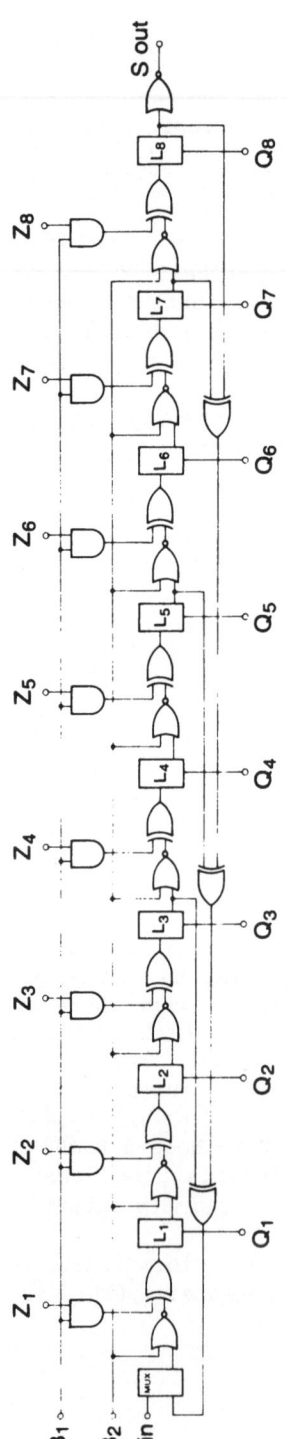

(a) General form of BILBO register

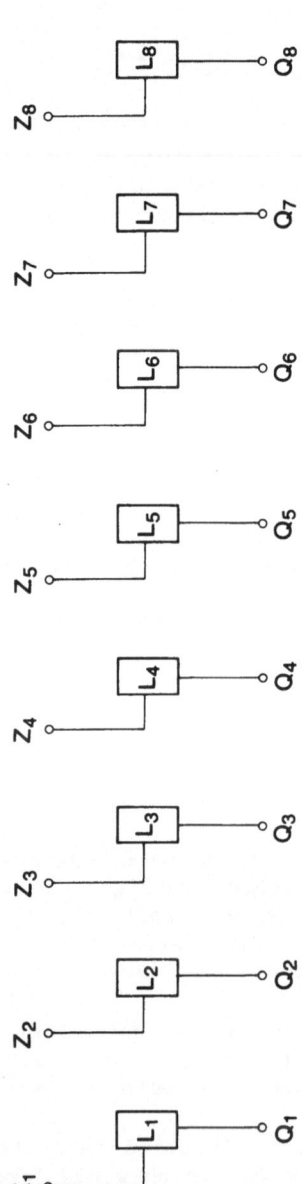

(b) $B_1 B_2 = 11$ System orientation mode

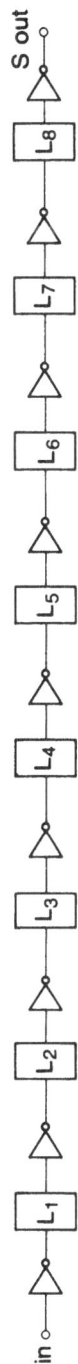

(c) $B_1 B_2 = 00$ Linear shift register mode

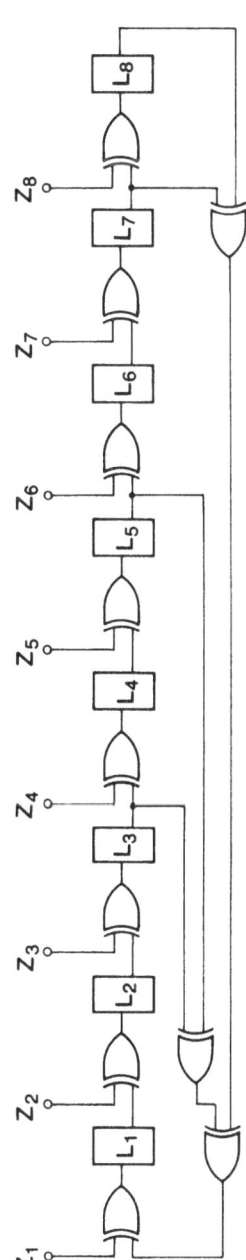

(d) $B_1 B_2 = 10$ Signature analysis register with multiple inputs ($Z_1, Z_2 \ldots Z_8$)

Figure 27. BILBO and its different modes

The BILBO registers are used in the system operation, as shown in Figure 28. Basically, a BILBO register with combinational logic and another BILBO register with combinational logic, as well as the output of the second combinational logic network can feed back into the input of the first BILBO register. The BILBO approach takes one other fact into account, and that is that, in general, combinational logic is highly susceptible to random patterns. Thus, if the inputs to the BILBO register, Z_1, Z_2, ..., Z_8, can be controlled to fixed values, such that the BILBO register is in the maximal length linear feedback shift register mode (Signature

Analysis) it will output a sequence of patterns which are very close to random patterns. Thus, random patterns can be generated quite readily from this register. These sequences are called Pseudo Random Patterns (PN).

If, in the first operation, this BILBO register on the left in Figure 28 is used as the PN generator--that is, its data inputs are held to fixed values--then the output of that BILBO register will be random patterns. This will then do a reasonable test, if sufficient numbers of patterns are applied, of the Combinational Logic Network 1. The results of this test can be stored in a Signature Analysis register approach with multiple inputs to the BILBO register on the right. After a fixed number of patterns have been applied, the signature is scanned out of the BILBO register on the right for good machine compliance. If that is successfully completed, then the roles are reversed, and the BILBO register on the right will be used as a PN sequence generator; the BILBO register on the left will then be used as a Signature Analysis register with multiple inputs from Combinational Logic Network 2, see Figure 29. In this mode, the Combinational Logic Network 2 will have random patterns applied to its inputs and its outputs stored in the BILBO register on the far left. Thus, the testing of the combinational logic networks 1 and 2 can be completed at very high speeds by only applying the shift clocks, while the two BILBO

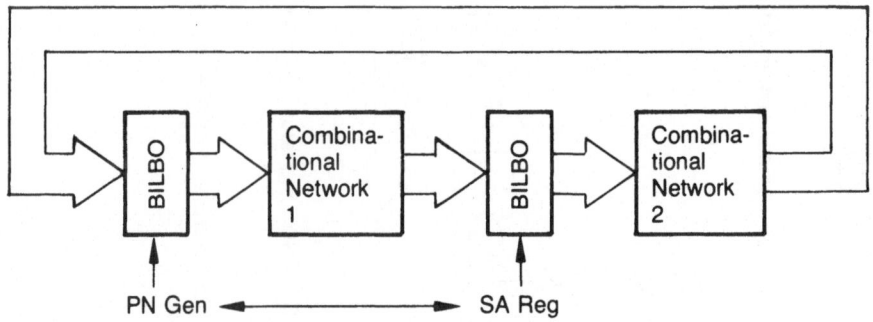

Figure 28. Use of BILBO registers to test Combinational Network 1

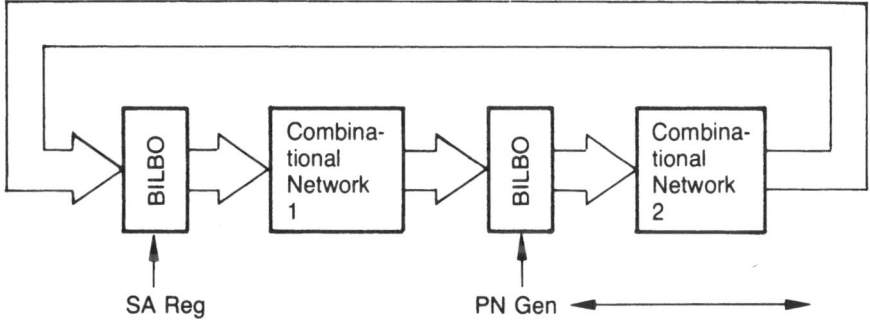

Figure 29. Use of BILBO registers to test Combinational Network 2

registers are in the Signature Analysis mode. At the conclusion of the tests, off-loading of patterns can occur, and determination of good machine operation can be made.

This technique solves the problem of test generation and fault simulation if the combinational networks are susceptible to random patterns. There are some known networks which are not susceptible to random patterns. They are Programmable Logic Arrays (PLAs), see Figure 30. The reason for this is that the fan-in in PLAs is too large. If and AND gate in the search array had 20 inputs, then each random pattern would have $\frac{1}{2}^{20}$ probability of coming up with the correct input pattern. On the other hand, random combinational logic networks with maximum fan-in of 4 can do quite well with random patterns.

The BILBO technique solves another problem and that is of test data volume. In LSSD, Scan Path, Scan/Set, or Random Access Scan, a considerable amount of test data volume is involved with the shifting in and out. With BIBLO, if 100 patterns are run between scan outs, the test data volume may be reduced by a factor of 100.

The overhead for this technique is higher than LSSD since about two Exclusive OR's must be used per latch position. Also, there is more delay in the system data path (one or two gate delays). If VLSI has the huge number of logic gates available than this may be a very efficient way to use them.

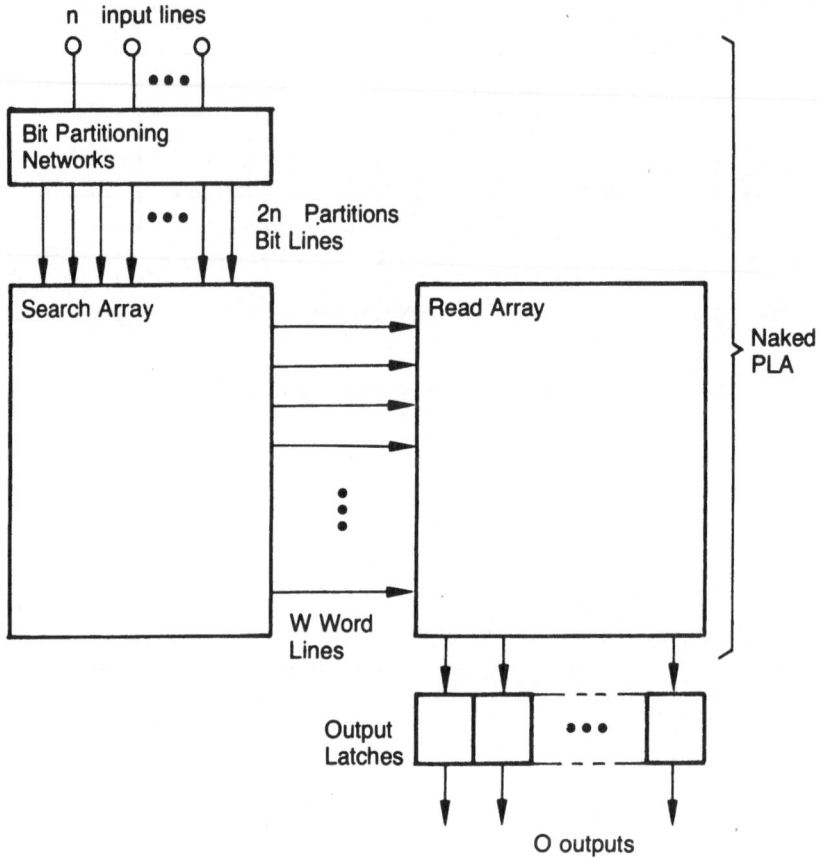

Figure 30. PLA Model

5.0 Conclusion

The area of Design for Testability is becoming a popular topic by necessity. Those users of LSI/VLSI which do not have their own captive IC facilities are at the mercy of the vendors for information. And, until the vendor information is drastically changed, the Ad Hoc approaches to design for testability will be the only answer.

In that segment of the industry which can afford to implement the Structured Design for Testability approach, there is considerable hope of getting quality test patterns at a very modest cost. Furthermore, many innovative techniques are appearing in the Structured Approach and probably will continue as we meander through VLSI and into more dense technologies.

6.0 Acknowledgements

The author wishes to thank Dr. K. P. Parker for his help in preparing this paper along with his support. The author also wishes to thank Mr. D. J. Brown for his helpful comments and suggestions. The assistance of Ms. B. Berger, Ms. C. Mendoza, and Mr. J. Smith in preparing this manuscript for publication was invaluable.

General References and Surveys

1. Breuer, M. A., (ed.), Diagnosis and Reliable Design of Digital Systems, Computer Science Press, 1976.

2. Chang, H. Y., E. G. Manning, and G. Metze, Fault Diagnosis of Digital Systems, Wiley Interscience, N.Y., 1970.

3. Friedman, A. D., and P. R. Menon, Fault Detection in Digital Circuits, Prentice Hall, New Jersey, 1971.

4. Hennie, F. C., Finite State Models for Logical Machines, J. Wiley & Sons, N. Y., 1968.

5. Kovijanic, P. G., "A New Look at Test Generation and Verification," Proc. 14th Design Automation Conference, June 1977, 77CH1216-1C, pp. 58-63.

6. Muehldorf, E. I., "Designing LSI Logic for Testability," Digest of Papers, 1976 Annl. Semiconductor Test Symp., Oct. 1976, 76CH1179-1C, pp. 45-49.

7. Susskind, A. K., "Diagnostics for Logic Networks," IEEE Spectrum, Oct. 1973, pp 40-47.

8. Williams, T. W., and K. P. Parker, "Testing Logic Networks and Design for Testability," Computer, Oct. 1979, pp. 9-21.

9. IEEE Standard Dictionary of Electrical and Electronics Terms, IEEE Inc., Wiley Interscience, N. Y., 1972.

References on Designing for Testability

10. "A Designer's Guide to Signature Analysis," Hewlett Packard Application Note 222, Hewlett Packard, 5301 Stevens Creek Blvd., Santa Clara, CA 95050.

408

11. Akers, S. B., "Partitioning for Testability," Journal of
    Design Automation & Fault-Tolerant Computing, Vol. 1,
    No. 2, Feb. 1977.

12. Ando, H., "Testing VLSI with Random Access Scan," Digest
    of Papers Compcon 80, Feb. 1980, 80CH1491-0C, pp. 50-52.

13. Bottorff, P., and E. I. Muehldorf, "Impact of LSI on Complex
    Digital Circuit Board Testing," Electro 77, New York,
    N.Y., April, 1977.

14. DasGupta, S., E. B. Eichelberger, and T. W. Williams, "LSI
    Chip Design for Testability," Digest of Technical Papers,
    1978 International Solid-State Circuits Conference,
    San Francisco, Feb., 1978, pp. 216-217.

15. "Designing Digital Circuits for Testability," Hewlett
    Packard Application Note 210-4, Hewlett Packard,
    Loveland, CO 80537.

16. Eichelberger, E. B., and T. W. Williams, "A Logic Design
    Structure for LSI Testability," Journal of Design
    Automation & Fault-Tolerant Computing, Vol 2., No. 2,
    May 1978, pp. 165-178.

17. Eichelberger, E. B., and T. W. Williams, "A Logic Design
    Structure for LSI Testing," Proc. 14th Design Automation
    Conf., June 1977, 77CH1216-1C, pp. 462-468.

18. Eichelberger, E. B., E. J. Muehldorf, R. G. Walter, and
    T. W. Williams, "A Logic Design Structure for Testing
    Internal Arrays," Proc. 3rd USA-Japan Computer Conference,
    San Francisco, Oct. 1978, pp. 266-272.

19. Funatsu, S., N. Wakatsuki, and T. Arima, "Test Generation
    Systems in Japan," Proc. 12th Design Automation Symp.,
    June 1975, pp. 114-22.

20. Godoy, H. C., G. B. Franklin, and P. S. Bottoroff, "Automatic
    Checking of Logic Design Structure for Compliance with
    Testability Groundrules," Proc. 14th Design Automation
    Conf., June 1977, 77CH1216-1C, pp. 469-478.

21. Hayes, J. P., "On Modifying Logic Networks to Improve their
    Diagnosability," IEEE-TC, Vol. C-23, Jan. 1974, pp. 56-62.

22. Hayes, J. P., and A. D. Friedman, "Test Point Placement to
    Simplify Fault Detection," FTC-3, Digest of Papers, 1973
    Symposium on Fault-Tolerant Computing, June 1973,
    pp. 73-78.

23. Koenemann, B., J. Mucha, and G. Zwiehoff, "Built-In Logic Block Observation Techniques," Digest of Papers, 1979 Test Conference, Oct. 1979, 79CH1509-9C, pp. 37-41.

24. Lippman, M. D., and E. S. Donn, "Design Forethought Promotes Easier Testing of Microcomputer Boards," Electronics, Jan. 18, 1979, pp. 113-119.

25. Nadig, H. J., "Signature Analysis-Concepts, Examples, and Guidelines," Hewlett Packard Journal, May 1977, pp. 15-21.

26. Neil, M., and R. Goodner, "Designing a Serviceman's Needs into Microprocessor Based Systems," Electronics, March 1, 1979, pp. 122-128.

27. Stewart, J. H., "Future Testing of Large LSI Circuit Cards," Digest of Papers 1977 Semiconductor Test Symp., Oct. 1977, 77CH1261-7C, pp. 6-17.

28. Toth, A., and C. Holt, "Automated Data Base-Driven Digital Testing," Computer, Jan. 1974, pp. 13-19.

29. White, E., "Signature Analysis, Enhancing the Serviceability of Microprocessor-Based Industrial Products," Proc. 4th IECI Annual Conference, March 1978, 78CH1312-8, pp. 68-76.

30. Williams, M. J. Y., and J. B. Angell, "Enhancing Testability of Large Scale Integrated Circuits via Test Points and Additional Logic," IEEE-TC, Vol. C-22, Jan. 1973, pp. 46-60.

31. Williams, T. W., "Utilization of a Structured Design for Reliability & Serviceability," Digest, Government Microcircuits Applications Conference, Monterey, Calif., Nov. 1978, pp. 441-444.

References on Faults and Fault Modeling

32. Boute, R., and E. J. McCluskey, "Fault Equivalence in Sequential Machines," Proc. Symp. on Computers and Automata, Polytechnic Inst. of Brooklyn, April 13-15, 1971, pp. 483-507.

33. Boute, R. T., "Optimal and Near-Optimal Checking Experiments for Output Faults in Sequential Machines," IEEE-TC, Vol.C-23, No. 11, Nov. 1974, pp. 1207-1213.

34. Boute, R. T., "Equivalence and Dominance Relations Between Output Faults in Sequential Machines," Tech. Report No. 38, SU-SEL-72-052, Nov. 1972, Stanford University, Stanford, Calif.

410

35. Dias, F. J. O., "Fault Masking in Combinational Logic
    Circuits," IEEE-TC, Vol. C-24, May 1975, pp. 476-482.

36. Hayes, J. P., "A NAND Model for Fault Diagnosis in
    Combinational Logic Networks," IEEE-TC, Vol. C-20, Dec.
    1971, pp. 1496-1506.

37. McCluskey, E. J., and F. W. Clegg, "Fault Equivalence in
    Combinational Logic Networks," IEEE-TC, Vol. C-20,
    Nov. 1971, pp. 1286-1293.

38. Mei, K. C. Y., "Fault Dominance in Combinational Circuits,"
    Technical Note No. 2, Digital Systems Laboratory, Stanford
    University, Aug. 1970.

39. Mei, K. C. Y., "Bridging and Stuck-At Faults," IEEE-TC,
    Vol. C-23, No. 7, July 1974, pp. 720-727.

40. Ogus, R. C., "The Probability of a Correct Output from a
    combinational Circuit," IEEE-TC, Vol. C-24, No. 5, May 1975,
    pp. 534-544.

41. Parker, K. P., and E. J. McCluskey, "Analysis of Logic
    Circuits with Faults Using Input Signal Probabilities,"
    IEEE-TC, Vol. C-24, No. 5, May 1975, pp. 573-578.

42. Schertz, D. R., and D. G. Metze, "A New Representation for
    Faults in Combinational Digital Circuits," IEEE-TC,
    Vol. C-21, No. 8, Aug. 1972, pp. 858-866.

43. Shedletsky, J. J., and E. J. McCluskey, "The Error Latency
    of a Fault in a Sequential Digital Circuit," IEEE-TC,
    Vol. C-25, No. 6, June 1976, pp. 655-659.

44. Shedletsky, J. J., and E. J. McCluskey, "The Error Latency
    of a Fault in a Combinational Digital Circuit," FTCS-5,
    Digest of Papers, Fifth International Symposium on Fault
    Tolerant Computing, Paris, France, June 1975, pp. 210-214.

45. To, K., "Fault Folding for Irredundant and Redundant
    Combinational Circuits," IEEE-TC, Vol. C-22, No. 11,
    Nov. 1973, pp. 1008-1015.

46. Wang, D. T., "Properties of Faults and Criticalities of
    Values Under Tests for Combinational Networks," IEEE-TC,
    Vol. C-24, No. 7, July 1975, pp. 746-750.

References on Testing and Fault Location

47. Batni, R. P., and C. R. Kime, "A Module Level Testing
    Approach for Combinational Networks," IEEE-TC, Vol. C-25,
    No. 6, June 1976, pp. 594-604.

48. Bisset, S., "Exhaustive Testing of Microprocessors and
    Related Devices: A Practical Solution," Digest of Papers,
    1977 Semiconductor Test Symp., Oct. 1977, pp. 38-41.

49. Czepiel, R. J., S. H. Foreman, and R. J. Prilik, "System for
    Logic, Parametric and Analog Testing," Digest of Papers,
    1976 Semiconductor Test Symp., Oct. 1976, pp. 54-69.

50. Frohwerk, R. A., "Signature Analysis: A New Digital Field
    Service Method," Hewlet Packard Journal, May 1977, pp. 2-8.

51. Gimmer, B. A., "Test Techniques for Circuit Boards Containing
    Large Memories and Microprocessors," Digest of Papers, 1976
    Semiconductor Test Symp., Oct. 1976, pp. 16-21.

52. Groves, W. A., "Rapid Digital Fault Isolation with FASTRACE,"
    Hewlett Packard Journal, March 1979, pp. 8-13.

53. Hayes, J. P., "Rapid Count Testing for Combinational Logic
    Circuits," IEEE-TC, Vol. C-25, No. 6, June 1976,
    pp. 613-620.

54. Hayes, J. P., "Detection of Pattern Sensitive Faults in
    Random Access Memories," IEEE-TC, Vol. C-24, No. 2,
    Feb. 1975, pp. 150-160.

55. Hayes, J. P., "Testing Logic Circuits by Transition Counting,"
    FTC-5, Digest of Papers, 1975 Symposium of Fault Tolerant
    Computing, Paris, France, June 1975, pp. 215-219.

56. Healy, J. T., "Economic Realities of Testing Microprocessors,"
    Digest of Papers, 1977 Semiconductor Test Symp., Oct. 1977,
    pp. 47-52.

57. Lee, E. C., "A Simple Concept in Microprocessor Testing,"
    Digest of Papers, 1976 Semiconductor Test Symp., Oct. 1976,
    76CH1179-1C, pp. 13-15.

58. Losq, J., "Referenceless Random Testing," FTCS-6, Digest of
    Papers, Sixth Int'l Symp. on Fault-Tolerant Computing,
    Pittsburgh, Penn., June 21-23, 1976, pp. 81-86.

59. Palmquist, S., and D. Chapman, "Expanding the Boundaries of
     LSI Testing with an Advanced Pattern Controller," Digest
     of Papers, 1976 Semiconductor Test Symp., Oct. 1976,
     pp. 70-75.

60. Parker, K. P., "Compact Testing: Testing with Compressed
     Data," FTCS-6, Digest of Papers, Sixth Int'l. Symp. on
     Fault-Tolerant Computing, Pittsburgh, Penn., June 21-23,
     1976.

61. Shedletsky, J. J., "A Rationale for the Random Testing of
     Combinational Digital Circuits," Digest of Papers, Compcon
     75 Fall, Washington, D.C., Sept. 9-11, 1975, pp. 5-9.

62. Strini, V. P., "Fault Location in a Semiconductor Random
     Access Memory Unit," IEEE-TC, Vol. C-27, No. 4, April 1978,
     pp. 379-385.

63. Weller, C. W., "An Engineering Approach to IC Test System
     Maintenance," Digest of Papers, 1977 Semiconductor Test
     Symp., Oct. 1977, pp. 144-145.

References on Testability Measures

64. Dejka, W. J., "Measure of Testability in Device and System
     Design," Proc. 20th Midwest Symp. Circuits Syst., Aug. 1977,
     pp. 39-52.

65. Goldstein, L. H., "Controllability/Observability Analysis of
     Digital Circuits," IEEE Trans. Circuits Syst., Vol. CAS-26,
     No. 9, Sept. 1979, pp. 685-693.

66. Keiner, W. L., and R. P. West, "Testability Measures,"
     presented at AUTOTESTCON '77, Nov. 1977.

67. Kovijanic, P. G., "Testability Analysis," Digest of Papers,
     1979 Test Conference, Oct. 1979, 79CH1509-9C, pp. 310-316.

68. Stephenson, J. E., and J. Grason, "A Testability Measure for
     Register Transfer Level Digital Circuits," Proc. 6th Fault
     Tolerant Computing Symp., June 1976, pp. 101-107.

References on Test Generation

69. Agrawal, V., and P. Agrawal, "An Automatic Test Generation
     System for ILLIAC IV Logic Boards," IEEE-TC, Vol. C-21,
     No. 9, Sept. 1972, pp. 1015-1017.

70. Armstrong, D. B., "On Finding a Nearly Minimal Set of Fault Detection Tests for Combinational Nets," IEEE-TC, EC-15, Vol. 13, No. 2, Feb. 1966, pp. 63-73.

71. Betancourt, R. "Derivation of Minimum Test Sets for Unate Logical Circuits," IEEE-TC, Vol. C-20, No. 11, Nov. 1973, pp. 1264-1269.

72. Bossen, D. C., and S. J. Hong, "Cause and Effect Analysis for Multiple Fault Detection in Combinational Networks," IEEE-TC, Vol. C-20, No. 11, Nov. 1971, pp. 1252-1257.

73. Bottorff, P. S., et al, "Test Generation for Large Networks," Proc. 14th Design Automation Conf., June 1977, 77CH1216-1C, pp. 479-485.

74. Edlred, R. D., "Test Routines Based on Symbolic Logic Statements," JACM, Vol. 6, No. 1, 1959, pp. 33-36.

75. Hsieh, E. P., et al., "Delay Test Generation," Proc. 14th Design Automation Conf., June 1977, 77CH1216-1C, pp. 486-491.

76. Ku, C. T., and G. M. Masson, "The Boolean Difference and Multiple Fault Analysis," IEEE-TC, Vol. C-24, No. 7, July 1975, pp. 691-695.

77. Muehldorf, E. I., "Test Pattern Generation as a Part of the Total Design Process," LSI and Boards: Digest of Papers, 1978 Annual Semiconductor Test Symp., Oct. 1978, pp. 4-7.

78. Muehldorf, E. I., and T. W. Williams, "Optimized Stuck Fault Test Patterns for PLA Macros," Digest of Papers, 1977 Semiconductor Test Symp., Oct. 1977, 77CH1216-7C, pp. 89-101.

79. Page, M. R., "Generation of Diagnositc Tests Using Prime Implicants," Coordinated Science Lab Report R-414, University of Illinois, Urbana, Il., May 1969.

80. Papaioannou, S. G., "Optimal Test Generation in Combinational Networks by Pseudo Boolean Programming," IEEE-TC, Vol. C-26, No. 6, June 1977, pp. 553-560.

81. Parker, K. P., "Adaptive Random Test Generation," Journal of Design Automation and Fault Tolerant Computing, Vol. 1, No. 1, Oct. 1976, pp. 62-83.

414

82. Parker, K. P., "Probabilistic Test Generation," Technical
    Note No. 18, Jan. 1973, Digital Systems Laboratory,
    Stanford University, Stanford, Calif.

83. Poage, J. F., and E. J. McCluskey, "Derivation of Optimum
    Tests for Sequential Machines," Proc. Fifth Annual Symp.
    on Switching Circuit Theory and Logic Design, 1964,
    pp. 95-110.

84. Poage, J. F., and E. J. McCluskey, "Derivation of Optimum
    Tests to Detect Faults in Combinational Circuits,"
    Mathematical Theory of Automation, Polytechnic Press,
    New York, 1963.

85. Putzolu, G. R., and J. P. Roth, "A Heuristic Algorithm for
    Testing of Asynchronous Circuits," IEEE-TC, Vol. C-20,
    No. 6, June 1971, pp. 639-647.

86. Roth, J. P., W. G. Bouricius, and P. R. Schneider,
    "Programmed Algorithms to Compute Tests to Detect and
    Distinguish Between Failures in Logic Circuits,"
    IEEE-TEC, EC-16, Oct. 1967, pp. 567-580.

87. Roth, J. P., "Diagnosis of Automata Failures: A Calculus
    and a Method," IBM Journal of Research and Development,
    No. 10, Oct. 1966, pp. 278-281.

88. Schneider, P. R., "On the Necessity to Examine D-chains in
    Diagnostic Test Generation-an Example," IBM Journal of
    Research and Development, No. 11, Nov. 1967, p. 114.

89. Schnurmann, H. D., E. Lindbloom, R. G. Carpenter, "The
    Weighted Random Test Pattern Generation," IEEE-TC,
    Vol. C-24, No. 7, July 1975, pp. 695-700.

90. Sellers, E. F., M. Y. Hsiao, L. W. Bearnson, "Analyzing
    Errors with the Boolean Difference," IEEE-TC, Vol. C-17,
    No. 7, July 1968, pp. 676-683.

91. Wang, D. T., "An Algorithm for the Detection of Test Sets
    for Combinational Logic Networks," IEEE-TC, Vol. C-25,
    No. 7, July 1975, pp. 742-746.

92. Williams, T. W., and E. E. Eichelberger, "Ransom Patterns
    within a Structured Sequential Logic Design," Digest of
    Papers, 1977 Semiconductor Test Symposium, Oct. 1977,
    77CH1261-7C, pp. 19-27.

93. Yau, S. S., and S. C. Yang, "Multiple Fault Detection for Combinational Logic Circuits, IEEE-TC, Vol. C-24, No. 5, May 1975, pp. 233-242.

References on Simulation

94. Armstrong, D. B., "A Deductive Method for Simulating Faults in Logic Circuits," IEEE-TC, Vol. C-22, No. 5, May 1972, pp. 464-471.

95. Breuer, M. A., "Functional Partitioning and Simulation of Digital Circuits," IEEE-TC, Vol. C-19, No. 11, Nov. 1970, pp. 1038-1046.

96. Chiang, H. Y. P., et al, "Comparison of Parallel and Deductive Fault Simulation," IEEE-TC, Vol. C-23, No. 11, Nov. 1974, pp. 1132-1138.

97. Eichelberger, E. B., "Hazard Detection in Combinational and Sequential Switching Circuits," IBM Journal of Research & Development, March 1965.

98. Manning, E., and H. Y. Chang, "Functional Technique for Efficient Digital Fault Simulation" IEEE Internat. Conv. Digest, 1968, p. 194.

99. Parker, K. P., "Software Simulator Speeds Digital Board Test Generation," Hewlett Packard Journal, March 1979, pp. 13-19.

100. Seshu, S., "On an Improved Diagnosis Program," IEEE-TEC, Vol. EC-12, No. 2, Feb. 1965, pp. 76-79.

101. Seshu, S., and D. N. Freeman, "The Diagnosis of Asynchronous Sequential Switching Systems," IRE Trans, Elec. Comp., Vol. EC-11, No. 8, Aug. 1962, pp. 459-465.

102. Storey, T. M., and J. W. Barry, "Delay Test Simulation," Proc. 14th Design Automation Conf., June 1977, 77CH1216-1C, pp. 491-494.

103. Szygenda, S. A., and E. W. Thompson, "Modeling and Digital Simulation for Design Verification Diagnosis," IEEE-TC, Vol. C-25, No. 12, Dec. 1976, pp. 1242-1253.

104. Szygenda, S. A., "TEGAS2-Anatomy of a General Purpose Test Generation and Simulation System for Digital Logic," Proc. 9th Design Automation Workshop, 1972, pp. 116-127.

105. Szygenda, S. A., D. M. Rouse, and E. W. Thompson, "A Model for Implementation of a Universal Time Delay Simulation for Large Digital Networks," <u>AFIPS Conf. Proc.</u>, Vol. 36, 1970, SJCC, pp. 207-216.

106. Ulrich, E. G., and T. Baker, "Concurrent Simulation of Nearly Identical Digital Networks," <u>Computer</u>, Vol. 7, No. 4, April 1974, pp. 39-44.

107. Ulrich, E. G., and T. Baker, "The Concurrent Simulation of Nearly Identical Digital Networks," <u>Proc. 10th Design Automation Workshop</u>, June 1973, pp. 145-150.

108. Ulrich, E. G., T. Baker, L. R. Williams, "Fault Test Analysis Techniques Based on Simulation," <u>Proc. 9th Design Automation Workshop</u>, 1972, pp. 111-115.

COMPUTER AIDS TO TESTING - AN OVERVIEW

Peter S. Bottorff

IBM System Products Division
Endicott
New York

This paper is intended to provide an overview of methods of Automatic Test Pattern Generation (ATPG) for digital logic networks. It will concentrate on methods which the author has found useful in his own experience; a good general review is given in the book by Breuer and Friedman [1].

1.0 COST CONSIDERATIONS FOR AUTOMATIC TEST GENERATION SYSTEMS

Software systems for automatic test generation are expensive to develop and maintain: development of a reasonably effective system will require 10-15 person-years of engineering labor plus associated computer charges. Thus, it may be most economical to purchase test generation services or programs from a vendor, or to rely exclusively on functional testing at the system/box level. Some of the factors which figure in the decision to develop an ATPG system are:

1. Open Vs. Closed Component Sets

If the number of different networks to be tested is large, the cost of generating tests for each network may exceed the cost of developing an ATPG system.

2. Small Vs. Large Number of Products

3. Number of Parts to be Manufactured and Tested

4. Quality Level Required

A good ATPG system will generally provide better tests
than can be obtained from hand-generated "functional"
patterns or from signature-analysis methods. This can
significantly influence the cost of system testing and
field repair. A rule of thumb is that the cost to find
and fix a defect increases by a factor of ten at each
level of test, from chip to board, box, system and
field.

5. Availability of Engineering-Programming Skills

The development personnel for an ATPG system must have
most of the skills of the logic/hardware system design
engineer and must also be first class programmers.
Such people are hard to find.

6. Total Investment the Organization Wishes to Make
   in Testing

Use of an ATPG system implies a large investment in
expensive stored program Automatic Test Equipment (ATE).
The software systems which control the ATE, and aid its
operators in fault diagnosis will require an investment
of the order of magnitude of the ATPG system.

The method of test pattern generation is primarily a
business decision. The objective is to minimize the
total testing and repair costs in the development, man-
ufacture, and field service of the product.

2.0 DEFINITIONS

The logic network consists of primitives (which may be
Boolean gates or more complicated functional elements)
interconnected by wires called nets. Logic signals
enter the network via terminals called primary inputs
(PI) and leave via primary outputs (PO). Networks with
memory elements are sequential; otherwise, they are
combinational. To test the network, a sequence of logic
signals is applied as a stimulus to the primary inputs,
and an expected response is measured at the primary outputs. A fault is a defect in the network which causes

the response to differ from that expected.

The ATPG system has 3 functions:

1. Find the set of test patterns to be applied to the PI, and the expected fault-free responses at the PO which will detect as many of the faults as possible.

2. Evaluate the effectiveness of the tests.

3. Provide fault diagnostics which relate the differences from the expected response to the tests to the faults which cause them to occur.

The fault model used must be realistic. It should correspond to the actual types of manufacturing defects which occur in the process of physically building the network. However, the fault model must not be so complex that an excessive amount of computation is needed to calculate the test patterns. This tradeoff is most important to the design of the ATPG system.

An important class of faults is stuck-at faults, in which signals at the inputs or outputs of the primitives are assumed to be fixed at logical-0 (s-a-0) or logical-1 (s-a-1). The stuck-at fault assumption is a model that is computationally efficient and can be related to real manufacturing defects.

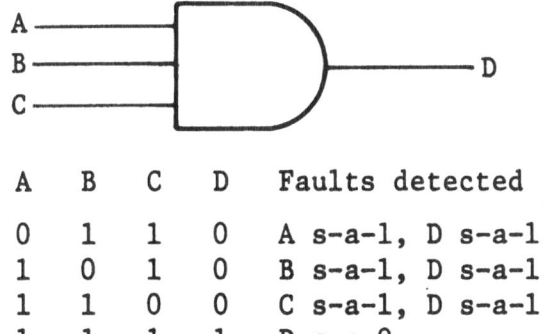

A	B	C	D	Faults detected
0	1	1	0	A s-a-1, D s-a-1
1	0	1	0	B s-a-1, D s-a-1
1	1	0	0	C s-a-1, D s-a-1
1	1	1	1	D s-a-0

Figure 1. Model for Stuck-at Faults

420

Another class of faults is <u>shorts</u> or <u>bridging</u> faults. This type of defect is caused by an unwanted short between two nets. Figure 2 illustrates this condition. Unwanted opens in nets usually correspond to stuck-at faults. In some cases, shorts faults may convert combinational networks to sequential networks.

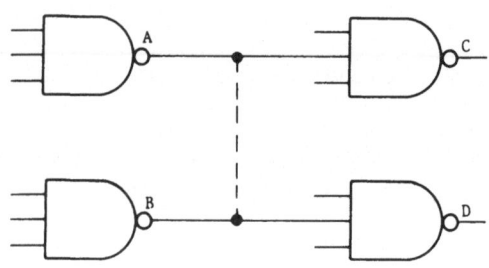

a) Bridging fault

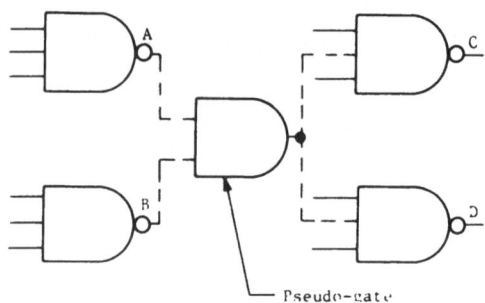

Test for either input of pseudo-gate s-a-1 to detect presence of bridging fault.

b) Model and test method

Figure 2. Model for Bridging Faults

<u>Delay faults</u> cause signals to propagate too slowly (or too fast) through the network. Stuck, shorts, and open faults can be detected by tests which are applied much more slowly than the normal rate of operation of the network. These are called <u>static</u> or <u>DC</u> tests. Delay faults will usually require testing at a rate approaching the normal rate of operation of the network; these tests are called <u>dynamic</u> or <u>AC</u> tests. Tests which measure voltage, current, or impedance values, as opposed to digital logic states, are called <u>parametric</u> or <u>analog</u> tests. ATPG for analog networks is in its infancy, and will not be considered here [see 2,3].

The test coverage is the ratio of the number of faults detected by the test patterns, to the total number of faults in the network.

There has been much theoretical work on networks containing multiple stuck-at faults. [See 4, 5, or 6 for typical papers.] The author regards multiple fault detection as a problem of little practical interest. It is difficult to obtain high test coverage in VLSI network with single stuck faults, and a good test for these will detect most multiple stuck faults [1, 7]. Further discussion will assume that there is only one fault present in the network under test.

The fault free network is often called the good machine. The network containing the fault is called the error machine or faulty machine.

In describing the net values present during test generation for a fault, it is useful to have a notation which represents a logic value which is 1 in the good machine, and 0 in the error machine, or vice versa. D will stand for the 1/0 (good value/faulty value) case, and $\overline{D}$ for 0/1. X will represent a state that is undefined or unknown. Figure 3 shows how to compute gate values for AND, OR, and INVERT gates.

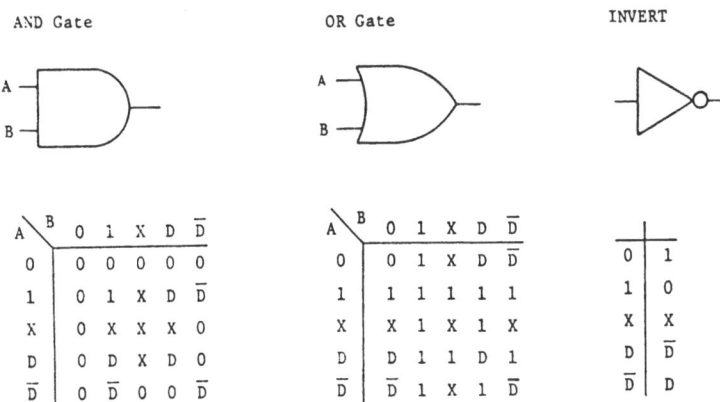

Figure 3. Calculation of Gate Values.

3.0 AUTOMATIC TEST GENERATION SYSTEMS

It is highly desirable for the ATPG system to have a
common data base with other programs in the logic design
automation system, because this will insure that physical layout data and test data are derived from the same
source description. A typical logic design automation
system is shown in Figure 4. This system includes means
for the logic designer to enter a description of the
network into the data base, a simulator for design verification, a physical design subsystem, a program for est-
imating wire and gate delays, and the ATPG system. A
macro expander is also included. This program uses the
information stored in a rules library to convert functional elements in the design, such as an ALU, into
smaller functional models or gate level descriptions.

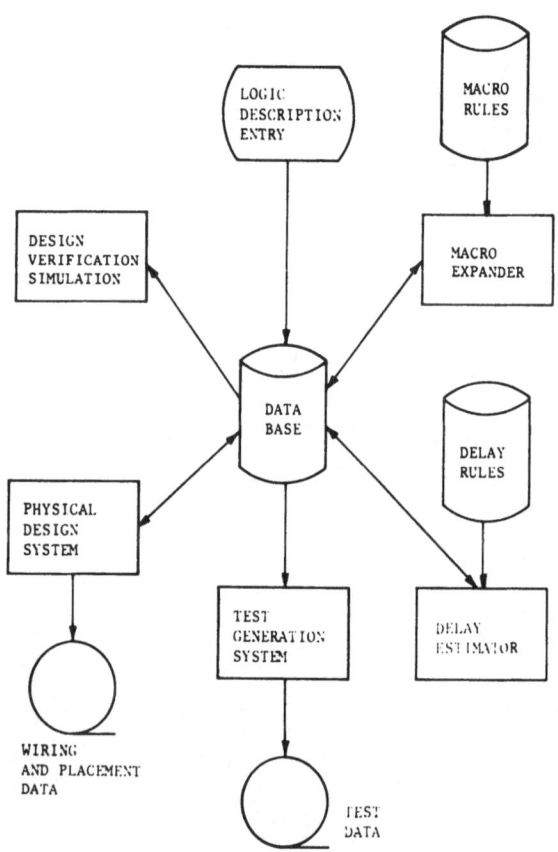

Figure 4. Logic Design Automation System

Information on failure modes of network elements may be placed in the macro rules library. The ATPG system uses results from macro expansion and delay estimation to do its job.

3.1   Programs in the ATPG system

Figure 5 shows the major programs and data files in the ATPG system.

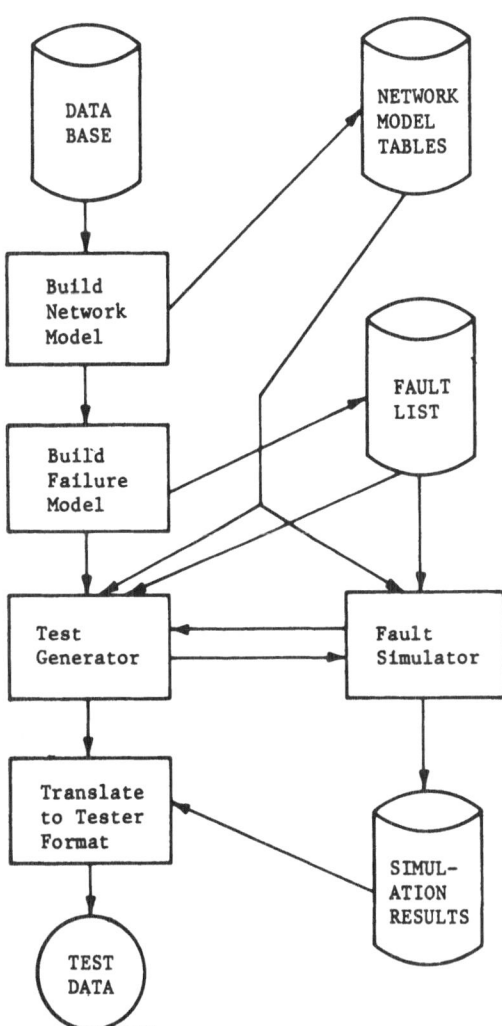

Figure 5.   Programs in an ATPG System.

3.1.1   Network Model Build

This program takes the logic network description stored
in the design automation system data base and converts
it into a set of network model tables which will be
loaded into main storage. These tables describe the
network layout to the test generation and fault simulation programs, and contain the following information:

1.   The logic function of each gate in the network.
      (e.g., PI, AND, NOR.)

2.   The gates which drive the inputs to a gate G are
      the fanin gates, or predecessors of G. The model
      tables contain pointers to the predecessors of
      each gate in the network.

3.   The gates which are driven by the output of a gate
      G are the fanout gates or successors of G. The
      model tables contain pointers to the successors
      of each gate in the network.

4.   Other information may be stored in the network
      model tables, such as gate and net delays, point-
      ers to routines which may be used to calculate
      gate values, and fields or data for use by the
      test generation and fault simulation programs.

Figure 6 illustrates the format of two typical network
model table structures. The structure shown in Figure
6a may be advantageous when the computer used to run
the programs has an instruction set oriented to fixed
length, aligned words.

Each gate in the network is designated by a number
called the gate index. In the fixed entry length model,
the gate index is the same as the number of the gate's
entry in the gate function table. Pointers to the lists
of fanin and fanout gates are in 1 to 1 correspondence
with the gate function table. The fanin and fanout
lists contain the indices of the fanin and fanout gates
of each gate. Other tables required for test generation
and fault simulation, such as a table of logic values
at gate outputs are put in 1 to 1 correspondence with
the function table, so that entries may be located by
gate index.

The variable structure shown in Figure 6b is more
efficient in machines with a flexible instruction set,
such as IBM System/370. All the information about

each gate is contiguous, and there are direct pointers
to the data for fanin and fanout gates. A separate
pointer field within the gate data is used to locate gate
related information not stored in the model tables.

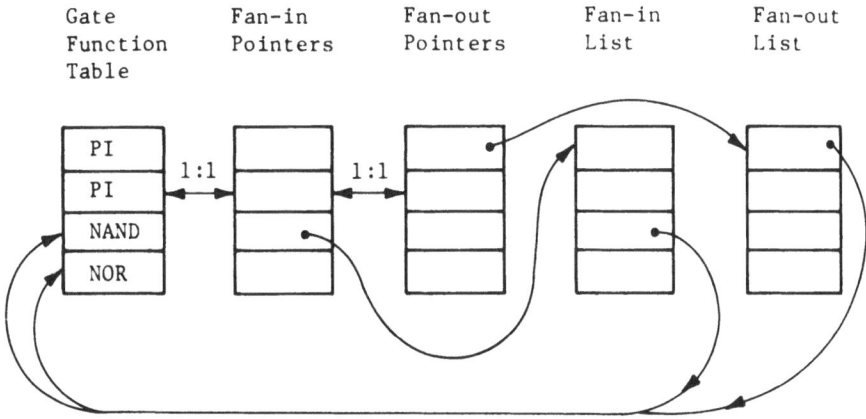

a) Fixed entry length model tables

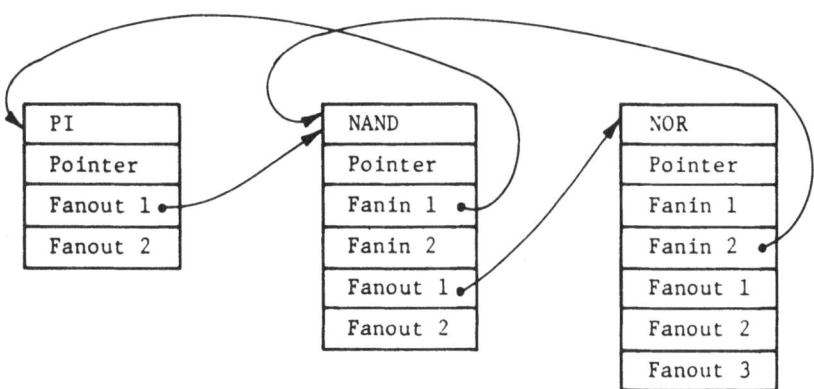

b) Variable length model tables

Figure 6.   Data Structure for the Network Model

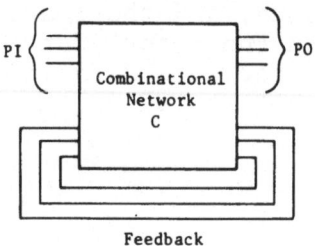

Feedback

a) Model of Sequential Network

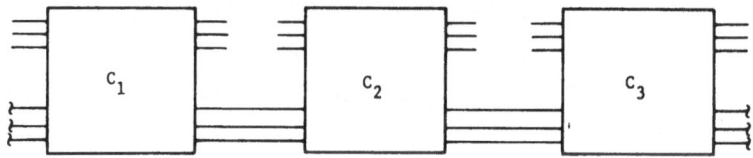

b) Model for Test Generation with Three Time Frames

Figure 7. Time Frame Model of Sequential Network

3.1.2   Other Functions of the Model Build Program

Test generation for sequential networks, discussed below in Section 5.0, is difficult. One approach that
has been used to simplify the problem is to model the
sequential network as an iterated series of combinational networks, as shown in Figure 7 [8]. Each replication of the combinational network is a time frame.
To construct this model, feedback loops within the network must be found and cut.

It is highly desirable for the test generation program
to have some guide to the relative difficulty of setting
gate inputs to a specified value. This may be done by
associating a number called a fanin weight with each
gate input. The fanin weight depends on the complexity
of the logic driving each input, as measured by distance
from a primary input in gate levels, and the total fan
out of gates in the driving network. (See [8] for a
simple procedure to compute weights. A more detailed
discussion of net controllability is found in [9].)

A method for assigning levels to gates and for cutting
feedback loops is shown in Figure 8. This procedure
requires an area for a work stack and a table area to
mark off gate inputs as they are processed. Points
at which loops are cut are indicated by flags in the
model tables.

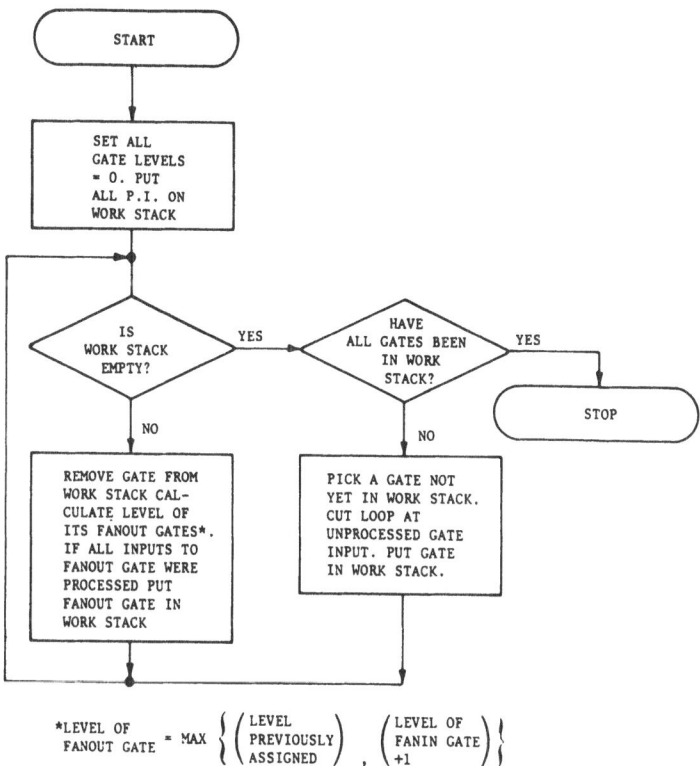

Figure 8. Method for Levelizing and Loop-cutting.

3.1.3   Other Programs in the ATPG System

The failure model building program constructs the fault
list. For stuck-at faults each list entry contains the
gate index, the gate input or output which is assumed
to be stuck, the stuck-at value, and an area reserved
for flags to be set by the test generator and fault
simulator. For bridging faults, the entry contains the
net indices of the nets assumed to be shorted, and the
logical function (AND, OR) of the short. Other fault
types may be handled similarly. Stuck-at faults are
usually determined by gate function: AND inputs s-a-1,
output s-a-0; OR inputs s-a-0, output s-a-1; but stuck-
at fault location may also be specified by the macro
rules. The points at which bridging faults are placed
are determined from the physical layout.

An important function of the fault model build program
is to identify faults which are equivalent to or dominate other faults [30]. Two faults are <u>equivalent</u> if
a test for one detects the other. For example, the
s-a-1 fault on the input of an INVERT gate is equivalent

to s-a-0 at its output. A fault f_1 is dominant over another fault f_2 if a test for f_1 detects f_2. In a combinational network, a set of tests which detects the stuck-at faults on primary inputs and on all branches of nets with multiple fanout will detect all other stuck-at faults in the network. The input and fanout branches of the network are called checkpoints, and their stuck-at faults are the prime faults of the network. Checkpoints may not be valid in sequential networks. The identification of equivalent and dominant faults usually reduces the number of faults to be considered in test generation and fault simulation by about 30%.

Fault simulation and test generation programs are discussed in detail in Sections 4.0 and 5.0.

A program is needed to translate the simulation results into the commands used to operate the automatic test equipment. This program replaces the logic 0-1 values with the voltages applied or measured at the network inputs and outputs, translates the detected fault indices to designations of physical components or circuits, and adds parametric tests. This program is oriented to the characteristics of the test equipment used, and may be extremely sophisticated when high speed, computer controlled testers are involved.

4.0 FAULT SIMULATION

Fault simulation provides the good machine response to the tests, test coverage, and diagnostic information. The diagnostics relate the primary outputs which show failing responses after each test pattern to the faults, and are used to locate process defects or to guide the repair of faulty assemblies.

For large networks, it is not practical to simulate all the faults against all the test patterns. The usual practice is to stop simulation of a fault when it is first detected. (This is called stop on first error, or SOFE.) As network sizes increase, stuck fault diagnostics are becoming less useful; the author recently observed a case in which 8000 faults were detected by the first test pattern.

Fault simulation saves time in the ATPG process because many test patterns will accidentally detect faults. By

interleaving the operation of the test generation and
fault simulation programs, and by marking off detected
faults, the number of faults for which tests must be
algorithmically generated is greatly reduced.

This section will only consider simulation methods which
use the significant events procedure. An event is significant only if it causes a change in the good machine
state of a net, or some change in fault machine status.
Usually only 2-10% of all nets will be active in the
simulation of a test; thus, the significant events procedure is much less expensive than simulation of all
gates.

4.1 Unit Delay Simulation Procedure

In unit delay simulation, all gates are assumed to have
an equal, fixed delay. To perform the simulation, two
work stacks (W_1 and W_2), a table of gate values (G),
and a list of updated gate values (U) are needed. Then
proceed as follows:

1.   Put the new values of any changed primary inputs
     into G. Enter the indices of the gates driven by
     these inputs into W_1.

2.   Fetch a gate index from W_1 and calculate the gate
     value using the input values currently in G. If
     the logic value calculated on an output differs
     from its current value, add the indices of all
     gates driven by this output net to W_2 and store
     the new value of the output in U. Continue until
     all gates in W_1 have been calculated.

3.   If W_2 is empty, quit. If not U $\rightarrow$ G, 0 $\rightarrow$ U,
     W_2 $\rightarrow$ W_1, 0 $\rightarrow$ W_2, and return to step 2.

Simple, isn't it.

A slight variation on this procedure can be used to
model the effects of delay uncertainty during input or
gate transitions. Set to X the values of any inputs
that are changing in step 1, and after n cycles of steps
2 and 3, repeat step 1 to change the X value to logic
0 or 1. This is n-X simulation. 1-X simulation is fast
and accurate for networks where timings are not critical
(such as LSSD design [10]) and is very useful for bipolar logic technologies.

4.2 Nominal-Delay Simulation Procedure

For MOSFET gates and other technologies where delays
depend on whether transitions are rising or falling,
or on the gate input which is changing, unit delay
simulation is unsatisfactory, unless hazard free design
techniques such as LSSD are employed. Nominal delay
runs about 3 times more slowly than unit delay.

In nominal delay simulation, each gate is assigned a
rising $(0 \rightarrow 1)$ delay Δ_R and falling $(1 \rightarrow 0)$ delay Δ_F
(These delays may be assigned separately to each input-
output pair of the gate.) Paths between the gates may
be given path delays Δ_P if their delays are not small
compared to Δ_R and Δ_F. Delays are estimated from the
characteristics of the devices used to build the gates,
gate loading, and wire lengths between gates. Accurate
delay estimation requires that the physical characteristics of the network be known.

The procedure for nominal delay simulation is a generalization of the procedure for unit delay. Future calculation events are scheduled at a time $\Delta_R + \Delta_P$ (or
$\Delta_F + \Delta_P$) after the present time, and update events
occur after Δ_R or Δ_F. There must be a separate new
work list W_2 and update table U for each future time
which is significant. The delays used always refer to
the gate being calculated at the current time and its
output net(s).

In addition, some method must be used to filter out
spikes which are shorter than the gate delay. The
following procedure [11] works well:

1. Establish a new table, called the <u>pipeline</u> table,
    with entries in 1:1 correspondence with the
    gate output nets. The pipeline table contains
    the <u>last</u> value calculated for each net.

2. The rules for updating G are:

    a) Do not update G to logic 0 or 1 if pipeline
        value $\neq$ update table value.

    b) Do not update to X if pipeline value =
        current value in G.

This check can be performed quickly while the gate
value table is being updated and avoids the difficulty
of searching the work stacks to unschedule events.

4.3   Fault Simulation Techniques

The most commonly used techniques are parallel, deductive and concurrent fault simulation. As its name implies, in parallel simulation, all the fault machines are simulated in parallel with the good machine. The deductive and concurrent techniques make use of fault lists which propagate along with signal flows through the network.

4.3.1   Parallel Fault Simulation

All general purpose computers have a standard word size (usually a multiple of 16 bits). In three valued (0-1-X) parallel simulation, the gate value table G uses two words for each gate output net (Figure 9). The first bit in each word is assigned to the good machine and the remainder to fault machines. To calculate the value of an AND (OR) gate, form the logical AND (OR) of Word 1 values from all its input nets and the logical OR (AND) of Word 2 values. The INVERT operation simply exhanges Word 1 and Word 2. The initial insertion of the fault machine values into G may be easily accomplished by using masks to force bits in the words to the fault machine state.

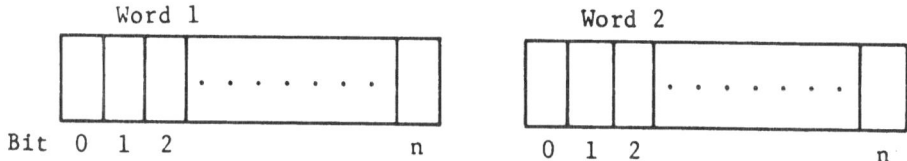

Bit 0 = Good machine state
Bits 1-n = Fault machine states

Bit Values		Logic
Word 1	Word 2	State
0	1	0
1	0	1
1	1	X

Figure 9. Gate Value Table Organization for Parallel
           Fault Simulation

To perform the simulation, all words in G are initialized to the X state, the faults on gate outputs are
forced into G, and any nets permanently tied to a value
are forced to that value. Gates fed by faulty or tied
nets are put on work stack W_1 and simulation using the
procedure of Section 4.1 or 4.2 is performed until the
network stabilizes. Then the input stimuli are applied;
when the network stabilizes after each input change is
simulated, the primary output entries in G are tested
to see if any faults have been detected. Faults detected are marked off, and good and faulty output states
with diagnostics are recorded on an output file. Because the number of faults which can be simulated is
limited by the number of bits in the machine word, repetitive passes are need to simulate all faults in the
network. The bit assignments in G are periodically compressed to eliminate detected faults from further calculation if the SOFE procedure is being used.

An efficient way to calculate gate values is to use
compiled subroutines for both good and faulty gates.
Other methods, such as table look-ups, may be used for
complex primitives. The generation of functional subroutines for high-level primitives is fairly complex,
because the manipulation of all the fault machine bits
in the gate value tables must be properly handled.

Dr. P. Goel has shown [12] that the cost to run parallel simulation varies as the cube of the number of gates
in the network. (The author's experience is that SOFE
will reduce the cost relationship to a 2.5 power rule.)
This makes parallel simulation prohibitively expensive
for large networks. Performance can be improved by increasing the number of bits in the gate value table
words [13], but this limits the maximum number of gates
which can be handled. Thus, the deductive and concurrent techniques are superior for VLSI networks.

4.3.2   Deductive Fault Simulation

Deductive fault simulation was independently invented
by Armstrong [14] and Godoy and Vogelsberg [15]. In
deductive simulation, a fault list is associated with
each gate output. Gates are scheduled for calculation
when there is a change in good machine value or in the
fault list content of their fanin nets. The latter type
of change is called a list event. Fault list content
at gate outputs is deduced from the states and fault
lists at the gate inputs.

Let $L_i = \{f_1, f_2, \ldots, f_K\}$ denote the set of faults at gate input i. If we do not consider X states, the rules for fault list calculation on AND and OR gate outputs are:

1. If all inputs are at a non-controlling value (0 for OR, 1 for AND)

$$\begin{array}{l}\text{Output} \\ \text{fault} \\ \text{list}\end{array} = \left(\bigcup_{\substack{\text{All gate}\\\text{inputs}}} L_i\right) \bigcup \left(\begin{array}{l}\text{Output stuck}\\\text{at controlling}\\\text{value}\end{array}\right)$$

2. If some inputs are at a controlling value (1 for OR, 0 for AND)

$$\begin{array}{l}\text{Output} \\ \text{fault} \\ \text{list}\end{array} = \left(\bigcap_{\substack{\text{Controlling}\\\text{inputs}}} L_i\right) \bigcap \overline{\left(\bigcup_{\substack{\text{Non-Controlling}\\\text{inputs}}} L_i\right)}$$

$$\bigcup \left(\begin{array}{c}\text{Output stuck}\\\text{at non-controlling}\\\text{value}\end{array}\right)$$

where $\overline{L}$ = set of faults not in L.

Figure 10a illustrates this fault list calculation procedure.

When X values are allowed, calculation of the fault lists becomes more complicated. One way to handle the problem is by keeping separate fault lists for the two states which differ from the good machine. This doubles gate processing time. Another method is to ignore faults for which the good machine state is X. If the good machine state is 0 or 1 and the fault state X, denote the fault by f*. The rules above now apply if the union and intersection operations are as shown in Figure 11. Figure 10b illustrates a typical list calculation.

Unfortunately, ignoring fault machines when the good machine is X causes problems. Some tested faults may not be marked as detected, and glitches of the form 0-X-0 will occur and set off fault machine oscillations in the sequential elements of the network. This causes a dramatic increase in run time. One way of eliminating most of the oscillations is to retain or add faults to a list even when the good machine changes to X, if the previous good machine state was 0 or 1 [16]. This pre-

434

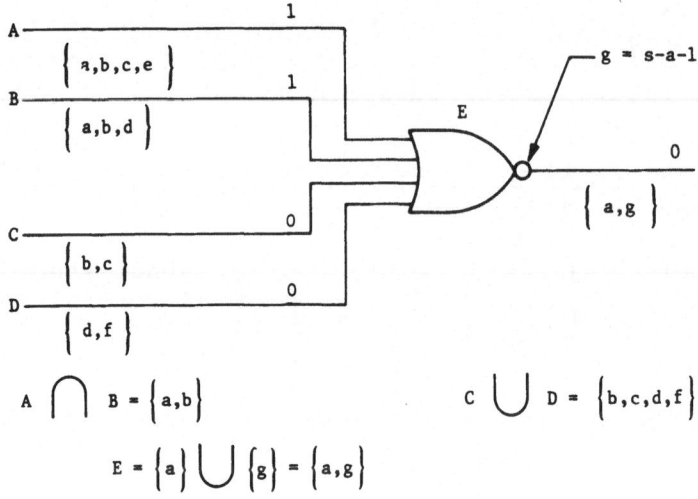

a) Calculation of new fault list for a NOR gate

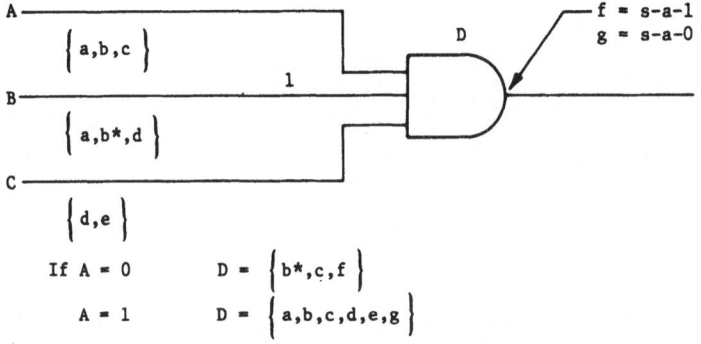

b) Calculation of fault lists with X value allowed.

Figure 10.   Examples of Fault List Calculation in Deductive Simulation.

A and B are two fault lists to be combined

0 indicates absence of f or f*

A	B	A ∪ B		A ∩ B		A ∩ B̄		B ∩ Ā	
f*	0	f*		0		f*		0	
f*	f	f		f*		0		f*	
f*	f*	f*		f*		f*		f*	

Figure 11.   Rules for List Operations with Faults at X State.

vents glitch generation during good machine transition.
Another alternative is to use special latch models to
handle fault list calculation in the sequential parts
of the network [14].

In simulating very large networks, it is possible that
the storage required for the fault lists may overflow
the maximum machine storage available. If this happens,
it is necessary to truncate the fault lists and resimu-
late the current test pattern against the remaining
faults. Normally overflow will occur on the first few
patterns only.

Run times for deductive simulation are proportional to
the square of the number of gates in the network [12].
This is a considerable improvement over the parallel
method. However, the difficulties associated with calculation of the fault lists and with scheduling events
for nominal delay simulation may make the concurrent
method more attractive.

4.3.3   Concurrent Fault Simulation

Concurrent simulation [17] combines the features of
parallel and deductive simulation. In parallel simulation primitives in both good and faulty networks are
simulated even when the logic values are equal; in deductive simulation only the good machine is simulated
and fault machine behavior is deduced from the good
machine state. In concurrent simulation, faulty gates
are simulated, but only when the fault causes some gate
input or output to differ from the good machine. This
significantly reduces the number of gate calculations
compared to the parallel method.

Figure 12 illustrates the concurrent calculation of a
gate for good and fault machines.

Input A to the gate has a s-a-1 fault. Because there
is a difference in the logic state of the good and
faulty gate on nets A and C, both the good and faulty
copies of the gate will be simulated. If net A switches
to a 1, simulation of the faulty copy will stop. Note
that the fault input B s-a-1 is not simulated, because
its state is identical to the good machine.

A fault list is associated with each gate, as in deductive simulation. An entry in the fault list contains
the index of the fault, and the gate input-output state

436

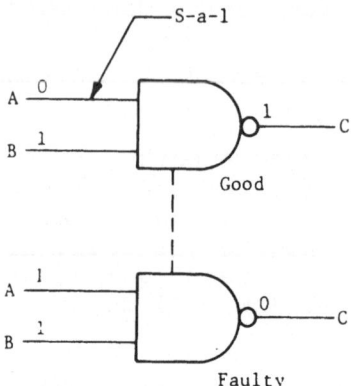

Figure 12. Conditions for Gate Calculation in a Concurrent Simulator

induced by that fault. (Fault state must differ from good machine state.) The list entries are ordered to provide for an easy search for a specific fault.

Gate calculations are scheduled as a result of significant events in the good machine or in the faulty machines in the fault lists. Good machine and fault machine calculations are scheduled with separate entries in the work stacks. An event affecting a fault machine causes an entry to be made in the work stack only for the same fault machine. The new logic values and index of the gate which caused the event are put into the work stack to eliminate look up at gate calculation time.

Calculation of a gate scheduled by a previous event takes place in four steps:

1. If the gate was scheduled by a good machine event, calculate new good machine output values. If there is a change in output state, schedule successor gates of the changed outputs for good machine calculations.

2. If any fault machines are scheduled for calculation at the gate, the action taken depends on whether the fault is already in the list at the gate being calculated.

   a. Fault is in the list. Update the input state in the list entry for the fault machine. Recalculate the output state for the fault machine, and schedule the fault for calcula-

tion at successor gates if the output state
changes.

    b.   Fault is not in the list. (This includes
faults at inputs of the gate being processed
which currently are stuck at a value different
from good machine value.) Add a new fault entry to the list. If the new entry has an output state different from good machine, schedule
the fault for calculation at successor gates.

3.   If a good machine calculation was performed (Step
1), update the input values of any fault entries
not processed in Step 2. Recalculate any entries
which have a changed input state, and schedule
the fault machine for calculation at successors if
there are any events at the outputs.

4.   Remove any fault entries which now have all input-
output states equal to good machine.

As an example, consider gate A in Figure 13a. The good
machine input state is 00, and the output net c is 1.
There are 3 faults which produce a different input-
output state from the good machine: net a s-a-1, net b
s-a-1, and net c s-a-0. Note that these 3 faults are
the same ones that would be considered to be different
by a parallel simulation at gate A. Note also that the
fault c s-a-0 will propagate to the input of gate C.

Now suppose that nets a and f are changed from 0 to 1.
The good machine output of gate A does not change.
Fault b_1 is recalculated at gate A, and the output value
changes from 1 to 0. This is a list event, and causes
fault b_1 to be scheduled for calculation at gate B. The
output state for fault c_0 does not change. Fault a_0 is
added to the list at gate A, but does not schedule
further events because its output value is the same as
the good machine. Fault a_1 now has all values equal to
the good machine. It is dropped from the list. Figure
13b shows the final result of the simulation in the
example network.

Figure 14 shows the concurrent simulation of a network
with feedback (or reconvergent fanout). On the first
cycle net b changes from 1 to 0, and faults a_1, b_0 are
replaced by a_0, b_1. All of these effects propagate to
gate B. On the second cycle, net d changes from 0 to 1,
fault a_0 is added to the list at gate B, and faults b_0,
d_1 are replaced by b_1, d_0. These changes propagate back

438

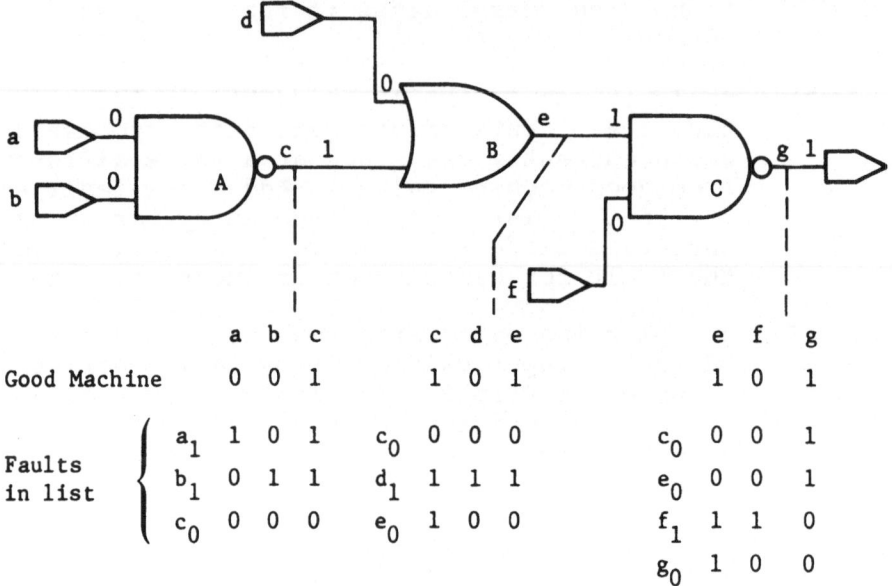

	a	b	c		c	d	e		e	f	g
Good Machine	0	0	1		1	0	1		1	0	1

Faults in list		a	b	c		c	d	e		e	f	g
	a_1	1	0	1	c_0	0	0	0	c_0	0	0	1
	b_1	0	1	1	d_1	1	1	1	e_0	0	0	1
	c_0	0	0	0	e_0	1	0	0	f_1	1	1	0
									g_0	1	0	0

a) Initial state of network and fault lists

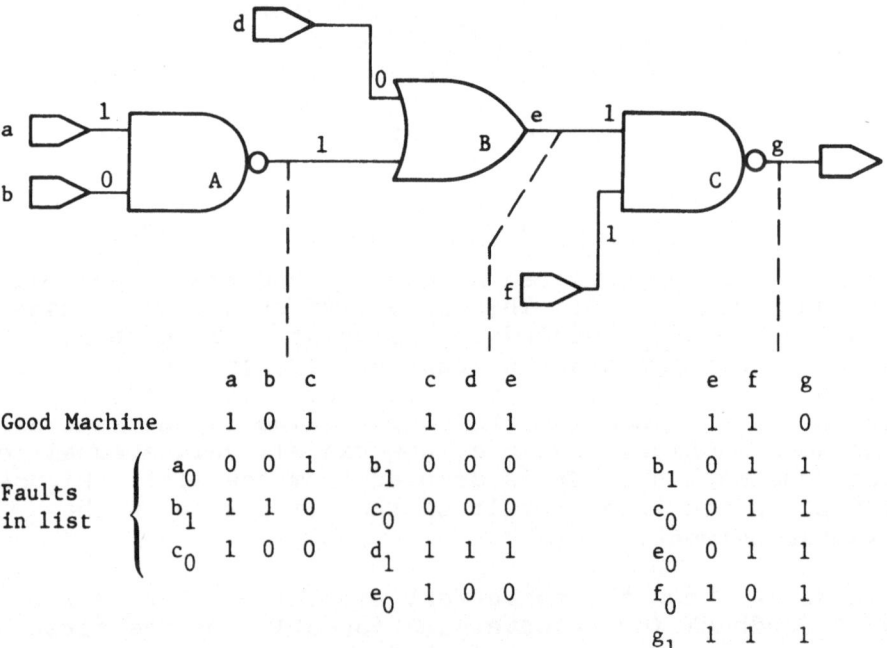

	a	b	c		c	d	e		e	f	g
Good Machine	1	0	1		1	0	1		1	1	0

Faults in list		a	b	c		c	d	e		e	f	g
	a_0	0	0	1	b_1	0	0	0	b_1	0	1	1
	b_1	1	1	0	c_0	0	0	0	c_0	0	1	1
	c_0	1	0	0	d_1	1	1	1	e_0	0	1	1
					e_0	1	0	0	f_0	1	0	1
									g_1	1	1	1

b) Final state of network and fault lists

Figure 13. Example of Concurrent Fault Simulation

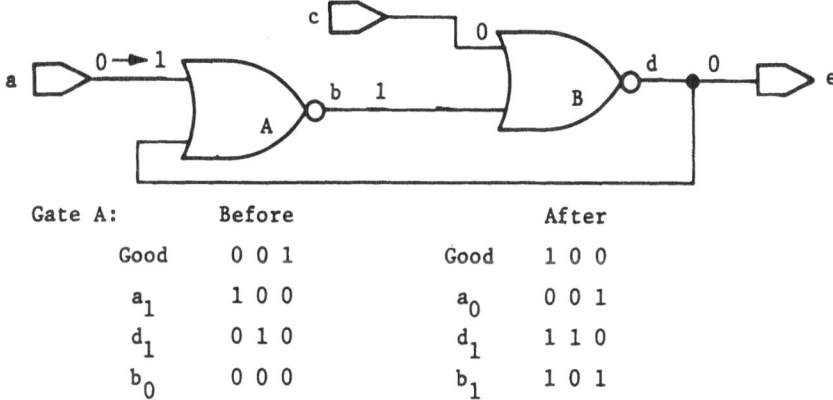

Gate A:

	Before		After	
Good	0 0 1	Good	1 0 0
a_1	1 0 0	a_0	0 0 1
d_1	0 1 0	d_1	1 1 0
b_0	0 0 0	b_1	1 0 1

a) First calculation cycle

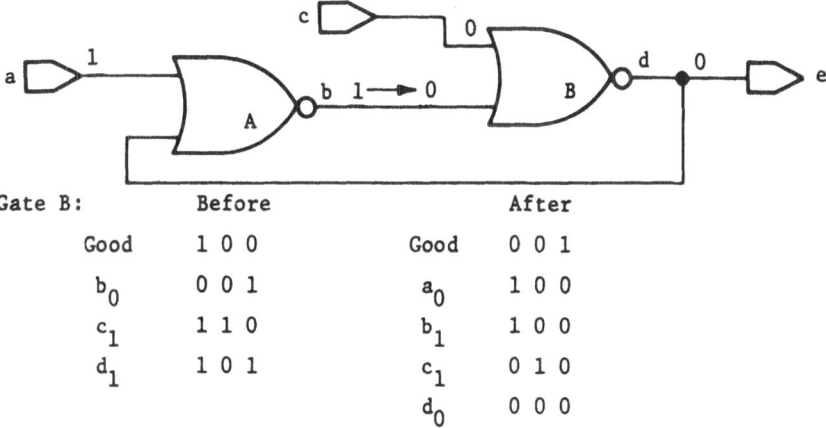

Gate B:

	Before		After	
Good	1 0 0	Good	0 0 1
b_0	0 0 1	a_0	1 0 0
c_1	1 1 0	b_1	1 0 0
d_1	1 0 1	c_1	0 1 0
		d_0	0 0 0

b) Second calculation cycle

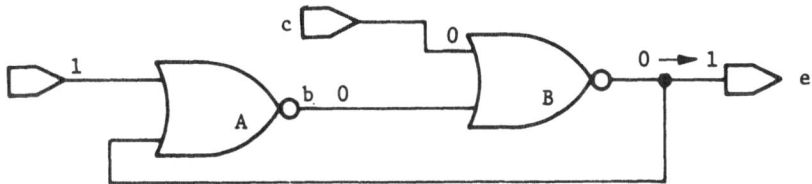

Gate A:

	Before		After	
Good	1 0 0	Good	1 1 0
a_0	0 0 1	a_0	0 1 0
b_1	1 0 1	b_1	1 1 1
d_1	1 1 1	d_0	1 0 0

c) Third calculation cycle

Figure 14. Network Illustrating Concurrent Simulation of a Feedback Loop.

to A. On the third cycle net b is steady, and d_1 is replaced by d_0. Fault a_0 is recalculated because it
changed at gate B on the previous cycle, but no new
entry is made in the list for a_0. Fault a_0 will continue
to propagate around the loop once more. Final entries
will be a_0 11 0 at gate A, and a_0 00 1 at gate B. Note
that when a fault machine is scheduled for calculation,
the fault lists at the successor gates must be scanned
to see if that fault is already present.

Because the fault lists in the concurrent simulator contain entries for all differences from the good machine
state as opposed to the deductive lists, which contain
only differences in gate output states, more storage
is required for the concurrent technique. The work lists
are also larger, because the fault machines are scheduled
separately from the good machine and fault values are
carried in the work list. The storage overhead may be
minimized by careful management and the same overflow
procedure used for the deductive case applies, but storage capacity could be a problem for very large networks.
Gate calculation is supposedly faster than deductive.
Overall simulation time varies as the square of the number of gates in the network.

4.3.4   Simulation by Fault Injection

This method is identical to concurrent simulation except
that only one fault is simulated at any time, and the
good machine states are determined separately and restored each time a new fault simulation begins. Thus
the "fault list" contains only one entry and propagation
stops if the fault machine state becomes equal to the
good machine state. This method requires minimal storage. It is faster than parallel simulation, but probably slower than concurrent for large networks.

4.4 Special Problems and Techniques

4.4.1   Oscillations

One of the most troublesome problems in fault simulation
is oscillation. Figure 15 illustrates an oscillating
condition in 1-X unit delay simulation. Good machine
oscillations are rare, and often indicate poor design
practice when they occur; but fault machine oscillations
are very common. Oscillations may increase the run
time by 50 to 100%.

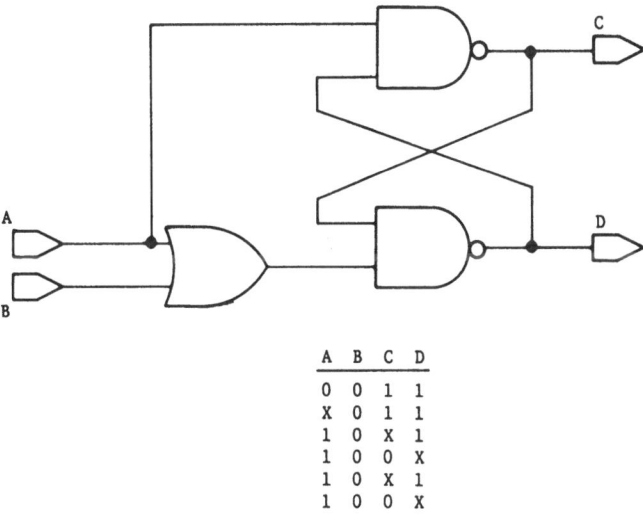

A	B	C	D
0	0	1	1
X	0	1	1
1	0	X	1
1	0	0	X
1	0	X	1
1	0	0	X

Figure 15.   Oscillation in 1-X Unit Delay Simulation

Oscillations are stopped by forcing the value of all
changing nets to X.   The number of simulation cycles
which must be completed before oscillation control
starts may be determined by a user parameter, or calculated from the number of logic levels and loops in
the network.   Oscillation control may also be invoked
if the number of changes on a feedback line exceeds a
predetermined value.   To save run time, the simulator
can detect and delete hyperactive faults from consideration.   For example, if any fault machine oscillates
more than 10 times, stop simulating it.

The author knows of no really good method to determine
when an oscillation has started.   This is a topic worth
investigating, because oscillations waste an enormous
amount of run time.

4.4.2   Functional vs. Gate Simulation

The number of gates in logic networks has increased by
an order of magnitude every 4-5 years.   This has brought
attention to the use of <u>functional primitives</u>, which
model collections of gates in simulation by the function
they perform.   Examples of functional primitives are
ROM and RAM arrays, shift registers, latches, and up-
down counters.   If the output behavior of a functional
primitive can be well defined for all expected input
sequences and internal states, it will provide as accurate a model for simulation as a gate level model.

The most common method of functional modeling is the use of subroutines which are executed by the simulator. Table look-up procedures are sometimes used. If the output values are a function of the internal states in addition to the inputs, extra storage must be provided. When the behavior is not specified for all input sequences and internal states, the X value may be used to represent indeterminate output states. Functional models may be coded to model internal fault machine behavior. This complicates the preparation of the model considerably; the decision on which types of faulty behavior to model is based on the physical characteristics of the device, an estimate of likelihood of the various modes of failure, and the sensitivity of the overall network to internal failures within the functional model. Gate level representations of the functional primitive may be helpful in the selection of internal faults, and in computing the simulus-response information to model faulty behavior.

Parallel simulation is not well suited to functional models, because it is usually necessary to unpack the input states and calculate responses one at a time. Regular structures such as RAM or ROM are an exception. Deductive simulation may not work well either, because simple rules for fault list calculation no longer apply. Concurrent simulation is more easily adapted because the same subroutine is used to calculate both good and fault machine responses [18, 19]. An excellent paper by Abramovici, Breuer, and Kumar provides a theoretical background for concurrent simulation of functional models with internal faults [20].

Properly used, functional primitives can save run time and storage. However, functional primitives will usually require more time to calculate than Boolean gates. For simple combinational models, gate level simulation is usually cost effective, and functional modeling should be considered only if more than 20 to 40 equivalent gates are contained in the functional primitive. Gate level models are also likely to be effective if the primitive has a large number of inputs, because a single input change usually will not generate many significant events, but one input change will force a complete recalculation of the functional model. Most networks will contain random logic as well as functional elements; thus, the simulator should be able to handle both gates and functional primitives together. Functional primitives should not be considered unless the behavior is very well specified; too many unspeci-

fied conditions will render a fault simulation virtually
useless.

The amount of effort needed to design, code, and test
an accurate functional model may be quite large. The
cost of model preparation may exceed the savings in simulation run time.

4.5   Conclusions on Fault Simulation

Parallel simulation is too slow for VLSI networks. Nominal delay simulation and functional primitives are more
difficult to apply to deductive simulation than concurrent, but concurrent requires more storage. A run time
edge is claimed for concurrent simulation by its users,
but there has been no clear cut demonstration of higher
speed for either method.

Future variations and syntheses of the known methods will
surely be found. Fault simulation is necessary, but
expensive; the return on investment spent in research can
be very high. Just as important as the algorithm used is
the care and skill spent in the programming and testing
process. The personnel selected to implement the simulator should be high performers, and the use of programmer
teams and design and code inspections is highly recommended. Interactive facilities for test and operation
are very desirable, and a troubleshooting package is
well worth the extra coding effort spent on it.

5.0   TEST GENERATION

Most test generation algorithms work in a similar way.
First, an untested fault is chosen from the fault list
and a test is set up at the faulty gate. Next, a
sensitized path is established from the faulty gate to
a primary output. Finally, the primary inputs are set
to logic states needed to produce values in the network
that are consistent with the first two steps. The algorithms differ in the method and order in which the
three steps are performed.

5.1   Algorithmic Test Generation Programs

5.1.1   FALTGEN

The most widely used procedure for test generation is
the D-algorithm, described in a famous paper by Roth
[21]. FALTGEN is a test generation program based on the
D-algorithm [46]. FALTGEN selects untested faults from
the fault list, one at a time, and attempts to generate
a test using a four step procedure:

Step 1   Setup. Assign the logic signal values on in-
           put nets to the faulty gate which will put a
           test value (D or $\overline{D}$) on its output(s).

           Each time a logic value is assigned to a net,
           a process called implication is performed.
           Implication may be described by an example:
           If the assigned value is 1, and the net is
           driven by an AND gate, all inputs of the AND
           are assigned to 1. Or, if the value assigned
           is 0, and the net drives an AND gate, the AND
           gate's output will be assigned to 0. When it
           is invoked, the implication routine continues
           until there are no more nets whose value can
           be unequivocally determined. All nets as-
           signed values are recorded in an assignment
           list. If it is found that a value to be as-
           signed is in conflict with a previously as-
           signed value (for example, previous value = 0,
           new value = 1), perform Step 4.

Step 2.   Propagation. Starting with a gate G driven
           by the faulty gate, assign the inputs of G to
           propagate the D (or $\overline{D}$) value to the output(s)
           of G. The implication routine is invoked
           each time a net value is assigned. Repeat

Step 2 until a D (or $\overline{D}$) value reaches a primary output. Note that Step 2 sets a sensitized path from the faulty gate to a network output.

Step 3.   Justification. If there are gates with an output assigned to a value, but have insufficient input values assigned to produce that output value, select such a gate and pick an unassigned input. Assign this input to the value needed to produce the value assigned at the output. The gate must be a decision gate; its index and the index of the input net selected are recorded in a decision list. Step 3 continues until no more gates with assigned outputs and incompletely assigned inputs are found. As before, the implication process is performed every time a new net value is assigned. When Step 3 is completed, the primary input state is a test for the fault under consideration.

Step 4.   Conflict resolution. If a conflict occurs, the decision list is used to choose a new course of action. This consists of selecting a previously untried input at a decision gate, or in selecting a new fanout path for D propagation. The entries in the assignment list are used to remove values assigned after the decision which is to be retried, and the entries in the assignment and decision lists made after the retried decision are deleted. Processing reverts to the step in which the retried decision was originally made. If no decision is found which can be retried, the attempt to generate a test for the fault fails.

The small network shown in Figure 16 illustrates the operation of FALTGEN. Suppose the fault selected is the output of gate F s-a-1:

Setup.   Put a 0 on the first input of gate J and a 1 on the output of gate G. This sets J's output to D. PI's A and B are set to 1 by implication.

Propagation.   Put a 1 on the output of gate K. This puts a $\overline{D}$ on the primary output fed by gate L.

Justification. There are two gates to justify: G and K.

446

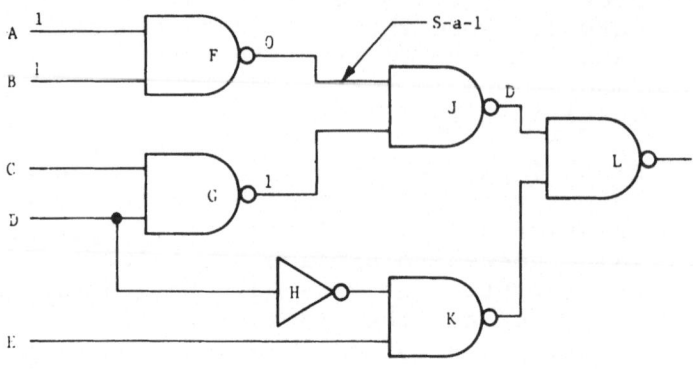

a) Setup.

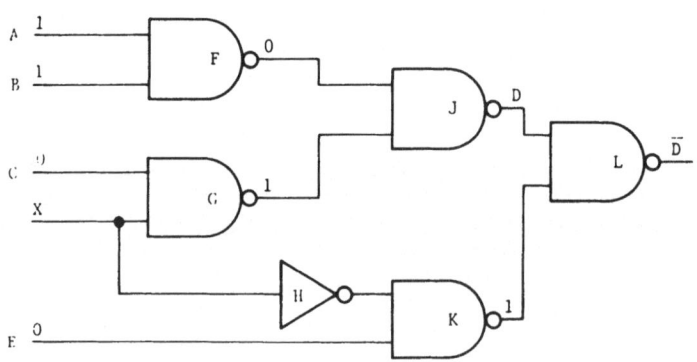

b) Propagation and Justification.

Figure 16. Example of FALTGEN Operation.

> This is accomplished by setting the PI's C
> and E to 0. (Note that if PI D and gate
> H had been used in an attempt to justify
> gates G and K, a conflict would occur.)

FALTGEN uses the fanin weights discussed in Section
3.1.2 to select paths for propagation and justification
in order to minimize the number of incorrect decisions;
but the conflict resolution procedure is FALTGEN's weak
point. Some faults in VLSI networks require millions
of retried decisions on one fault before the procedure
terminates. The running times become impossibly long
if there is no limit placed on the number of retries.
In practice, a limit of 25 retries gives satisfactory
results; however, some tests are missed.

FALTGEN can generate tests for simple sequential networks
using the model of Figure 7. On complex sequential net-

works containing counters and shift registers, FALTGEN
is useless. For combinational networks of 4000 gates
FALTGEN has been reasonably effective, achieving test
coverages in the high 90% range.

5.1.2   SOFTG

The Simulator Oriented Fault Test Generator (SOFTG)
[22, 23] is an effective algorithm for sequential networks. It uses a fault simulator to perform the propagation function. A primary objective of SOFTG was to
generate test sequences with no arbitrary restriction on
sequence length; this enables SOFTG to generate tests
for counters and shift registers.

Figure 17 is an overview of SOFTG. The procedure begins
by setting all primary inputs to 0 (or some other arbitrary state), and simulating all faults. If any faults
are not tested, SOFTG selects an untested fault to work
on. Fault selection is done using the following rules
in descending priority order:

Rule 1.   Fault has a D or $\bar{D}$ value at some gate in the
            network.

            a.   If SOFTG has previously attempted and
                  failed to find a test for the fault and
                  the D or $\bar{D}$ is at a lower number of logic
                  levels from a PO compared to the previous
                  attempt, select it.

            b.   Select a previously untried fault.

            If more than one fault exists which meets
            condition a or b, select the one with D or $\bar{D}$
            nearest to a PO.

Rule 2.   If no fault has a D or $\bar{D}$, select the next un-
            tested fault in the fault list.

The fault selection procedure attempts to capitalize on
the fact that many D and $\bar{D}$ values occur in the network
by "chance". This heuristic is extremely valuable.

When a fault has been selected, the next action depends
on whether Rule 1 or Rule 2 was used. If Rule 1 was
applied, then there must be some gate G in the network
which has D or $\bar{D}$ on one or more of its inputs, but does
not have D or $\bar{D}$ on its output. In this case, the initial

448

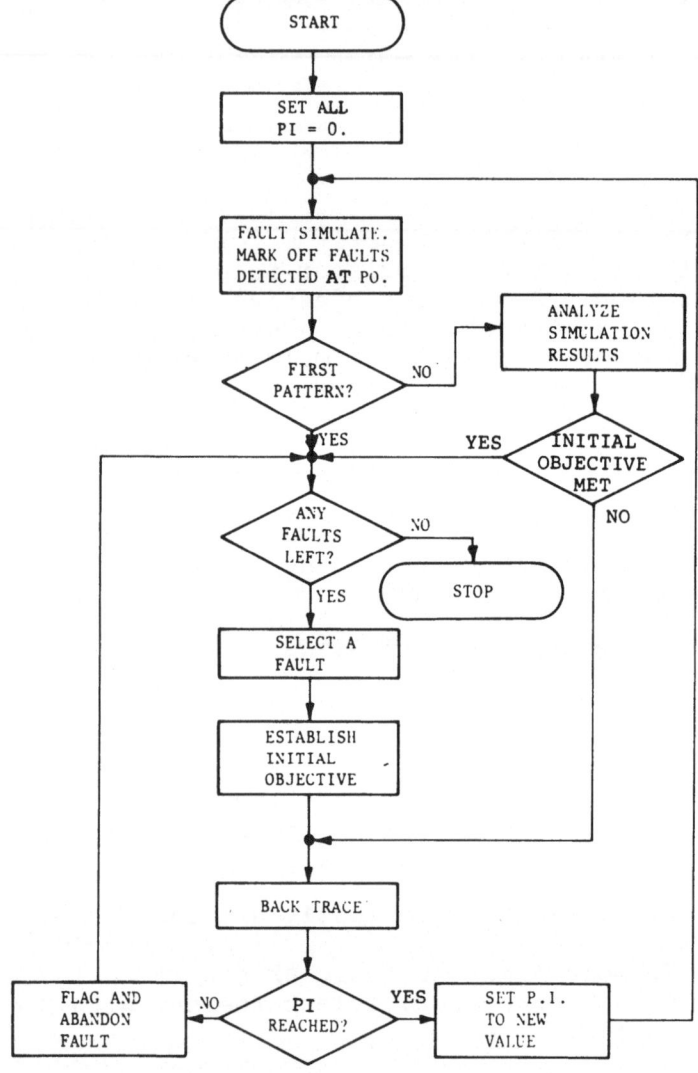

Figure 17. Overview of SOFTG.

objective of SOFTG will be to set the inputs of G so that D or $\bar{D}$ will propagate to the output of G. If Rule 2 was used, the objective will be to set up the inputs of the faulty gate to produce D or $\bar{D}$ at its output. Note that the gate G in the Rule 1 case may not be the faulty gate.

Objectives are recorded in an objective list (Figure 18a). This list contains the gate index, the desired state of the gate inputs, a flag that indicates whether

Gate Index	Desired Input State	Imply- Decision Flag	Number of Inputs Processed

a) Objective List Entry Format

Gate Index	Output Value Assigned	Good- Faulty Flag	Pointer to Objective List

b) Assignment List Entry Format

Figure 18. Tables Used in SOFTG

all input values are implied by the desired output value
(imply gate) or not (decision gate), and the number of
inputs previously processed. Suppose that the initial
gate to be processed is a NAND that has a $\overline{D}$ value on one
input and does not have D on its output. Because $\overline{D}$
means 0 in the good machine, 1 in the faulty machine,
the good machine value at the NAND gate output must be
1. (The failing machine output state must be either 1
or X, because D is not the value at the gate output.)
Therefore, the initial objective will be to set all
the inputs to the NAND gate to 1 in the faulty machine.
This will place a $\overline{D}$ on the output. Because SOFTG
operates with a fault simulator, both good and faulty
machine states are available.

Note that if the good machine 0 part of the $\overline{D}$ value at
the NAND input changes to 1 or X, it will not be possible to set the output to D. To prevent this, the good
machine 0 is assigned. An assignment list (Figure 18b)
records all logic values which are critical to satis-
faction of objectives in the objective list. Entries
in the assignment list contain: the index of the gate
which drives a net to the assigned value, the assigned
value, a flag which indicates whether the value relates
to good or faulty machine, and a pointer to the associated entry in the objective list. For the example of the
NAND gate the initial objective is to set all inputs to
1 in the faulty machine (imply gate), and the first
assignment is to hold a good machine 0 value at the input which has current value $\overline{D}$.

When the initial objective list entry has been established, further objective and assignment list entries
are made by the backtrace procedure shown in Figure 19.
The fanin weights discussed in Section 3.1.2 are used

450

to determine the priority order in which gate inputs
are processed. If a gate input is already at the required value, it is assigned and another input selected.

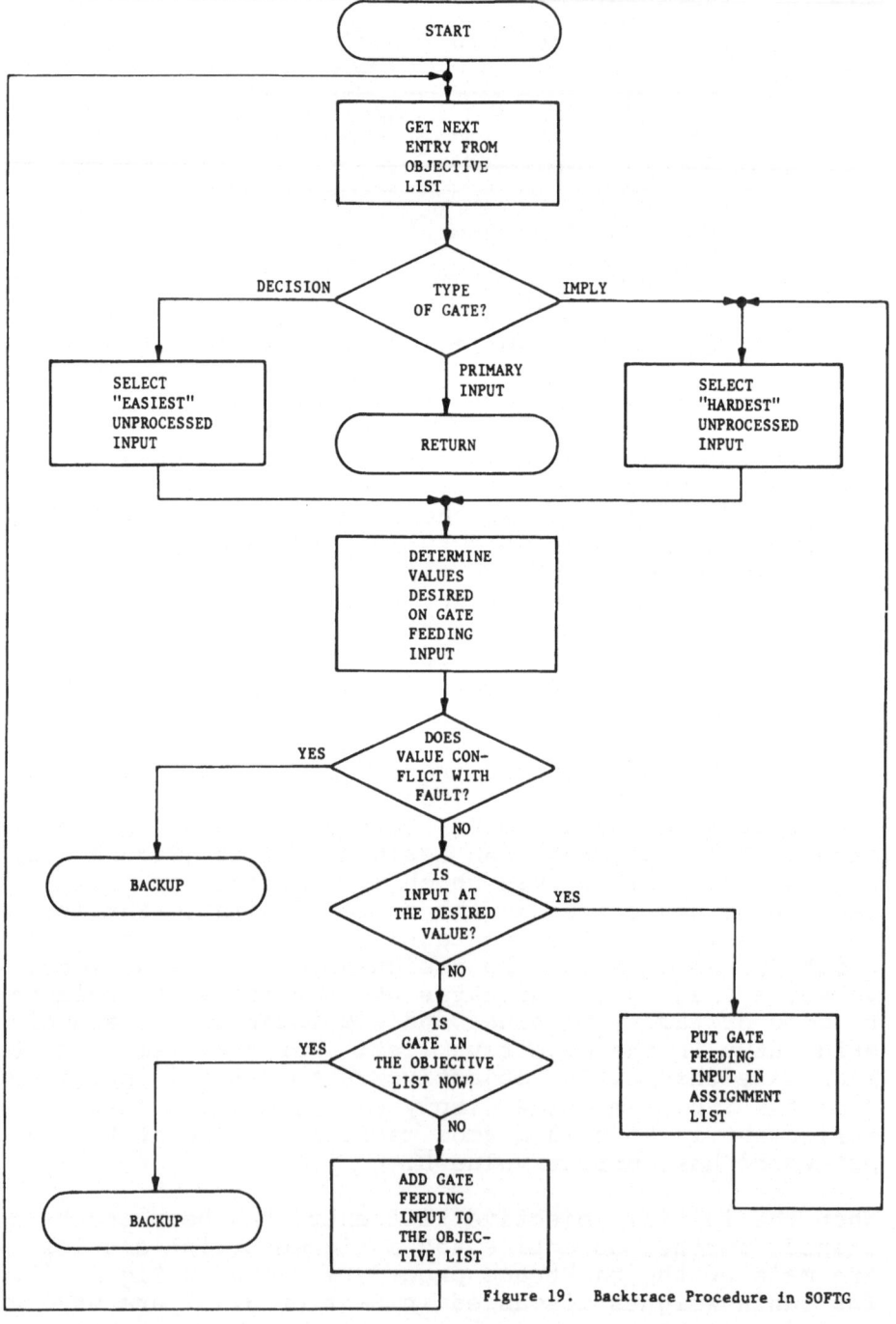

Figure 19. Backtrace Procedure in SOFTG

Two types of conflict may occur in the backtrace procedure. If the net to be set to a fault machine value is
the faulty net, the value required must be equal to the
stuck-at value; and, if the next gate to be added to the
objective list is already in the objective list, there
is a conflict condition. The second test prevents the
backtrace from running endlessly around a loop.

When a conflict occurs in the backtrace, processing reverts to the last decision gate entered on the objective
list. If this gate has an unprocessed input, all subsequent objective and assignment list entries are removed,
and the unprocessed input is tried. If no such decision
gate is in the objective list, the attempt to generate
a test for the selected fault fails.

If no conflicts occur, the backtrace procedure calculates
the values required at the selected predecessor gate
and adds it to the objective list. This gate becomes
the next to be processed, and the process iterates until
a primary input is reached. This primary input is
changed to a new value and the fault simulator is called.

If the initial objective is met after simulation, the
fault selection routine is invoked. Usully, the same
fault will be selected again, and the D value will move
toward an output one gate level at a time. If the initial objective was not met, the simulation results
analysis routine tests the objective list entries and
associated assignments until an objective which is met
is found. If any assigned values associated with unmet
objectives are not equal to the simulated values, the
last input change is discarded and the procedure for
handling conflicts is invoked. Otherwise, backtracing
resumes at the last unsatisfied entry on the objective
list. The process continues until the initial objective
is met or the fault selected is abandoned.

SOFTG has much less trouble with retried decisions than
FALTGEN, because the only decision gates are the predecessors of the initial objective gate. Usually very few
retried decisions are necessary to establish that it is
impossible to propagate a D value. Many faults are detected "accidentally" by the propagation of other faults.

A disadvantage of SOFTG is that only one primary input
is switched at a time. The number of patterns generated
is usually large.

SOFTG has produced high test coverages when applied to

complex sequential networks containing 2000 gates. For larger networks, coverage begins to fall off, and other methods must be used.

5.1.3 PODEM

The Path Oriented Decision Making (PODEM) algorithm [24] is an offshoot of SOFTG which has been effective when applied to combinational networks containing over 20,000 gates. The high-level flow of PODEM (Figure 20) is very similar to that of SOFTG. However, PODEM does not simulate all the faults; it selects one fault at a time, initializes all gate values to X, and works with the selected fault until a test is found or it is determined that no test is possible.

The backtracing procedure in PODEM is the same as that employed in SOFTG, except that no assignment list is used. The establishment of the initial objective is also similar, but does have an important difference shown in Figure 21. Once a D value is established for a fault, PODEM tries to find the gate G nearest to a primary output with a D value on an input and no D value on the output. This is identical to fault selection Rule 1b in SOFTG. Unlike SOFTG, when the gate G is found, a check is made to see if it is possible to create a sensitized path from G to a primary output. This is not possible unless there is some path in which all gates are still at the X state from G to a primary output. If the X-path test fails at G, an attempt is made to select another gate for propagation of the D value.

The major novelty in PODEM is in the method used to record and retry decisions. FALTGEN and SOFTG select a new input on a decision gate in the network; PODEM changes values at the primary inputs. Figure 22 shows the decision list structure used by PODEM. The index and assigned logic value of each primary input reached by backtracing are recorded at the bottom of the list. When a backtrace conflict or failure in the X-path check occurs, primary inputs in the decision list are removed and their values set to X in a last in - first out manner, until an input which has not been retried is found. The state of this input is switched to the opposite value, and the remade decision flag bit is turned on. Processing resumes with simulation of the new primary input state. If there is no primary input to retry, there is no test for the fault under consideration.

453

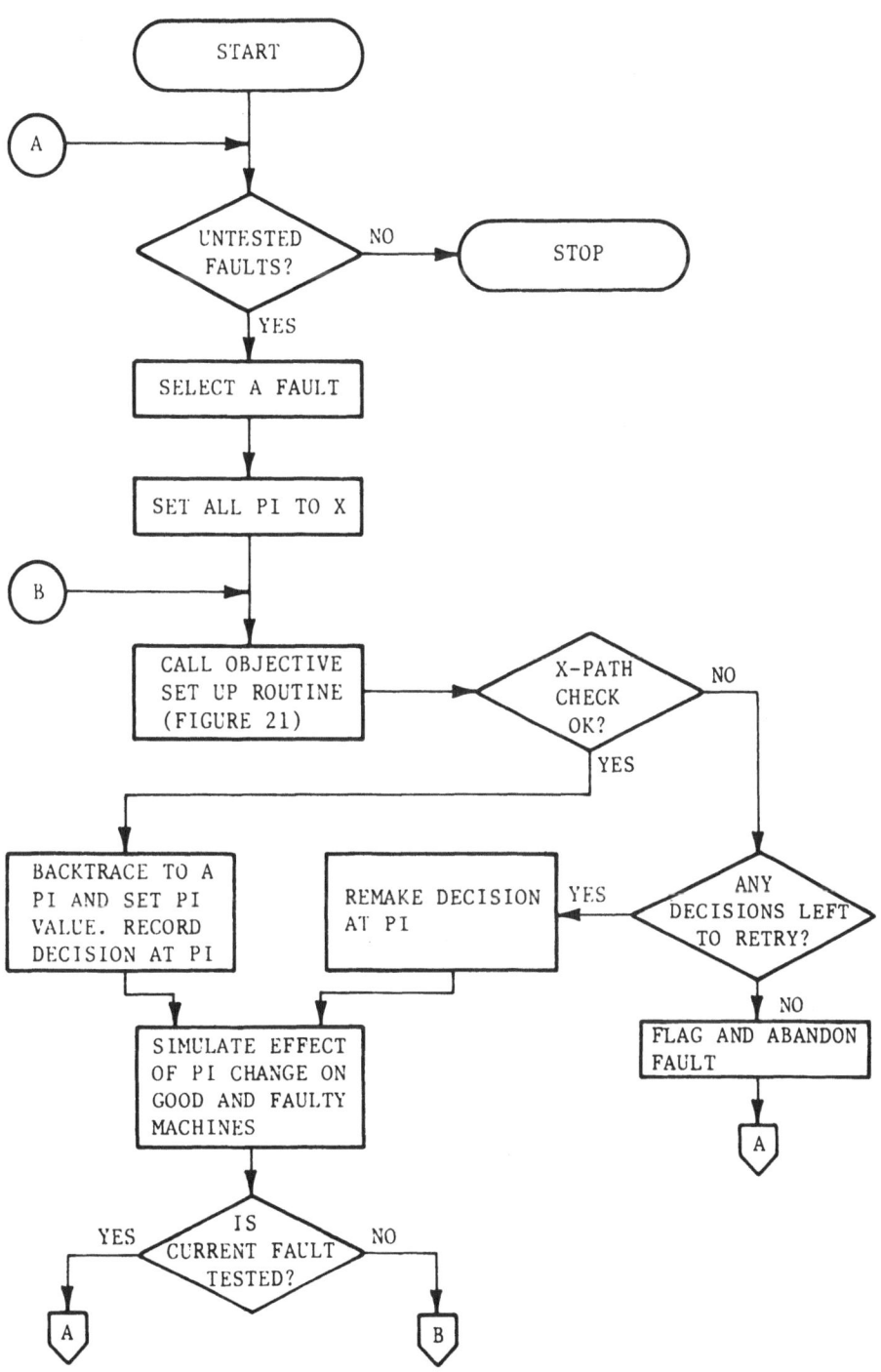

FIGURE 20. PODEM FLOW

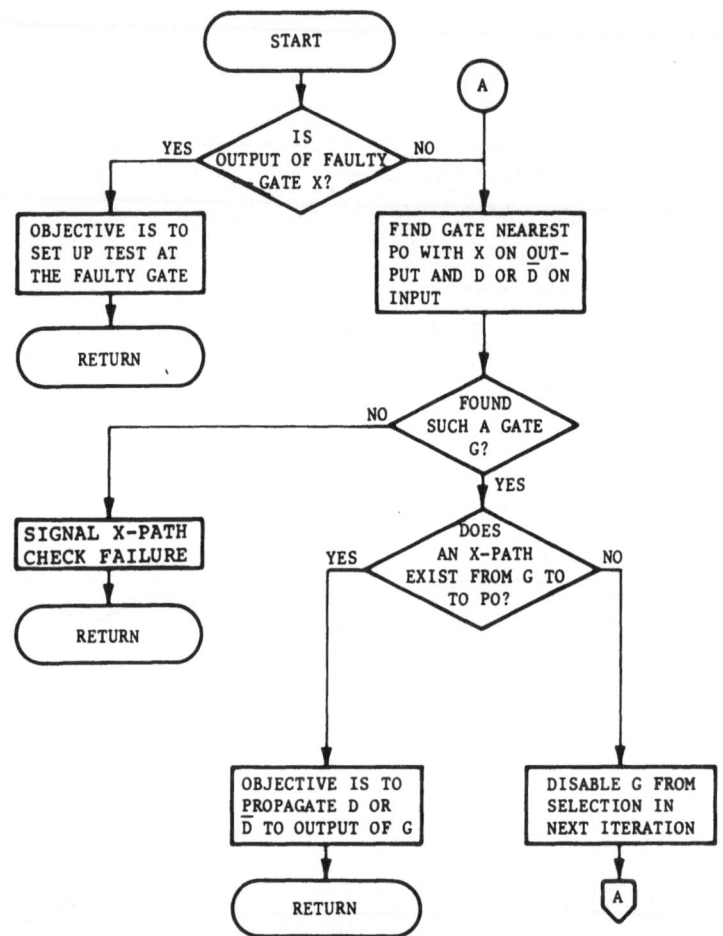

FIGURE 21.   DETERMINATION OF INITIAL OBJECTIVE IN PODEM

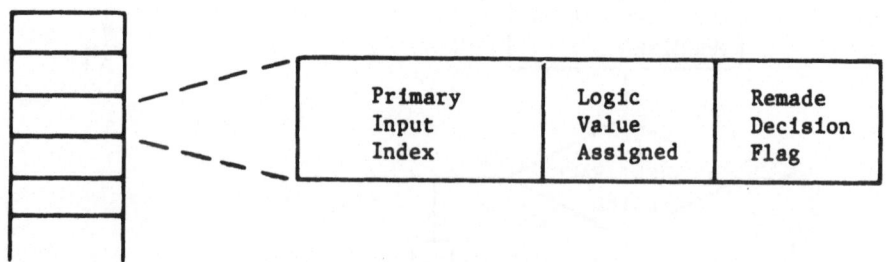

Figure 22.   PODEM Decision List Structure

If a test for the fault under consideration exists, this
procedure will always find it if no limit is placed on
the number of retries. The backtrace and retry procedures are compact and efficient, and account for the
successful use of PODEM on very large networks.

5.1.4   Path Sensitizing

If it is possible to sensitize a path from a primary
input to a primary output such that a primary input
change will switch all the gates on the path, every
stuck fault in the path will be detected. This is the
basis for the commercially available LASAR [25, 26] and
TESTAID [27] programs. Path sensitizing is effective
on moderately sequential networks of up to 4000 gates.
It has the disadvantage of producing a very large number of test patterns.

5.1.5   Algebraic Methods

Methods have been devised for generation of stuck fault
tests from Boolean equations, such as those based on
Boolean Difference [28]. They are worthless for large
networks.

5.2   Random Pattern Test Generation

Because programming and operating algorithmic test generators is expensive, test pattern generation based on
random numbers [30, 31] has become popular. The effectiveness of random patterns decreases as the number of
fanins to gates increases and as the number of logic
levels in the network increases [32, 33]. Both of these
are likely to be large in a VLSI network. Responses
must be obtained from good machine simulation or comparison with a known good device. Fault simulation of
random patterns is extremely expensive, but no other
measure of test effectiveness is available. This author
believes that exclusive reliance on random pattern testing is unwise.

5.3   Test Generation Using Functional Primitives

As was the case with fault simulation, consideration
has been given to using functional primitives in test
generation. This is very difficult; simulation requires

456

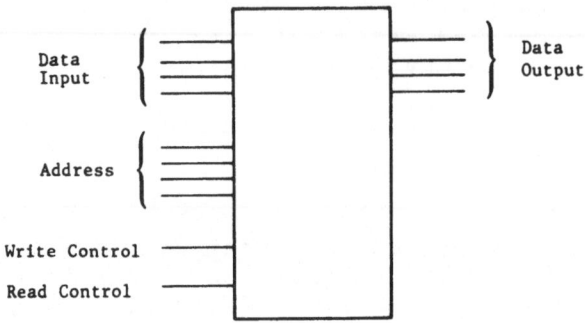

Figure 23.   RAM Model

only forward propagation of logic signals, while test
generation requires both forward propagation and reverse
justification. The use of functional models has been
successful only for primitives which have a high degree
of structural regularity or perform simple, well defined
operations. Functional models are usually assumed to
be internally fault-free.

5.3.1   Random Access Memories

RAMs may be represented by the model shown in Figure 23.
Logic values are propagated or justified from the data
inputs to the data outputs by pulsing the write control
and read control in succession. D values on the address
inputs may be propagated to the data outputs with the
following sequence:

Step 1.   Initialize the value at the faulty address
           input to 1 (or 0) if the faulty value is D
           (or $\overline{D}$). Initialize all other address inputs
           to known states (0 or 1). Put a known value
           on one of the data input nets. Pulse the
           write control net.

Step 2.   Switch the good machine values at the data
           input selected in Step 1 and at the faulty
           address input net. Pulse the write control
           net.

Step 3.   Restore the address inputs to the state set in
           Step 1. Pulse the read control net. A $\overline{D}$
           (or D) value will now appear on the data out-
           put corresponding to the selected data input.

Similar tests may be devised for faults on the control
inputs.

5.3.2 Other Functional Primitives

The RAM is easy to model because it has a regular structure and well-defined operation. Other regular structures, such as Programmable Logic Arrays, may be handled
functionally [34, 35, 36, 45]. A recent article by
Breuer and Friedman discusses methods for modeling shift
registers and up-down counters [37]. Extension of the
models to include more complex primitives, such as a
microprocessor, or to include internal faults, is a
formidable problem.

6.0 TEST GENERATION FOR VLSI NETWORKS

Despite the success of SOFTG, and claims made for programs like LASAR, it is unlikely that automatic test
generation programs will be able to reliably achieve
test coverages of better than 90% on complex sequential
networks containing more than 5000 gates unless some
form of design for testability is employed. There are
four design principles which will improve the performance
of the ATPG system:

1.   Add additional primary inputs and outputs to the
      network.

2.   Make sequential logic appear combinational to the
      ATPG system.

3.   Insure that a large network may be subdivided into
      smaller networks which may be individually tested.

4.   Include special hardware in the network so that it
      can test itself.

6.1 Level Sensitive Scan Design (LSSD)

LSSD [10] employs the first three design principles in
combination. Figure 24 illustrates an LSSD network.
The latches in the shift registers act as extra primary
inputs and outputs to the combinational network.

Test generation for an LSSD network is a three-step
process:

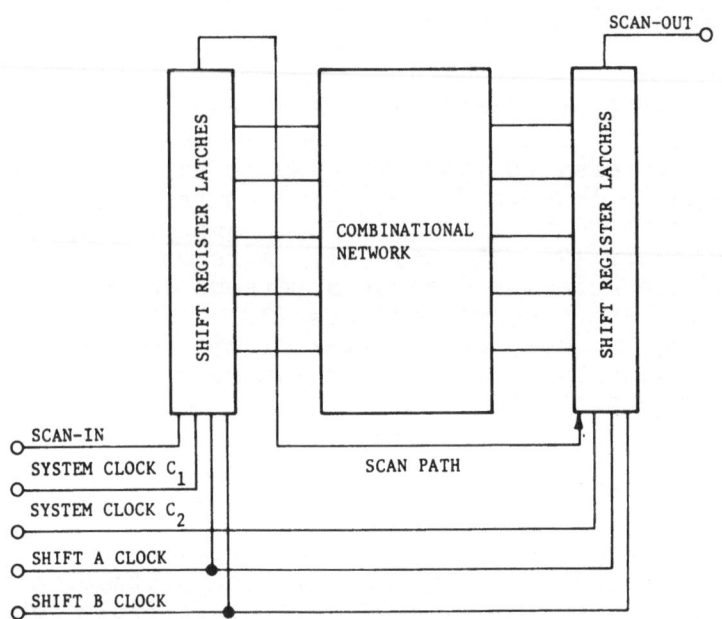

Figure 24. LSSD Network

1. The shift registers are functionally tested to insure that the scan-in and scan-out of bits in the
   shift registers works correctly. The shift register test also detects many faults in the combinational network.

2. 50 to 100 random patterns are applied to the combinational network to test "easy" faults.

3. The PODEM program is run.

The tests are applied as follows:

1. Scan the input state for the combinational network
   into the shift register.

2. Apply logic values to the network primary inputs.

3. Measure responses at the network primary outputs.

4. Pulse the system and shift B clocks to capture
   test results in the shift register latches.

5. Shift out and measure latch states at the scan out
   primary output.

The LSSD tests for separate faults are combined into an
efficient test set before application [38].

6.2 Subdividing Large LSSD Networks for Test Generation

LSSD makes subdivision of large networks possible [39,
40]. This is done as follows:

1. Initialize a <u>backtrace</u> <u>list</u> <u>set</u> L to 0.

2. Back trace through the network starting from each
    primary output and shift register latch (SRL).
    Stop tracing at primary inputs and SRLs. Record
    the indices of all gates traced in a list L_i. Add
    L_i to L. Figure 25 illustrates this step.

3. Repeat Step 2 for all PO and SRLs.

4. Obtain an area for subnetwork gate lists S_j. Set
    j = 1.

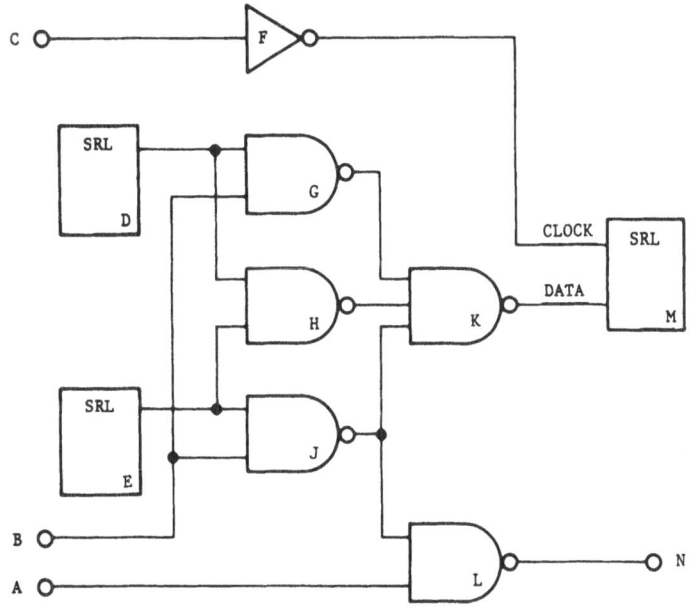

BACKTRACE LIST FOR SRL M=

{M, K, F, G, H, J, D, E, B. + SHIFT CLOCK
AND SCAN PATH LOGIC }

BACKTRACE LIST FOR P.O.N=

{ N, L, J, E, B, A}

Figure 25. Example of Backtrace

5. S_j = member of L with the largest number of gates. Delete the selected member from L.

6. If the number of gates in $S_j \geq$ desired subnetwork size, go to Step 8.

7. Select the member L_k of L such that $S_j \cap L_k$ contains the largest number of gates. Set $S_j = S_j \cup L_k$, delete L_k and go to Step 6. If all L_i have now been processed, go to Step 8.

8. Save the list S_j. If there are members L_i which have not been processed, set j = j + 1 and return to Step 5. Otherwise, stop.

When this procedure is completed, the gate indices in the subnetwork gate lists may be used to build the network model tables for the subnetworks. Test generation is done on all subnetworks separately and the results are combined.

6.3 Other Methods for ATPG of Large Networks

LSSD network subdivision is one example of a "divide and conquer" approach to testing VLSI. It is possible, by careful design, to electrically isolate the parts of a network that are implemented on separate chips or modules. Each part and the inter-part wiring may then be separately tested. This is useful as long as the number of gates on a chip is small enough to be handled by the ATPG system.

A good general review of the technique and status of design for testability is found in a paper by Williams and Parker [41].

7.0 CONCLUSIONS

There has been some success in automatic test generation for sequential networks, but this problem will become unsolvable with the rapid increase in the number of gates and gate/pin ratios on chips. The outlook for LSSD combinational networks is much brighter; ATPG was used on a 30,000 gate board in IBM System 38. More efficient algorithmic procedures for test generation of LSSD networks will probably be found. The key seems to lie in minimizing or eliminating retried decisions.

Overall system run time varies approximately as the square of the number of gates in the network [12]. This sets an ultimate upper limit to the size of the network which can be economically processed, and emphasizes the importance of "divide and conquer" approaches to testing. Work needs to be done to find efficient ATPG procedures for networks of 100,000 gates or more. Test generation and fault simulation using functional primitives is still in the early stages. There is a need to broaden the fault model to include delay and intermittent faults; not much work has been done in this area [42, 43, 44, 47].

This author is optimistic about the future of ATPG. The silicon and wiring overheads which have discouraged logic designers from using design for testability are becoming less significant. Design for testability will become mandatory with the expected increases in gate density on chips, and there will be further progress made in improving the capacity and performance of ATPG systems. The challenge of ATPG for VLSI networks will be met.

8.0 ACKNOWLEDGEMENTS

The author wishes to thank G. West, H. Godoy, R. MacLean, T. Snethen, P. Shearon, and H. Li for their helpful comments and suggestions. The assistance of Mrs. N. M. Henson in preparing the manuscript for publication was invaluable.

REFERENCES

1. M. Breuer and A. Friedman, Diagnosis and Reliable Design of Digital Systems, Computer Science Press, Potomac, MD, 1976.
2. W. Plice, 1979 Test Conference, IEEE Computer Society, 128-136.
3. T. Trick, E. El-Masry, A. Sakla, B. Inn, 1979 Test Conference, IEEE Computer Society, 137-142.
4. M. Du and C. Weiss, IEEE Trans. on Computers, (C-22), 235-240 (1973).
5. I. Kohavi and Z. Kohavi, IEEE Trans. on Computers (C-21), 556-567 (1972).
6. D. Bossen and S. Hong, IEEE Trans. on Computers (C-20), 1252-1257 (1971).
7. V. Agarwal and G. Masson, IEEE Trans. on Computers (C-29), 288-299 (1980).

8. G. Putzolu and J. Roth, IEEE Trans. on Computers (C-20), 639-646 (1971).

9. L. Goldstein, IEEE Trans. on Circuits and Systems (CAS-26), 685-693 (1979).

10. E. Eichelberger and T. Williams, Proceedings 14th Design Automation Conf., IEEE Computer Society, 462-468 (1977).

11. R. MacLean, Private communication.

12. P. Goel, Proceedings 17th Design Automation Conf., IEEE Computer Society (1980).

13. H. Chang, S. Chappel, C. Elmendorf, and L. Schmidt, IEEE Trans. on Computers (C-23), 1132-1138 (1974).

14. D. Armstrong, IEEE Trans. on Computers (C-21), 464-471 (1972).

15. H. Godoy and R. Vogelsberg, IBM Technical Disclosure Bulletin (13), 3343 (1971).

16. H. Godoy, Private communication.

17. E. Ulrich and T. Baker, Computer (7), 39-44 (1974).

18. D. Schuler, T. Baker, R. Fisher, S. Hirchhorn, M. Hommel, H. McGinness, R. Bosslet, 1979 Test Conference, IEEE Computer Society, 203-207.

19. D. Schuler and R. Cleghorn, Proceedings 14th Design Automation Conf., IEEE Computer Society, 230-238 (1977).

20. M. Abramovici, M. Breuer, K. Kumar, Proceedings 14th Design Automation Conference, IEEE Computer Society, 128-137 (1977).

21. J. Roth, IBM Journal of Research and Development (10), 278-291 (1966).

22. T. Snethen, Proceedings 14th Design Automation Conference, IEEE Computer Society, 88-93 (1977).

23. T. Snethen, U. S. Patent 3,961,250 (1976).

24. P. Goel, Fault Tolerant Computing Symposium (1980).

25. K. Bowden, 1975 IEEE Intercon Conference Record, Session 15.

26. R. Heckelman, USAF Report RADC-TR-7B-233 (1978).

27. P. Accampo, Electronic Packaging and Production, 186-191 (April, 1978).

28. F. Sellers, M. Hsiao, and C. Bearnson, IEEE Trans. on Computers (C-17), 676-683 (1968).

29. D. Schertz and G. Metze, IEEE Trans. on Computers (C-21), 858-866 (1972).

30. H. Schnurmann, E. Lindbloom, and E. Carpenter, IEEE Trans. on Computers (C-24), 695-700 (1975).

31. J. Hayes, IEEE Trans. on Computers (C-26), 613-619 (1976).

32. P. Agrawal and V. Agrawal, IEEE Trans. on Computers (C-25), 691-694 (1975).

33. J. Shedletsky, Proc. 7th Annual Conf. on Fault Tolerant Computing, 175-179 (1977).

34. C. Cha, <u>Proc. 16th Design Automation Conf.</u>, IEEE Computer Society, 326-334 (1979).

35. J. Smith, <u>IEEE Trans. on Computers</u> (C-28), 845-853 (1979).

36. E. Eichelberger and E. Lindbloom, <u>IBM Journal of Research and Development</u> (24), 15-22 (1980).

37. M. Breuer and A. Friedman, <u>IEEE Trans. on Computers</u> (C-29), 223-234 (1980).

38. P. Goel and B. Rosales, <u>1979 Test Conference</u>, IEEE Computer Society, 189-192.

39. P. Bottorff, R. France, N. Garges, and E. Orosz, <u>Proc. 14th Design Automation Conf.</u>, IEEE Computer Society, 479-485 (1977).

40. A. Yamada, N. Wakatsuki, T. Fukui, and S. Funatsu, <u>Proc. 15th Design Automation Conf.</u>, IEEE Computer Society, 347-352 (1978).

41. T. Williams and K. Parker, <u>Computer</u> (12), <u>10</u>, 9-21 (October, 1979).

42. R. Rasmussen, E. Hsieh, L. Vidunas, and W. Davis, <u>Proc. 14th Design Automation Conf.</u>, IEEE Computer Society, 486-491 (1977).

43. J. Lesser and J. Shedletsky, <u>IEEE Trans. on Computers</u> (C-29), 235-248 (1980).

44. M. Breuer, <u>IEEE Trans. on Computers</u> (C-22), 241-245 (1973).

45. D. Ostapko and S. Hong, <u>IEEE Trans. on Computers</u> (C-28), 617-626 (1979).

46. FALTGEN was programmed by R. A. Rasmussen (1966).

47. S. Kamal and C. Page, <u>IEEE Trans. on Computers</u> (C-23), 713-719 (1974).

LAYOUT AUTOMATION BASED ON PLACEMENT AND ROUTING ALGORITHMS.

W. Sansen, W. Heyns[*], H. Beke[**]

Katholieke Universiteit Leuven
Departement Elektrotechniek, afdeling ESAT
Kardinaal Mercierlaan 94, 3030 Heverlee, Belgium

ABSTRACT. The advent of VLSI has necessitated the use of layout automation to generate the masks. Whereas in earlier small and medium scale circuits, automatic layout merely served to shorten the layout process from months to days, it now has become a real necessity to cope with the ever increasing complexity of VLSI.

The automatic generation of masks heavily depends on the technology and the design philosophy. The automatic layout of a linear bipolar chip is a planarity problem whereas for MOS-LSI a normal placement and routing task has to be carried out. Present day master slice layouts on the other hand require a placement with a very limited number of alternatives in combination with a conventional routing. Regular MOS-structures do not require automated routing at all. The first paragraph deals with all intermediate philosophies of layout automation from manual layout to fully automated layout and provides a general introduction on placement and routing algorithms.

In order to demonstrate the capabilities of placement and routing algorithms, an analysis is presented of the CALMOS program [1] and the CAL-MP program [2] which is the result of further development of the CALMOS program. These software packages have proven to be very successful for the layout of MOS-LSI as they are in use in several companies in Europe, USA and Japan as well. Although both programs make use of a quadratic placement algorithm followed by a fast channel router which guarantees full 100 % completeness of all interconnections, the many differences do not allow to consider CAL-MP as just an extension of CALMOS. Both pro-

---

[*] IWONL Fellow.
[**] Leuven Industrial Software Company.

grams and the algorithms that have been selected for them, are discussed in the second paragraph.

The third and last paragraph will focus on layout automation for VLSI. It will discuss the hierarchical design philosophy and introduce general cells. It will also discuss some new algorithms which were developed especially to handle general cell assemblies. These algorithms can be used in most future automated layout systems for VLSI.

I. GENERAL INTRODUCTION ON LAYOUT AUTOMATION AND PLACEMENT AND ROUTING ALGORITHMS.

I.1. Types of layout automation.

The term layout automation has been abused by lack of insight in the degree of assistance provided by the computer. Manual layout, one extreme, does not require any computer. Automatic layout, the other extreme, allows no human interaction. In the latter all decisions are taken by computer software which provides well-defined and predictable decisions. Between manual layout and automatic layout exists a whole range of layout philosophies in which the degree of assistance and intelligence by the computer are of increasing order.

The evolution from manual layout to automatic layout can be marked by the following types :

a. Manual layout : The designer carries out the drawings himself. The functionally related blocks are put together. The interconnection is not given priority. Very dense layouts result but at the expense of an enormous design time and high risk of errors. The use of a digitizer and/or graphic system will always shorten the design time and is thus highly recommandable.

b. Manual layout with the use of a digitizer : The coordinates of the figures are manually read in a computer which provides a precise drawing on a plotter. The drawings are precise but the conversion is again time consuming and prone to errors.

c. Manual layout on a graphic system : The system is used as a drawing aid such that a manual layout can be realized in a relatively short time with much less errors than before. Some degree of design verification (e.g. tolerance checking) can be built in. The correction of errors is easy and takes little time.

This type of layout is still a full manual layout with as much computer support as possible. It still provides the densest layout at a considerable design time cost. For large product quantities this type of layout is preferred because of the lowest chip area obtained.

d. Interactive automatic layout (computer aided layout) : This

type of layout involves the use of a graphic system to which automatic layout routines (e.g. placement and routing) are added to
aid and even to replace the designer. This system is interactive
because the designer can always repeat any step of the layout.
However the designer still (interactively) supervises the automatic (placement and routing) procedures. On the other hand he is
relieved of much of the repetitive and routine tasks associated
with the design of a complex layout.

This type of layout is the best compromise between the added
computer intelligence and the flexibility kept with the designer.
On the other hand extensive design rule checking is required after the interaction between designer and computer.

    e. Automatic layout on mainframe computers : Procedures are
devised to carry out an automatic layout. These procedures are
based on trial-and-error techniques and thus require considerable
computing time. The user does not interact with the system but
either accepts the output or has to try a different input. The
computer time is kept short by use of only very simple procedures,
which are repeated a large number of times.

    f. Automatic layout : In this type of layout full use is made
of the analyzing and decision making capabilities of the computer.
Dedicated software is applied to automatically process the design
through each step from raw logic data to final routing. The output is quite predictable in performance. Layout rules are never
violated. Checking is thus superfluous.

I.2. Methodology of layout automation.

    Introduction. Layout automation is a direct consequence of
the ever increasing design cost with circuit complexity. This cost
is based on two factors : the design time and the number of errors per design cycle. Therefore layout automation is aimed at
the reduction of both.

The design time has been shortened by :
- the use of standard cells with elementary logical functions
  [1], [2], [3].
- the implementation of layout automation on several hierarchical
  levels [4], [5].
- the use of highly regular structures such as PLA's, ULA's, etc.
  [6].

These different approaches will be commented on. A discussion
on the efficiency of layout automation in general is also presented.

Different approaches towards layout automation.

Standard cell approach. For designs using the same basic building blocks over and over again, automatic placement and routing

have provided drastic reductions in design time by introducing
the notion of standard cells. Most of the time is actually spent
in the design and maintenance of the standard cell library. As a
consequence this approach is the right one only if many designs
have to be realized in the same technology. If not the cells can
better be replaced by their symbolic equivalent. The translation
into a physical layout by means of the design rules is then carried out after the application of layout automation. Although
this latter approach seems to be advantageous it poses much more
problems with respect to design rule checking. The use of standard cells followed by full automatic placement and routing makes
design rule checking superfluous. Manual (interactive) intervention or symbolic layout do necessitate design rule checking.

On the other hand the degree of interactivity or manual intervention depends on the degree of personalisation by the designers. High level designers with a long tradition cannot accept
the result of a fully automatic software package. They personali-
ze i.e. they improve the design at the risk of more errors i.e.
additional expenses.

Whether physical standard cells are to be used or symbolic
cells is clearly a point of discussion as well as whether fully
automatic layout is preferred above interactive layout. All combinations are in use depending on the specific design task. The
second paragraph will focus on fully automated layout of physical
standard cell designs in which any level of interactivity can be
introduced.

Hierarchical layout. For VLSI, where the number of gates can
be as high as a hundred thousand, the humain mind has to resort
to tools of simplification. Of course he could rely on fully automated computer power but he prefers to keep a grasp on the design
process in order to apply his cleverness (and/or experience) wherever possible. The most obvious method of simplification is the
subdivision into several hierarchical levels of layout.

The circuit is partitioned on a functional basis into smaller
subcircuits or blocks. Each of them then can be laid out separately or eventually partitioned into yet smaller subcircuits. If the
blocks are always equivalent to standard cells on the higher level, then the same methods can be applied every time on each level to obtain the layout of the full circuit [4]. In this way
complex circuits can be laid out relatively easy. Since on each
level however the layout will be organized in rows, as is usual
in the standard cell approach, this still imposes some rather severe restrictions on the layout. A more general procedure will be
discussed in the third paragraph.

Regularly structured layout. The use of regular structures
(e.g. logic arrays) can be very efficient with respect to design
time and obtained density. The synthesis of these arrays in com-

bination with a design rule checking is in full development. Structured arrays however will never make up for a full design. They will rather form a block on the lowest hierarchical level to include most of the logic. As a consequence there will always be a need for placement and routing procedures to layout the several blocks with respect to one another. The placement and routing algorithms however are vastly different from the ones used in programs such as CALMOS and CAL-MP. The blocks can take any size and shape. They can have connections on all sides. Also routing is no more confined to channels between rows of cells. The interaction between placement and routing is much stronger, which complicates the predictability of the result. Some algorithms to handle these problems will also be discussed in the third paragraph.

The efficiency of layout automation. Layout automation is especially efficient for plain digital circuitry where :
- no analog parts or interface circuits are present.
- several reruns can be anticipated (e.g. because of changes in specifications) and the last one can be finished in an interactive way.
- violation of the design rules is never accepted.
- similar designs have been realized before, such that the output is predictable in performance.

Full design automation can then be realized if the digital circuit is made up of standard cells consisting of elementary logical functions. The best examples are n-MOS and CMOS logic, for which cell libraries exist. The cells are interconnected on two levels which are usually preferential i.e. all vertical lines are made up with aluminum whereas all horizontal lines are made up of a low-resistivity diffusion.

For a very large number of cells (> 500) the ratio of interconnectivity area to the total area increases. Therefore for large circuits it is preferable to layout a larger number more complex cells. This situation is depicted in Fig. 1.

A circuit of given complexity can be laid out by means of ten large cells with total area BC. The cells have all been laid out by hand. They are relatively dense but have required a long development time. The silicon area required to interconnect the ten cells is very small (AB). The total chip area is AC. If the ten large cells are split up into elementary cells such that one thousand of them are needed to realize the same circuit functions, then much more area is required to carry out the interconnections (A'B'). On the other hand the cells can be laid out by hand and be made very dense. The total cell area is therefore much smaller (B'C'). The optimal solution however is somewhere in between. Solution ABC is a typical manual solution whereas solution A'B'C' is a typical design automation solution. Solution ABC takes much more development time than solution A'B'C'. Moreover solution ABC remains unique : no designer would try e.g. a solution with 20

cells in search for the optimum. By means of design automation
only slightly more complex cells have to be designed or some of
the cells have to be taken together in order to provide a smaller
layout. Thisonly requires a minor design time. Proceeding in this
way will lead to the optimal solution.

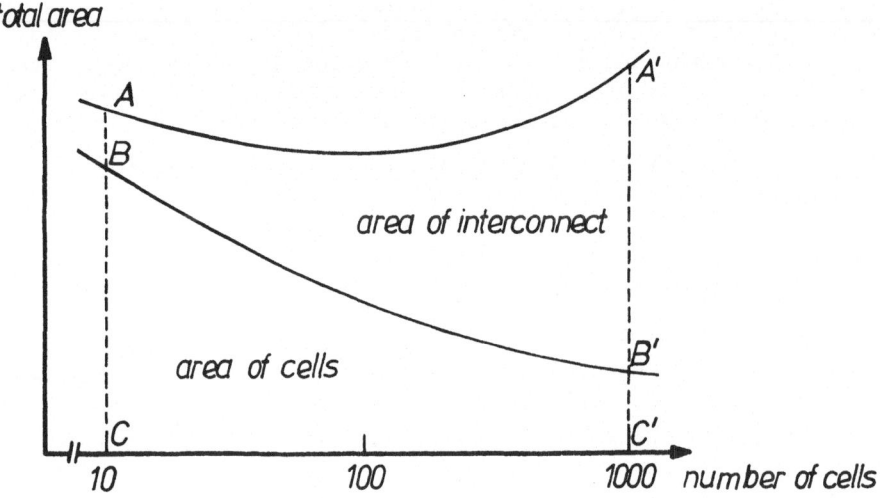

Fig. 1. The change in total chip area (specified as the sum of
cell area and interconnect area) in function of the number of cells for a circuit of given complexitiy.

This figure also illustrates the advantages of hierarchical
layout and structured layout. In a hierarchical layout philosophy
the large cells are laid out by use of design automation themselves, which leads to a reduction in BC. In structured layout the
interconnections are superimposed or laid out first in an array,
which leads to a reduction in A'B'. How the total area changes is
an open question. No optimum may exist.

I.3. Layout automation algorithms.

Introduction. We will restrict ourselves to placement and routing algorithms since these are of central importance in finding
an optimal layout. Topics that are not going to be discussed here
are for instance :
- partitioning of the circuit in several functionally related
  blocks.
- circuit verification in which the logic diagram is reconstruc-

ted from the layout and compared with the original logic diagram.
- design rule checking in which the geometrical design rules are
verified.

In the standard cell approach the partitioning has been carried out to an extreme point such that only elementary logic
functions result, the circuit is to be verified by means of the
network description and normally no design rules are ever violated.

Placement algorithms. Placement or mapping is the placement
of the different modules on a chip. It is the first design step
which involves the actual physical layout. The main criterion is
ideally the minimization of the difficulty in subsequent routing.
However this goal is rarely met in practice. Moreover other factors such as crosstalk between routing channels, capacitive loading, etc., often complicate the placement function.

In a linear placement procedure [7], all modules are put on
one row such that the total one-dimensional length of all interconnection links is minimal. In quadratic mapping [8], [13] the
total sum of two-dimensional distances between a module and all
his interconnected neighbours, has to be minimal. Since the latter
procedure is used in CALMOS and CAL-MP, it is discussed more in
detail. The min-cut placement [9], also discussed here, distinguishes itself from the previous procedures in that not the interconnection lenght is minimized but the number of "signal-cuts".

Placement by force-directed pairwise interchange. The placement problem can actually be defined as follows. Given a number
of modules (blocks, cells, ...) with signal nets interconnecting
them and a number of slots with the rectilinear distance defined
over all pairs of slots, the placement then consists of the assignment of the modules to the slots such that the sum of the minimum spanning tree distance over all the signal sets is minimum.
The quadratic assignment problem differs from the placement problem in that the distance measure is taken on pairs of points instead of sets of points (i.e. the signal sets). It is defined by
the minimum of :

$$\sum_{ij} c_{ij}\, d_{p(i),p(j)}$$

over all permutations p where $C = [c_{ij}]$ is a cost matrix and $D =
[d_{kl}]$ is a distance matrix. The simplest way to transform the
placement problem into an associated quadratic assignment problem
is to choose c_{ij} equal to the sum of the signal sets common to modules i and j for $i \neq j$ and $c_{ii} = \emptyset$, and choose the distance matrix D equal to the distance matrix of the original placement problem. The computation time for an algorithm for the quadratic assignment problem will be considerably less than the corresponding
time for the same algorithm running on the placement problem.

For large systems as in VLSI, it was shown [8] that the only viable combination of placement algorithms is based on the force-directed pairwise relaxation algorithm. It not only achieves the best placement but it also achieves it in the shortest computer time.

Pairwise interchange is a simple algorithm consisting of a trial interchange of a pair of modules after an initial (constructive) placement. If this interchange results in a reduction of the distance, the interchange is accepted. Otherwise the modules are returned to their initial positions. All $n(n-1)/2$ possible pairs of modules are trial interchanged in a complete cycle (if there are n modules and n slots). In the force-directed pairwise relaxation algorithm however for each module M a force vector is computed

$$\overline{F}_M = \sum_i c_{Mi} \, \overline{s}_{Mi}$$

where c_{Mi} is the edge weight connecting modules M and i, and $\overline{s}_{Mi}$ is the vector distance from M to i. This force vector allows to compute the target point i.e. the point where the sum of the forces on module M is zero. The closest slot to this point is then the target location. It is a kind of gravity center of the modules considered for a trial interchange.

A module A is now chosen for a trial interchange with a module B located in the neighbourhood of the target location of A, only if the target location of B is in the neighbourhood of A. Again the trial interchange is accepted if it results in a reduction in distance (cost). This procedure clearly limits the number of trial interchanges. Because of the computation of the force vector only a limited set of modules is considered for an interchange with a selected module. Of this set only these modules are trial interchanged of which the target point is located in the neighbourhood of the module selected for an interchange. In fact we try to optimize the placement of two modules at once. Computational speed is therefore enhanced considerably.

The placement algorithm also shows some disadvantages. The resulting placement is such that all larger modules with many interconnections are clustered together. The algorithm even does not perform at all if the sizes of the modules are largely different. The first problem can be avoided by manual intervention, the latter problem can be solved by replacement of a large module into smaller ones.

Min-cut placement. In the min-cut algorithm the classical objective functions based upon distance are abandoned in favour of an objective function based upon "signal cuts". This objective was motivated by two observations, namely :
- successfully routing is dependent on the density of interconnections.

- some areas are usually more dense than others.
The min-cut algorithm tries to minimize the number of interconnections crossing a cut-line and hence the density in that region.

Consider a substrate represented in fig. 2 by block B and a
set of modules which are to be assigned to slots. We first process cut-line c_1 which divides block B in the blocks B_1 and B_2.
If a signal is connected to a module to the left of c_1 and to a
module to the right of c_1, then this signal must cross cut-line
c_1. By choosing a proper assignment of the modules to the blocks
B_1 and B_2 the number of signals that crosses the cut-line can be
minimized. This procedure can be repeated for cut-line c_2, which
divides B_1 in the blocks B_1 (a new B_1) and B_3, and B_2 in the
blocks B_2 (a new B_2) and B_4. Note that once a module has been assigned to the left (right) of c_1, the module can never be moved
to the other side of c_1 later on. The algorithm stops when each
block contains just one component i.e. when the components can be
assigned to a specific slot instead of a set of slots as during
the min-cut procedure.

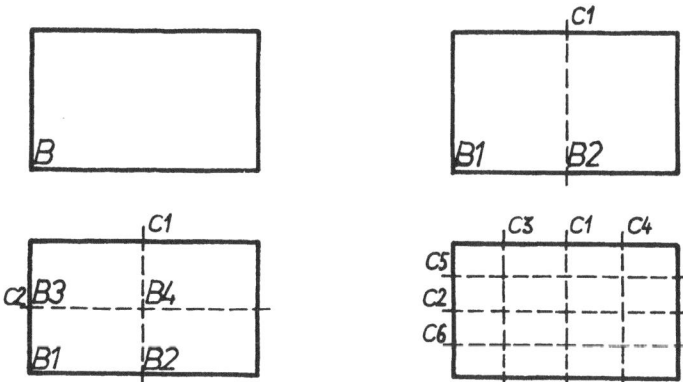

Fig. 2. Min-cut algorithm.

Routing algorithms. Routing is the hardware realization of
all the interconnections between the modules. This step thus
leads to the layout of the conductor paths (aluminum) to achieve
the listed connections subject to the imposed constraints. These
constraints are related to the number of interconnection levels,
restrictions in minimum line width and spacing, number and location of vias (feedthroughs), available routing space, etc.

The greatest challenge of a DA system is a 100 % routing
without time consuming manual intervention. This is why routing
has intensively been studied. The most successful routers are
those of Lee, Hightower and Hashimoto-Stevens. The wavefront or
grid expansion router of Lee [10] is the most general one but it
is time consuming. It routes sequentially which reduces its capability to route all the interconnections. Hightowers router is a

line search router [11] which also routes sequentially but consu-
mes much less CPU time. The router of Hashimoto and Stevens [14]
is a channel router which is directly aimed to perform a 100 %
complete routing. It achieves this by trying a global solution.
All later channel routers (branch and bound, [16], dogleg, [17],
etc.) are derived from it.

All these routers are meant to solve an interconnection problem on two interconnection levels. They are the most frequent
ones. Single level interconnections (e.g. in linear IC's) hit
planarity problems whereas three and more levels are technologi-
cally less attractive.

The routers described here (because of their importance) are
Lee's algorithm and Highthowers line search algorithm.

Grid expansion algorithm of Lee. In the Lee algorithm a series of successive wavefronts are generated from the starting
point towards the target point. (Fig. 3). If the target point is
reached it is easy to trace back the solution.

7	6	7	8					
6	5	6	7	8				
5	4	5	6	7	8	x		
4	3	4	X	8				
3	2	3	X					
2	1	2	X	X	X	X	X	
1	x	1	2	3	4	5	6	7
2	1	2	3	4	5	6	7	8
3	2	3	4	5	6	7	8	

Fig. 3. Wavefront generation in the Lee algorithm and trace-back
when the target point is reached.

The algorithm is so popular because it always guarantees to find
a solution if one exists. It is therefore often used to complete
the last remaining interconnections that were left over by some
ohter routing algorithm. It is only used afterwards because it is
very memory and time consuming. Many routing algorithms were derived from the original algorithm. They all try to speed up the
search for an interconnection by enhancing the search in the direction towards the target point. In general the cost-function is
considered in the Lee-algorithm can be a weighted sum

aL + bA + cK

where a, b and c are nonnegative coefficients, L is the path
length, A is an adjacency penalty (for wire-spreading purposes)

and K is the number of wire crossovers. If b = c = $\emptyset$ then the shortest path will be found.

Hightowers line search algorithm. The main advantage of the line search routing algorithm developed by D. Hightower is that it is very fast compared to the grid expansion technique of Lee. It does not guarantee to find the shortest path available or even to find a path at all. It somehow tends to find a path with the minimal number of bends.

The algorithm works as follows. A path has to be found between points A and B (fig. 4). Two perpendicular line segments are first constructed through point A extending in each direction from boundary to boundary. The algorithm then tries to find an escape point such that an escape line, which is constructed through it, will

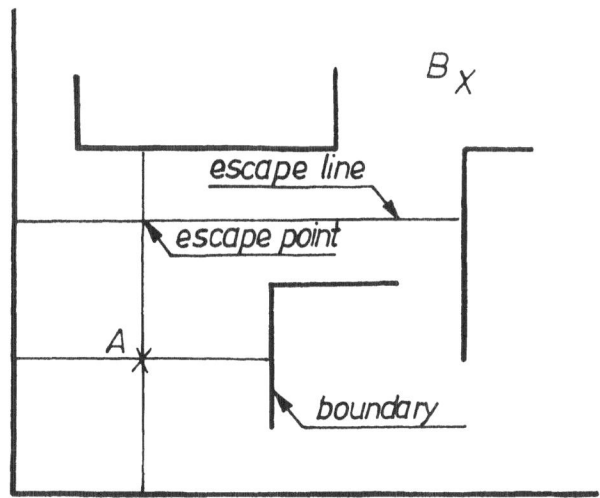

Fig. 4. Line search algorithm based on escape point.

extend beyond one of the previous boundaries of point A. If such a point is found, it becomes the new point A and the process can be repeated. The algorithm alternatively starts from points A and B until a line segment associated with A crosses one from point B. A path is then reconstructed consisting of all the line segments connecting the escape points from A to the intersection and from the intersection to B. A suitable data structure is to be found which makes the number of searches minimal and allows the algorithm to run at very high speed.

The same philosophy of the escape point can be applied to a full escape line as in the line-expansion algorithm [12]. This provides an even higher efficiency in computer use. In fact this algorithm guarantees that a solution will be found (if one exists) and that this solution is the one with the least bends.

In one of the earlier MOS-LSI layout programs [15] a line-search router was used to route the channel between the rows of cells. If at some point no path could be found between some points of a net, a track was added to the channel. In this way a 100 % complete solution could be achieved. Although nowadays most MOS-LSI layout programs use a channel-router, the latter principle of adjusting the width of the channel to its needs, has remained. The channel-routing algorithm will be discussed later.

II. ANALYSIS OF CALMOS AND CAL-MP.

II.1. Introduction.

CALMOS and CAL-MP are computer systems for the layout of custom MOS/LSI circuits. Starting with a standard cell library and a simple circuit connectivity description, the programs perform various automatic and/or interactive procedures such as initial placement, assignment of equivalent and equipotential pins, optimization of the placement, prerouting, routing, routing compression, fan-in fan-out and circuit verification.

The database has reentrant properties such that a designer can step through the system and try out different possibilities. After each step the results are immediately available and can be compared with previous outputs. This on-line optimization avoids multiple rerunning of the task and will also yield a better chip minimization.

The program can be run in both batch and interactive mode. In batch mode the interactive commands are still acceptable as batch-commands. The program provides interaction in editing an existing placement.

Thanks to the use of a MORTRAN preprocessor (giving a FORTRAN program as output), program limits are adjustable to the memory size of the computer on which the program has to be installed, e.g. on VAX 11/780 the maximum number of cells is 9000.

CALMOS/CAL-MP can easily be installed on almost any computer system with a FORTRAN compiler.

II.2. Advantages.

a) of both programs. The efficiency of CALMOS/CAL-MP is based on some specific properties and several new algorithms which are listed below.
- It contains a "quadratic placement" algorithm and provides easy preplacement (absolute or relative) in macro.
- The channel routing algorithm always provides a 100 % complete routing (cyclic constraints are removed by a set of algorithms or by the user).
- Several algorithms are implemented to measure and optimize the

"quality" of placement and routing.
- Automatic routing compression is provided by a separate post processor.
- The interactive mode allows easy and immediate correction, avoiding superfluous computations and resulting in a better minimization of the silicon area.
- Program limits are adjustable to computer memory size.
- The definition of macro-cells still extends the maximum number of simple cells, and simplifies the network description.
- CALMOS/CAL-MP is maintained, supported and continuously extended and updated by an industrial software house.

   b) of CAL-MP vs. CALMOS. As in CALMOS, cells must be rectangular and of nearly equal height. But whereas I/O-pins had to come out at the bottom side of the cell, in CAL-MP they may also come out at the top.

   This extension has many advantages :
1. The cell designer can draw smaller cells.
2. Often it doesnot require extra space to let a I/O-pin come out both at the top and at the bottom of a cell, thus creating an "internal feed-through". This feature decreases the need for space consuming extra connections through cell rows (i.e. external feed-throughs).
3. If cells are designed in a symmetrical way (symmetrical power and clock lines), they can be swapped, i.e. mirrored around the X-axis. Both mirroring (around the Y-axis) and swapping can be allowed or inhibited by simple parameter setting. If both are allowed, the program will automatically choose the best of the four possible positions, thus minimizing routing area.

   The use of standard cells with I/O-pins at two sides results in 5 to 10 % less chip area compared to CALMOS.

II.3. System overview.

   Library creation. Using CALMOS/CAL-MP requires the creation of standard cells on a graphic system. Although a complete internal description is necessary for the final artwork CALMOS-CAL-MP needs to be fed with the external characteristics of a cell. (e.g. width, height, number and definition of IO-pins, fan-in fan-out specifications, ...). An interactive program LIBBLD allows the user to set up such an "external library".

   The same program allows for the creation of three different types of MACRO cells out of existing basic cells. A BLOCK macro behaves as a new basic cell. A CHAIN macro behaves as one cell during placement but allows feedthrough interconnections to pass between basic cells (during routing). A CLUSTER macro behaves as a chain macro but the basic cells of the macro can change their relative position in the cluster (during placement).

Input.CALMOS basically needs two inputs : an "external library" and a network description in terms of cells belonging to that library. The network is cell oriented. Assignment of pins to specific equivalent or equipotential pins is possible. Without user specifications, the assignment is optimized automatically.

Placement of the cells. In order to get enough computational speed, the quadratic placement problem is divided into 3 consecutive steps : global clustering, local placement within every cluster and overall optimization of cell placement, net-to-channel assignment and pin assignment. The global placement [3][13] distributes the cells over a limited set of clusters. Each cluster is defined by its physical location on the chip, its capacity (maximum cell area available in the cluster) and its accessability for different cell types. Since CALMOS/CAL-MP uses channel routing, the clusters must be positioned in rows. The local placement calculates the relative positioning of all cells belonging to one cluster. It can take into account relative prepositioning of cells within a row. [5]

Both the global clustering and the local placement are realized with an initial constructive placement followed by a force directed pairwise interchange algorithm [8], minimizing total netlength.

The initial constructive placement can be done manually or automatically, taking into account the interconnectivity of the cells. If automatically the cells are not placed all at once, but gradually, each time followed by an optimization step. This decreases the chance for obstruction of a cluster. Obstruction is the situation where a cell cannot be placed in the optimal cluster, because of a shortage of space in this cluster.

The force used for the optimization algorithm (a force-directed pairwise relaxation algorithm) is not a real force but derived from a direct calculation of the change in total routing length by moving a cell from one position to another. The length of a net is defined as half the perimeter of the circumscribing rectangle.

An example is given in Fig. 5. We have drawn the evolution of total routing length resulting from a move of cell C along the X and Y axis. It is assumed that cells A,B and C are interconnected with a net. As long as C remains in the circumscribing rectangle the netlength is not changed. When the circumscribing rectangle decreases (increases) by a move of C, then, by definition, the netlength decreases (increases).

This calculation is done for every net of a cell. For each cell, the results of all nets are added together. This defines three regions:when moved to one of these regions the total routing length will increase, stay unchanged or decrease. These regions are used in the force directed pairwise relaxation algorithm to control the pairwise interchange.

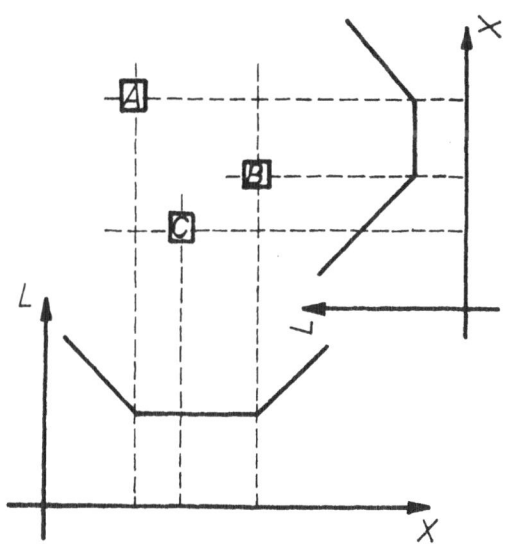

Fig. 5. Evolution of routing length.

The placement ends with an overal optimization of cell placement and the net-to-channel assignment and pin assignment which are done taking into account :
- the minimization of the use of feedthrough-cells.
- the optimization of the place of these feedthrough-cells.
- the minimization of the track density in each channel.

The user can continuously store and edit placement configurations, and change the default starting conditions of the algorithms. Any modification (by hand or through algorithms) of a previous placement is automatically followed by a new estimation of final chipsize for that placement. This allows the designer to combine interactive and automatic sessions towards a final optimal placement solution.

Routing. The routing of interconnections is achieved with a heuristic routing algorithm [14], [16], [17]. Routing is done on two distinct levels with all horizontal paths being assigned to one level and all vertical paths to the other. Connections between the levels are made through contact windows. A single net may result in many horizontal and vertical segments. By restricting ourselves to a gridded channel, we have been able to introduce a special database with ringstructures, making all characteristics of each net easily accessible, and thereby making routing very fast.

480

<u>Grid calculation</u>. An important factor in using a gridded router is the choice of the grid.

The channel router works on a grid in both the vertical and horizontal directions. The gridsize is defined by the layoutrules. Normal routers have to work with symmetrical contact windows. In that case we have a minimal gridsize (Fig. 6.a).

$$GS = Max \quad / \quad xTOx + \frac{xCON}{2} + \frac{xCON}{2} \quad /$$

(x = metal, poly, dif.).

CALMOS and CAL-MP allow for the use of asymmetrical windows (Fig. 6.b). Now we can work with a grid

$$GS = Max \quad / \quad xTOx + \frac{xMIN}{2} + \frac{xCON}{2} \quad /$$

xMIN is always less than xCON. Thus the last gridsize is smaller. Using this grid for the layout of the basic cells and for the interconnections will already yield smaller chip sizes. Moreover, a special routing compression algorithm will reduce the total channel height by using non gridded horizontal tracks, causing a further decrease of total chip size. The effect of the compression is dependent on the used layout-rules. For some technologies however the decrease of routing area can reach values of 25 %.

Since the channel router is heuristic and only looks at the total horizontal track count, a second post processor will run a routine to minimize the total vertical interconnection length (for the same horizontal track count and the same horizontal interconnection length).

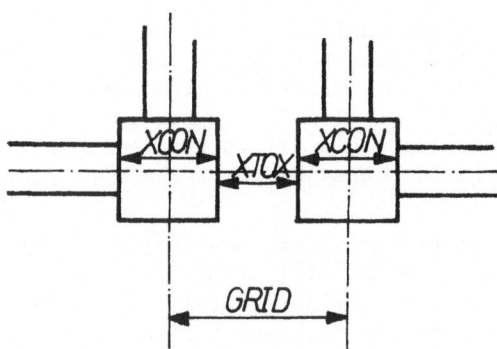

Fig. 6.a. : Minimum grid for symmetrical windows.

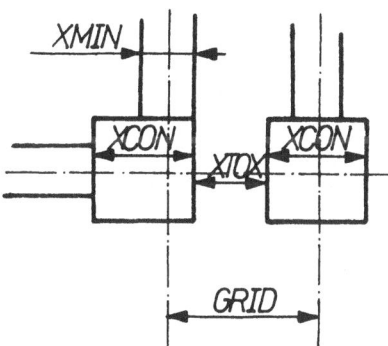

Fig. 6.b. : Minimum grid for assymmetrical windows.

The channel-routing algorithm. Assume that the channel to be routed is rectangular, and that all input-output pins are situated on grid points along the upper or lower side of the channel. A net consists of a certain number of pins that must be interconnected via some routing path.

Horizontal routing segments are placed on one level (usually metal) while vertical segments occur on another level (usually polysilicon and/or diffusion). In order to interconnect a horizontal and vertical segment, a contact must be placed at the intersection point.

The easiest way to solve such a routing problem is by the so-called "Left-Edge"-channel routing algorithm [14]. One starts searching from the left side of the channel for a starting pin of a net. In Fig.7.a the first starting pin we encounter is of net A (at the bottom). We now assign net A to the first track in the channel, and layout the complete net. (Fig. 7.b).

We then continue the search for a starting pin from the point where the first net ended, and encounter the starting pin of net C next. If we layout C in the way the dotted line shows, we prohibit net B from being routed. In trying to route B the program will notice this vertical constraint, and it will conclude that another order must be tried. We rip off net C, and continue. Since no net starts to the right of C, we restart from the left side of the channel for the second track.

The first starting pin we encounter is now of net B, therefore net B is assigned to the second track and laid out. (Fig. 7.c).

At last, on the third track, net C is laid out. (Fig. 7.d).

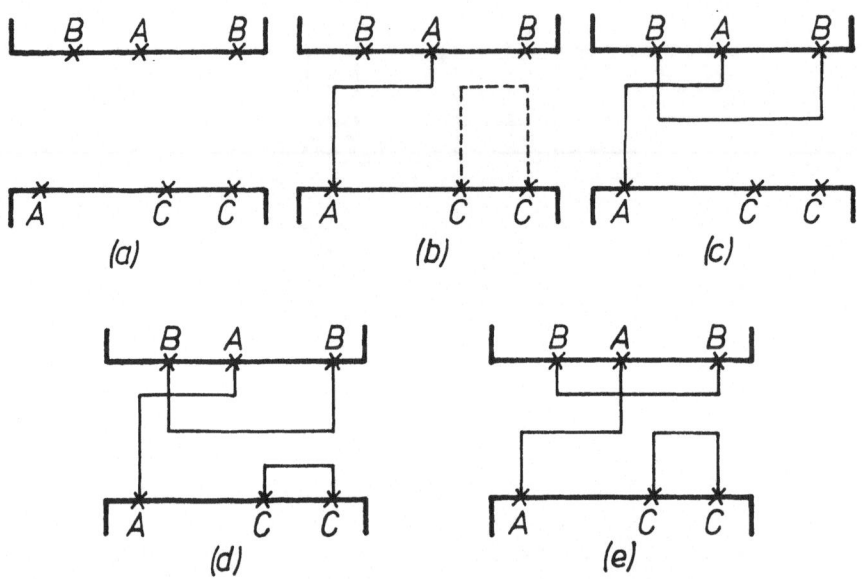

Fig. 7. : The channel routing algorithm.

The important part of the algorithm is the order in which the nets are laid out : for the same problem, a solution requiring only two horizontal tracks is found when starting from the right side of the channel (Fig. 7.e).

There are two practical possibilities :
- Branch & bound (trying every possible arrangement and compare it to the best result so far) gives always the best solution, if run to the end, but usually this is so time-consuming that it is stopped after a given number of trials.
- The heuristic approach (trying a limited number of strategies, such as upper left edge, upper right edge, lower left edge, etc.) has a very good chance to find a better solution than the branch & bound algorithm, in the same time. The channel router implemented in CALMOS-CAL-MP is a heuristic one.

Cyclic constraints. A typical difficulty for channel routers is the solution of "cyclic constraints" [14], [16]. Such constraints, if not removed, force the channel to unroutability. For instance if nets AD and BC (Fig. 8) are to be connected as shown, one of the nets can never be routed with a single horizontal section.

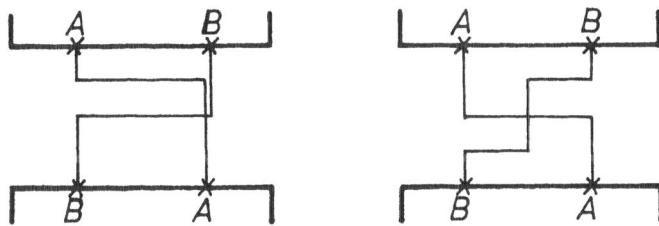

Fig. 8. : A cyclic constraint and its solution.

As clearly shown, it is sufficient to break the horizontal section of AD in the constraint cycle into two horizontal segments to be able to carry out the routing. The result is an additional vertical jump in some column. The two horizontal segments can be treated as two independent nets as long as they are connected to the same column jump.

CALMOS and CAL-MP contain four algorithms to remove cyclic constraints. Without user intervention they are executed in the given order :
1. Interchange of equivalent and/or equipotential pins.
2. Mirroring of cells.
3. Interchange of cells within a row.
4. Forcing the net to split into several horizontal pieces (dog-legging).
5. Editing by the user i.e. the user can in all ways change the cell placement to overcome persistent constraints.

Outputs. The ultimate of CALMOS/CAL-MP is a "general magtape", containing all necessary information to create a set of masks (on the graphic system). The output consists of readable (ASCII)for-matted records and is translated into the graphic systems databa-se with an interface program.

If the system includes a plotter, this general output file can be used to produce a quick checkplot. If a plotter is not available, symbolic layouts of all the routing channels can be out-put to the systems line printer.

These layouts not only visualize placement and routing but al-so some typical factors that are important for iterative and in-teractive optimization as for instance the critical zones where we have routing congestion. Normally the user will run CALMOS/CAL-MP several times. After every run he carefully studies the symbolic

484

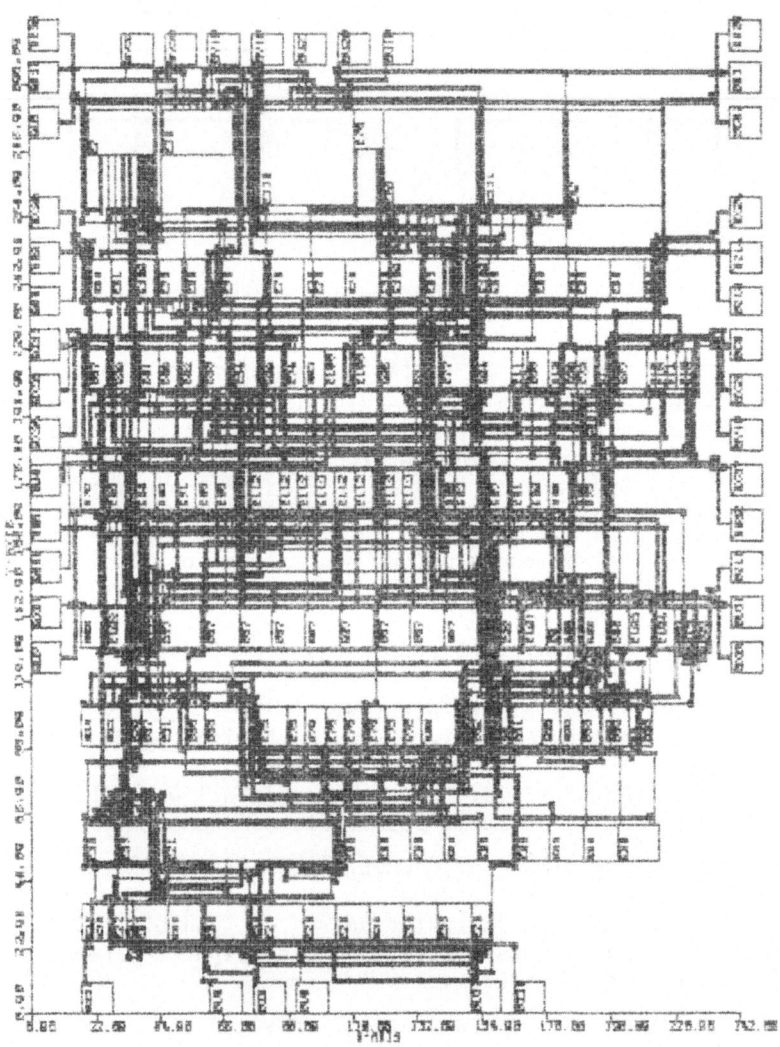

Fig. 9. : Example of CAL-MP plot output.

layouts and tries out possible editings of the placement. Every modification is automatically evaluated through an estimation of final chipsize. After some major editing, the router recreates the total chip layout and the associated new symbolic or graphic layouts.

Another important result of CALMOS is the calculation of capacitive loads. Actually this data can be used offline as input for a logic simulator. An online coupling between CALMOS/CAL-MP and the hybrid simulation program DIANA [24], [25] is under development. Such coupling will complete the interactive optimization of the layout (minimizing the chipsize) with a new and higher level iteration process (layout-simulation-layout ...) to optimize the performance of the chip.

Another important factor is the runtime of CALMOS/CAL-MP. The example of Fig. 9, containing 200 cells, was realized in 5 iterations on a VAX 11/780. A first batch run (no user intervention) took about 2 min. of CPU time. Afterwards the designer performed 4 interactive sessions to minimize chipsize. The complete design cycle required about 10 min. (total) of CPU time.

Although the program can handle cells of varying height, the use of big building blocks (ROM, RAM, PLA) as basic cells will highly degrade the routing efficiency and the overall chipsize. If such blocks are necessary, practice has proven it is better to preplace them in the corners of the chip and to reorganize the definitions of global clusters for the placement of normal cells. The routing towards these cells also causes some difficulties : the router actually only handles IO-pins coming out at the top and bottom side.

III. LAYOUT AUTOMATION FOR VLSI.

In the standard cell approach towards automated MOS-LSI layout it is always assumed that the cells have equal heights and that the input-output pins are located on the top and/or bottomside of the cells. These restrictions make it possible to perform a placement of the cells in rows with the pins facing clearly defined routing channels. Optimization of the placement can be done very easily by interchanging cells in different rows or in the same row. Routing can be done very effectively with a channel router. A channel router is very fast and can guarantee a 100 % solution. Because of all this programs as for instance CALMOS and CAL-MP can achieve very efficient layouts for designs with a few hundred to some thousand cells in a very short time.

In structured design however large blocks, typically memory structures and PLA's, which were laid out once, are used. It is difficult to use them in a regular standard cell approach. Furthermore the complexity of circuits is still increasing. As we talk of designs with ten thousand to a hundred thousand cells, we will have to introduce new methods and algorithms to handle the enormous amount of data.

In this part we will discuss the hierarchical design methodology and some algorithms for the placement and the routing in general cell assemblies.

3.1. Hierarchical design.

Hierarchical design itself is not such a new idea. Complex circuits can be several orders of magnitude larger than circuits which can be easily comprehended. Therefore the design of such circuits usually starts with a block diagram. Every block in this diagram in its turn is then specified more in detail on a lower abstraction level. This process is continued until the lowest level, for instance the transistor level, is reached. At every level the designer reduced the amount of detail to a manageable size. Much in the same way works the hierarchical design principle for VLSI automated layout.

In the standard cell approach much of the actual structure of the circuit is lost in the final layout. The circuit is designed in terms of functions but implemented by cells. This decrepancy will be even more evident in VLSI designs if the standard cell approach was to be used. In hierarchical design however it is possible to structure the layout so that it is still transparent to the designer.

The functional blocks in the circuit can be thought of as general cells (as opposed to standard cells) of which neither width nor height are restricted and of which the input-output pins are not restricted to any particular side. How the functions are realized is not important at this point, only how they are to be interconnected. The designer now can approximate the size of each (general) cell, see how they fit together and eventually adjust the dimensions until finally the global chip area is optimized. The algorithms to automatically perform a placement of general cell assemblies will be discussed later on.

However we can note here that it would be nice if programs using these algorithms made a distinction between the following classes of general cells :
- cells of which the dimensions are fixed.
- cells for which some alternatives are specified.
- cells for which the minimum area is specified together with the minimum width or height.
- cells which are flexible in one or both directions i.e. when they are gridded inside (as for instance memorie structures and PLA's).

Particularly the last class is of much interest. By adjusting the grid of these cells it is possible to straighten-out busstructures i.e. to aligne the pins on different cells so that they can be interconnected very neatly by straight lines instead of space consuming staircase structures.

Once the layout on some hierarchical level is acceptable, the designer can go on to the next level in the hierarchical design. Every cell on the higher hierarchical level is now to be implemented by lower level cells. Eventually these can be standard cells. In general for very complex chip layouts such as those encountered in VLSI circuits, the designer will have to distinct several levels in his hierarchical decomposition in order to reduce the complexity on each level so as to make it consistent with the available design capability and, what is even more important, to keep the design comprehensive. The hierarchical structure of a design can be represented by a structure tree (fig. 10).

 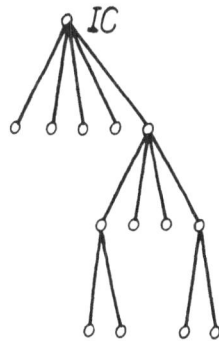

Fig. 10. : Entire layout for an IC and the corresponding structure tree.

Once arrived at the lowest level in this top-down design, the designer has to go bottom-up again to construct the chip. The chip design is complete when the top level is reached. In general however he will have to go through same iterative cycles. It is possible that still during the top-down procedure it becomes evident that he had underestimated the size of a certain cell for one reason or another. It suffices that some output buffers have to be added or replaced by types that can drive a higher load although this could have been anticipated when laying out the higher level. Then he has to apply the top-down procedure again starting from the corresponding node in the structure tree.

It is believed that it is profitable to start with a top-down design instead of immediately going bottom-up. The reason is that optimization of a higher level is more profitable than optimization of a lower level because it affects the total chip layout more directly. In the top-down design for instance it becomes possible to influence the pin positioning on a general cell by the interconnections with other general cells. In this way we can very easily optimize for busstructures. If the pin positioning was dictated because the general cell was to be optimized internally we could end up with weird busses. If however we dictate the pin positions when we layout the cell, the layout of the cell will be automatically adjusted so as to be optimal for that pin positioning.

III.2. Algorithms for placement and routing in general cell assemblies.

Introduction. By using a hierarchical design method we can reduce the complexity on each level to be consistent with the available design capability. We introduced the "general cell" i.e. a cell of which neither the dimensions nor the positioning of the pins are restricted in any way. The algorithms developed in the standard cell approach for the placement and the routing can not be used as such for general cells. In [18] an interactive system is described for the computer aided layout of thick film circuits. The problems encountered there look very much the same as for general cells. There the placement is implemented as an interactive program in which some force-directed placement algorithms are available to the user. These result in a relative placement from which the user can interactively obtain a final placement. For the routing a line-expansion algorithm is implemented. This algorithm routes over the whole substrate at once. Therefore it is very difficult to optimize routing locally. Furthermore a 100 % complete routing cannot be guaranteed because it is assumed that the placement of the components is fixed. As a result the routing area is limited and cannot be adjusted dynamically. In the standard cell approach a 100 % complete routing can be achieved because the width of the routing channel can be adjusted.

In recent years new algorithms have been devised which can generate completely automatically a placement and a 100 % complete routing. The latter is rather important because otherwise the user is left with the difficult task to draw the last remaining interconnections. These are usually the difficult ones. Sometimes it can be required to reroute a large subset of the interconnections that were routed. This is a very time consuming job and not totally error-free. It has to be avoided if possible.

Most of the algorithms discussed further on are based on a proper representation of the substrate which makes it possible to have a good global oversight during the placement and the initial phase of the routing, and to optimize the routing locally during the final phase. Furthermore it is possible to fix any coordinates only at the last moment to allow spreading of the cells in order to achieve 100 % complete routing.

Systems using graph representations.

Graph representations. The idea to represent an arrangement of
non overlapping rectangles by a graph has been applied already to
both the layout of integrated circuits and hybrid layout. However
originally all work was based on the concept of planarity. In [19]
[20] independently from the planarity concept, graphs were used to
represent the placement of MOS building blocks.

The layout (Fig. 11.a.) is represented by a pair of mutually
dual graphs $G_x = (V_x, E_x)$ and $G_y = (V_y, E_y)$ (Fig. 11. b. and c.).
These graphs are called respectively the vertical and the horizontal channel positioning graph. They are planar, acyclic directed
graphs containing one source and one sink.

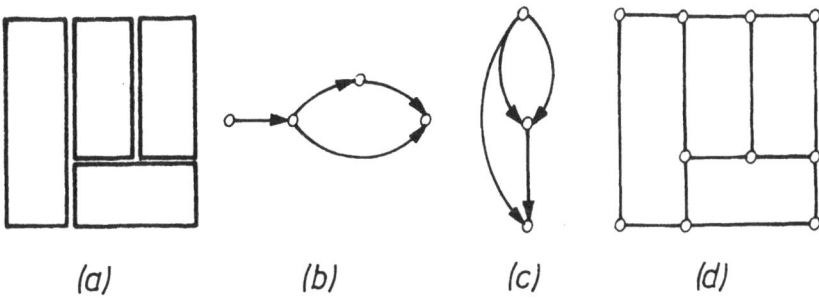

(a)   (b)   (c)   (d)

Fig. 11. : A placement of general building blocks (a) and the corresponding vertical channel positioning graph (b), horizontal channel positioning graph (c) and channel intersection graph (d).

Parallel edges are permitted. There is a one to one correspon-
dance between the edges of G_x and G_y. Each pair of edges (e_x^1, e_y^1)
represents a rectangle with x-dimension 1 (e_x^1) and y-dimension
1 (e_y^1) where 1 denotes the length associated with edge e. Each ver-
tex (or node) V_x corresponds to a vertical channel, while each ver-
tex V_y corresponds to a horizontal channel. Identical to the way
we assign weights to edges to represent the dimensions of the
building blocks, we can assign weights to vertices to represent
the width of a channel. Sometimes [5] channels are also represented by edges. There is however no substantial difference between
both methods.

The graphs G_x and G_y represent only the relative positioning of
the building blocks to each other. The coordinates are not really
fixed until all weights are assigned. Then to find the total di-

490

mensions of the substrate (and idem for the proper position of
the blocks and/or channels) longest path calculations can be used.

There is still a third representation of a layout by means of
a graph (Fig. 11.d.). The channel intersection graph is a more detailed representation of the channel intersections. The refinement
for the representation of the channels in the graphs discussed in
[21] is in fact automatically achieved in this last graph representation. It can be seen that every edge in Fig. 12.a. corresponds to an edge in Fig. 12.b.

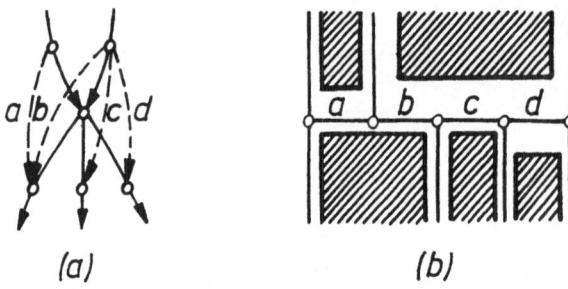

(a)   (b)

Fig. 12. : Refined representation of the channel widths (a) and
          the corresponding channel intersection graph (b).

After this introduction on graph representations, we can now
focus on how they are constructed i.e. how the placement is done,
and how they are used for the routing.

Placement of general cell assemblies. There are two possible
ways to handle the placement of general cells using graph representations. You can either start with a complete substrate and
divide it into pieces or build it up piece by piece. Both alternatives will be discussed.

a. Placement by division of the substrate. Starting with the total cell area and the desired shape of the final substrate, it is
possible to calculate the expected dimensions of the substrate.
The substrate is then represented in a graph as just one edge
(Fig. 13). In the next step the substrate will be divided in two
parts of approximately equal area. In [21] a min-cut algorithm is
then applied to assign the cells to the left or the right part of
the substrate. In the graph representation one edge is to be added to represent the new situation. This procedure can be repeated
until every block contains exactly one cell i.e. when all cells
are assigned to a specific block. In the final graph representations the dimensions of the blocks are to be replaced by the dimensions of the cells. Note that until now we only have a relative positioning of the cells as represented by the edges in a graph.
To find the actual coordinates of the cells we still have to apply
some longest path calculations.

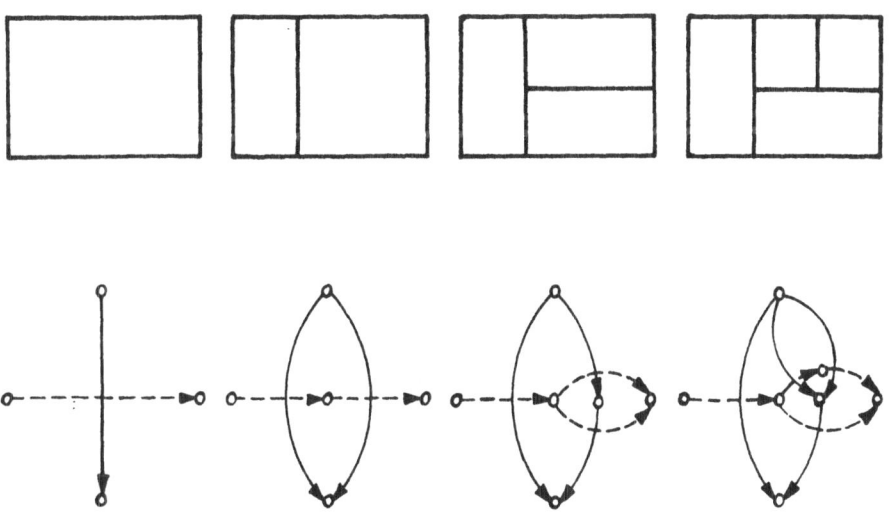

Fig. 13. : The different stages in finding a placement of general
cells by division of the substrate.

b. Placement by build up of the substrate. As opposed to the top-
down procedure of the previous method, the second approach will be
bottom-up i.e. start from one cell and add the other cells one by
one until all cells are placed [22]. This procedure is demonstrated in Fig. 14. In the graph representation of the placement we
start with one edge and add one for every cell added. Note that an
edge now always represents a single cell while in the previous method an edge usually represented a block containing a set of cells.
Furthermore here we have to consider for the placement of a cell
every possible position in which it can be placed. For the first
cell to be added this is respectively above, under, to the left
and to the right of the starting cell. These positions can be derived from the (partial) channel intersection graph. The costfunction used to select the best placement can be a combination of minimum interconnection length and minimal area. It is also necessary to check if the placement will result in a chip of acceptable
shape and size.

In the first method, since we start from an expected result,
we know where we will end up if we divide the substrate carefully.
In the second method we know where we want to end up but it is
difficult to see at every stage if we actually will. For both methods it holds that it is not just enough to find a dense placement

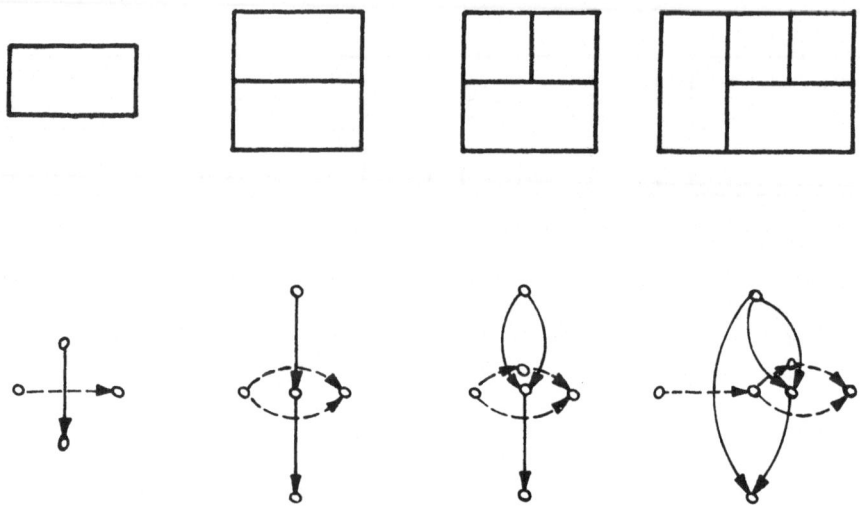

Fig. 14. : The different stages in finding a placement of general cells by build up of the substrate.

of the cells. It is absolutely necessary to include some expected routing area. If we did not include some expected routing area, this could result in an irregularly shaped chip after the routing when at certain positions the cells are to be spread further away to allow a 100 % complete routing then in other positions. The expected routing area can be better calculated in the second method because we immediately construct the channel intersection graph i.e. if we want to be very precise we can perform a complete routing at every stage (for the partial placement of course) at the cost of excessive computer time.

Routing in general cell assemblies. The graphs constructed during the placement will now be used for the routing. Routing will be done in two steps. In the first step we will assign the nets loosely to the routing channels. This step is commonly indicated as global routing. It is only in the second step that the nets are finally assigned to a specific track in the routing channel.

a. Global routing.The loose routing phase finds a strategic route on the channel intersection graph for each interconnection. After the insertion of the points to be interconnected by some net in the channel intersection graph, the Lee-algorithm can be used to find a path between these points (Fig. 15). Note that the maximum number of grid points to be considered in the Lee-algorithm

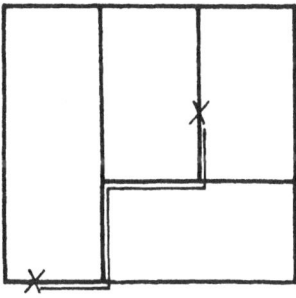

Fig. 15. : Loose determination a path i.e. assignment of the net
to certain routing channels.

i.e. the number of nodes in the channel intersection graph, is
equal to 4 times the number of cells. This is considerably less
then the total number of grid points of the substrate (usually
one has to store information about any one of these) and therefore
the algorithm will perform here very effectively. Moreover we can
use a costfunction which will account not only for length but also for channel congestion.

Often the notion of a critical channel is introduced. A channel is called a critical channel if, when a net is assigned to
the channel, this results in an enlargement of the total chip area.
A longer route can be preferred over a shorter one because it may
not add any incremental area to the chip and this because it does
not pass through a position of maximum track density on a critical
channel.

Since the channels still have infinite track capacity at this
stage, a 100 % route completion is guaranteed. Rip-up and reroute
of nets is easy because the nets are only assigned to channels and
not yet to specific tracks. Therefore a constructive initial solution can be followed by iterative improvement.

b. Channel_routing. When all nets have been assigned to channels,
the channels can be routed one by one. This means that we have reduced the routing task from routing over a complex substrate to
routing in a channel. Thus a channel router can be used and hence
routing is very fast and 100 % complete. We only have to take care
that at the intersections the channels fit together. It is because
of these intersections that the channels have to be routed in a
specific order. An intersection as in Fig. 16.a. imposes a T-con-

straint i.e. the horizontal channel has to be routed before the
vertical channel is routed. It is clear that there can be cyclic
T-constraints as is shown in Fig. 16.b. This cyclic T-constraints

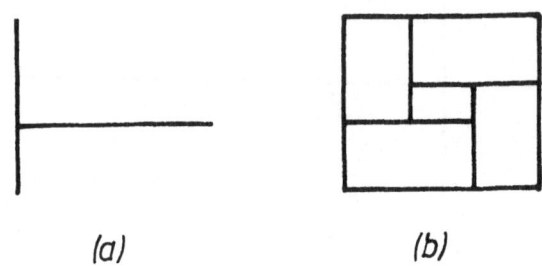

(a)   (b)

Fig. 16. : A simple T-constraint (a) and a set of cyclic T-constraints (b).

can be broken by assuming a certain width for a channel in the cycle, performing the routing for the other channels and then see if
the first channel can be routed. In general a 100 % routing completion can always be achieved, eventually after some iterations.

Systems not using graph representations. There are some systems [23] for the layout of general cell assemblies which do not
use graph representations. In these systems placement and routing
are performed in parallel. A chip layout is built in a series of
successive levels working from one edge (bottom) of the chip pro-
gressively towards the opposite edge. At each level, an area (slot)
at a time is locally optimized (Fig. 17). Components are chosen
for each slot which are most closely connected to the existing
portion of the layout and to any other components already selected for the current slot.

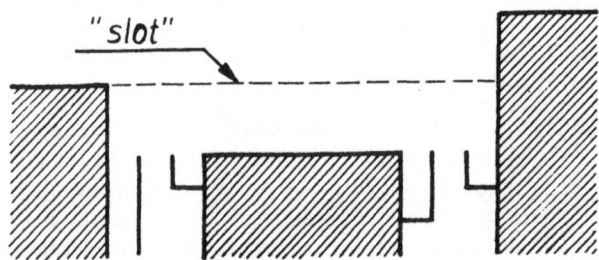

Fig. 17. : Example of a "slot".

Routes that cannot be completed in a certain slot are floated upwards. This procedure also guarantees a 100 % completion rate.

Although this approach looks very promising (there seems to be no wasted area because all slots are packed as tight as possible), it lacks the global overview obtained in the previous methods. The same remarks are valid as for the bottom-up placement approach discussed earlier but there optimization is still possible because you have a lot of information about your layout and it is important that nothing is fixed until the end.

CONCLUSION.

Many placement and routing algorithms have been proposed over the years for varying technologies and design philosophies. Some of these have been discussed. An analysis was presented of the CALMOS and CAL-MP programs to demonstrate the usefulness of layout automation in MOS-LSI design. The turn-around time for a design can be decreased drastically and although highly automated these programs still allow the designer a large flexibility. For VLSI it seems quite natural to complete a layout on several hierarchical levels. It was mentioned that the philosophies of the standard cell approach could still be used in hierarchical designs however it also seems more apt to go to general cell assemblies. Finally it was demonstrated that already several algorithms exist to succesfully deal with the problems imposed by general cell assemblies.

Those new layout techniques impose the need for hierarchical description languages and flexible data base systems. The latter is of primary importance if we think of the amount of data that is going to be stored. Furthermore nothing was said on the way bus-structures are to be treated. As future layouts will be more and more structured i.e. with a lot of bus-alike interconnections, this can become very important. Another important concept are integrated systems [25]. In its most extreme form all data concerning a specific design from system description to masks is stored in a common data base. Centered around this data base then are a number of programs ranging from system simulation to design rule verification. It is important to note that in this concept there exists an immediate link between for instance simulation programs and layout programs. Output data from the layout program can be fed back immediately to the simulation program to verify the proper functioning of the circuit. Also because all data is centralized the need to transport and convert data from one program to another is no longer existing. Everyone has access to the same data and this is very important to coordinate a design in the future.

496

REFERENCES.

[1] H. BEKE, W. SANSEN, "CALMOS, a portable software system for the automatic and interactive layout of MOS/LSI", Proc. 16 th Design Automation Conf., June 1979, pp. 102-108.

[2] H. BEKE, G. PATROONS, W. SANSEN, "CAL-MP : An advanced computer aided layout program for MOS-LSI", Proc. ESSCIRC 1980.

[3] G. PERSKY, D. DEUTSCH, D. SCHWEIKERT, "LTX - A system for the directed automated design of LSI circuits", Proc. 13 th Design Automation Conf., June 1976, pp. 399-408.

[4] K. SATO, T. NAGAI, H. SHIMOYAMA, T. YAHARA, "Mirage - A simple-model routing program for the hierarchical design of IC masks", Proc. 16th Design Automation Conf., July 1979, pp. 297-304.

[5] B. PREAS, C. GWYN, "Methods for hierarchical automatic layout of custom LSI circuit masks", Proc. 15th Design Automation Conf., June 1978, pp. 206-212.

[6] C. MEAD, L. CONWAY, "Introduction to VLSI systems", Addison-Wesley Publ. Co., Mass., 1979.

[7] D. SCHULER, E. ULRICH, "Clustering and linear placement", Proc. 9th Design Automation Workshop, June 1972, pp. 50-56.

[8] M. HANAN, P. WOLFF, B. AGULE, "A study of placement techniques", Proc. 13th Design Automation Conf., June 1976, pp. 27-61.

[9] M. BREUER, "Min-cut placement", Design Automation and fault-tolerant computing, october 1977, pp. 343-362.

[10] C.Y. LEE, "An algorithm for path connections and its applications", IEEE Trans. on Electronic Computers, Sept. 1961, pp. 346-365.

[11] D. HIGHTOWER, "A solution to the routing problems on the continuous plane", Proc. 6th Design Automation Workshop, June 1969, pp. 1-24.

[12] W. HEYNS, W. SANSEN, H. BEKE, "A line-expansion algorithm for the general routing problem with a guaranteed solution", Proc. 17th Design Automation Conf., June 1980.

[13] D. SCHWEIKERT, "A 2-dimensional placement algorithm for the layout of electronic circuits", Proc. 13th Design Automation Conf., June 1976, pp. 408-416.

[14] A. HASHIMOTO, J. STEVENS, "Wire-routing by optimizing channel-assignment within large apertures", Proc. 8th Design Automation Workshop, June 1971, pp. 155-169.

[15] R. MATTISON, "A high quality, low cost router for MOS/LSI", Proc. 9th Design Automation Workshop, June 1972, pp. 94-103.

[16] B. KERNIGHAN, D. SCHWEIKERT, G. PERSKY, "An optimum channel-routing algorithm for polycell layouts of integrated circuits", Proc. 10th Design Automation Conf., June 1973, pp. 50-59.

[17] D. DEUTSCH, "A dogleg channel router", Proc. 13th Design Automation Conf., June 1976, pp. 425-433.

[18] W. SANSEN, R. GOVAERTS, W. HEYNS, H. BEKE, "A minicomputer interactive system for hybrid automation", Proc. 1st Int. Microelectronics Conf., 1980, Japan.

[19] Y. SUGIYAMA, K. KANI, "A multichip LSI routing method",
Electronics and Communications in Japan, Vol. 58-C, No. 4,
1975, pp. 106-114.

[20] K. KANI, H. KAWANISHI, A. KISHIMOTO, "ROBIN, A building block
LSI routing program", IEEE Proc. ISCAS, 1976, pp. 658-661.

[21] U. LAUTHER, "A min-cut placement algorithm for general cell
assemblies based on a graph representation", Proc. 16th Design Automation Conf. , July 1979, pp. 1-10.

[22] B.T. PREAS, W.M. VAN CLEEMPUT, "Placement algorithms for arbitrarily shaped blocks", Proc. 16th Design Automation
Conf., July 1979, pp. 474-480.

[23] K.G. LOOSEMOORE, "IC layout - the automatic approach",
Proc. ESSCIRC, 1979, pp. 48-50.

[24] G. ARNOUT, H. DE MAN, "The use of threshold functions and
boolean controlled networks for macromodeling of LSI circuits", Trans. IEEE, Vol. SC-13, June 1978, pp. 326-332.

[25] E. LAPORTE, H. DE MAN, W. SANSEN, H. BEKE, "Towards an integrated minicomputer based LSI design system", Proc. of the
14th European Solid State Circuits Conference , Sept. 1978,
pp. 163-165.

SYMBOLIC LAYOUT COMPACTION

M. Y. Hsueh

IBM T. J. Watson Research Center
Yorktown Heights, NY 10598, USA

1. INTRODUCTION

Many experienced hand layout designers plan their layout by first rearranging the circuit topology in a way that they believe will result in the best final layout. The rearranged configuration is often drawn as a rough diagram indicating the desired placement of circuit elements and interconnection lines. Such a rough layout plan is then used as a guide during the tedious and error-prone actual layout process to remind the designer of the intended space utilization. In the past few years, layout plans of this type have become known as "stick diagrams" [1, 2] for they contain mostly simple line drawings.

It is more productive to use stick diagrams as a starting point for a computer program to generate the compact actual layout. Such layout minimization and generation systems have become available during recent years as a result of the research and development by separate groups and individuals [2-5]. In general, these layout minimization systems have been termed symbolic layout compaction systems because they reduce the total area of a layout represented by the symbolic description of the desired layout topology. Note that the word compaction refers to the minimization operation and, during the compaction process, certain areas in the layout may actually expand to satisfy, for example, a design-rule requirement. CABBAGE (for Computer-Aided Building-Block Artwork Generator and Editor) [5, 6] is such a compaction system and its organizational and algorithmic aspects are the subject of this paper. Fig. 1 gives a brief summary of the major functions of the CABBAGE system. It shows a stick diagram of an MOS T-flip flop being compacted and, in this case, a manually introduced jog being generated to produce a corresponding compact actual layout. At the present, only MOS circuit families are supported by CABBAGE.

A layout compaction system offers a number of advantages that are difficult to achieve with hand layout. First, it is now possible to incorporate technology design rules in the program and have the program apply all the rules during the generation of the layout. Thus, as long as the initial layout plan is consistent with the circuit or logic design, the output from the program would be a correct layout. Second, if the program is reasonably fast, it becomes practical to try out different types of topologies with the program and the best final layout can be selected from a group of candidates. In fact, experiments indicate that, for the same circuit, CABBAGE helps a user produce a smaller layout faster than what an experienced layout designer can do by hand [6]. In short, the use of a layout compaction system has divided the layout process into the two stages of planning the layout topology and generating the actual and correct layout geometry. The man-computer cooperation in this two-stage process is clear: the user decides the general organization of the layout, which is difficult for computer programs to perform, and the computer program does the uninteresting work of filling in all the details. As a result of this cooperation, it is possible to maintain or even improve the quality of the layout, as compared to layouts done by hand. At the same time it is possible to achieve a satisfactory reduction in the total design cost in terms of time and effort.

The capability to maintain layout quality and reduce design cost is important to meeting the challenge of the ever-increasing system complexity of VLSI chips. Other types of computer-based layout systems generally make less desirable compromises between quality and design cost. For example, with interactive layout drawing systems the user has all the freedom of generating the most special and sophisticated layout that he deems necessary. Such systems offer very limited de-

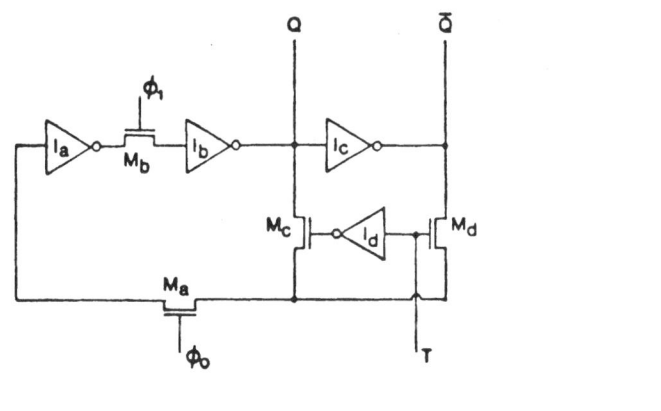

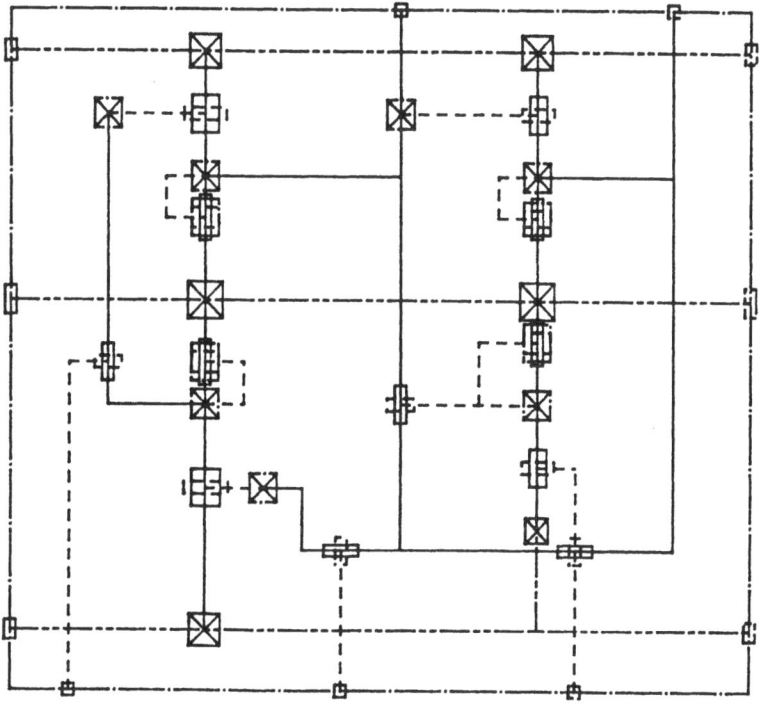

Fig. 1. The circuit schematic and the stick diagram of
an MOS T-flip flop. In the stick diagram, the
solid, dashed, and mixed long and short dashed
lines denote features on the diffusion, polysilicon, and metal layers, respectively. The symbols
made up of two overlapping rectangles represent MOS
transistors. The X's enclosed in boxes represent
contacts.

502

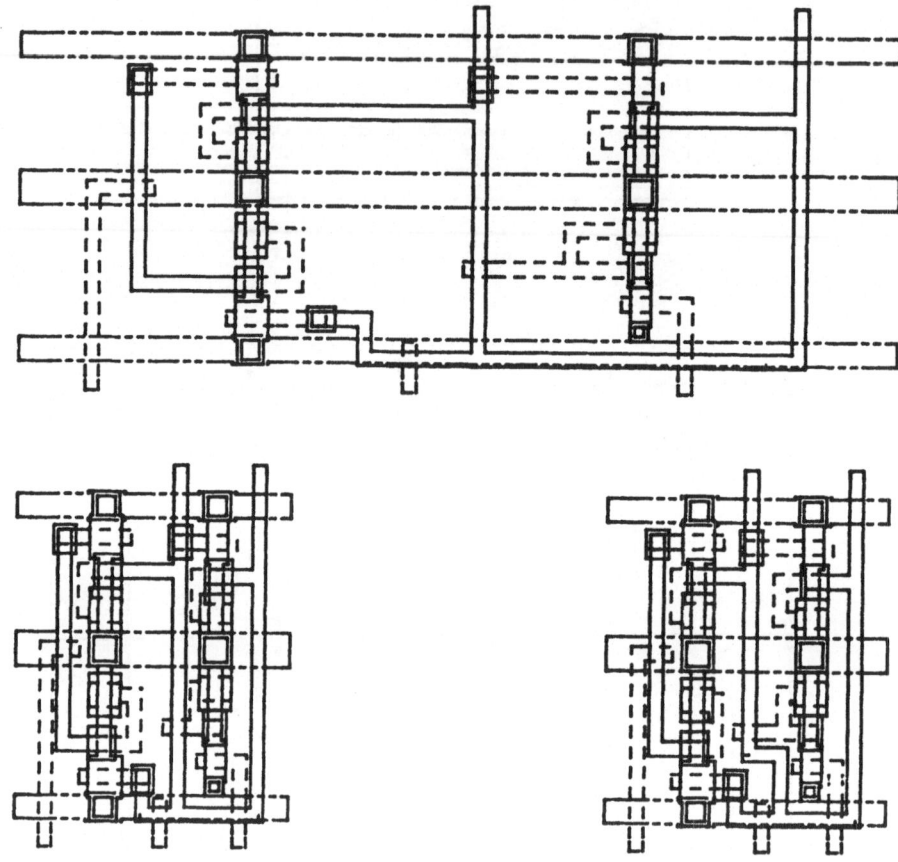

Fig. 1. (continued) Steps of layout compaction. The
sequence begins with the figure on the top and ends
with the figure at the lower right. First, the
layout is compacted in the vertical direction.
Next, the layout is compacted in the horizontal
direction. Finally, a jog is introduced selective-
ly in the vertical line in the center portion of
the layout.

sign cost improvement because the user is responsible for ensuring the correctness and the compactness of the layout. A higher degree of cost reduction can be obtained with symbolic layout translators [7, 8], which simply substitute symbols with actual layout features. With such translators the user must place symbols on a fixed coarse grid so that the translator can map the user's input into a correct layout directly. A still higher degree of cost reduction can be achieved with "automated" layout aids such as standard-cell layout systems [9, 10] and programmable logic array (PLA) synthesizers [11]. These layout aids convert circuit interconnection descriptions or logic equations directly into layouts. However, the user must settle for layouts that are loose and contain a family of predefined circuitry arranged in fixed formats. (Standard-cell type of layouts generally are arranged in a format that is similar to the arrangement of IC packages on a circuit board. PLA's are matrix-like structures that perform two or more levels of alternating AND and OR logic.)

It must be emphasized, however, that layout compaction systems are not the only solution for the VLSI layout problem. For some exploratory projects, standard-cell layout systems may be the most effective layout aid for getting the test chips made quickly. In contrast, hand or interactive drawing type of layout may be necessary for designs where the shape and size of nearly all features are critical, such as in sense amplifiers used in memory chips. Thus, the most successful layout aid for VLSI systems will evolve as designers' needs dictate and should incorporate the advantages of all layout systems available today. In short, the following presentation should serve as an exposition of a new type of layout aid that fills in the gap left uncovered by other layout aids. Also the presentation should offer a new perspective of the layout problem and provide a framework for research and development of more useful layout aids for VLSI systems.

2. THE CABBAGE SYSTEM

This section presents an overview that is intended to offer some insights into the organization and characteristics of the CABBAGE system. These insights may prove valuable in later sections where the algorithms and implementations are described.

CABBAGE is developed based on a circuit designer's need to streamline the hand layout process. This origin strongly influenced some important characteristics of the system. Initially, CABBAGE is designed as a tool to help the user evaluate different layout topologies by providing the user with compact actual layouts that correspond to those topologies. As such, CABBAGE does not attempt to make drastic alterations to a stick diagram. Instead, it lets the user do most of the major changes at the simple stick diagram level to direct the progress of the overall layout process. With such an uncomplicated approach, the program can work fast and stay lean, resulting in quick response, easy usage, and simple maintenance. In keeping with the initial objective, CABBAGE performs only those tasks that the user either is unwilling to do or cannot do well. First, CABBAGE enforces all design-rule requirements automatically during the layout compaction process. The guaranteed correctness with respect to design rules eliminates the need for repeated layout corrections. Second, CABBAGE collects all design-rule requirements in every local neighborhood of a layout plan before it begins to minimize the entire layout at once. In contrast, a human designer tends to localize his work space when it comes to detailed geometrical drawings, and he often over-refines a particular area unnecessarily.

Specifically, the CABBAGE system consists of two programs. The first program, GRLIC, is a simple interactive graphics editor which provides the user with a means for entering stick diagrams. The stick diagram is stored in a disk file as a collection of symbols without explicit connectivity information. Each record in the stick diagram file contains the type, orientation, location, size and name of each of the constituent elements, such as transistors and interconnection lines. The second program, PRSLI, performs layout compaction based on the stick diagram. It reads in the stick diagram from the disk file, derives the interconnection and adjacency relations of individual symbols, analyzes all design-rule requirements for each symbol, and determines the new location of each symbol based on these information to generate a correct and compact actual layout. From PRSLI the compacted layout can be displayed or stored in a file in the same format as that used by the graphics editor GRLIC. Thus the simple file serves as a link between the editor and the layout compactor. The existence of such a link allows the user to carry out repeated editing and compaction of a layout based on results obtained from previous compaction operations. The quality of the layout can be improved rapidly with this type of iterative process.

The graphics editor GRLIC presents the user with a set of predefined element symbols for constructing stick diagrams. An element symbol is selected via a command specifying its type, size and the optional name. The location of the element is specified with the graphics cursor and the orientation of the element can be adjusted at the cursor location. In its present form, the GRLIC program is a simple tool for the sole purpose of data entry and modification.

The PRSLI program lets the user issue compaction requests in the vertical and the horizontal directions separately. In addition, an automatic jog generation routine may be used to bend some elements to achieve a possibly more dense layout. Automatic reflection and rotation of elements have not been programmed into the system. However, as mentioned previously, the user can perform manual alteration of the compacted stick representation after each compaction operation to guide the progress of the layout compaction process. In the unfortunate event that a compacted layout must be expanded due to change in technology design rules for improving the chip yield, these interim manual modifications by the user are not totally lost because the compaction process employed in PRSLI makes both contractions and expansions to meet design-rule requirements. PRSLI also lets the user employ fixed constraints to set certain distances constant during the compaction. These fixed constraints can be used in the representation of compacted building blocks that are used in the hierarchical construction of a larger layout.

The similarities in data structures used by both programs make it possible to merge all the functions of the editor and the compactor into a single program. However, the CABBAGE system uses separate and loosely linked programs for two good reasons. First, such an arrangement provides modularity. Changes and improvements made in one program do not affect the operation of the other program, as long as the format of the

506

common file is not altered. Second, the modularity promotes the development of other related design aids, such as those shown in Fig. 2. The functions and usages of the related design aids will be described in Section 5, where possible extensions and enhancements to the CABBAGE system are considered.

The next two sections will concentrate on the algorithmic and organizational aspects of the layout compactor PRSLI. The graphics editing aspects of the CABBAGE system will not be considered any further because graphics editing is in general a relatively mature area. Furthermore, the modularity of the CABBAGE system allows any good graphics editors to be used in place of GRLIC.

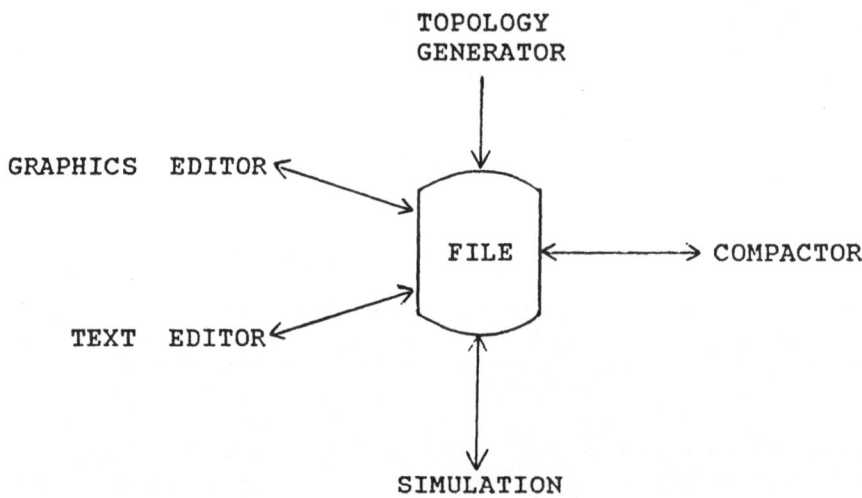

Fig. 2. Tools that may be interfaced to the stick diagram file used by CABBAGE.

3. METHODS FOR LAYOUT COMPACTION

Several types of formulations can be used for solving the layout compaction problem. Perhaps the most obvious formulation is to treat the compaction problem as a placement problem: the individual elements in the stick diagram are packed into the available space to construct a compact actual layout. It is possible to use classical placement methods such as force-directed placement, pair-wise interchange [13], etc. to generate a compaction solution. These placement methods typically assume that individual elements can be moved around separately and the proper interconnection may be restored in a subsequent wire routing phase. However, at the lowest level of layout compaction where basic elements such as MOS transistors are to be packed together, areas consumed by interconnection may be much larger in comparison to the aggregate of basic elements. Thus, at least for circuits containing basic elements, it is quite difficult to use separate placement and routing phases in the construction of a compact actual layout.

To avoid the difficulties of rearranging the layout configuration, a simple alternative to the classical placement approach is to strictly follow the interconnection configuration of the stick diagram during the packing process. For example, elements in a stick diagram can be sorted according to their coordinates and packed into the layout in that order. Here, basic elements as well as interconnection lines are placed in a uniform fashion [2, 12]. The orderly introduction of elements preserves the original connectivity and eliminates the need for a separate routing phase. In a way, such a packing process simulates what a layout designer does by hand. Because of the similarity between this type of compaction method and the hand layout process, the computer program may run into the same difficulty of over-compacting a certain area that hinders the satisfactory placement of subsequent elements.

Another layout compaction scheme is the "compression ridge" method proposed by Akers et al [14]. Here, bands of continuous unused area, called compression ridges, are removed from a loose layout as shown in the example in Fig. 3.1. Also indicated in that figure is the use of "rift lines" for joining unused areas to form effectively continuous bands. Since the compression ridge method cannot create new space, the starting

508

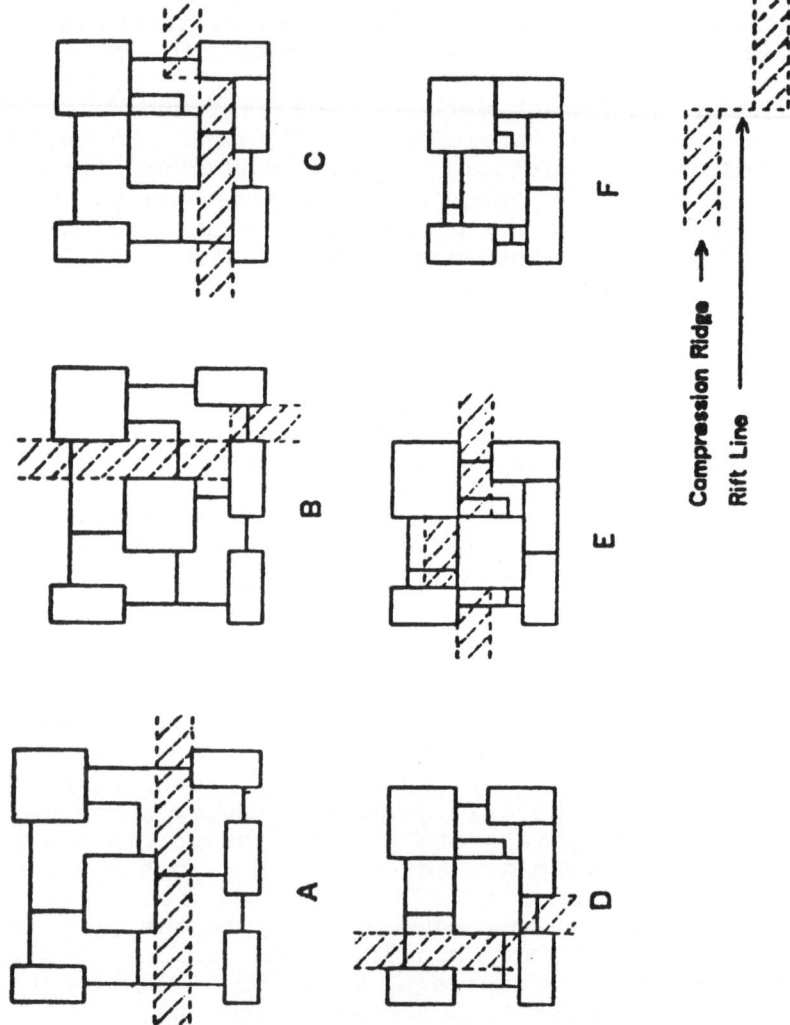

Fig. 3.1. An example showing the operation of the compression ridge method.

layout must be free of design-rule errors. Thus, in addition to the sequential removal of the unused area, an initial expansion may be required at the beginning of the layout compaction process to satisfy all design-rule requirements.

The compression ridge method can be more effective than the aforementioned sequential build-up method because it adjusts the locations of a group of elements at once. However, both methods have the same type of fundamental drawback: only a limited area in the layout is compacted at a time and such a compacted area may hamper the progress of subsequent compaction operations. In contrast, the compaction method used in CABBAGE evaluates all relevant constraints before it determines the final placement for all elements in either the horizontal or the vertical direction. Specifically, the compaction operation involves the collection of all relevant design-rule requirements in each local neighborhood and the selection of those "critical" requirements whose combined value limits the overall size of the final compacted layout. In the following, the principles of the compaction procedure used in CABBAGE are described. A simplified version of the complete procedure for compaction is given at the end of this section.

3.1. A graph representation for a stick diagram

A convenient way of organizing all relevant constraints in a computer program is the rectangle dissection technique described by Tutte et al [15]. Here, a graph is mapped into a dissected rectangle, as shown in Fig. 3.2, by considering such a graph as representing a resistive network whose branch voltages and currents map into the heights and widths, respectively, of the corresponding constituent rectangles.

The dissection-type mapping has been used by many researchers for handling the circuit layout problem at various levels [16-20]. Here the total area for the layout is dissected into many smaller areas, each of which contains a circuit element used in the layout. In contrast to the original dissection problem where the resistance of each branch is fixed, the layout problem generally requires that the area and aspect ratio of each corresponding rectangle be flexible. Typically, the solution process for circuit layout

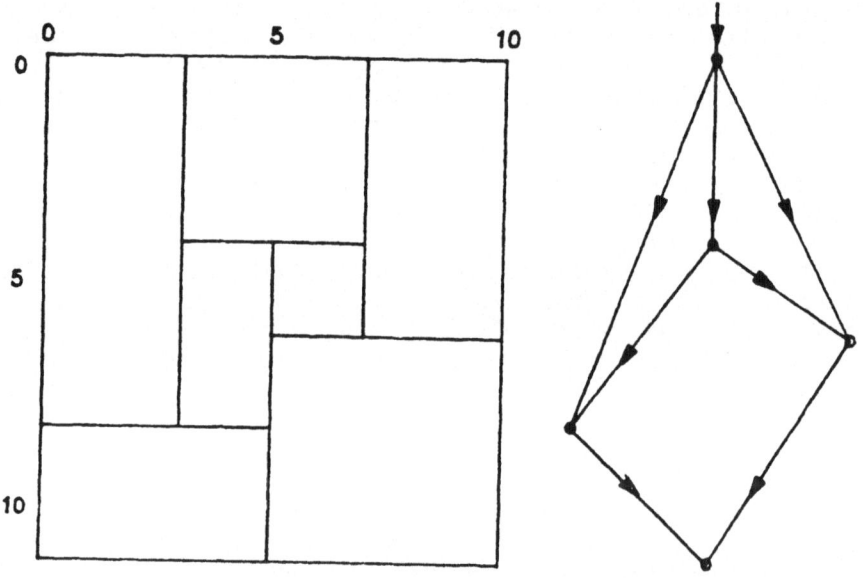

Fig. 3.2. The graph representation of a dissected rectangle.

problems consists of the construction of a suitable graph representation of the adjacency relationship of all circuit elements and the solution of the graph with mathematical programming techniques to generate actual rectangles in a minimum area.

For symbolic layout compaction, however, the stick diagram provides a good starting point and can be used directly to guide the construction of the adjacency graph. The conversion of a stick diagram into such a graph is a simple task and is described in Subsection 3.2 below. Furthermore, the overall compaction problem is divided into independent compaction operations in the horizontal and vertical directions, respectively, for simplicity. The disadvantage of such a division is that the compaction operation is no longer truly global and compaction in one direction may hamper the progress of a subsequent compaction in the orthogonal direction. However, such a division makes it possible to treat the graph-to-rectangle conversion operation as a linear programming problem since the constraints are now one-dimensional.

The linear programming problem may be simplified further. Assume for the moment that all constraints are of the lower-bound type; for example, the size of an element and the spacing between two elements must be greater than or equal to certain prescribed values, respectively. Specifically, the constraints can be written as

$$\mathbf{Ap} \geq \mathbf{h}$$

where $\mathbf{A}$ is the nodal incidence matrix of the adjacency graph shown in Fig. 3.2, $\mathbf{p}$ is a vector containing the locations of the cut lines in the corresponding dissected rectangle, and $\mathbf{h}$ is a vector containing the minimum element size and the associated clearances to be accommodated in each of the constituent rectangles. Since the constraint matrix in this linear programming problem is the incidence matrix representing the branch incidence relationship of the adjacency graph, the linear programming problem of minimizing either the horizontal or the vertical dimension is equivalent to finding the longest path through the graph in that dimension [21]. It is a simple matter to determine the longest path through the adjacency graph since the graph is directed and acyclic [6, 28]. Thus, CABBAGE uses the simpler operation of finding the longest path to determine element locations in a layout. The case of mixed lower- and upper-bound constraints is considered in Subsection 3.4.

Conceptually, the longest path corresponds to the aggregate of the most constraining elements in the layout. Thus, the overall size of the compact actual layout is determined once the longest path is found. Elements whose constraints do not contribute to the longest path are nonconstraining elements. A nonconstraining element does not use up the total area allocated to it based on the longest path calculation. The extra areas left in the vicinity of the nonconstraining elements appear as "dead areas" in a typical layout.

3.2. The graph construction

The constraint graph is constructed directly from the stick diagram. In order to facilitate such a translation and the subsequent compaction operations, the four rules listed below must be observed in drawing a stick diagram.

a) All line elements must be terminated. A line may be terminated by another perpendicular line or with a terminal symbol. In particular, a line element should not be terminated by any other point structure, such as a transistor, alone.

b) All point structures other than the terminal symbol must be placed at the intersection of lines. The function of a point structure is to indicate the type of the intersection where it is located.

c) The center locations of any two point structures must not overlap. In general, the user should save a reasonable amount of space around each element in the initial symbolic input. This may save the user from redoing the entire input when it is desired to modify a part of the symbolic input to achieve a better compaction result. Any violation of minimum spacing design-rules will be corrected by the compactor, provided the center lines of the elements in question do not overlap each other.

d) A border must be drawn around the stick diagram. A terminal symbol must be placed at the intersection where a line terminates on the border.

Although each element (a transistor, a contact or a line segment) may be enclosed in a rectangular area that is in turn represented by a branch in the constraint graph, as outlined in Fig. 3.2, it is difficult to preserve the connectivity of the original stick diagram with such a representation. The difficulty arises from the fact that the constraint graph indicates only the adjacency but not the connectivity information. Hachtel described a formulation in which an auxiliary set of connectivity equations is used in conjunction with the adjacency equations, depicted here as a directed graph, to overcome this difficulty [29]. Rather than using another set of equations to describe the proper connectivity, CABBAGE uses a simple alternative of grouping all topologically connected elements that share the same center line in the stick diagram as a connected unit. Such a grouping holds together connected elements in the direction perpendicular to the compaction direction. Fig. 3.3 gives an example of the grouping of elements in a layout.

The grouping operation constitutes a preprocessing step that reduces the size of the graph and simplifies the problem. However, the grouping operation also makes some element connections unnecessarily rigid that many smaller elements cannot fully utilize their surrounding areas because their movement is restricted by larger elements in the same group. Such a drawback can be remedied by breaking a group into two or more new groups appropriately and repeat the compaction with the new stick diagram. The breakage would appear as bent features or jogs after compaction. The automatic introduction of all possible break points is described in the next subsection.

In the actual implementation in CABBAGE, the center lines associated with each group are mapped to nodes and the spacing requirements among groups are represented by branches. Such a mapping is consistent with the mapping used in the rectangle dissection method; the center lines correspond to the cut lines in the dissected rectangle. The mapping used in CABBAGE provides conceptual simplicity because the adjacency graph can be viewed as a PERT chart, and the determination of group placement becomes equivalent to the task scheduling problem.

The spacing requirements among groups are examined by a design-rule analysis routine. This routine in-

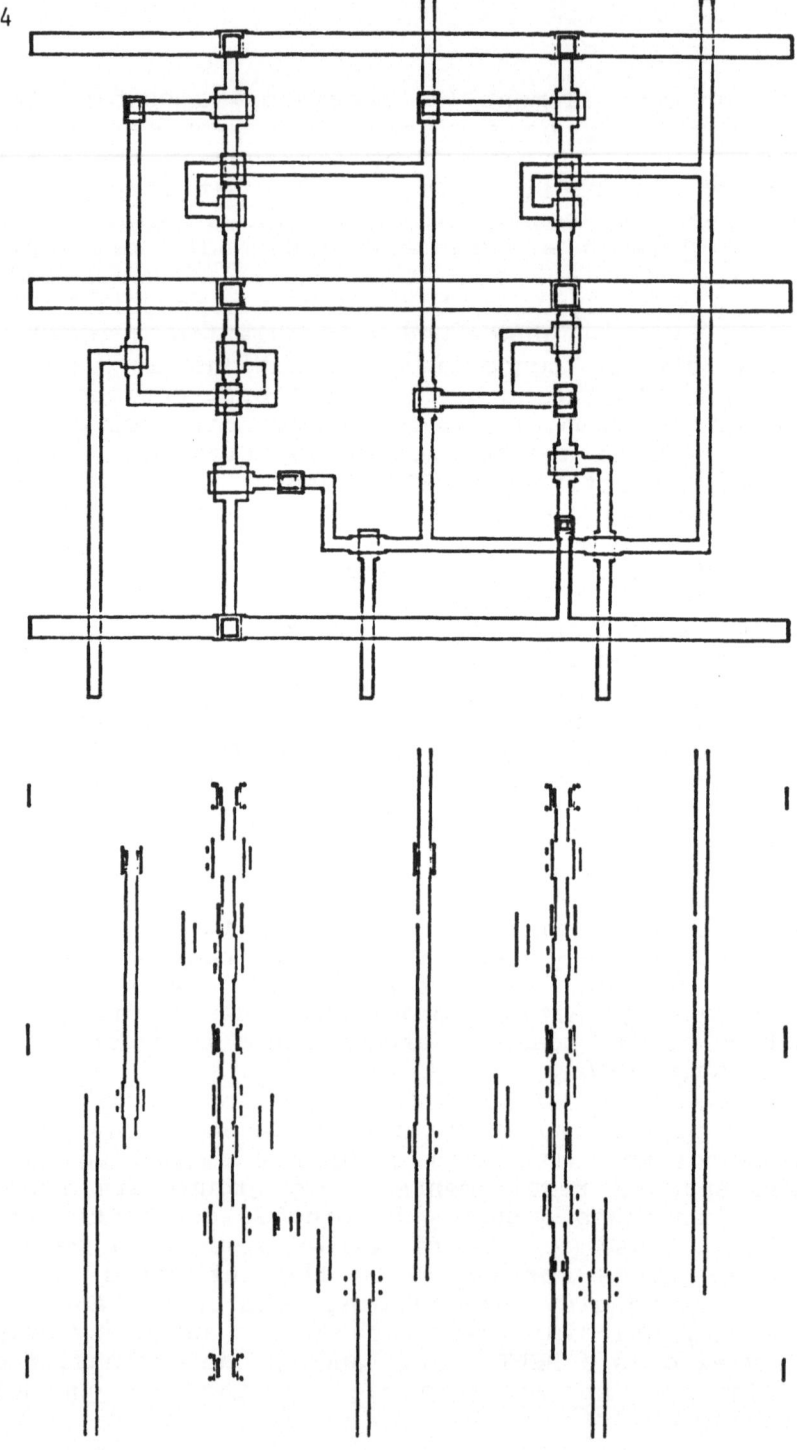

Fig. 3.3. The initial layout of the T-flip flop (top) and its vertical groups used in a horizontal compaction operation (bottom).

serts a branch between two nodes if the corresponding element groups must be set apart. Note that, among other criteria, two groups must be set apart if they are electrically unconnected. Thus, it is necessary to have the electrical connectivity available to the design-rule analyzer. The maximum of all the required separation determined by the design-rule analysis is used as the branch weight. By definition, the branch points away from the groups toward which the rest of the groups are to be moved. Fig. 3.4 shows the constraint graph for the groups shown in Fig. 3.3.

3.3. The determination of element placement

The longest path length to each node is computed during the analysis process for determining the longest path through the constraint graph. The path lengths to individual nodes can be used as center line locations of the corresponding groups to construct the compact actual layout. The algorithm for determining the longest path is given in Section 4.

However, the placement based on the longest path calculation alone is not entirely satisfactory because element groups not in the longest path are pulled towards the starting groups. In other words, the entire extra space available to such a nonconstraining element group appears on the side between it and its following groups. A more satisfactory placement can be obtained by adjusting the allocation of the extra space based on the interconnection configuration in the neighborhood of the nonconstraining element groups. A simple technique based on the force-directed placement principle is used in CABBAGE to perform the space redistribution.

The new location of each group is used to determine the new end points (and the length) of line segments linking the groups. These line segments lie in the direction of compaction and can be viewed as rubber bands; their end points may move towards the origin of compaction and their length may shrink each time a more compact layout is realized. Note that one of the functions of the aforementioned drawing rules is to provide a means through which every linking segment can find out where its two ends are supposed to be.

Unused space may exist around groups in the longest path as well. Such a situation occurs, for example,

516

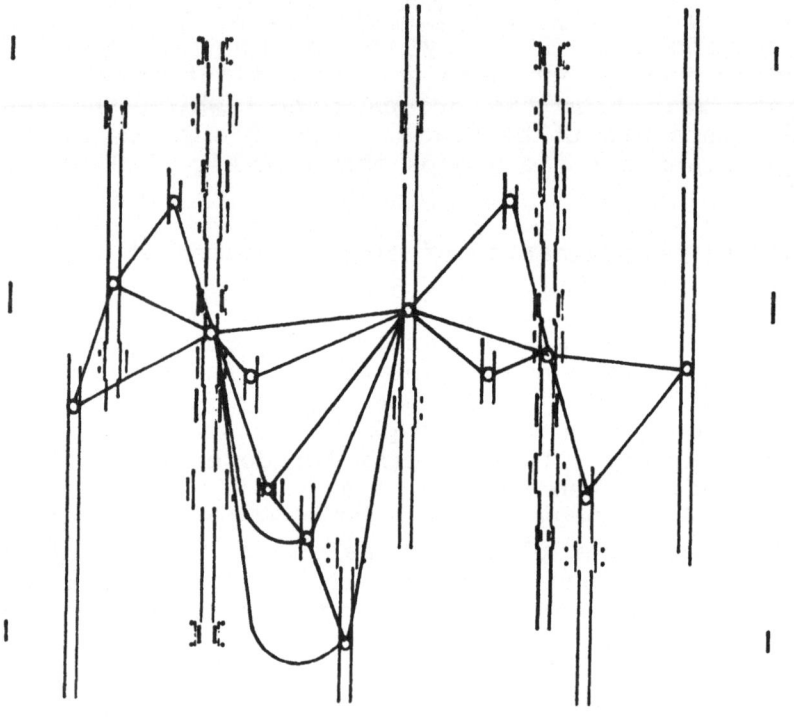

Fig. 3.4. The constraint graph for the groups in Fig. 3.3.

when a long group is constrained by a short group on the one end and the long group is in turn constraining another short group on the other end, as shown in Fig. 3.5a. If the ranges of the two short groups do not overlap, then it may be advantageous to bend the long group in the middle to achieve a possible reduction of the size of the layout, as shown in Fig. 3.5b. Thus, break points for jog generation can be derived directly from information about the longest path and the group size.

3.4. Handling mixed constraint types

Most layout rules indicate the lower-bound constraints within or among elements. For example, typical layout rules specify the minimum size and spacing of features on a given mask layer. Occasionally, however, an upper bound may be specified also to limit the size or spacing of elements. For example, it may be desirable to restrict the length of a connection so that the wire capacitance can be kept within a reasonable limit. Another common use of upper bounds is to keep signal lines a fixed distance apart so that they can make direct contact with continuing lines in other parts of a separately compacted layout. (A fixed constraint is represented by equal lower and upper bounds.) In general, most upper bounds appear as user-defined constraints that augment the existing design-rule set.

The use of the longest path technique for determining the element placement is predicated on the assumption that only lower-bound constraints are present. In the presence of upper-bound constraints, however, the longest path technique is still applicable if a two-pass placement strategy is used. In the first pass, only the lower bounds are used to construct the adjacency graph and the longest path analysis is applied as before to determine the tentative placement. In the second pass, the upper bounds are examined against the element locations determined during the first pass. Thus, any violation of the upper-bound requirements can be detected. The drawback of this approach is that violations of upper bounds can be found only after the complete graph is traversed once. However, the alternative of considering both types of constraints in a linear programming problem may lead to cycles in the solution process.

518

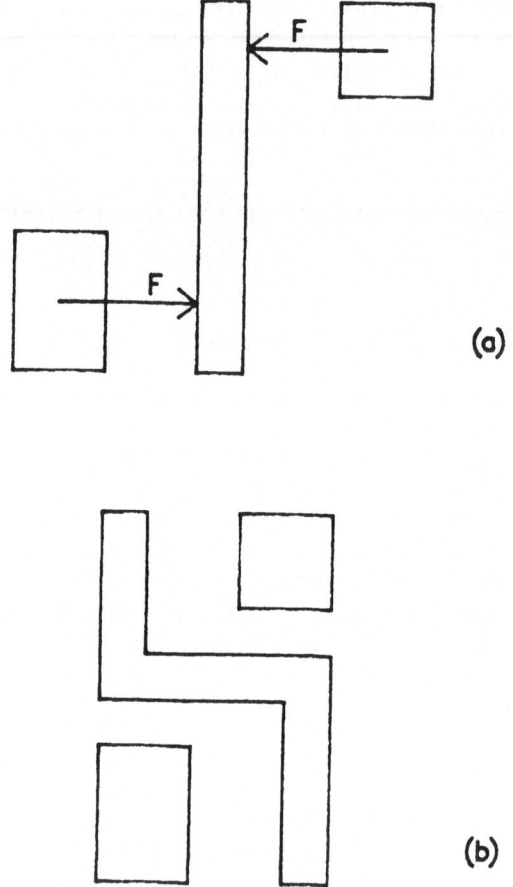

(a)

(b)

Fig. 3.5. A simple example of the use of a jog. The
    "torque" applied by neighboring groups shown in (a)
    bends the middle group as in (b).

3.5 The compaction procedure

The last several subsections described the major components of the layout compaction procedure used in CABBAGE. These major components as well as other supporting operations are combined in the following sequence to perform layout compaction:

a) sort the elements by their coordinate values for ease of reference. Since the layout compaction is carried out as separate horizontal and vertical operations, separate orderings by X and Y coordinates are used.

b) group topologically connected elements in both compaction directions.

c) develop electrical connectivity for proper design-rule analysis.

d) perform design-rule analysis in the direction of compaction to build up the constraint graph.

e) find the longest path through the constraint graph and report violations of upper-bound constraints.

f) adjust group locations based on the longest path calculation. Also adjust the terminal locations of the interconnection lines (the rubber bands) in the direction of compaction.

g) use the longest path information to generate break points in groups if the user uses the automatic jog generation feature.

h) if necessary, go back to step d and ready for the next round of compaction.

The detailed algorithms used in these steps are the subject of the next section.

4. ALGORITHMS AND THEIR IMPLEMENTATION

The algorithms used in the PRSLI compactor are described and explained in this section. These algorithms are presented in the general sequence of the layout compaction steps outlined at the end of the last section. This section is intended to bring out some of the subtle but important details that must be skipped in the earlier presentation to avoid possible confusions.

For clarity, some of the algorithms are described in a language that closely resembles the computer programming language RATFOR [24], which is the programming language used in PRSLI. In this algorithm description language, statements are terminated with semicolons and assignments are denoted with the equal sign. Several statements may be grouped together with a pair of braces to form a compound statement. The "for" statement signifies a loop in which all arguments satisfying a given condition are examined sequentially. As an example, the statement "for each element j connected to the element i {...}" means that each one of the elements connected to the element i is to go through the actions specified in the pair of braces.

It is useful at this point to get an overview of the data structure used in PRSLI. The data structure is built upon element symbol description blocks (SDB's) that constitute the aforementioned disk file for recording a stick diagram. The SDB's have the structure shown in Fig. 4.1. An element may be a transistor, a contact, or a line segment. Also recorded as elements are line terminals and user-specified constraints that present additional information to the compactor but do not appear as actual elements in the layout. The actual location of an element as it appears in a stick diagram is recorded as the X and Y center locations. The size of an element is recorded as half-widths from the corresponding center line locations and are termed "offsets" in the SDB's. Note that a distinction is made between the line segment element and the rest of the elements. A line segment has variable length and can be placed anywhere in the layout. As such, only the X center location of a vertical line needs be recorded. Similarly only the Y center location of a horizontal line is recorded. Other elements such as transistors are considered as "point structures" and must be placed at intersections of line segments. Both their X and Y center locations are defined and recorded.

When PRSLI reads in such a file, it stores all real elements and line terminals in a linear array called

element type
orientation
x center location
x left (offset)
x right (offset)
y center location
y bottom (offset)
y top (offset)
name (pointer)

Fig. 4.1. The data stored in a Symbol Description Block.

the LSG array. The user-specified constraints are stored elsewhere in memory. Preliminary setup procedures such as grouping elements and tracing the electrical connectivity can be performed at this point. Specifically, the following operations are performed.

4.1. Sort elements

Since the SDB's in the input file may be recorded in any order, it is necessary to rearrange their sequence for easy reference in the subsequent steps of the compaction process. A linear array called LREF is set up for this purpose. Each block in the LREF array corresponds to a center location in either the X or the Y direction and serves as the basic data area for the element groups that are to be developed shortly. As shown in Fig. 4.2, each LREF block contains information about the element group's locations before and after compaction, a pointer to a LIST block which contains pointers to SDB's located at the center location represented by this LREF block, as well as other facts that are used in the design-rule analysis and path search. Thus, elements are sorted and collected by their respective X and Y center locations. Furthermore, center locations in each of the SDB's are substituted with pointers to the corresponding LREF blocks to complete the pointer chain; each element can find out about the rest of the elements that share the same center location through this pointer chain.

The LREF blocks must be sorted as well. Separate arrays LXRF and LYRF are used to record the ordering of LREF blocks that represent either X or Y center locations. (The LXRF and LYRF arrays are omitted from Fig. 4.2.) A rudimentary data structure is now available for the design-rule analyzer to examine the layout sequentially. The electrical connectivity of a layout is also needed by the design-rule analyzer and are derived with the algorithms in the next subsection.

It should be noted that the data structure developed so far is similar to what the interactive graphics editor GRLIC uses for all its operations.

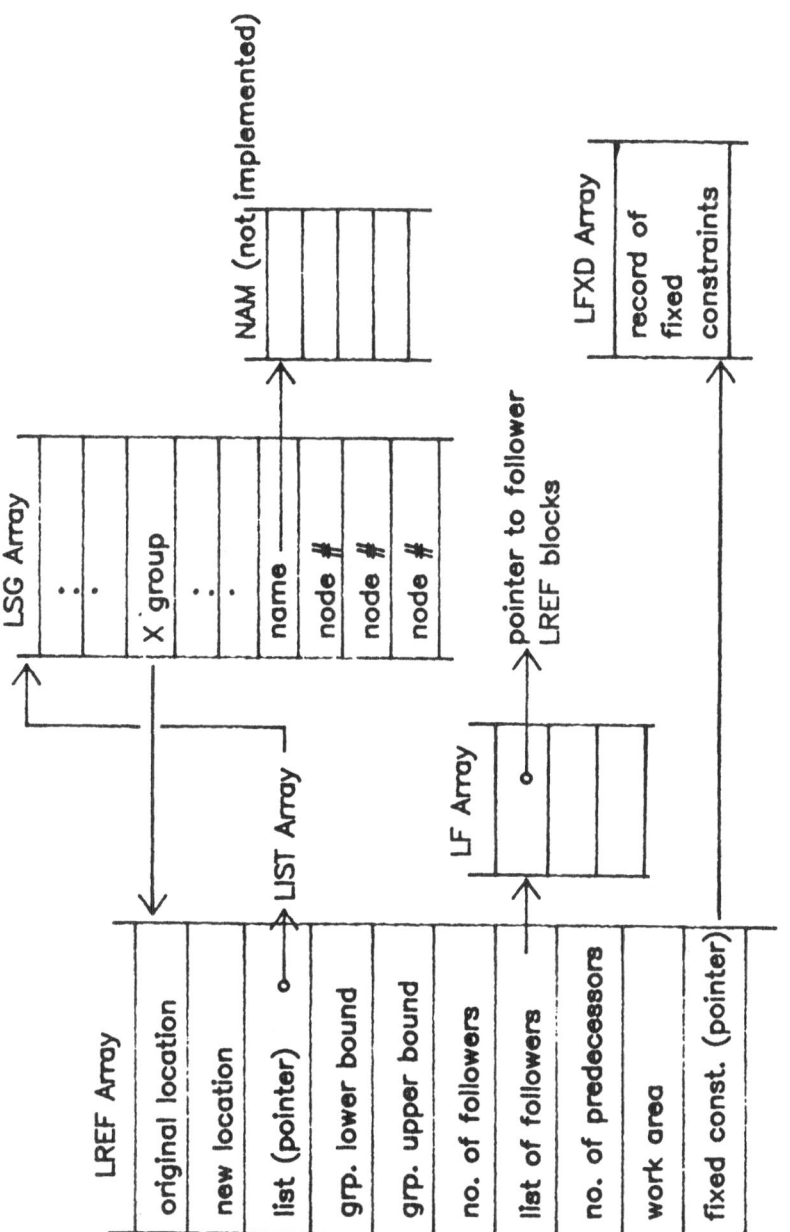

Fig. 4.2. An outline of the data structure used in PRSLI.

4.2. Trace the electrical connectivity

Recall that one of the drawing conventions used in the CABBAGE system requires that transistors be placed at intersections of lines. Thus, a line must be separated into two electrically unconnected segments if the source and drain terminals of a transistor fall on top of it, as shown in Fig. 4.3. A specialized routine SPLIT is used to perform this separation by cutting such a line into two halves at the center of the transistor. The separation of drain and source lines is just one example where technology-specific routines must be written to help the compaction program understand the electrical characteristics of the stick diagram. The alternative in this case would be to ask the user to draw separate drain and source lines. Another such technology-specific routine that may prove useful is one that tells the user about all unmarked line intersections where either a transistor or a contact must exist.

The next step is to group topologically connected elements. Since elements at the same center location are collected in the same LREF block already, this grouping amounts to dividing a given LREF block into several new LREF blocks each of which containing only topologically connected elements.

The electrical connectivity is represented by node numbers assigned to elements. The node number of a given element may be determined with a path-finding scheme [22]. For simplicity, assume that each element is tied to only one node. The node numbering algorithm used in PRSLI begins by assigning a distinct, monotonically increasing temporary node number to each element sequentially. The elements are then examined in the same order to deduce the real node number as follows.

```
for each element i in the layout {
    for each element j connected to it {
    N = smaller of the two node numbers;
    L = larger of the two node numbers;
        assign N to all elements whose node number
            equals L;
        decrement node numbers above L by 1;
    }
}
```

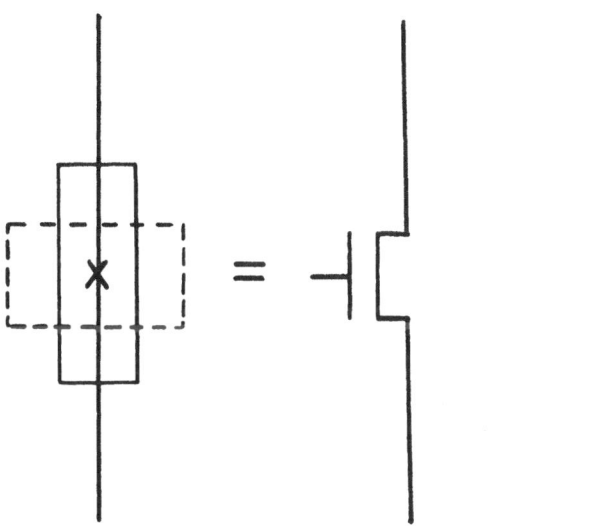

Fig. 4.3. The splitting of a diffusion line at the center of an MOS transistor.

DIF	PLY	MET	RNX	ACT	CNT	BUR	IMP	TRM	
5	2	X	5	X	6	4	X	X	DIFFUSION
	5	X	5	X	6	4	X	X	POLYSILICON
		6	X	X	X	X	X	X	METAL
			5	5	5	5	5	5	RUNX
				5	5	6	4	4	X	ACTIVE AREA
					10	2	X	X	CONTACT WINDOW
						4	X	X	BURRIED CONTACT
							X	X	IMPLANT REGION
								X	TERMINAL POINT

Fig. 4.4. The table of spacing design rules used by CABBAGE.

Since point structures such as transistors must be
placed at the intersection of line segments, it is
sufficient to examine only line elements in a given
group to discover all other elements connected to them.
Thus the first line of the algorithm can be reduced to
"for each line element i". Note that two touching elements may be unconnected. For example, a metal line
may overlap a polysilicon line in the layout but the
two are unconnected if they are not hooked up with a
contact. Thus, a technology-dependent routine is needed again to provide information regarding the possible
connectivity between any two masks.

Finally, it is useful to build up a node table that
groups elements by node numbers. In fact, the last
statement in the above algorithm is included to produce
continuous node numbers which simplify the construction
of a node table. Additionally, the node numbers may be
used in conjunction with other circuit parameters that
can be derived from the stick diagram to feed simulation programs such as SPLICE [23].

4.3. Apply design rules and build the constraint graph

The graph representation of the stick diagram is to
be generated in this step. Since nodes and branches of
a constraint graph represent group center lines and
constraints among groups, respectively, information
about the connectivity of a node is stored in the LREF
block that represents the corresponding center line.
Specifically, the number of follower nodes, a pointer
to a list of these follower nodes, and the number of
predecessor nodes are stored in the LREF block as shown
in Fig. 4.2. Note that branch weights are stored in
the list of followers, LF, as lower- and upper-bound
constraints against the respective follower groups.

The graph construction operation essentially involves filling in these data areas with appropriate
pointers and numbers. To begin with, user-specified
constraints are placed into the appropriate LF blocks.
Next, a design-rule analysis is carried out in the
compaction direction to complete the constraint graph.
At the present, PRSLI compacts all elements towards the
lower left corner. Thus, the design-rule analyzer
starts with the group at the extreme left or the bottom
of the layout and works across to the other end of the
layout.

Since the design-rule analysis must deal with the exact geometry of the stick diagram, a set of routines is used to convert a given element group into actual geometrical features. Specifically, the output from these routines is a collection of description blocks each of which contains the location, length, offset from the group center, mask layer, and node number of an external edge contributed by an element of a group. Let EDGE(i, j, d) represent the edge description block of the j-th element of the i-th group when the group is viewed from the d direction (left, right, top, or bottom). It is clear that the geometrical layout associated with a stick diagram can be obtained directly by enumerating through all possible combinations of i, j, and d. In fact, the partial layout shown in the lower part of Fig. 3.3 is generated by plotting out only edge blocks containing left and right edges.

An outline of the design-rule analysis algorithm that uses the edge blocks is given below. For simplicity, the outline addresses the horizontal compaction only.

```
for each group i starting with the left-most group {
   develop its left edge blocks EDGE(i, *, L);
   for each group j sequentially to the right {
      develop its right edge blocks EDGE(j, *, R);
    maxsp = -1;
      inspect spacing requirements between any 2 blks
         e.g. EDGE(i, m, L) and EDGE(j, n, R);
      if a spacing requirement exists then {
      maxsp = MAX ( maxsp, this requirement );
         delete the inspected portion of EDGE(i, m, L);
         }
      quit this loop when all EDGE(i, *, L) are deleted;
      }
   if maxsp = -1 then
      groups i and j need no separation
   else
      use maxsp as the lower-bound constraint
         between groups i and j;
   }
```

Two of the important features of the above algorithm are

a) no spacing requirement is inserted between two mergeable groups, and

b) the design-rule analysis for a given group is carried out)/ only until that group is fully fenced off
by its immediate neighbors.

The first feature maximizes space utilization in the
compacted layout, while the second feature enhances the
program execution.

For the efficient use by a computer program, most
spacing rules may be organized in a matrix-like table
whose indices include regular mask layers as well as
logical layers, such as that containing the active
channel region of a transistor formed by the intersection of the diffusion and polysilicon layers in a thin
oxide region. An example of such a table is given in
Fig. 4.4. Rules regarding the minimum size, enclosure,
and overlap of elements are enforced at the time the
stick diagram is drawn since the program-supplied symbols are made to meet these rules. But there are more
involved rules that need special handling. For example, the extra overlap of a buried contact window is
often keyed to the direction of the incoming diffusion
line. As the compaction process changes the shape of
the diffusion line, the contact window must also change
in shape. Since the direction of a line may change
after each compaction operation, the directional enlargement of a contact window must be performed on the
compacted layout. The net result is that more than one
compaction operation in each direction is needed to get
the layout into its final correct form. Fig. 4.5 provides an example in which the result after two compac-
tions (shown in the lower left corner) contains design-
rule violations.

4.4. Search the longest path

The constraint graph obtained with the mapping described above is directed and acyclic. Moreover, since
branches are put in with a top-down, left-to-right
sequential design-rule analysis, the predecessor-
descendant relationship of each node is easily obtained
and is recorded in the LF blocks.

It is a simple matter to determine the longest path
through such a directed graph with known descendants
for each node. (The determination of the longest path
through such a graph is known as the critical path
method for analyzing the project network in the field

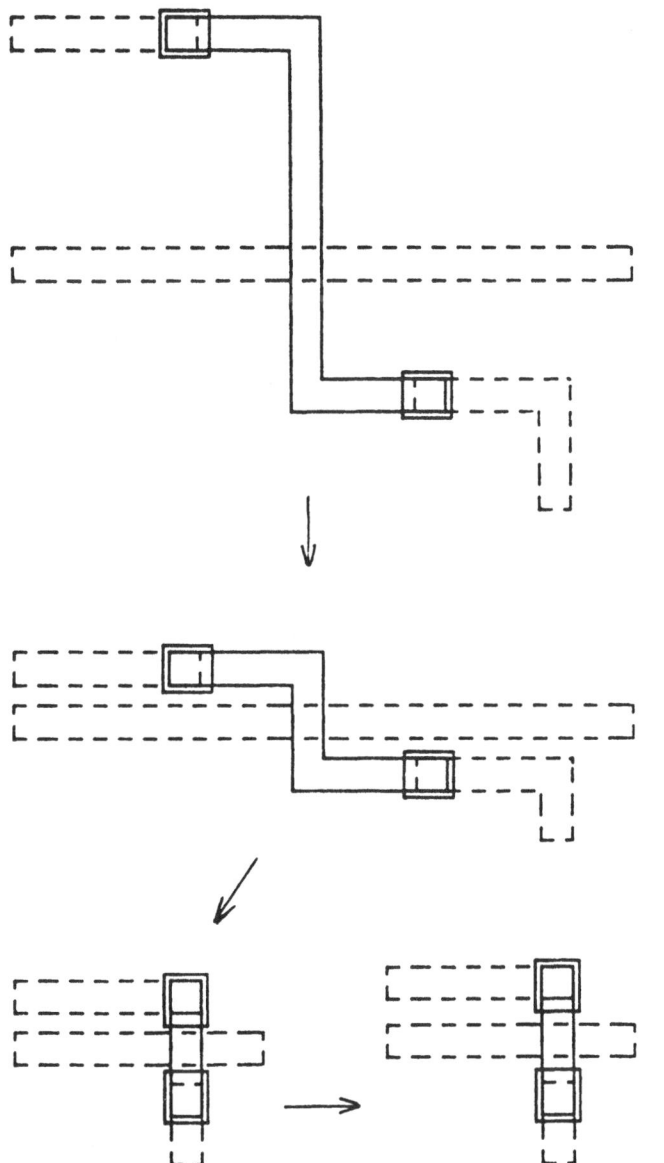

Fig. 4.5. An example showing that the directional enlargement of a buried contact window causes design-rule violation after the second compaction operation.

of operations research [25].) Informally, the longest
path to a given node is taken as the maximum of

a) the present path length to the node, and

b) the sum of the value of the lower-bound constraint
   between the present node and one of its predecessor
   nodes, and the longest path length to that predecessor node.

A formal description of the longest path algorithm
is given below. Before the presentation of the algorithm, however, a few shorthand notations must be defined. Items in the LREF block are referred to with
the prefix LR.. Specifically, the items involved are
the new location, the number of follower nodes, the
list of followers, and the number of predecessors and
they are abbreviated as LR.new, LR.nf, LR.lf, and LR.np,
respectively. Similarly, a lower-bound spacing requirement is referred to as LF.lb. The longest path length
to each node can be computed and recorded in the LR.new
item of the corresponding LREF block as follows:

```
initialize LR.new to 0 for all nodes;
put those nodes whose LR.np equals 0 in a queue;
while the queue is not empty {
    take the first node in the queue as node i;
    for each of its descendant nodes j {
    LR.new of j = MAX ( LR.new of j,
       LF.lb between i and j + LR.new of i );
    decrement LR.np of j by 1;
    if LR.np of j equals 0 then
        add node j to the end of the queue;
    }
    pop the queue (thus deleting node i from it);
}
```

A second pass through the graph is then taken to
detect violations of upper-bound constraints.

The center locations of the most-constraining groups
are given by the path lengths LR.new to their corresponding nodes. The slack or extra space surrounding a
non-constraining group can be determined by reversing
the direction of the branches and searching for the
longest paths from the (originally) last node towards
the (originally) starting node. Note that for a path
containing consecutive non-constraining groups the
slack thus obtained for each group is equal to the

total slack available to that entire path. At the present, PRSLI takes one more pass through such a path to redistribute the total slack equally to each member group. Subsequently, PRSLI applies a simple force-directed placement technique [13] to these non-constraining groups to put proportional unoccupied space on their two sides. Specifically, the linear dimension available to a group in the compaction direction is distributed according to the ratio of LR.np / (LR.np + LR.nf). Such a crude approximation works out well for most examples.

4.5. Display results

The new placement just obtained necessitates some changes in the stick diagram. First, line segments in the direction of compaction must be moved and their lengths must be adjusted according to the new center locations of the groups that they link to. Second, some regrouping may be needed to reflect the merger of elements resulting from the compaction operation. The worst changes, however, are those related to special design-rule requirements such as the directional enlargement of contact windows mentioned earlier. Such changes may introduce design-rule violations in an otherwise correct layout. Rather than performing such changes, PRSLI gives the user the old placement (which is still available from the LREF blocks) if it meets the new constaints (which are in the LF blocks). If the user chooses to use the new placement or if the old placement does not satisfy the new constraints, the user is advised that another compaction operation is necessary.

For the graphic output, edge blocks of all elements in this newly placed and grouped stick diagram are used to draw the compact actual layout.

4.6 Introduce break points for jogs

Jog points are locations at which a line can be bent and then continued a short distance away. Such a bending may result in a more area-efficient layout as shown by the example in Figure 4.6. (A higher degree of area efficiency can not be guaranteed unconditionally since the bent line must occupy some space in the direction of the bend.)

532

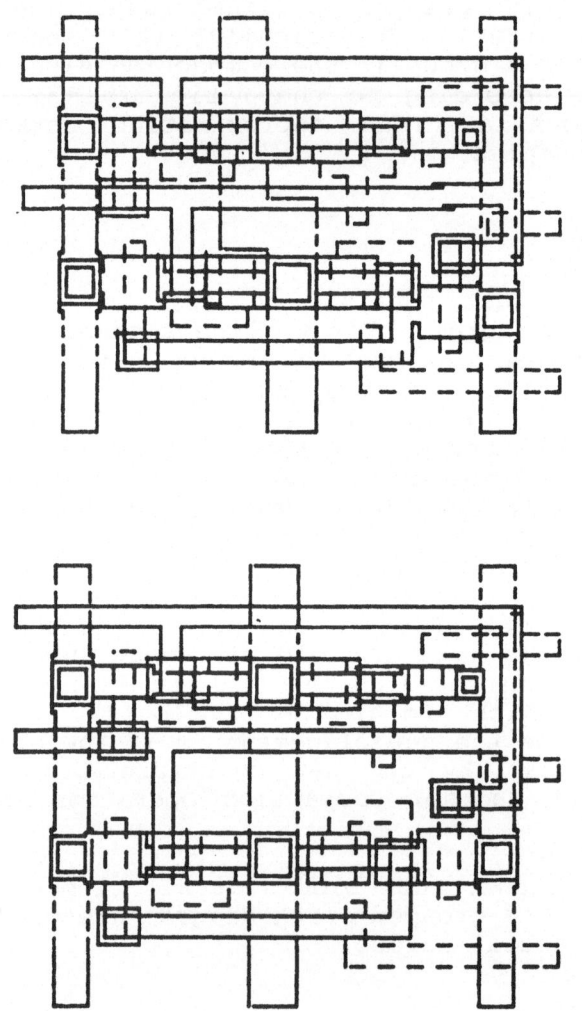

Fig. 4.6. The application of automatic jog generation on the compacted layout that was shown in Fig. 1. The new version with jogs is shown on the right and is 9% smaller than the original version on the left.

Although jog points may be put into the symbolic layout plan manually at any stage of the compaction operation, the longest path information derived during the initial layout compaction operation may be used to determine all possible jog points at once. Specifically, a jog point may be included in a group where the longest path goes into and comes out from two different elements of the group.

In the actual implementation, the element intervals over which the maximum separation requirement between two groups exists are recorded during the design-rule analysis and the record is accessible to the descendant group. Thus, it is simple to find three consecutive groups where the longest path is due to different elements in the middle group. Break points for jogs can be inserted in the middle group at appropriate locations between these constraining elements. The inclusion of a jog point involves the splitting of the group into two parts at the point of the jog and the addition of a line element at that point to connect the new group to the original group. The actual jogs are brought out with a subsequent compaction in the direction of interest.

5. AREAS FOR IMPROVEMENT

The CABBAGE system in its present form is a useful layout tool. However, there are a number of features that may make it much more efficient and versatile. These features are either unavailable now due to an oversight in the initial planning of the system or are hard to use because, in the first round of development, they were deemed less essential to the successful operation of the system and were not emphasized during the development. In this section, these areas for major improvements of the system are considered.

5.1. User control of the compaction process

CABBAGE offers the user a good control of the overall topology of the compacted layout because it does not make drastic changes to the stick diagram supplied by the user. Furthermore, the user can give one dimension more freedom of space utilization by compacting in that dimension first. However, these controls are often insufficient for handling production-oriented layout problems. In many chip designs, each building block has a prescribed size which the layout designer must observe. With CABBAGE, the designer has as much control over the final size of the layout as he does with hand layout; in both cases he is likely to have to work on the size of the layout several times before the layout can fit into the prescribed area.

The problem with the prescribed area is one of the inherent deficiencies of the fixed-topology compaction approach used by CABBAGE. The separation of the compaction problem into two one-dimensional minimization operations also contributes to the lack of control of the overall size. The development of a simplified scheme that is based on the quadratic programming for solving the two-dimensional layout minimization problem will certainly provided the user with more control of this nature. Additionally, good heuristics that use the longest path information to manipulate the layout topology can be useful too.

A second type of deficiency in control arises from the fact that PRSLI compacts a layout towards the lower left corner only. With such an arbitrary choice of origin, any user-devised symmetry about other points are lost after compaction. Most experienced layout designer use some type of symmetry in their layout to achieve good space usage. A satisfactory mechanism for preserving symmetry would be to let the user specify symmetry points and axes in the stick diagram. The inclusion of this feature entails changes to the structure of the constraint graph as well as the use of a different approximation for the force-directed placement of nonconstraining element groups.

In general, the user-specified constraints offer the user additional control over some layout parameters that are not specified in the design-rule set. However, these constraints must be specified in terms of fixed numerical quantities. Since the layout compaction process deals with a relative space, it is often difficult for the user to determine a priori the exact values for these constraints. In many cases the user would be much happier if he can specify simply that, for example, the distance A must be equal to the distance B in the compacted actual layout. Examples in which this type of relative constraint is useful include the design of balanced amplifiers and the construction of a building block whose input and output ports must be aligned to facilitate the stacking of the block.

5.2. Hierarchical layout construction

A hierarchical approach facilitates the construction of large layouts. An efficient macro representation of compacted building-blocks is essential to an effective hierarchical layout construction. The availability of fixed constraints and the use of element groups in PRSLI provide the basic components for a macro representation. At the present, the macro representation of a compacted layout consists of the peripheral element groups that fully enclose the layout. These groups are then separated with fixed constraints to maintain the size and shape of the layout that they represent. Since every detail of the peripheral group must be kept, the computational savings achieved with such a representation can be insignificant for small blocks. An alternative approach that increases computation performance is to replace the peripheral structure with a uniform safety zone, if the area lost to the safety zone is of less importance.

5.3. Organization of design rules

In the present version of PRSLI, layout design rules are scattered in many parts of the program. To begin with, the minimum size, overlap and enclosure rules are implemented as default parameters for the individual elements. The spacing design rules are more organized and are placed in a table. However, specialized routines are needed for accommodating other more involved rules and for indicating permissible electrical connections.

In order to streamline this complicated setup, it is first necessary to devise some algorithm that can translate most reasonable design rules into entries in a generalized design-rule table. Next, it is useful to organize the rules according to their complexity and apply them in a hierarchical manner. In particular, the hierarchy should be set up such that the use of more refined rules would not cause design-rule violations in a compacted layout obtained with a set of simpler rules. Such a hierarchy enables the user to generate first-cut compaction outputs quickly and assures him that the final layout always can be obtained from the first-cut results.

5.4. Integration with other tools

The simple disk file used by CABBAGE to record stick diagrams may serve as the basic data structure for other tools that cooperate with or enhance the usefulness of the CABBAGE system. Several such tools have been shown in Fig. 2 earlier. As mentioned before, it is possible to derive circuit connectivity and device parameters from such a file to serve as input to simulation programs. Another useful tool is a topology generator that supplies the user with a series of topological configurations derived from logic equations. These recommended topologies may be used as the preliminary stick diagrams and they guarantee the consistency between the circuit logic and the final layout. Finally, a language for describing items of the file is useful for applications such as the replication of a building block to construct a functional unit.

6. SUMMARY

The complexity of VLSI circuits makes it necessary to carry out the layout process hierarchically for the construction of compact, well-organized and error-free layouts. The symbolic layout compaction method described in this chapter is most useful for the efficient generation of the actual cells and building-blocks at the lowest level of the hierarchy. The generation of actual layouts from symbolic descriptions allows the user to direct his attention to the development of good layout topologies that lead to geometrically simpler, more structured and more compact layouts. The compaction program is used only as an aid to the user and performs the repetitive and tedious tasks of determining and comparing geometric constraints among the symbolic elements. It provides the user with an almost immediate feedback in the form of a compact actual layout based on the layout topology selected by the user.

The fundamental algorithm for layout compaction used in the layout generation system described here is a rectangle compaction algorithm related to the rectangle dissection method developed by Tutte et al. [15]. A

directed graph summarizing the adjacencies and constraints among elements in the stick diagram is used to
provide the compaction algorithm with a global view of
its topological structure in the direction of the compaction operation. Based on this graph, the compaction
program can determine the most advantageous location
for placing each element to construct a compact geometric layout. The compaction process is carried out
iteratively in the horizontal and the vertical directions until the layout is free of violations of layout
design-rules and the user is satisfied with the overall
size and organization of the layout. Further research
is needed to provide the user with more control over
the final size and aspect ratio of the compacted layout.

For algorithmic simplicity and in the interest of
providing the user with control over the layout topology, the compaction program does not attempt to make
drastic changes to the organization of the stick diagram specified by the user. The automatic jog introduction is the only major modification performed by the
compaction program and is controlled by the outcome of
the previous layout compaction operations. The
computer-aided generation of good layout topologies is
best implemented with a separate program that tackles
the problem with both rigorous mathematical methods,
such as the graph planarization methods [26, 27], and
heuristic approaches to topology planning.

At all stages during the layout generation and compaction process, the user has access to the most recent
layout topology with an interactive graphics editor.
Thus, the user may guide the progress of the layout
compaction process by modifying the interim results and
by restricting the movement of certain elements with
user-defined constraints. Further, information about
the circuit interconnection and composition is available from the compactor throughout the compaction process. This information can be used to derive the necessary input for circuit simulation programs for the
examination of the electrical performance of the particular geometric layout.

At the present time, the compaction program handles
only the generation of n-channel polysilicon-gate MOS
transistor circuit layouts. This limitation arises
from the lack of design-rule analysis capabilities for
other integrated circuit fabrication process families.

(The design-rule analysis is essential to the construction of the constraint graph.) Preliminary results from experiments indicate that this design-rule analysis technique can be applied to other MOS process families. The application of this technique to bipolar junction transistor processes must be determined. For all circuit families, a design-rule translator must be developed to convert the rules specified by the process developer into forms readily usable by the analyzer. The grouping of rules according to their level of importance and complexity should be considered also for the more efficient hierarchical design-rule analysis.

Finally, it should be recognized that the most significant contribution made by layout compaction systems such as CABBAGE is the fact that they make the circuit layout process efficient through an intelligent rearrangement of the classical layout procedure. In the classical approach, a detailed layout is generated before its geometries are examined by a design-rule checker. In addition, the topology of the layout is extracted for simulation and archiving purposes. Iterations through these steps are inevitable if the layout is done in this sequence. Layout compaction systems reverse this sequence and begin with the layout topology to generate the correct geometrical layout in one pass. It would be a useful exercise to determine if many of the bottlenecks in today's VLSI design process may be removed with a similar reformulation of the problem.

REFERENCES

[1] C. Mead and L. Conway, Introduction to VLSI Systems, Addison-Wesley, 1979.

[2] J. D. Williams, "STICKS - A Graphical Compiler for High-Level LSI Design," AFIPS Conference Proceedings, Vol. 47, June 1978, pp. 289-295.

[3] Y. E. Cho, A. J. Korenjak and D. E. Stockton, "FLOSS: An Approach to Automated Layout for High-Volume Designs," Proc. 14th Design Automation Conference, June 1977, pp. 138-141.

[4] A. E. Dunlop, "SLIP: Symbolic Layout of Integrated Circuits with Compaction," Computer-Aided Design, Vol. 10, No. 6, Nov. 1978, pp. 387-391.

[5] M. Y. Hsueh and D. O. Pederson, "Computer-Aided Layout of LSI Circuit Building-Blocks," Proc. 1979 IEEE International Symposium on Circuits and Systems, pp. 474-477.

[6] M. Y. Hsueh, "Symbolic Layout and Compaction of Integrate Circuits," Electronics Research Laboratory Memorandum No. UCB/ERL M79/80, Dec. 1979.

[7] R. P. Larson, "Versatile Mask Generation Techniques for Custom Microelectronic Devices," Proc. 15th Design Automation Conference, June 1978, pp. 193-198.

[8] D. Gibson and S. Nance, "SLIC - Symbolic Layout of Integrated Circuits," Proc. 13th Design Automation Conference, June 1976, pp. 434-440.

[9] G. Persky, D. N. Deutsch and D. G. Schweikert, "LTX - A Minicomputer-Based System for Automatic LSI Layout," Journal of Design Automation and Fault-Tolerant Computing, Vol. 1, No. 3, May 1977, pp. 217-255.

[10] A. Feller, "Automatic Layout of Low-Cost Quick-Turnaround Random-Logic Custom LSI Devices," Proc. 13th Design Automation Conference, June 1976, pp. 79-85.

[11] S. Hong, R. G. Cain and D. L. Ostapko, "MINI: A Heuristic Approach for Logic Minimization," IBM Journal of Research and Development, Vol. 18, No. 5, Sept 1974, pp. 443-458. Also, R. Ayers, "Silicon Compilation - A Hierarchical Use of PLAs," Proc. 16th Design Automation Conference, June 1979, pp. 314-326.

[12] L. Abraitis, J. Blonskis and W. Kuzmicz, "The Method of Computer- Aided Layout Design for Integrated Circuits," J. Design Automation & Fault-Tolerant Computing, Vol. 3, No. 3,4, Winter 1979, pp. 191-209.

[13] M. Hanan, P. K. Wolff and B. J. Agule, "Some Experimental Results on Placement Techniques," Proc. 13th Design Automation Conference, June 1976, pp. 214-224.

[14] S. B. Akers, J. M. Geyer and D. L. Roberts, "IC Mask Layout with a Single Conductor Layer," Proc. 7th Design Automation Conference, June 1970, pp. 7-16.

[15] R. L. Brooks, C. A. B. Smith, A. H. Stone and W. T. Tutte, "The Dissection of Rectangles into Squares," Duke Math. Journal, Vol. 7, 1940, pp. 312-340.

[16] T. Ohtsuki, N. Sugiyama and H. Kawanishi, "An Optimization Technique for Integrated Circuit Layout Design," Proc. International Conference on Circuit and System Theory, Kyoto, Sept. 1970, pp. 67-68.

[17] K. D. Brinkmann and D. A. Mlynski, "Computer-Aided Chip Minimization for IC Layout," Proc. 1976 IEEE International Symposium on Circuits and Systems, pp. 650-653.

[18] R. H. J. M. Otten and M. C. van Lier, "Automatic IC Layout: The Geometry of the Islands," Proc. 1975 IEEE International Symposium on Circuits and Systems, pp. 231-234.

[19] K. Zibert and R. Saal, "On Computer-Aided Hybrid Circuit Layout," Proc. 1974 IEEE International Symposium on Circuits and Systems, pp. 314-318.

[20] U. Lauther, "A Min-cut Placement Algorithm for General Cell Assemblies Based on a Graph Representation," Proc. 16th Design Automation Conference, June 1979, pp. 1-10.

[21] G. D. Hachtel, Private communication.

[22] A. V. Aho, J. E. Hopcroft and J. D. Ullman, The Design and Analysis of Computer Algorithms, Addison-Wesley, 1974, Chapter 5.

[23] A. R. Newton, "Techniques for the Simulation of Large-Scale Integrated Circuits," IEEE Transaction on Circuits and Systems, Vol. CAS-26, No. 9, Sept. 1979, pp. 741-749.

[24] B. W. Kernighan and P. J. Plauger, Software Tools, Addison-Wesley, 1976.

[25] A. Thesen, Computer Methods in Operations Research, Academic Press, 1978, Chapter V.

[26] J. Hopcroft and R. E. Tarjan, "Efficient Planarity Testing," J. ACM, Vol. 21, No. 4, 1974, pp. 549-568.

[27] W. L. Engl, D. A. Mlynski and P. Pernards, "Computer-Aided Topological Design for Integrated Circuits," IEEE Transaction on Circuit Theory, Vol. CT-20, No. 6, Nov. 1973, pp. 717-725.

[28] E. Lawler, "Combinatorial Optimization: Networks and Matroids," Holt, Rinehart and Winston, 1976, Chapter 3.

[29] G. D. Hachtel, "On the Sparse Tableau Approach to Optimal Layout," Proc. First IEEE International Conf. on Circuits and Computers, Oct. 1980, pp. 1019-1022.